Third Edition

COMPUTATIONAL FLUID DYNAMICS
VOLUME II

KLAUS A. HOFFMANN

STEVE T. CHIANG

A Publication of Engineering Education System™, Wichita, Kansas, 67208-1078, USA

ISBN 0-9623731-2-5
First Print: August 1998

This book is typeset by Jeanie Duvall dba SciTech Computer Typesetting of Austin, Texas.

To purchase additional copies, please write to:

Engineering Education System™
P.O. Box 20078
Wichita, KS 67208-1078
USA

CONTENTS

Preface

Chapter Ten: A Review 1

10.1 Introductory Remarks 1
10.2 Classification of Partial Differential Equations 1
  10.2.1 Linear and Nonlinear PDEs 1
  10.2.2 Classification Based on Characteristics 2
10.3 Boundary Conditions 3
10.4 Finite Difference Equations 4
  10.4.1 Parabolic Equations 4
    10.4.1.1 One-Space Dimension 4
    10.4.1.2 Multi-Space Dimensions 7
  10.4.2 Elliptic Equations 8
  10.4.3 Hyperbolic Equations 10
    10.4.3.1 Linear Equations 10
    10.4.3.2 Nonlinear Equations 13
10.5 Stability Analysis 14
10.6 Error Analysis 15
10.7 Grid Generation – Structured Grids 16
10.8 Transformation of the Equations from the Physical Space to Computational Space 19

Chapter Eleven: Transformation of the Equations of Fluid Motion from Physical Space to Computational Space 21

11.1 Introductory Remarks 21
11.2 Generalized Coordinate Transformation 22
11.2.1 Equations for the Metrics 23
11.3 Nondimensionalization of the Equations of Fluid Motion 25
11.4 Navier-Stokes Equations 28
11.4.1 Linearization 31
11.4.2 Inviscid and Viscous Jacobian Matrices 33
11.5 Thin-Layer Approximation 57
11.6 Parabolized Navier-Stokes Equations 60
11.7 Two-Dimensional Planar or Axisymmetric Formulation 69
11.8 Incompressible Navier-Stokes Equations 85
11.8.1 Inviscid and Viscous Jacobian Matrices 89
11.8.2 Two-Dimensional Incompressible Navier-Stokes Equations 91
11.9 Problems 94

Chapter Twelve: Euler Equations 97

7.1 Introductory Remarks 97
12.2 Euler Equations 98
12.3 Quasi One-Dimensional Euler Equations 99
12.3.1 Numerical Issues 100
12.3.2 Explicit Formulations 107
12.3.2.1 Steger and Warming Flux Vector Splitting 107
12.3.2.2 Van Leer Flux Vector Splitting 108
12.3.2.3 Modified Runge-Kutta Formulation 112
12.3.2.4 Second-Order TVD Formulation 112

12.3.2.4.1 Harten-Yee Upwind TVD 113
12.3.2.4.2 Roe-Sweby Upwind TVD 114
12.3.2.4.3 David-Yee Symmetric TVD 115
12.3.2.5 Modified Runge-Kutta Scheme with TVD 115
12.3.3 Implicit Formulations 115
12.3.3.1 Steger and Warming Flux Vector Splitting 116
12.4 Boundary Conditions 118
12.5 Application 1: Diverging Nozzle Configuration 121
12.5.1 Supersonic Inflow, Supersonic Outflow 124
12.5.1.1 Analytical Solution 124
12.5.2.2 Numerical Solutions 125
12.5.2 Supersonic Inflow, Subsonic Outflow 133
12.5.2.1 Analytical Solution 133
12.5.2.2 Numerical Solutions 134
12.6 Grid Clustering 144
12.7 Global Time Step and Local Time Step 146
12.8 Application 2: Shock Tube or Riemann Problem 152
12.8.1 Problem Description 152
12.8.2 Analytical Solution 153
12.8.3 Numerical Solution 157
12.9 Two-Dimensional Planar and Axisymmetric Euler Equations 162
12.9.1 Numerical Considerations 163
12.9.2 Explicit Formulations 170
12.9.2.1 Steger and Warming Flux Vector Splitting 170
12.9.2.1.1 Matrix Manipulations 171
12.9.2.1.2 Existence of Zero Metrics Within the Domain 175
12.9.2.1.3 Eigenvector Matrices 177
12.9.2.2 Van Leer Flux Vector Splitting 178
12.9.2.3 Modified Runge-Kutta Formulation 180
12.9.2.4 Second-Order TVD Formulation 181
12.9.2.4.1 Harten-Yee Upwind TVD 181
12.9.2.4.2 Roe Sweby Upwind TVD 183
12.9.2.4.3 David-Yee Symmetric TVD 185
12.9.2.5 Modified Runge-Kutta Scheme with TVD 185

12.9.3 Boundary Conditions 186
12.9.3.1 Body Surface 186
12.9.3.2 Symmetry 189
12.9.3.3 Inflow 190
12.9.3.4 Outflow 190
12.9.3.5 Boundary Characteristics Based on Characteristics 191
12.9.3.5.1 Inflow Boundary 192
12.9.3.5.2 Outflow Boundary 192
12.9.3.5.3 Determination of Flow Variables 192
12.9.4 Implicit Formulations 193
12.9.4.1 Steger and Warming Flux Vector Splitting 193
12.9.4.1.1 Computation of the Jacobian Matrices 197
12.9.4.1.2 Boundary Conditions 197
12.10 Application: Axisymmetric/Two-Dimensional Problems 200
12.10.1 Supersonic Channel Flow 200
12.10.1.1 Grid Generation 201
12.10.1.2 Numerical Scheme 203
12.10.1.3 Analytical Solution 204
12.10.1.4 The Physical Domain and Flow Conditions 205
12.10.1.5 Initial Conditions and Time Step 206
12.10.1.6 Results 206
12.10.2 Axisymmetric Blunt Body 210
12.11 Concluding Remarks 215
12.12 Problems 216

Chapter Thirteen: Parabolized Navier-Stokes Equations 218

13.1 Introductory Remarks 218
13.2 Governing Equations of Motion 221
13.3 Streamwise Pressure Gradient 223
13.4 Numerical Algorithm 225
13.5 Boundary Conditions 235

13.6 Extension to Three Dimensions 244
13.6.1 Numerical Algorithm 244
13.7 Numerical Damping Terms 249
13.8 Shock Fitting Procedure 250
13.8.1 Extension to Three-Dimensions 257
13.9 Application 261
13.10 Summary Objectives 264
13.11 Problems 265

Chapter Fourteen: Navier-Stokes Equations 266

14.1 Introductory Remarks 266
14.2 Navier-Stokes Equations 267
14.3 Thin-Layer Navier-Stokes Equations 268
14.4 Numerical Algorithms 269
14.4.1 Explicit Formulations 269
14.4.1.1 MacCormack Explicit Formulation 270
14.4.1.2 Flux Vector Splitting 272
14.4.1.3 Modified Runge-Kutta Scheme 275
14.4.2 Boundary Conditions 276
14.4.3 Implicit Formulations 277
14.4.3.1 Flux Vector Splitting 279
14.4.3.2 Higher-Order Flux-Vector Splitting 288
14.4.3.3 Second-Order Accuracy in Time 290
14.4.3.4 LU Decomposition 291
14.5 Extension to Three-Dimensions 294
14.5.1 Explicit Flux Vector Splitting Scheme 295
14.5.2 Implicit Formulation 303
14.6 Concluding Remarks 304
14.7 Problems 306

Chapter Fifteen: Grid Generation—Unstructured Grids 307

15.1 Introductory Remarks 307
15.2 Domain Nodalization 308
15.3 Domain Triangulation 311
15.3.1 The Advancing Front Method 312
15.3.1.1 Simply-Connected Domain 313
15.3.1.2 Multiply-Connected Domain 316
15.3.2 The Delaunay Method 320
15.3.2.1 Geometrical Description 320
15.3.2.2 Outline of the Algorithm 321
15.3.2.3 An Illustrative Example 329
15.4 Concluding Remarks 333
15.6 Problems 334

Chapter Sixteen: Finite Volume Method 336

16.1 Introductory Remarks 336
16.2 General Description of the Finite Volume Method 338
16.2.1 Cell Centered Scheme 339
16.2.2 Nodal Point Scheme 340
16.3 Two-Dimensional Heat Conduction Equation 341
16.3.1 Interior Triangles 342
16.3.2 Boundary Triangles 346
16.3.2.1 Dirichlet Type Boundary Condition 347
16.3.2.2 Neumann Type Boundary Condition 347
16.4 Flux Vector Splitting Scheme 353
16.4.1 Interior Triangles 354
16.4.2 Boundary Triangles 359

16.5 Concluding Remarks 361

16.6 Problems 362

Chapter Seventeen: Turbulent Flow and Turbulence Models 369

17.1 Introductory Remarks 369
17.2 Fundamental Concepts 369
17.3 Modifications of the Equations of Fluid Motion 372
17.4 Reynolds Averaged Navier-Stokes Equations and Turbulence Models 379
17.4.1 General Remarks 379
17.4.2 Turbulent Shear Stress and Heat Flux 380
17.4.3 Reynolds Averaged Navier-Stokes Equations 383
17.4.4 Zero-Equation Turbulence Models 384
17.4.5 One-Equation Turbulence Models 387
17.4.5.1 Baldwin-Barth One-Equation Turbulence Model 388
17.4.5.1.1 Nondimensional Form 389
17.4.5.1.2 Baldwin-Barth Turbulence Model in Computational Space 390
17.4.5.2 Spalart-Allmaras One-Equation Turbulence Model 393
17.4.5.2.1 Nondimensional Form 395
17.4.5.2.2 Spalart-Allmaras Turbulence Model in Computational Space 396
17.4.6 Two-Equation Turbulence Models 398
17.4.6.1 k-ε Two-Equation Turbulence Model 399
17.4.6.2 k-ω Two-Equation Turbulence Model 401
17.4.6.3 Combined k-ε/k-ω Two-Equation Turbulence Model 402
17.4.6.3.1 Baseline Model 403
17.4.6.3.2 Shear-Stress Transport Model 404
17.4.6.3.3 Nondimensional Form 404
17.4.6.3.4 k-ε/k-ω Turbulence Model in Computational Space 405
17.5 Numerical Considerations 407
17.6 Finite Difference Formulations 408

17.6.1 Baldwin-Barth One-Equation Turbulence Model 408
17.6.1.1 Implicit Formulation 409
17.6.1.2 Explicit Formulation 414
17.6.2 Spalart-Allmaras One-Equation Turbulence Model 416
17.6.2.1 Implicit Formulation 416
17.6.2.2 Explicit Formulation 423
17.6.3 k-ε/k-ω Two-Equation Turbulence Model 424
17.7 Concluding Remarks 430

Chapter Eighteen:
An Introduction to High Temperature Gases 432

18.1 Introductory Remarks 432
18.2 Fundamental Concepts 432
18.2.1 Real Gas and Perfect Gas 433
18.2.2 Partial Pressure 434
18.2.3 Frozen Flow 434
18.2.4 Equilibrium Flow 435
18.2.5 Nonequilibrium Flow 435
18.2.6 Various Modes of Energy 435
18.2.7 Reaction Rates 437
18.2.8 Five-Species Model 439
18.3 Quasi One-Dimensional Flow/Equilibrium Chemistry 440
18.4 Quasi One-Dimensional Flow/Nonequilibrium Chemistry 441
18.4.1 Species Continuity Equation 442
18.4.2 Coupling Schemes 443
18.4.3 Numerical Procedure for the Loosely-Coupled Scheme 443
18.5 Applications 446
18.5.1 Quasi One-Dimensional Flow 446
18.5.2 Two-Dimensional Axisymmetric Flow 448
18.6 Concluding Remarks 451

Appendices

Appendix G: An Introduction to Theory of Characteristics: Euler Equations 452
Appendix H: Computation of Pressure at the Body Surface 469
Appendix I: Transformation of Turbulence Models from Physical Space to Computational Space 475
Appendix J: Rate of Formation of Species 480

References 483

Index 486

PREFACE

The fundamental concepts of computational schemes established in the first volume are extended to the solution of Euler equations, Parabolized Navier-Stokes equations, and Navier-Stokes equations in this volume. Additional topics such as unstructured grids, finite volume schemes, turbulent flows, and chemically reacting flows at the introductory level are included as well.

This volume begins with a review of the basic concepts which is presented in Chapter 10.

Subsequently, the transformation of the equations of fluid motion from physical space to computational space is provided in Chapter 11. This chapter also includes the linearization of the equations as well as the derivation of the Jacobian matrices. Chapter 12 presents numerical schemes for the solution of the Euler equations for inviscid flowfields. Specifications of the boundary conditions, along with illustrated examples, are provided in this chapter. Chapter 13 presents Parabolized Navier-Stokes (PNS) equations and a numerical algorithm for solution. The shock fitting procedure is discussed in this chapter as well. The Navier-Stokes equations and various numerical schemes for solutions are discussed in Chapter 14. To familiarize the reader with unstructured grids which are used in conjunction with finite volume and finite element schemes, it is introduced in Chapter 15. It developes some fundamental concepts and explores two techniques for generation of unstructured grids in two-dimensions. Finite volume schemes are developed in Chapter 16. Finally, in Chapters 17 and 18, some special topics which include consideration for turbulence and chemistry are discussed.

Finally, our sincere thanks and appreciation are extended to all individuals acknowledged in the preface of the first volume. Thank you all very much for your friendship and encouragement.

Klaus A. Hoffmann
Steve T. Chiang

Chapter 10

A Review

10.1 Introductory Remarks

The fundamental concepts of computational fluid dynamics were introduced in the previous chapters. Various aspects of numerical schemes were explored with regard to simple partial differential equations. In all cases up to Chapter 8, the investigations were limited to a single equation. In the upcoming chapters the concepts are extended to systems of equations. Before proceeding further, however, it is beneficial to review and summarize the content of the previous chapters.

10.2 Classification of Partial Differential Equations

Partial differential equations (PDEs) can be classified into different categories, where within each category they may be classified further into subcategories. The numerical procedure used to solve a partial differential equation very much depends on the classification of the governing equation. A brief review of the classification of partial differential equations is provided in the following subsections.

10.2.1 Linear and Nonlinear PDEs

(a) Linear PDE: There is no product of the dependent variable and/or product of its derivatives within the equation.

(b) Nonlinear PDE: The equation contains a product of the dependent variable and/or a product of the derivatives.

10.2.2 Classification Based on Characteristics

(I) First-order PDE: Almost all first-order PDEs have real characteristics, and therefore behave much like hyperbolic equations of second order.

(II) Second-order PDE: A second-order PDE in two independent variables, x and y, may be expressed in a general form as

$$A\frac{\partial^2 \phi}{\partial x^2} + B\frac{\partial^2 \phi}{\partial x \partial y} + C\frac{\partial^2 \phi}{\partial y^2} + D\frac{\partial \phi}{\partial x} + E\frac{\partial \phi}{\partial y} + F\phi + G = 0 \tag{10-1}$$

The equation is classified according to the expression $(B^2 - 4AC)$ as follows :

$$\left(B^2 - 4AC\right) \begin{cases} < 0 \rightarrow \text{elliptic equation} \\ = 0 \rightarrow \text{parabolic equation} \\ > 0 \rightarrow \text{hyperbolic equation} \end{cases}$$

The following criteria may be stated with regard to each category defined above:

(a) Elliptic equations

- No real characteristic lines exist
- A disturbance propagates in all directions
- Domain of solution is a closed region
- Boundary conditions must be specified on the boundaries of the domain

(b) Parabolic equations

- Only one characteristic line exists
- A disturbance propagates along the characteristic line
- Domain of solution is an open region
- An initial condition and two boundary conditions are required

(c) Hyperbolic equations

- Two characteristic lines exist
- A disturbance propagates along the characteristic lines
- Domain of solution is an open region
- Two initial conditions along with two boundary conditions are required

(III) System of First-Order PDEs

A system of first-order PDEs may be expressed in a vector form as

$$\frac{\partial \boldsymbol{\Phi}}{\partial t} + [A]\frac{\partial \boldsymbol{\Phi}}{\partial x} + [B]\frac{\partial \boldsymbol{\Phi}}{\partial y} + \boldsymbol{\Psi} = 0 \tag{10-2}$$

where the vector $\boldsymbol{\Phi}$ contains the dependent variables. The system is classified according to the eigenvalues of coefficient matrices $[A]$ and $[B]$. If the eigenvalues of matrix $[A]$ are all real and distinct, the system is classified as hyperbolic in t and x. If the eigenvalues of $[A]$ are complex, the system is elliptic in t and x. Similarly, the system is classified with respect to the independent variables t and y based on the eigenvalues of matrix $[B]$.

For a steady equivalent of (10-2), given by

$$[A]\frac{\partial \boldsymbol{\Phi}}{\partial x} + [B]\frac{\partial \boldsymbol{\Phi}}{\partial y} + \boldsymbol{\Psi} = 0 \tag{10-3}$$

the classification is as follows:

$$H \begin{cases} < 0 \rightarrow \text{elliptic} \\ = 0 \rightarrow \text{parabolic} \\ > 0 \rightarrow \text{hyperbolic} \end{cases}$$

where

$$H = R^2 - 4PQ$$

and

$$P = |A|, \quad Q = |B|$$

For a system composed of two equations, R is given by

$$R = \begin{vmatrix} a_1 & a_4 \\ b_1 & b_4 \end{vmatrix} + \begin{vmatrix} a_3 & a_2 \\ b_3 & b_2 \end{vmatrix}$$

where

$$[A] = \begin{bmatrix} a_1 & a_2 \\ b_1 & b_2 \end{bmatrix} \quad \text{and} \quad [B] = \begin{bmatrix} a_3 & a_4 \\ b_3 & b_4 \end{bmatrix}$$

(IV) System of Second-Order PDEs

The classification of a system of second-order PDEs is facilitated if the second-order PDEs are reduced to their equivalent first-order PDEs. Subsequently, the system is classified as previously seen. The procedure could be cumbersome. For specific details and examples, Section 1.9 should be reviewed.

10.3 Boundary Conditions

A set of specific information with regard to the dependent variable and/or its derivative must be specified along the boundaries of the domain. This set of information is known as the *boundary condition* and may be categorized as follows.

(a) The Dirichlet boundary condition: The value of the dependent variable along the boundary is specified.

(b) The Neumann boundary condition: The normal gradient of the dependent variable along the boundary is specified.

(c) The Mixed boundary condition: A combination of the Dirichlet and the Neumann type boundary conditions is specified.

10.4 Finite Difference Equations

The partial derivatives appearing in the differential equations are replaced by approximate algebraic expressions to provide an equivalent algebraic equation known as the *finite difference equation.* Subsequently, the finite difference equation is solved within a domain which has been discretized into equally spaced grids. Finite difference equations commonly used for the solution of parabolic, elliptic, and hyperbolic equations are reviewed in this section.

10.4.1 Parabolic Equations

Various finite difference formulations are reviewed for the one-dimensional parabolic equations initially and, subsequently, extended to multi-dimensional problems.

10.4.1.1 One-Space Dimension

The simple diffusion equation is used in this section to review various finite difference equations. The model equation is given by

$$\frac{\partial u}{\partial t} = \alpha \frac{\partial^2 u}{\partial x^2}$$

where α is assumed to be a constant and hence a linear equation. To facilitate the review process, various aspects of each finite difference formulation such as the order of accuracy, amplification factor, stability requirement, and the corresponding modified equation are summarized. In the formulations to follow, the diffusion number is designated by d, which is defined by

$$d = \alpha \frac{\Delta t}{(\Delta x)^2}$$

Scheme: FTCS explicit

Formulation: $u_i^{n+1} = u_i^n + d\,(u_{i+1}^n - 2u_i^n + u_{i-1}^n)$

Order: $O\left[(\Delta t)\,,\ (\Delta x)^2\right]$

Amplification Factor: $G = 1 + 2d(\cos\theta - 1)$

Stability Requirement: $d \leq \frac{1}{2}$

Modified Equation:

$$\frac{\partial u}{\partial t} = \alpha\frac{\partial^2 u}{\partial x^2} + \left[-\frac{1}{2}\alpha^2(\Delta t) + \frac{1}{12}\alpha(\Delta x)^2\right]\frac{\partial^4 u}{\partial x^4} + \left[\frac{1}{3}\alpha^3(\Delta t)^2\frac{\partial^4 u}{\partial x^4} - \frac{1}{12}\alpha^2(\Delta t)\,(\Delta x)^2 + \frac{1}{360}\alpha(\Delta x)^2\right]\frac{\partial^6 u}{\partial x^6} + \ldots$$

Special Considerations: None

Scheme: DuFort-Frankel explicit

Formulation: $u_i^{n+1} = \frac{1-2d}{1+2d}u_i^{n-1} + \frac{2d}{1+2d}\,(u_{i+1}^n + u_{i-1}^n)$

Order: $O\left[(\Delta t)^2\,,\ (\Delta x)^2\,,\ \left(\frac{\Delta t}{\Delta x}\right)^2\right]$

Amplification Factor: $G = \frac{1}{1+2d}\left[2d\,\cos\theta \pm (1 - 4d^2\mathrm{Sin}^2\theta)^{\frac{1}{2}}\right]$

Stability Requirement: Unconditionally stable

Modified Equation:

$$\frac{\partial u}{\partial x} = \alpha\frac{\partial^2 u}{\partial x^2} + \left[\frac{1}{12}\alpha(\Delta x^2) - \alpha^3\left(\frac{\Delta t}{\Delta x}\right)^2\right]\frac{\partial^4 u}{\partial x^4} + \left[-\frac{1}{3}\alpha^3(\Delta t)^2 + \frac{1}{360}\alpha(\Delta x)^4 + 2\alpha^5\left(\frac{\Delta t}{\Delta x}\right)^4\right]\frac{\partial^6 u}{\partial x^6} + \ldots$$

Special Considerations: Requires two sets of data to proceed

Scheme: Laasonen implicit

Formulation: $du_{i-1}^{n+1} - (1+2d)u_i^{n+1} + du_{i-1}^{n} = u_i^n$

Order: $O\left[(\Delta t)\,,\ (\Delta x)^2\right]$

Amplification Factor: $G = \dfrac{1}{1+2d(1-\text{Cos}\theta)}$

Stability Requirement: Unconditionally stable

Modified Equation:

$$\frac{\partial u}{\partial t} = \alpha\frac{\partial^2 u}{\partial x^2} + \left[\frac{1}{2}\alpha^2(\Delta t) + \frac{1}{12}\alpha(\Delta x)^2\right]\frac{\partial^4 u}{\partial x^4}$$
$$+\left[\frac{1}{3}\alpha^3(\Delta t)^2 + \frac{1}{12}\alpha^2(\Delta t)\,(\Delta x)^2\right.$$
$$\left.+\frac{1}{360}\alpha(\Delta x)^4\right]\frac{\partial^6 u}{\partial x^6} + \ldots$$

Special Considerations: Requires solution of a tridiagonal system

Scheme: Crank-Nicolson implicit

Formulation:

$$\frac{1}{2}du_{i+1}^{n+1} - (1+d)u_i^{n+1} + \frac{1}{2}du_{i-1}^{n+1} = -\frac{1}{2}u_{i+1}^n$$
$$+\,(d-1)u_i^n - \frac{1}{2}du_{i-1}^n$$

Order: $O\left[(\Delta t)^2\,,\ (\Delta x)^2\right]$

Amplification Factor: $G = \dfrac{1-d(1-\text{Cos}\theta)}{1+d(1-\text{Cos}\theta)}$

Stability Requirement: Unconditionally stable

Modified Equation:

$$\frac{\partial u}{\partial x} = \alpha\frac{\partial^2 u}{\partial x^2} + \left[\frac{1}{12}\alpha(\Delta x)^2\right]\frac{\partial^4 u}{\partial x^4}$$
$$+\left[\frac{1}{12}\alpha^3(\Delta t)^2 + \frac{1}{360}\alpha(\Delta x)^4\right]\frac{\partial^6 u}{\partial x^6} + \ldots$$

Special Considerations: Requires solution of a tridiagonal system

10.4.1.2 Multi-Space Dimensions

The review of multi-dimensional problems will be limited to two-space dimensions. The procedure to three-space dimensions is similar. However, it is cautioned that the extension may not be trivial, and certain formulations which may have been unconditionally stable in two-space dimensions may become only conditionally stable in three-space dimensions. The model equation used is the diffusion equation in two-space dimensions given by

$$\frac{\partial u}{\partial t} = \alpha\left(\frac{\partial^2 u}{\partial x^2} + \frac{\partial^2 u}{\partial y^2}\right)$$

Scheme: FTCS explicit

Formulation:

$$u_{i,j}^{n+1} = u_{i,j}^n + d_x(u_{i+1,j}^n - 2u_{i,j}^n + u_{i-1,j}^n) + d_y\left(u_{i,j+1}^n - 2u_{i,j}^n + u_{i,j-1}^n\right)$$

Order: $O\left[(\Delta t),\ (\Delta x)^2,\ (\Delta y)^2\right]$

Stability Requirement: $(d_x + d_y) \le \frac{1}{2}$

Scheme: ADI

Formulation:

$$-\left(\frac{1}{2}d_x\right)u_{i-1,j}^{n+\frac{1}{2}} + (1+d_x)u_{i,j}^{n+\frac{1}{2}} - \left(\frac{1}{2}d_x\right)u_{i+1,j}^{n+\frac{1}{2}} = \left(\frac{1}{2}d_y\right)u_{i,j+1}^n + (1-d_y)u_{i,j}^n + \left(\frac{1}{2}d_y\right)u_{i,j-1}^n$$

and

$$-\left(\frac{1}{2}d_y\right)u_{i,j-1}^{n+1} + (1+d_y)u_{i,j}^{n+1} - \left(\frac{1}{2}d_y\right)u_{i,j+1}^{n+1} = \left(\frac{1}{2}d_x\right)u_{i+1,j}^{n+\frac{1}{2}} + (1-d_x)u_{i,j}^{n+\frac{1}{2}} + \left(\frac{1}{2}d_x\right)u_{i-1,j}^{n+\frac{1}{2}}$$

Order: $O\left[(\Delta t)^2,\ (\Delta x)^2,\ (\Delta y)^2\right]$

Amplification Factor:

$$G = \frac{[1 - d_x(1-\mathrm{Cos}\theta_x)]\ [1 - d_y(1-\mathrm{Cos}\theta_y)]}{[1 + d_x(1-\mathrm{Cos}\theta_x)]\ [1 + d_y(1-\mathrm{Cos}\theta_y)]}$$

Stability Requirement: Unconditionally stable

Scheme:	Fractional step
Formulation:	$d_x u_{i+1,j}^{n+\frac{1}{2}} - (1+2d_x) u_{i,j}^{n+\frac{1}{2}} + d_x u_{i-1,j}^{n+\frac{1}{2}}$
	$= -d_x u_{i+1,j}^{n} + (2d_x - 1) u_{i,j}^{n} - d_x u_{i-1,j}^{n}$
and	$d_y u_{i,j+1}^{n+1} - (1+2d_y) u_{i,j}^{n+1} + d_y u_{i,j-1}^{n+1}$
	$= -d_y u_{i,j+1}^{n+\frac{1}{2}} + (2d_y - 1) u_{i,j}^{n+\frac{1}{2}} - d_y u_{i,j-1}^{n+\frac{1}{2}}$
Order:	$O\left[(\Delta t)^2, (\Delta x)^2, (\Delta y)^2\right]$
Stability Requirement:	Unconditionally stable

10.4.2 Elliptic Equations

The model equation utilized to represent various finite difference equations is the Poisson equation expressed in two-space dimensions given by

$$\frac{\partial^2 u}{\partial x^2} + \frac{\partial^2 u}{\partial y^2} = 0$$

Only iterative schemes which are usually the most efficient schemes to solve elliptic equations are reviewed in this section. In the formulations to follow, the ratio of stepsizes is designated by β, i.e., $\beta = \dfrac{\Delta x}{\Delta y}$.

Scheme:	Point Gauss-Seidel (PGS)
Formulation:	$u_i^{k+1} = \dfrac{1}{2(1+\beta^2)} \left[u_{i+1,j}^{k} + u_{i-1,j}^{k+1} + \beta^2 \left(u_{i,j+1}^{k} + u_{i,j-1}^{k+1}\right)\right]$
Order:	$O\left[(\Delta x)^2, (\Delta y)^2\right]$

Scheme:	Line Gauss-Seidel (LGS)
Formulation: (x direction)	$u_{i-1,j}^{k+1} - 2(1+\beta^2) u_{i,j}^{k+1} + u_{i+1,j}^{k+1} = -\beta^2 u_{i,j+1}^{k} - \beta^2 u_{i,j-1}^{k+1}$
Order:	$O\left[(\Delta x)^2, (\Delta y)^2\right]$
Modified Equation:	$\dfrac{\partial^2 u}{\partial x^2} + \dfrac{\partial^2 u}{\partial y^2} = -\dfrac{1}{12}(\Delta x)^2 \dfrac{\partial^4 u}{\partial x^4} - \dfrac{1}{12}(\Delta y)^2 \dfrac{\partial^4 u}{\partial y^4} + \ldots$

Scheme: Point Successive Over-Relaxation (PSOR)

Formulation:

$$u_{i,j}^{k+1} = (1-\omega)u_{i,j}^{k} + \frac{\omega}{2(1+\beta^2)}\left[u_{i+1,j}^{k} + u_{i-1,j}^{k+1} + \beta^2\left(u_{i,j+1}^{k} + u_{i,j-1}^{k+1}\right)\right]$$

Order: $O\left[(\Delta x)^2\,, (\Delta y)^2\right]$

Special Considerations: The range of relaxation parameter is $1 \le \omega < 2$

Scheme: Line Successive Over-Relaxation (LSOR)

Formulation:
(x direction)

$$\omega u_{i-1,j}^{k+1} - 2\left(1+\beta^2\right)u_{i,j}^{k+1} + \omega u_{i+1,j}^{k+1}$$
$$= -(1-\omega)\left[2(1+\beta^2)\right]u_{i,j}^{k} - \omega\beta^2\left(u_{i,j+1}^{k} + u_{i,j-1}^{k+1}\right)$$

Special Considerations: The range of relaxation parameter is $1 \le \omega < 2$

Scheme: ADI

Formulation:

$$u_{i-1,j}^{k+\frac{1}{2}} - 2(1+\beta^2)u_{i,j}^{k+\frac{1}{2}} + u_{i+1,j}^{k+\frac{1}{2}} = -\beta^2\left(u_{i,j+1}^{k} + u_{i,j-1}^{k+\frac{1}{2}}\right)$$

and

$$\beta^2 u_{i,j-1}^{k+1} - 2(1+\beta^2)u_{i,j}^{k+1} + \beta^2 u_{i,j+1}^{k+1} = -u_{i+1,j}^{k+\frac{1}{2}} - u_{i-1,j}^{k+1}$$

Scheme: ADI with relaxation parameter

Formulation:

$$\omega\, u_{i-1,j}^{k+\frac{1}{2}} - 2(1+\beta^2)u_{i,j}^{k+\frac{1}{2}} + \omega\, u_{i+1,j}^{k+\frac{1}{2}}$$
$$= -(1-\omega)\;\left[2\left(1+\beta^2\right)\right]u_{i,j}^{k} - \omega\,\beta^2\left(u_{i,j+1}^{k} + u_{i,j-1}^{k+\frac{1}{2}}\right)$$

and

$$\omega\,\beta^2 u_{i,j-1}^{k+1} - 2(1+\beta^2)u_{i,j}^{k+1} + \omega\,\beta^2 u_{i,j+1}^{k+1}$$
$$= -\,(1-\omega)\;\left[2\left(1+\beta^2\right)\right]u_{i,j}^{k+\frac{1}{2}} - \omega\;\left(u_{i+1,j}^{k+\frac{1}{2}} + u_{i-1,j}^{k+1}\right)$$

10.4.3 Hyperbolic Equations

Investigation of various finite difference equations is easily accomplished with regard to linear hyperbolic equations. Subsequently, the conclusions may be extended to nonlinear hyperbolic equations. With that in mind, the review of the formulations is performed sequentially in two parts.

10.4.3.1 Linear Equations

The wave equation given by

$$\frac{\partial u}{\partial t} = -a\frac{\partial u}{\partial x} \; , \quad a > 0$$

is used to review linear hyperbolic equations. Note that the speed of sound, a, in the equation above is assumed to be a constant and, hence, a linear equation. The parameter, $a\Delta t/\Delta x$, defined as the Courant number and designated by c, will be used in the formulations to follow.

Scheme:	First upwind differencing
Formulation:	$u_i^{n+1} = u_i^n - c\left(u_i^n - u_{i-1}^n\right)$
Order:	$O\left[(\Delta t)\, ,\ (\Delta x)\right]$
Amplification Factor:	$G = 1 - c(1 - \text{Cos}\,\theta) - i(c\text{Sin}\,\theta)$
Stability Requirement:	$c \leq 1$
Modified Equation:	$\frac{\partial u}{\partial t} = -a\frac{\partial u}{\partial x} + \frac{1}{2}a\Delta x(1-c)\frac{\partial^2 u}{\partial x^2} - \frac{1}{6}a(\Delta x)^2(2c^2 - 3c + 1)\frac{\partial^3 u}{\partial x^3} + \ldots$

Scheme:	Lax
Formulation:	$u_i^{n+1} = \frac{1}{2}\left(u_{i+1}^n + u_{i-1}^n\right) - \frac{1}{2}c\left(u_{i+1}^n - u_{i-1}^n\right)$
Order:	$O\left[(\Delta t)\, ,\ \frac{(\Delta x^2)}{(\Delta t)}\right]$
Amplification Factor:	$G = \text{Cos}\,\theta - i(c\text{Sin}\,\theta)$
Stability Requirement:	$c \leq 1$
Modified Equation:	$\frac{\partial u}{\partial t} = a\frac{\partial u}{\partial x} + \frac{1}{2}a(\Delta x)\left(\frac{1}{c} - c\right)\frac{\partial^2 u}{\partial x^2} + \frac{1}{3}a(\Delta x)^2(1-c^2)\frac{\partial^3 u}{\partial x^3} + \ldots$

Scheme:	Midpoint Leapfrog
Formulation:	$u_i^{n+1} = u_i^{n-1} - c\,(u_{i+1}^n - u_{i-1}^n)$
Order:	$O\left[(\Delta t)^2,\ (\Delta x)^2\right]$
Amplification Factor:	$G = \pm\left[1 - c^2 \text{Sin}^2\,\theta\right]^{\frac{1}{2}} - i(c\text{Sin}\,\theta)$
Stability Requirement:	$c \leq 1$
Modified Equation:	$\frac{\partial u}{\partial t} = -a\frac{\partial u}{\partial x} - \frac{1}{6}a(\Delta x)^2\left(1 - c^2\right)\frac{\partial^3 u}{\partial x^3} + \ldots$
Special Considerations:	Requires two sets of data for the solution to proceed

Scheme:	Lax-Wendroff
Formulation:	$u_i^{n+1} = u_i^n - \frac{1}{2}c\,(u_{i+1}^n - u_{i-1}^n) + \frac{1}{2}c^2\,(u_{i+1}^n - 2u_i^n + u_{i-1}^n)$
Order:	$O\left[(\Delta t)^2,\ (\Delta x)^2\right]$
Amplification Factor:	$G = 1 - c^2\,(1 - \text{Cos}\,\theta) - i(c\,\text{Sin}\,\theta)$
Stability Requirement:	$c \leq 1$
Modified Equation:	$\frac{\partial u}{\partial t} = -a\frac{\partial u}{\partial x} - \frac{1}{6}a(\Delta x)^2\left(1 - c^2\right)\frac{\partial^3 u}{\partial x^3} - \frac{1}{8}a(\Delta x)^3 c(1 - c^2)\frac{\partial^4 u}{\partial x^4} + \ldots$

Scheme:	BTCS implicit
Formulation:	$\frac{1}{2}c\,u_{i-1}^{n+1} - u_i^{n+1} - \frac{1}{2}c\,u_{i+1}^{n+1} = -u_i^n$
Order:	$O\left[(\Delta t)\ ,\ (\Delta x)^2\right]$
Amplification Factor:	$G = \dfrac{1 - i(c\,\mathrm{Sin}\,\theta)}{1 + c^2\,(\mathrm{Sin}^2\,\theta\,)}$
Stability Requirement:	*None*
Modified Equation:	$\frac{\partial u}{\partial t} = -a\frac{\partial u}{\partial x} + \frac{1}{2}a^2(\Delta t)\frac{\partial^2 u}{\partial x^2} - \left[\frac{1}{6}a(\Delta x)^2 + \frac{1}{3}a^3(\Delta t)^2\right]\frac{\partial^3 u}{\partial x^3} + \ldots$

Scheme:	Crank-Nicolson
Formulation:	$\frac{1}{4}c\,u_{i+1}^{n+1} - u_i^{n+1} - \frac{1}{4}c\,u_{i-1}^{n+1} = u_i^n - \frac{1}{4}c\,(u_{i+1}^n - u_{i-1}^n)$
Order:	$O\left[(\Delta t)^2\ ,\ (\Delta x)^2\right]$
Amplification Factor:	$G = \dfrac{1 - 0.5i(c\,\mathrm{Sin}\,\theta)}{1 + 0.5i\,(c\,\mathrm{Sin}\,\theta\,)}$
Stability Requirement:	None
Modified Equation:	$\frac{\partial u}{\partial t} = -a\frac{\partial u}{\partial x} - \frac{1}{6}a\,(\Delta x)^2\left(1 + \frac{1}{2}c^2\right)\frac{\partial^3 u}{\partial x^3}\cdots$

Scheme:	Lax-Wendroff
Formulation:	$u_{i+\frac{1}{2}}^{n+\frac{1}{2}} = \frac{1}{2}\,(u_{i+1}^n + u_i^n) - \frac{1}{2}c\,(u_{i+1}^n - u^{n_i})$
and	$u_i^{n+1} = u_i^n - c\left(u_{i+\frac{1}{2}}^{n+\frac{1}{2}} - u_{i-\frac{1}{2}}^{n+\frac{1}{2}}\right)$
Order:	$O\,[(\Delta t)^2\ ,\ (\Delta x)^2]$
Stability Requirement:	$c \leq 1$

Scheme:	MacCormack
Formulation:	$u_i^* = u_i^n - c\,(u_{i+1}^n - u_i^n)$
and	$u_i^{n+1} = \frac{1}{2}\left[u_i^n + u_i^* - c\,(u_i^* - u_i^*)\right]$
Order:	$O\left[(\Delta t)^2\ ,\ (\Delta x)^2\right]$
Stability Requirement:	$c \leq 1$

10.4.3.2 Nonlinear Equations

The inviscid Burgers equation is used to review various schemes for the solution of hyperbolic equations. Recall that the model equation is given by

$$\frac{\partial u}{\partial t} = -u\frac{\partial u}{\partial x} = -\frac{\partial E}{\partial x}$$

where $E = \frac{1}{2}u^2$. Now, the Courant number is defined as $c = u\Delta t/\Delta x$.

Scheme:	Lax
Formulation:	$u_i^{n+1} = \frac{1}{2}\left(u_{i+1}^n + u_{i-1}^n\right) - \frac{1}{2}\frac{\Delta t}{\Delta x}\left(E_{i+1}^n - E_{i-1}^n\right)$
Order:	$O\left[(\Delta t)\ ,\ (\Delta x)^2\right]$
Amplification Factor:	$G = \text{Cos}\,\theta - i(c\,\text{Sin}\,\theta)$
Stability Requirement:	$\left\lvert u_{\max}\frac{\Delta t}{\Delta x}\right\rvert \leq 1$

Scheme:	Lax-Wendroff
Formulation:	$u_i^{n+1} = u_i^n - \frac{1}{2}\frac{\Delta t}{\Delta x}\left(E_{i+1}^n - E_{i-1}^n\right) + \frac{1}{4}\left(\frac{\Delta t}{\Delta x}\right)^2$
	$\left[\left(u_{i+1}^n + u_i^n\right)\left(E_{i+1}^n - E_i^n\right) - \left(u_i^n + u_{i-1}^n\right)\left(E_i^n - E_{i-1}^n\right)\right]$
Amplification Factor:	$G = 1 - 2c^2(1 - \text{Cos}\,\theta) - 2i(c\,\text{Sin}\,\theta)$
Stability Requirement:	$\left\lvert u_{\max}\frac{\Delta t}{\Delta x}\right\rvert \leq 1$

Scheme:	MacCormack
Formulation:	$u_i^* = u_i^n - \frac{\Delta t}{\Delta x}\left(E_{i+1}^n - E_i^n\right)$
and	$u_i^{n+1} = \frac{1}{2}\left[u_i^n + u_i^* - \frac{\Delta t}{\Delta x}\left(E_i^* - E_{i-1}^*\right)\right]$
Order:	$O\left[(\Delta t)^2\,,\ (\Delta x)^2\right]$

10.5 Stability Analysis

The error introduced in the finite difference equations due to the truncation of the higher order terms in the Taylor series expansion may grow unbounded, producing an unstable solution. The control of errors within the solution is of primary concern for any numerical scheme. To establish the necessary requirements, a stability analysis must be performed. Among various methods available for stability analysis are: (1) The discrete perturbation stability analysis, (2) The von Neumann (or Fourier) stability analysis, and (3) The matrix method. It should be noted that direct stability analysis of a nonlinear, multi-dimensional, coupled system of equations is usually cumbersome. In most cases, expressions are proposed which are based on the stability analysis of simple model equations complimented and reinforced by numerical experimentation. Thus, one encounters the suggested stability requirement for a particular scheme which resembles those of simple model equations, but includes some modifications based on numerical experimentations.

To review the limitations and conclusions provided from the von Neumann stability analysis, the summary stated in Chapter 4 is repeated at this point.

1. The von Neumann stability analysis can be applied to linear equations only.

2. The influence of the boundary conditions on the stability of the solution is not included.

3. For a scalar PDE which is approximated by a two-level FDE, the mathematical requirement is imposed on the amplification factor G as follows:

   (a) if G is real, then $|G| \leq 1$

   (b) if G is complex, then $|G|^2 \leq 1$, where $|G|^2 = G\overline{G}$

4. For a scalar PDE which is approximated by a three-level FDE, the amplification factor is a matrix. In this case, the requirement is imposed on the eigenvalues of G as follows:

(a) if λ is real, then $|\lambda| \leq 1$

(b) if λ is complex, then $|\lambda|^2 \leq 1$

5. The method can be easily extended to multi-dimensional problems.

6. The procedure can be used for stability analysis of a system of linear PDEs. The requirement is imposed on the largest eigenvalue of the amplification matrix.

7. Benchmark values for the stability of unsteady one-dimensional problems may be established as follows:

   (a) For most explicit formulations:

   I. Courant number, $c \leq 1$

   II. Diffusion number, $d \leq \frac{1}{2}$

   III. Cell Reynolds number, $Re_c \leq (2/c)$

   (b) For implicit formulation, most are unconditionally stable.

8. For multi-dimensional problems with equal grid spacing in all spatial directions, the stated benchmark values are adjusted usually by dividing them by the number of spatial dimensions.

9. On occasions where the amplification factor is a difficult expression to analyze, graphical solution along with some numerical experimentation will facilitate the analysis.

10.6 Error Analysis

The truncation of terms in the approximation of a partial derivative could begin from an odd-order or an even-order derivative term. For example, one may approximate a first-order derivative by either

$$\frac{\partial u}{\partial x} = \frac{u_{i+1} - u_i}{\Delta x} + \frac{(\Delta x)}{2!}\frac{\partial^2 u}{\partial x^2} + \frac{(\Delta x)^2}{3!}\frac{\partial^3 u}{\partial x^3} + \ldots \tag{10-4}$$

or

$$\frac{\partial u}{\partial x} = \frac{u_{i+1} - u_{i-1}}{2\Delta x} + \frac{(\Delta x)^2}{3!}\frac{\partial^3 u}{\partial x^3} + \ldots \tag{10-5}$$

The approximation (10-4) may be truncated and expressed as

$$\frac{\partial u}{\partial x} = \frac{u_{i+1} - u_i}{\Delta x} + O(\Delta x)$$

where the dominant (or leading) error term includes a second-order derivative, i.e., even. The second expression given by (10-5) is written as

$$\frac{\partial u}{\partial x} = \frac{u_{i+1} - u_{i-1}}{2\Delta x} + O(\Delta x)^2$$

where the dominant error term now includes an odd derivative. The behavior of error associated with finite difference equations is strongly influenced by the dominant error term. To clarify the types of error introduced to the finite difference equations, a convective dominated equation, where physical viscosity is absent, will be used. Thus, consider the wave equation and two different finite difference equations given by

$$u_i^{n+1} = u_i^n - c(u_i^n - u_{i-1}^n) \tag{10-6}$$

and

$$u_i^{n+1} = u_i^{n-1} - c(u_{i+1}^n - u_{i-1}^n) \tag{10-7}$$

The FDEs are recognized as the first upwind differencing scheme and the mid-point leapfrog method. To identify the dominant error term of an FDE, the modified equation must be investigated. The corresponding modified equations for the FDEs given by (10-6) and (10-7) are, respectively:

$$\frac{\partial u}{\partial t} = -a\frac{\partial u}{\partial x} + \frac{1}{2}a(\Delta x)(1-c)\frac{\partial^2 u}{\partial x^2} - \frac{1}{6}a(\Delta x)^2\left(2c^2 - 3c + 1\right)\frac{\partial^3 u}{\partial x^3} + \ldots \tag{10-8}$$

and

$$\frac{\partial u}{\partial t} = -a\frac{\partial u}{\partial x} + \frac{1}{6}a(\Delta x)^2\left(c^2 - 1\right)\frac{\partial^3 u}{\partial x^3} + \ldots \tag{10-9}$$

Observe that the dominant error terms in Equations (10-8) and (10-9) include second-order derivative and third-order derivative, respectively. Recall that, from a physical point of view, a second-order derivative is associated with diffusion. Indeed, the coefficient of the second-order derivative in Equation (10-8) is known as the *numerical viscosity.* Thus, it is obvious that the error associated with Equation (10-8) is dissipative and, hence, it is called *dissipation error.* On the other hand, an FDE scheme, where its corresponding modified equation possesses an odd-order derivative as the lead term in error, is associated with oscillations within the solution. Such an error is called *dispersion error.*

10.7 Grid Generation-Structured Grids

Finite difference equations are most efficiently solved in a rectangular domain (for 2-D applications and an equivalent hexahedral domain for 3-D applications) with equal grid spacings. Unfortunately, the majority of physical domains encountered are nonrectangular in shape. Thus, it is necessary to transform the nonrectangular physical domain to a rectangular computational domain where grid points are distributed at equal spacings. It is also important to note that the transformation allows the alignment of one of the coordinates along the body, thus facilitating the implementation of the boundary conditions. The objective of grid generation is then to identify the location of the grid points in the computational domain and the location of the corresponding grid points in the physical space. Furthermore, the metrics and Jacobian of transformation which are required for the solution of flow equations are computed within the grid generation routine.

Typically, grid generation schemes may be categorized as algebraic methods or differential methods. In the latter case, the scheme is based on the solution of a set of PDEs and may be subcategorized as either an elliptic, parabolic, or hyperbolic grid generation. Either category of grid generation scheme should include the following considerations.

1. A mapping which guarantees one-to-one correspondence ensuring grid lines of the same family do not cross each other;
2. Smoothness of the grid distribution;
3. Orthogonality or near orthogonality of the grid lines;
4. Options for grid clustering.

A brief summary of the advantages and disadvantages of each method is provided below.

1. Algebraic grids

   The advantages of this category of grid generators are:

   (a) They are very fast computationally;
   (b) Metrics may be evaluated analytically, thus avoiding numerical errors;
   (c) The ability to cluster grid points in different regions can be easily implemented.

   The disadvantages are:

   (a) Discontinuities at a boundary may propagate into the interior region which could lead to errors due to sudden changes in the metrics;
   (b) Smoothness and skewness may be difficult to control.

2. Elliptic grids

   The advantages of this class of grid generators are:

(a) Will provide smooth grid point distribution, i.e., if a boundary discontinuity point exists, it will be smoothed out in the interior domain;

(b) Numerous options for grid clustering and surface orthogonality are available;

(c) Method can be extended to 3-D problems.

The disadvantages of the method are:

(a) Computation time is large (compared to algebraic methods or hyperbolic grid generators);

(b) Specification of the forcing functions P and Q (or the constants used in these functions) is not easy;

(c) Metrics must be computed numerically.

3. Hyperbolic grids

The advantages of hyperbolic grid generators are:

(a) The grid system is orthogonal in two dimensions;

(b) Since a marching scheme is used for the solution of the system, computationally they are much faster compared to elliptic systems;

(c) Grid line spacing may be controlled by the cell area or arc-length functions.

The disadvantages are:

(a) Boundary discontinuity may be propagated into the interior domain;

(b) Specifying the cell-area or arc-length functions must be handled carefully. A bad selection of these functions easily leads to undesirable grid systems.

Finally, the metrics and Jacobian of tranformation are given by the following expressions.

1. Two dimensions

$$\begin{aligned}
\xi_x &= J y_\eta \\
\xi_y &= -J x_\eta \\
\eta_x &= -J y_\xi \\
\eta_y &= J x_\xi
\end{aligned}$$

where

$$J = \frac{1}{x_\xi y_\eta - y_\xi x_\eta}$$

2. Three dimensions

$$\xi_x = J(y_\eta z_\zeta - y_\zeta z_\eta)$$
$$\xi_y = J(x_\zeta z_\eta - x_\eta z_\zeta)$$
$$\xi_z = J(x_\eta y_\zeta - x_\zeta y_\eta)$$
$$\eta_x = J(y_\zeta z_\xi - y_\xi z_\zeta)$$
$$\eta_y = J(x_\xi z_\zeta - x_\zeta z_\xi)$$
$$\eta_z = J(x_\zeta y_\xi - x_\xi y_\zeta)$$
$$\zeta_x = J(y_\xi z_\eta - y_\eta z_\xi)$$
$$\zeta_y = J(x_\eta z_\xi - x_\xi z_\eta)$$
$$\zeta_z - J(x_\xi y_\eta - x_\eta y_\xi)$$
$$\xi_t = -(x_\tau \xi_x + y_\tau \xi_y + z_\tau \xi_z)$$
$$\eta_t = -(x_\tau \eta_x + y_\tau \eta_y + z_\tau \eta_z)$$
$$\zeta_t = -(x_\tau \zeta_x + y_\tau \zeta_y + z_\tau \zeta_z)$$

where

$$J = \frac{\partial(\xi\,,\eta\,,\zeta)}{\partial(x\,,y\,,z)} = \frac{1}{x_\xi(y_\eta z_\zeta - y_\zeta z_\eta) - x_\eta(y_\xi z_\zeta - y_\zeta z_\xi) + x_\zeta(y_\xi z_\eta - y_\eta z_\xi)}$$

10.8 Transformation of the Equations From the Physical Space to Computational Space

The partial derivatives expressed in the physical space are related to the partial derivatives in the computational space by the following relations:

$$\frac{\partial}{\partial t} = \frac{\partial}{\partial \tau} + \xi_t \frac{\partial}{\partial \xi} + \eta_t \frac{\partial}{\partial \eta} + \zeta_t \frac{\partial}{\partial \zeta}$$
$$\frac{\partial}{\partial x} = \xi_x \frac{\partial}{\partial \xi} + \eta_x \frac{\partial}{\partial \eta} + \zeta_x \frac{\partial}{\partial \zeta}$$
$$\frac{\partial}{\partial y} = \xi_y \frac{\partial}{\partial \xi} + \eta_y \frac{\partial}{\partial \eta} + \zeta_y \frac{\partial}{\partial \zeta}$$
$$\frac{\partial}{\partial x} = \xi_z \frac{\partial}{\partial \xi} + \eta_z \frac{\partial}{\partial \eta} + \zeta_z \frac{\partial}{\partial \zeta}$$

The Navier-Stokes equations in a flux vector form may be expressed in the physical space by

$$\frac{\partial Q}{\partial t} + \frac{\partial E}{\partial x} + \frac{\partial F}{\partial y} + \frac{\partial G}{\partial z} = \frac{\partial E_v}{\partial x} + \frac{\partial F_v}{\partial y} + \frac{\partial G_v}{\partial z}$$

can be transformed to the computational space and expressed by

$$\frac{\partial \overline{Q}}{\partial \tau} + \frac{\partial \overline{E}}{\partial \xi} + \frac{\partial \overline{F}}{\partial \eta} + \frac{\partial \overline{G}}{\partial \zeta} = \frac{\partial \overline{E}_v}{\partial \xi} + \frac{\partial \overline{F}_v}{\partial \eta} + \frac{\partial \overline{G}_v}{\partial \zeta}$$

where

$$\overline{Q} = \frac{Q}{J}$$

$$\overline{E} = \frac{1}{J}(\xi_t Q + \xi_x E + \xi_y F + \xi_z G)$$

$$\overline{F} = \frac{1}{J}(\eta_t Q + \eta_x E + \eta_y F + \eta_z G)$$

$$\overline{G} = \frac{1}{J}(\zeta_t Q + \zeta_x E + \zeta_y F + \zeta_z G)$$

$$\overline{E}_v = \frac{1}{J}(\xi_x E_v + \xi_y F_v + \xi_z G_v)$$

$$\overline{F}_v = \frac{1}{J}(\eta_x E_v + \eta_y F_v + \eta_z G_v)$$

$$\overline{G}_v = \frac{1}{J}(\zeta_x E_v + \zeta_y F_v + \zeta_z G_v)$$

The inviscid and viscous Jacobian matrices which are produced in the process of linearization of the equations are given in Chapter 11 for the Navier-Stokes, Thin-Layer Navier-Stokes, Euler, and Parabolized Navier-Stokes equations.

Chapter 11

Transformation of the Equations of Fluid Motion from Physical Space to Computational Space

11.1 Introductory Remarks

To enhance the efficiency and accuracy of a numerical scheme and to simplify implementation of boundary conditions, a transformation from physical space to computational space is performed. This transformation allows clustering of grid points in regions where flow variables undergo high gradients and grid point motion when required. The computational domain is a rectangular shape which is divided into an equally spaced grid system. In order to solve the governing equations of motion in the computational space, a transformation of the equations from physical space into computational space is required. Any assumption on the simplification of the equations of motion is imposed on the transformed equations. For example, to reduce computational time and required storage, the full Navier-Stokes equations may be simplified by neglecting the circumferential and streamwise gradient of stresses, while retaining only the normal gradient of the stresses. The resulting equations are known as the Thin-Layer Navier-Stokes equations. The reduction of equations is performed on the transformed equations.

This chapter investigates generalized coordinate transformation of the governing equations of fluid motion expressed in the Cartesian coordinate system (x, y, z) from physical space to computational space (ξ, η, ζ). Various formulations of the equations which are discussed include full Navier-Stokes (NS), Euler, Thin-Layer Navier-Stokes (TLNS), and Parabolized Navier-Stokes (PNS) equations. A summary of the assumptions which are imposed in the reduction process of the equations is presented. The formulations include three-dimensional flows as well as two-dimensional planar or axisymmetric flow fields.

In order to include the capability for shock capturing, the equations of motion are written in conservative form. Any set of equations is then approximated by finite difference formulation, which is solved in the rectangular grid system. Furthermore, in the linearization process Jacobian matrices are produced, which are included in this chapter.

This chapter summarizes the equations of fluid motion in computational space and presents them in concise and orderly manner. Thus, the objectives of this chapter are summarized as follows:

(1) Define the metrics and the Jacobian of transformation;

(2) Express the equations of fluid motion in a generalized coordinate system for NS, TLNS, Euler, and PNS equations; and

(3) Derive the Jacobian matrices for each set of equations which are used in various numerical algorithms.

11.2 Generalized Coordinate Transformation

The equations of motion are transformed from the physical space (x, y, z) to the computational space (ξ, η, ζ) by the following relations:

$$\tau = t \tag{11-1}$$

$$\xi = \xi(t, x, y, z) \tag{11-2}$$

$$\eta = \eta(t, x, y, z) \tag{11-3}$$

$$\zeta = \zeta(t, x, y, z) \tag{11-4}$$

The chain rule of partial differentiation provides the following expressions for the Cartesian derivatives:

$$\frac{\partial}{\partial t} = \frac{\partial}{\partial \tau} + \xi_t \frac{\partial}{\partial \xi} + \eta_t \frac{\partial}{\partial \eta} + \zeta_t \frac{\partial}{\partial \zeta} \tag{11-5}$$

$$\frac{\partial}{\partial x} = \xi_x \frac{\partial}{\partial \xi} + \eta_x \frac{\partial}{\partial \eta} + \zeta_x \frac{\partial}{\partial \zeta} \tag{11-6}$$

$$\frac{\partial}{\partial y} = \xi_y \frac{\partial}{\partial \xi} + \eta_y \frac{\partial}{\partial \eta} + \zeta_y \frac{\partial}{\partial \zeta} \tag{11-7}$$

$$\frac{\partial}{\partial z} = \xi_z \frac{\partial}{\partial \xi} + \eta_z \frac{\partial}{\partial \eta} + \zeta_z \frac{\partial}{\partial \zeta} \tag{11-8}$$

11.2.1 Equations for the Metrics

From Equations (11-5) through (11-8), it is obvious that the value of the metrics ξ_t, η_t, ζ_t, ξ_x, η_x, ζ_x, ξ_y, η_y, ζ_y, ξ_z, η_z, and ζ_z must be provided in some fashion. In most cases the analytical determination of the metrics is not possible and, therefore, they must be computed numerically. Since the stepsizes in the computational domain are equally spaced, x_ξ, x_η, x_ζ, etc., can be computed by various finite difference approximations. Thus, if the metrics appearing in Equations (11-5) through (11-8) can be expressed in terms of these derivatives, the numerical computation of metrics is completed. To obtain such relations, the following differential expressions are considered:

$$dt = \frac{\partial t}{\partial \tau} d\tau + \frac{\partial t}{\partial \xi} d\xi + \frac{\partial t}{\partial \eta} d\eta + \frac{\partial t}{\partial \zeta} d\zeta$$

But according to (11-1),

$$\frac{\partial t}{\partial \tau} = 1 \quad \text{and}$$

$$\frac{\partial t}{\partial \xi} = \frac{\partial t}{\partial \eta} = \frac{\partial t}{\partial \zeta} = 0 \quad \text{thus}$$

$$dt = d\tau \tag{11-9}$$

Similarly,

$$dx = x_\tau d\tau + x_\xi d\xi + x_\eta d\eta + x_\zeta d\zeta \tag{11-10}$$

$$dy = y_\tau d\tau + y_\xi d\xi + y_\eta d\eta + y_\zeta d\zeta \tag{11-11}$$

$$dz = z_\tau d\tau + z_\xi d\xi + z_\eta d\eta + z_\zeta d\zeta \tag{11-12}$$

Equations (11-9) through (11-12) are expressed in a matrix form as

$$\begin{bmatrix} dt \\ dx \\ dy \\ dz \end{bmatrix} = \begin{bmatrix} 1 & 0 & 0 & 0 \\ x_\tau & x_\xi & x_\eta & x_\zeta \\ y_\tau & y_\xi & y_\eta & y_\zeta \\ z_\tau & z_\xi & z_\eta & z_\zeta \end{bmatrix} \begin{bmatrix} d\tau \\ d\xi \\ d\eta \\ d\zeta \end{bmatrix} \tag{11-13}$$

Reversing the role of the independent variables,

$$d\tau = dt \tag{11-14}$$

$$d\xi = \xi_t dt + \xi_x dx + \xi_y dy + \xi_z dz \tag{11-15}$$

$$d\eta = \eta_t dt + \eta_x dx + \eta_y dy + \eta_z dz \tag{11-16}$$

$$d\zeta = \zeta_t dt + \zeta_x dx + \zeta_y dy + \zeta_z dz \tag{11-17}$$

which are expressed as

$$\begin{bmatrix} d\tau \\ d\xi \\ d\eta \\ d\zeta \end{bmatrix} = \begin{bmatrix} 1 & 0 & 0 & 0 \\ \xi_t & \xi_x & \xi_y & \xi_z \\ \eta_t & \eta_x & \eta_y & \eta_z \\ \zeta_t & \zeta_x & \zeta_y & \zeta_z \end{bmatrix} \begin{bmatrix} dt \\ dx \\ dy \\ dz \end{bmatrix} \tag{11-18}$$

Comparing Equations (11-13) and (11-18), one concludes that

$$\begin{bmatrix} 1 & 0 & 0 & 0 \\ \xi_t & \xi_x & \xi_y & \xi_z \\ \eta_t & \eta_x & \eta_y & \eta_z \\ \zeta_t & \zeta_x & \zeta_y & \zeta_z \end{bmatrix} = \begin{bmatrix} 1 & 0 & 0 & 0 \\ x_\tau & x_\xi & x_\eta & x_\zeta \\ y_\tau & y_\xi & y_\eta & y_\zeta \\ z_\tau & z_\xi & z_\eta & z_\zeta \end{bmatrix}^{-1}$$

From which,

$$\xi_x = J(y_\eta z_\zeta - y_\zeta z_\eta) \tag{11-19}$$

$$\xi_y = J(x_\zeta z_\eta - x_\eta z_\zeta) \tag{11-20}$$

$$\xi_z = J(x_\eta y_\zeta - x_\zeta y_\eta) \tag{11-21}$$

$$\eta_x = J(y_\zeta z_\xi - y_\xi z_\zeta) \tag{11-22}$$

$$\eta_y = J(x_\xi z_\zeta - x_\zeta z_\xi) \tag{11-23}$$

$$\eta_z = J(x_\zeta y_\xi - x_\xi y_\zeta) \tag{11-24}$$

$$\zeta_x = J(y_\xi z_\eta - y_\eta z_\xi) \tag{11-25}$$

$$\zeta_y = J(x_\eta z_\xi - x_\xi z_\eta) \tag{11-26}$$

$$\zeta_z = J(x_\xi y_\eta - x_\eta y_\xi) \tag{11-27}$$

$$\xi_t = -(x_\tau \xi_x + y_\tau \xi_y + z_\tau \xi_z) \tag{11-28}$$

$$\eta_t = -(x_\tau \eta_x + y_\tau \eta_y + z_\tau \eta_z) \tag{11-29}$$

$$\zeta_t = -(x_\tau \zeta_x + y_\tau \zeta_y + z_\tau \zeta_z) \tag{11-30}$$

After substitution of Equations (11-19) through (11-27) into Equations (11-28), (11-29), and (11-30),

$$\xi_t = J[x_\tau(y_\zeta z_\eta - y_\eta z_\zeta) + y_\tau(x_\eta z_\zeta - x_\zeta z_\eta) + z_\tau(x_\zeta y_\eta - x_\eta y_\zeta)] \tag{11-31}$$

$$\eta_t = J[x_\tau(y_\xi z_\zeta - y_\zeta z_\xi) + y_\tau(x_\zeta z_\xi - x_\xi z_\zeta) + z_\tau(x_\xi y_\zeta - x_\zeta y_\xi)] \tag{11-32}$$

$$\zeta_t = J[x_\tau(y_\eta z_\xi - y_\xi z_\eta) + y_\tau(x_\xi z_\eta - x_\eta z_\xi) + z_\tau(x_\eta y_\xi - x_\xi y_\eta)] \tag{11-33}$$

where J is the Jacobian of transformation defined by

$$J = \frac{\partial(\xi,\eta,\zeta)}{\partial(x,y,z)} = \frac{1}{x_\xi(y_\eta z_\zeta - y_\zeta z_\eta) - x_\eta(y_\xi z_\zeta - y_\zeta z_\xi) + x_\zeta(y_\xi z_\eta - y_\eta z_\xi)} \tag{11-34}$$

11.3 Nondimensionalization of the Equations of Fluid Motion

Equations of fluid motion may be nondimensionalized to achieve certain objectives. For one, it would provide conditions upon which dynamic and energetic similarity may be obtained for geometrically similar situations. Second, the solution of such equations would usually provide values within limits between zero and one. Generally a characteristic dimension, such as the chord of an airfoil or the length of a vehicle, is selected to nondimensionalize the independent spatial variables. Freestream conditions are used to nondimensionalize the dependent variables. Among many choices available, the following will be used in this and in the subsequent chapters:

$$t^* = \frac{tu_\infty}{L}, \quad x^* = \frac{x}{L}, \quad y^* = \frac{y}{L}, \quad z^* = \frac{z}{L}$$

$$\mu^* = \frac{\mu}{\mu_\infty}, \quad u^* = \frac{u}{u_\infty}, \quad v^* = \frac{v}{u_\infty}, \quad w^* = \frac{w}{u_\infty}$$

$$\rho^* = \frac{\rho}{\rho_\infty}, \quad T^* = \frac{T}{T_\infty}, \quad p^* = \frac{p}{\rho_\infty u_\infty^2}, \quad e_t^* = \frac{e_t}{u_\infty^2}$$

The nondimensional parameters are defined as:

Reynolds number: $Re_\infty = \dfrac{\rho_\infty u_\infty L}{\mu_\infty}$

Prandtl number: $Pr = \dfrac{\mu c_p}{k}$

Using the nondimensional variables defined above, the equations of fluid motion in the Cartesian coordinate system are expressed as:

1. Continuity:

$$\frac{\partial \rho^*}{\partial t^*} + \frac{\partial}{\partial x^*}(\rho^* u^*) + \frac{\partial}{\partial y^*}(\rho^* v^*) + \frac{\partial}{\partial z^*}(\rho^* w^*) = 0 \tag{11-35}$$

2. X-component of the momentum equation:

$$\frac{\partial}{\partial t^*}(\rho^* u^*) + \frac{\partial}{\partial x^*}(\rho^* u^{*^2} + p^*) + \frac{\partial}{\partial y^*}(\rho^* u^* v^*) + \frac{\partial}{\partial z^*}(\rho^* u^* w^*) =$$

$$\frac{\partial}{\partial x^*}(\tau_{xx}^*) + \frac{\partial}{\partial y^*}(\tau_{xy}^*) + \frac{\partial}{\partial z^*}(\tau_{xz}^*) \tag{11-36}$$

For a Newtonian fluid,

$$\tau^*_{xx} = \frac{1}{Re_\infty}\left[2\mu^* \frac{\partial u^*}{\partial x^*} + \lambda^* \nabla^* \cdot \vec{V}^*\right] \tag{11-37}$$

$$\tau^*_{xy} = \tau^*_{yx} = \frac{1}{Re_\infty}\left[\mu^* (\frac{\partial u^*}{\partial y^*} + \frac{\partial v^*}{\partial x^*})\right] \tag{11-38}$$

$$\tau^*_{xz} = \tau^*_{zx} = \frac{1}{Re_\infty}\left[\mu^* (\frac{\partial w^*}{\partial x^*} + \frac{\partial u^*}{\partial z^*})\right] \tag{11-39}$$

3. Y-component of the momentum equation:

$$\frac{\partial}{\partial t^*}(\rho^* v^*) + \frac{\partial}{\partial x^*}(\rho^* u^* v^*) + \frac{\partial}{\partial y^*}(\rho^* v^{*^2} + p^*) + \frac{\partial}{\partial z^*}(\rho^* v^* w^*) =$$

$$\frac{\partial}{\partial x^*}(\tau^*_{xy}) + \frac{\partial}{\partial y^*}(\tau^*_{yy}) + \frac{\partial}{\partial z^*}(\tau^*_{yz}) \tag{11-40}$$

where

$$\tau^*_{yy} = \frac{1}{Re_\infty}\left[2\mu^* \frac{\partial v^*}{\partial y^*} + \lambda^* \nabla^* \cdot \vec{V}^*\right] \tag{11-41}$$

$$\tau^*_{yz} = \tau^*_{zy} = \frac{1}{Re_\infty}\left[\mu^* (\frac{\partial v^*}{\partial z^*} + \frac{\partial w^*}{\partial y^*})\right] \tag{11-42}$$

and τ^*_{xy} is given by (11-38).

4. Z-component of the momentum equation:

$$\frac{\partial}{\partial t^*}(\rho^* w^*) + \frac{\partial}{\partial x^*}(\rho^* u^* w^*) + \frac{\partial}{\partial y^*}(\rho^* v^* w^*) + \frac{\partial}{\partial z^*}(\rho^* w^{*^2} + p^*) =$$

$$\frac{\partial}{\partial x^*}(\tau^*_{xz}) + \frac{\partial}{\partial y^*}(\tau^*_{yz}) + \frac{\partial}{\partial z^*}(\tau^*_{zz}) \tag{11-43}$$

where

$$\tau^*_{zz} = \frac{1}{Re_\infty}\left[2\mu^* \frac{\partial w^*}{\partial z^*} + \lambda^* \nabla^* \cdot \vec{V}^*\right] \tag{11-44}$$

and τ^*_{xz} and τ^*_{yz} are given by Equations (11-39) and (11-42), respectively.

5. Energy:

$$\frac{\partial}{\partial t^*}(\rho^* e_t^*) + \frac{\partial}{\partial x^*}(\rho^* u^* e_t^* + p^* u^*) + \frac{\partial}{\partial y^*}(\rho^* v^* e_t^* + p^* v^*) +$$

$$\frac{\partial}{\partial z^*}(\rho^* w^* e_t^* + p^* w^*) = \frac{\partial}{\partial x^*}[u^*\tau_{xx}^* + v^*\tau_{xy}^* + w^*\tau_{xz}^* - q_x^*] + \tag{11-45}$$

$$\frac{\partial}{\partial y^*}[u^*\tau_{yx}^* + v^*\tau_{yy}^* + w^*\tau_{yz}^* - q_y^*] + \frac{\partial}{\partial z^*}[u^*\tau_{zx}^* + v^*\tau_{zy}^* + w^*\tau_{zz}^* - q_z^*]$$

where

$$q_x^* = -\frac{\mu^*}{Re_\infty Pr\,(\gamma - 1)\,M_\infty^2}\frac{\partial T^*}{\partial x^*} \tag{11-46}$$

$$q_y^* = -\frac{\mu^*}{Re_\infty Pr\,(\gamma - 1)\,M_\infty^2}\frac{\partial T^*}{\partial y^*} \tag{11-47}$$

$$q_z^* = -\frac{\mu^*}{Re_\infty Pr\,(\gamma - 1)\,M_\infty^2}\frac{\partial T^*}{\partial z^*} \tag{11-48}$$

The nondimensional equations of fluid motion may be expressed in a flux vector form as:

$$\frac{\partial Q^*}{\partial t^*} + \frac{\partial E^*}{\partial x^*} + \frac{\partial F^*}{\partial y^*} + \frac{\partial G^*}{\partial z^*} = \frac{\partial E_v^*}{\partial x^*} + \frac{\partial F_v^*}{\partial y^*} + \frac{\partial G_v^*}{\partial z^*} \tag{11-49}$$

where

$$Q^* = \begin{bmatrix} \rho^* \\ \rho^* u^* \\ \rho^* v^* \\ \rho^* w^* \\ \rho^* e_t^* \end{bmatrix} \tag{11-50}$$

$$E^* = \begin{bmatrix} \rho^* u^* \\ \rho^* u^{*^2} + p^* \\ \rho^* u^* v^* \\ \rho^* u^* w^* \\ (\rho^* e_t^* + p^*)u^* \end{bmatrix} \tag{11-51}$$

$$E_v^* = \begin{bmatrix} 0 \\ \tau_{xx}^* \\ \tau_{xy}^* \\ \tau_{xz}^* \\ u^*\tau_{xx}^* + v^*\tau_{xy}^* + w^*\tau_{xz}^* - q_x^* \end{bmatrix} \tag{11-52}$$

$$F^* = \begin{bmatrix} \rho^* v^* \\ \rho^* v^* u^* \\ \rho^* v^{*^2} + p^* \\ \rho^* v^* w^* \\ (\rho^* e_t^* + p^*)v^* \end{bmatrix} \tag{11-53}$$

$$F_v^* = \begin{bmatrix} 0 \\ \tau_{yx}^* \\ \tau_{yy}^* \\ \tau_{yz}^* \\ u^*\tau_{yx}^* + v^*\tau_{yy}^* + w^*\tau_{yz}^* - q_y^* \end{bmatrix} \tag{11-54}$$

$$G^* = \begin{bmatrix} \rho^* w^* \\ \rho^* w^* u^* \\ \rho^* w^* v^* \\ \rho^* w^{*^2} + p^* \\ (\rho^* e_t^* + p^*) w^* \end{bmatrix} \quad (11\text{-}55) \qquad G_v^* = \begin{bmatrix} 0 \\ \tau_{zx}^* \\ \tau_{zy}^* \\ \tau_{zz}^* \\ u^* \tau_{zx}^* + v^* \tau_{zy}^* + w^* \tau_{zz}^* - q_z^* \end{bmatrix} \quad (11\text{-}56)$$

The notation "*" which is used to designate nondimensional quantities will be dropped for the remainder of this chapter and those following. Thus, all the equations will be in nondimensional form unless otherwise specified.

11.4 Navier-Stokes Equations

The equations of fluid motion in complete form which include the conservation of mass, conservation of momentum and conservation of energy are referred to as the Navier-Stokes equations. The nondimensional Navier-Stokes equations given in the previous section in the Cartesian coordinate system are now transformed to the computational space by the following. Due to similarity of the left- and right-hand sides of Equation (11-49), the mathematical details are carried out for the left-hand side (LHS) of Equation (11-49) only. Equation (11-49), which is repeated here for convenience, is:

$$\frac{\partial Q}{\partial t} + \frac{\partial E}{\partial x} + \frac{\partial F}{\partial y} + \frac{\partial G}{\partial z} = \frac{\partial E_v}{\partial x} + \frac{\partial F_v}{\partial y} + \frac{\partial G_v}{\partial z}$$

The Cartesian derivatives are replaced by Equations (11-5) through (11-8) to yield:

$$LHS = \frac{\partial Q}{\partial \tau} + \xi_t \frac{\partial Q}{\partial \xi} + \eta_t \frac{\partial Q}{\partial \eta} + \zeta_t \frac{\partial Q}{\partial \zeta} + \xi_x \frac{\partial E}{\partial \xi} + \eta_x \frac{\partial E}{\partial \eta} +$$

$$\zeta_x \frac{\partial E}{\partial \zeta} + \xi_y \frac{\partial F}{\partial \xi} + \eta_y \frac{\partial F}{\partial \eta} + \zeta_y \frac{\partial F}{\partial \zeta} + \xi_z \frac{\partial G}{\partial \xi} + \eta_z \frac{\partial G}{\partial \eta} + \zeta_z \frac{\partial G}{\partial \zeta} \quad (11\text{-}57)$$

It is recognized that this equation is no longer in a conservative form. To recast the equation in a conservative form, some manipulation must be performed. To do so, Equation (11-57) is first divided by J, and then a combination of terms which sums up to zero is added. In the following, only four terms are considered to show the required mathematical steps. The conclusion is then extended to the remaining terms. The first four terms of Equation (11-57) divided by J and with the added zeros shown as the brackets are

$$\frac{1}{J}\frac{\partial Q}{\partial \tau} + \frac{1}{J}\xi_t\frac{\partial Q}{\partial \xi} + \frac{1}{J}\eta_t\frac{\partial Q}{\partial \eta} + \frac{1}{J}\zeta_t\frac{\partial Q}{\partial \zeta} + \left[Q\frac{\partial}{\partial \tau}(\frac{1}{J}) - Q\frac{\partial}{\partial \tau}(\frac{1}{J})\right]$$

$$+ \left[Q\frac{\partial}{\partial \xi}(\frac{\xi_t}{J}) - Q\frac{\partial}{\partial \xi}(\frac{\xi_t}{J})\right] + \left[Q\frac{\partial}{\partial \eta}(\frac{\eta_t}{J}) - Q\frac{\partial}{\partial \eta}(\frac{\eta_t}{J})\right]$$

$$+ \left[Q\frac{\partial}{\partial \zeta}(\frac{\zeta_t}{J}) - Q\frac{\partial}{\partial \zeta}(\frac{\zeta_t}{J})\right]$$

which may be rearranged as

$$\left[\frac{1}{J}\frac{\partial Q}{\partial \tau} + Q\frac{\partial}{\partial \tau}(\frac{1}{J})\right] + \left[\frac{\xi_t}{J}\frac{\partial Q}{\partial \xi} + Q\frac{\partial}{\partial \xi}(\frac{\xi_t}{J})\right]$$

$$+ \left[\frac{\eta_t}{J}\frac{\partial Q}{\partial \eta} + Q\frac{\partial}{\partial \eta}(\frac{\eta_t}{J})\right] + \left[\frac{\zeta_t}{J}\frac{\partial Q}{\partial \zeta} + Q\frac{\partial}{\partial \zeta}(\frac{\zeta_t}{J})\right]$$

$$- Q\left[\frac{\partial}{\partial \tau}(\frac{1}{J}) + \frac{\partial}{\partial \xi}(\frac{\xi_t}{J}) + \frac{\partial}{\partial \eta}(\frac{\eta_t}{J}) + \frac{\partial}{\partial \zeta}(\frac{\zeta_t}{J})\right] \quad (11\text{-}58)$$

The terms in the first four brackets may be combined, and by substitution of expressions (11-31) through (11-34) into the last bracket, it can be shown that it is zero. Therefore, expression (11-58) is combined as

$$\frac{\partial}{\partial \tau}(\frac{Q}{J}) + \frac{\partial}{\partial \xi}(\xi_t\frac{Q}{J}) + \frac{\partial}{\partial \eta}(\eta_t\frac{Q}{J}) + \frac{\partial}{\partial \zeta}(\zeta_t\frac{Q}{J})$$

Extending this conclusion to the remaining terms of Equation (11-57) results in the following expression:

$$\begin{aligned} LHS = {} & \frac{\partial}{\partial \tau}(\frac{Q}{J}) + \frac{\partial}{\partial \xi}(\xi_t\frac{Q}{J}) + \frac{\partial}{\partial \eta}(\eta_t\frac{Q}{J}) + \frac{\partial}{\partial \zeta}(\zeta_t\frac{Q}{J}) + \\ & \frac{\partial}{\partial \xi}(\xi_x\frac{E}{J}) + \frac{\partial}{\partial \eta}(\eta_x\frac{E}{J}) + \frac{\partial}{\partial \zeta}(\zeta_x\frac{E}{J}) + \frac{\partial}{\partial \xi}(\xi_y\frac{F}{J}) + \frac{\partial}{\partial \eta}(\eta_y\frac{F}{J}) + \\ & \frac{\partial}{\partial \zeta}(\zeta_y\frac{F}{J}) + \frac{\partial}{\partial \xi}(\xi_z\frac{G}{J}) + \frac{\partial}{\partial \eta}(\eta_z\frac{G}{J}) + \frac{\partial}{\partial \zeta}(\zeta_z\frac{G}{J}) \end{aligned} \quad (11\text{-}59)$$

The RHS of Equation (11-49) is reformulated in a similar fashion, and therefore

(11-49) is written as

$$\frac{\partial}{\partial \tau}(\frac{Q}{J}) + \frac{\partial}{\partial \xi}\left[\frac{1}{J}(\xi_t Q + \xi_x E + \xi_y F + \xi_z G)\right] +$$

$$\frac{\partial}{\partial \eta}\left[\frac{1}{J}(\eta_t Q + \eta_x E + \eta_y F + \eta_z G)\right] + \frac{\partial}{\partial \zeta}\left[\frac{1}{J}(\zeta_t Q + \zeta_x E + \zeta_y F + \zeta_z G)\right]$$

$$= \frac{\partial}{\partial \xi}\left[\frac{1}{J}(\xi_x E_v + \xi_y F_v + \xi_z G_v)\right] + \frac{\partial}{\partial \eta}\left[\frac{1}{J}(\eta_x E_v + \eta_y F_v + \eta_z G_v)\right] +$$

$$\frac{\partial}{\partial \zeta}\left[\frac{1}{J}(\zeta_x E_v + \zeta_y F_v + \zeta_z G_v)\right]$$

The terms in this equation are redefined such that it may be written as

$$\frac{\partial \bar{Q}}{\partial \tau} + \frac{\partial \bar{E}}{\partial \xi} + \frac{\partial \bar{F}}{\partial \eta} + \frac{\partial \bar{G}}{\partial \zeta} = \frac{\partial \bar{E}_v}{\partial \xi} + \frac{\partial \bar{F}_v}{\partial \eta} + \frac{\partial \bar{G}_v}{\partial \zeta} \tag{11-60}$$

where

$$\bar{Q} = \frac{Q}{J} \tag{11-61}$$

$$\bar{E} = \frac{1}{J}(\xi_t Q + \xi_x E + \xi_y F + \xi_z G) \tag{11-62}$$

$$\bar{F} = \frac{1}{J}(\eta_t Q + \eta_x E + \eta_y F + \eta_z G) \tag{11-63}$$

$$\bar{G} = \frac{1}{J}(\zeta_t Q + \zeta_x E + \zeta_y F + \zeta_z G) \tag{11-64}$$

$$\bar{E}_v = \frac{1}{J}(\xi_x E_v + \xi_y F_v + \xi_z G_v) \tag{11-65}$$

$$\bar{F}_v = \frac{1}{J}(\eta_x E_v + \eta_y F_v + \eta_z G_v) \tag{11-66}$$

$$\bar{G}_v = \frac{1}{J}(\zeta_x E_v + \zeta_y F_v + \zeta_z G_v) \tag{11-67}$$

The viscous shear stresses given by Equations (11-37), (11-38), (11-39), (11-41), (11-42) and (11-44) with Stoke's hypothesis ($\lambda = -\frac{2}{3}\mu$) in the transformed computational space are:

$$\tau_{xx} = \frac{\mu}{Re_\infty}\left[\frac{4}{3}(\xi_x u_\xi + \eta_x u_\eta + \zeta_x u_\zeta) - \frac{2}{3}(\xi_y v_\xi + \eta_y v_\eta + \zeta_y v_\zeta) - \frac{2}{3}(\xi_z w_\xi + \eta_z w_\eta + \zeta_z w_\zeta)\right] \quad (11\text{-}68)$$

$$\tau_{yy} = \frac{\mu}{Re_\infty}\left[\frac{4}{3}(\xi_y v_\xi + \eta_y v_\eta + \zeta_y v_\zeta) - \frac{2}{3}(\xi_x u_\xi + \eta_x u_\eta + \zeta_x u_\zeta) - \frac{2}{3}(\xi_z w_\xi + \eta_z w_\eta + \zeta_z w_\zeta)\right] \quad (11\text{-}69)$$

$$\tau_{zz} = \frac{\mu}{Re_\infty}\left[\frac{4}{3}(\xi_z w_\xi + \eta_z w_\eta + \zeta_z w_\zeta) - \frac{2}{3}(\xi_x u_\xi + \eta_x u_\eta + \zeta_x u_\zeta) - \frac{2}{3}(\xi_y v_\xi + \eta_y v_\eta + \zeta_y v_\zeta)\right] \quad (11\text{-}70)$$

$$\tau_{xy} = \tau_{yx} = \frac{\mu}{Re_\infty}(\xi_y u_\xi + \eta_y u_\eta + \zeta_y u_\zeta + \xi_x v_\xi + \eta_x v_\eta + \zeta_x v_\zeta) \quad (11\text{-}71)$$

$$\tau_{xz} = \tau_{zx} = \frac{\mu}{Re_\infty}(\xi_z u_\xi + \eta_z u_\eta + \zeta_z u_\zeta + \xi_x w_\xi + \eta_x w_\eta + \zeta_x w_\zeta) \quad (11\text{-}72)$$

$$\tau_{yz} = \tau_{zy} = \frac{\mu}{Re_\infty}(\xi_z v_\xi + \eta_z v_\eta + \zeta_z v_\zeta + \xi_y w_\xi + \eta_y w_\eta + \zeta_y w_\zeta) \quad (11\text{-}73)$$

where the heat conduction terms in the computational space are:

$$q_x = -\frac{\mu}{Pr\, Re_\infty(\gamma-1)M_\infty^2}(\xi_x T_\xi + \eta_x T_\eta + \zeta_x T_\zeta) \quad (11\text{-}74)$$

$$q_y = -\frac{\mu}{Pr\, Re_\infty(\gamma-1)M_\infty^2}(\xi_y T_\xi + \eta_y T_\eta + \zeta_y T_\zeta) \quad (11\text{-}75)$$

$$q_z = -\frac{\mu}{Pr\, Re_\infty(\gamma-1)M_\infty^2}(\xi_z T_\xi + \eta_z T_\eta + \zeta_z T_\zeta) \quad (11\text{-}76)$$

11.4.1 Linearization

In order to numerically solve Equation (11-60), a linearization procedure is introduced, and all flux vectors are expressed in terms of the flux vector $\bar{Q}$. The procedure was previously described in Chapter 8; however, due to its importance, it is reemphasized in this section.

Consider the following Taylor series expansion:

$$\bar{E}^{n+1} = \bar{E}^n + \frac{\partial \bar{E}}{\partial \tau}\Delta\tau + O(\Delta\tau)^2 \tag{11-77}$$

In order to rewrite $\partial\bar{E}/\partial\tau$ in terms of the gradient of flux vector $\bar{Q}$, recall that

$$\bar{E} = f(\bar{Q}, \xi_t, \xi_x, \xi_y, \xi_z)$$

The chain-rule of differentiation yields the following relation:

$$\frac{\partial \bar{E}}{\partial \tau} = \frac{\partial \bar{E}}{\partial \bar{Q}}\frac{\partial \bar{Q}}{\partial \tau} + \frac{\partial \bar{E}}{\partial \xi_t}\frac{\partial \xi_t}{\partial \tau} + \frac{\partial \bar{E}}{\partial \xi_x}\frac{\partial \xi_x}{\partial \tau} + \frac{\partial \bar{E}}{\partial \xi_y}\frac{\partial \xi_y}{\partial \tau} + \frac{\partial \bar{E}}{\partial \xi_z}\frac{\partial \xi_z}{\partial \tau} \tag{11-78}$$

For many applications, the grid is independent of time, and therefore time gradients of the metrics are zero. Hence, for a time independent grid system, Equation (11-78) is reduced to

$$\frac{\partial \bar{E}}{\partial \tau} = \frac{\partial \bar{E}}{\partial \bar{Q}}\frac{\partial \bar{Q}}{\partial \tau} \tag{11-79}$$

This equation is substituted into Equation (11-77) to yield:

$$\bar{E}^{n+1} = \bar{E}^n + \frac{\partial \bar{E}}{\partial \bar{Q}}\frac{\partial \bar{Q}}{\partial \tau}\Delta\tau + O(\Delta\tau)^2 \tag{11-80}$$

The partial derivative $\partial\bar{Q}/\partial\tau$ is approximated by a first-order backward difference expression as

$$\frac{\partial \bar{Q}}{\partial \tau} = \frac{\bar{Q}^{n+1} - \bar{Q}^n}{\Delta\tau} + O(\Delta\tau) = \frac{\Delta\bar{Q}}{\Delta\tau} + O(\Delta\tau)$$

Substitution of this equation into Equation (11-80) yields:

$$\bar{E}^{n+1} = \bar{E}^n + \frac{\partial \bar{E}}{\partial \bar{Q}}\left[\frac{\Delta\bar{Q}}{\Delta\tau} + O(\Delta\tau)\right]\Delta\tau + O(\Delta\tau)^2$$

or

$$\bar{E}^{n+1} = \bar{E}^n + \frac{\partial \bar{E}}{\partial \bar{Q}}\Delta\bar{Q} + O(\Delta\tau)^2 \tag{11-81}$$

If a time dependent grid system is used, the additional terms which appeared in Equation (11-78) would be included in Equation (11-81) as well. In either case, terms such as $\partial\bar{E}/\partial\bar{Q}$ and similar terms related to flux vectors $\bar{F}, \bar{G}, \bar{E}_v, \bar{F}_v$, and $\bar{G}_v$ will always appear in the linearization process. They are defined as the Jacobian matrices, which are the focus of the next section.

For steady-state problems such as Parabolized Navier-Stokes equations, the linearization process is similar. In that case, the Taylor series is expanded with respect to the streamwise space coordinate (marching direction). A detailed mathematical approach is shown in Chapter 13.

11.4.2 Inviscid and Viscous Jacobian Matrices

To linearize Equation (11-60), the following and similar approximations are employed:

$$\bar{E}^{n+1} = \bar{E}^n + \frac{\partial \bar{E}}{\partial \bar{Q}} \Delta \bar{Q} + O(\Delta \tau)^2$$

where $\partial \bar{E}/\partial \bar{Q}$ is defined as the flux Jacobian matrix. The remaining flux Jacobian matrices are $\partial \bar{F}/\partial \bar{Q}$ and $\partial \bar{G}/\partial \bar{Q}$. The viscous Jacobian matrices are $\partial \bar{E}_v/\partial \bar{Q}$, $\partial \bar{F}_v/\partial \bar{Q}$, and $\partial \bar{G}_v/\partial \bar{Q}$. Since flux vectors $\bar{Q}$ and $\bar{E}$ are 5×1 vectors, each one of the Jacobian matrices are 5×5 (for three-dimensional problems). The derivations of the Jacobians are considered next. In the following derivations we will assume perfect gas; therefore, the equation of state is expressed as

$$p = \rho e(\gamma - 1).$$

The inviscid Jacobian $\partial \bar{E}/\partial \bar{Q}$ is

$$\frac{\partial \bar{E}}{\partial \bar{Q}} = \frac{\partial(\bar{E}_1,\, \bar{E}_2,\, \bar{E}_3,\, \bar{E}_4,\, \bar{E}_5)}{\partial(\bar{Q}_1,\, \bar{Q}_2,\, \bar{Q}_3,\, \bar{Q}_4,\, \bar{Q}_5)}$$

$$= \begin{bmatrix} \frac{\partial \bar{E}_1}{\partial \bar{Q}_1} & \frac{\partial \bar{E}_1}{\partial \bar{Q}_2} & \frac{\partial \bar{E}_1}{\partial \bar{Q}_3} & \frac{\partial \bar{E}_1}{\partial \bar{Q}_4} & \frac{\partial \bar{E}_1}{\partial \bar{Q}_5} \\ \frac{\partial \bar{E}_2}{\partial \bar{Q}_1} & \frac{\partial \bar{E}_2}{\partial \bar{Q}_2} & \frac{\partial \bar{E}_2}{\partial \bar{Q}_3} & \frac{\partial \bar{E}_2}{\partial \bar{Q}_4} & \frac{\partial \bar{E}_2}{\partial \bar{Q}_5} \\ \frac{\partial \bar{E}_3}{\partial \bar{Q}_1} & \frac{\partial \bar{E}_3}{\partial \bar{Q}_2} & \frac{\partial \bar{E}_3}{\partial \bar{Q}_3} & \frac{\partial \bar{E}_3}{\partial \bar{Q}_4} & \frac{\partial \bar{E}_3}{\partial \bar{Q}_5} \\ \frac{\partial \bar{E}_4}{\partial \bar{Q}_1} & \frac{\partial \bar{E}_4}{\partial \bar{Q}_2} & \frac{\partial \bar{E}_4}{\partial \bar{Q}_3} & \frac{\partial \bar{E}_4}{\partial \bar{Q}_4} & \frac{\partial \bar{E}_4}{\partial \bar{Q}_5} \\ \frac{\partial \bar{E}_5}{\partial \bar{Q}_1} & \frac{\partial \bar{E}_5}{\partial \bar{Q}_2} & \frac{\partial \bar{E}_5}{\partial \bar{Q}_3} & \frac{\partial \bar{E}_5}{\partial \bar{Q}_4} & \frac{\partial \bar{E}_5}{\partial \bar{Q}_5} \end{bmatrix} \tag{11-82}$$

In order to determine the elements of matrix (11-82), the flux vectors E, F, and G are expressed in terms of the components of vector Q according to

$$E = \begin{bmatrix} \rho u \\ \rho u^2 + p \\ \rho uv \\ \rho uw \\ (\rho e_t + p)u \end{bmatrix} = \begin{bmatrix} Q_2 \\ \dfrac{Q_2^2}{Q_1} + (\gamma - 1)\left[Q_5 - \dfrac{1}{2}(\dfrac{Q_2^2}{Q_1} + \dfrac{Q_3^2}{Q_1} + \dfrac{Q_4^2}{Q_1})\right] \\ \dfrac{Q_2 Q_3}{Q_1} \\ \dfrac{Q_2 Q_4}{Q_1} \\ \left[\gamma Q_5 - \dfrac{\gamma - 1}{2}(\dfrac{Q_2^2}{Q_1} + \dfrac{Q_3^2}{Q_1} + \dfrac{Q_4^2}{Q_1})\right] \dfrac{Q_2}{Q_1} \end{bmatrix} \tag{11-83a}$$

$$F = \begin{bmatrix} \rho v \\ \rho uv \\ \rho v^2 + p \\ \rho vw \\ (\rho e_t + p)v \end{bmatrix} = \begin{bmatrix} Q_3 \\ \dfrac{Q_2 Q_3}{Q_1} \\ \dfrac{Q_3^2}{Q_1} + (\gamma - 1)\left[Q_5 - \dfrac{1}{2}(\dfrac{Q_2^2}{Q_1} + \dfrac{Q_3^2}{Q_1} + \dfrac{Q_4^2}{Q_1})\right] \\ \dfrac{Q_3 Q_4}{Q_1} \\ \left[\gamma Q_5 - \dfrac{\gamma - 1}{2}(\dfrac{Q_2^2}{Q_1} + \dfrac{Q_3^2}{Q_1} + \dfrac{Q_4^2}{Q_1})\right] \dfrac{Q_3}{Q_1} \end{bmatrix} \tag{11-83b}$$

$$G = \begin{bmatrix} \rho w \\ \rho wu \\ \rho wv \\ \rho w^2 + p \\ (\rho e_t + p)w \end{bmatrix} = \begin{bmatrix} Q_4 \\ \dfrac{Q_2 Q_4}{Q_1} \\ \dfrac{Q_3 Q_4}{Q_1} \\ \dfrac{Q_4^2}{Q_1} + (\gamma - 1)\left[Q_5 - \dfrac{1}{2}(\dfrac{Q_2^2}{Q_1} + \dfrac{Q_3^2}{Q_1} + \dfrac{Q_4^2}{Q_1})\right] \\ \left[\gamma Q_5 - \dfrac{\gamma - 1}{2}(\dfrac{Q_2^2}{Q_1} + \dfrac{Q_3^2}{Q_1} + \dfrac{Q_4^2}{Q_1})\right] \dfrac{Q_4}{Q_1} \end{bmatrix} \tag{11-83c}$$

Recall that, from Equation (11-62),

$$\bar{E}_1 = \frac{1}{J}[\xi_t Q_1 + \xi_x E_1 + \xi_y F_1 + \xi_z G_1] \tag{11-84}$$

With the components E_1, F_1, and G_1 from Equations (11-83a), (11-83b), and (11-83c), (11-84) may be written as

$$\bar{E}_1 = \frac{1}{J}[\xi_t Q_1 + \xi_x Q_2 + \xi_y Q_3 + \xi_z Q_4] \tag{11-85}$$

Now the first element of (11-82) is determined as

$$\frac{\partial \bar{E}_1}{\partial \bar{Q}_1} = \frac{\partial\{\frac{1}{J}[\xi_t Q_1 + \xi_x Q_2 + \xi_y Q_3 + \xi_z Q_4]\}}{\partial\{\frac{Q_1}{J}\}}$$

$$= \frac{\partial(\xi_t Q_1 + \xi_x Q_2 + \xi_y Q_3 + \xi_z Q_4)}{\partial Q_1} = \xi_t \tag{11-86}$$

The remaining elements of the first row are

$$\frac{\partial \bar{E}_1}{\partial \bar{Q}_2} = \xi_x \ , \qquad \frac{\partial \bar{E}_1}{\partial \bar{Q}_3} = \xi_y \ , \qquad \frac{\partial \bar{E}_1}{\partial \bar{Q}_4} = \xi_z \ ,$$

and $\dfrac{\partial \bar{E}_1}{\partial \bar{Q}_5} = 0$.

The elements of the second, third, fourth, and fifth rows are determined in a similar fashion. The resulting inviscid Jacobians are:

$$\frac{\partial \bar{E}}{\partial \bar{Q}} = \left[\begin{array}{c:c:c:c:c}
\xi_t & \xi_x & \xi_y & \xi_z & 0 \\ \hdashline
\begin{array}{l} -u(\xi_x u + \xi_y v + \xi_z w) \\ +\xi_x\left[\frac{1}{2}(\gamma-1)(u^2+v^2+w^2)\right] \end{array} & \begin{array}{l} \xi_t + \xi_x(2-\gamma)u \\ +(\xi_x u + \xi_y v + \xi_z w) \end{array} & \xi_y u - (\gamma-1)\xi_x v & \xi_z u - (\gamma-1)\xi_x w & (\gamma-1)\xi_x \\ \hdashline
\begin{array}{l} -v(\xi_x u + \xi_y v + \xi_z w) \\ +\xi_y\left[\frac{1}{2}(\gamma-1)(u^2+v^2+w^2)\right] \end{array} & \xi_x v - (\gamma-1)\xi_y u & \begin{array}{l} \xi_t + \xi_y(2-\gamma)v \\ +(\xi_x u + \xi_y v + \xi_z w) \end{array} & \xi_z v - (\gamma-1)\xi_y w & (\gamma-1)\xi_y \\ \hdashline
\begin{array}{l} -w(\xi_x u + \xi_y v + \xi_z w) \\ +\xi_z\left[\frac{1}{2}(\gamma-1)(u^2+v^2+w^2)\right] \end{array} & \xi_x w - (\gamma-1)\xi_z u & \xi_y w - (\gamma-1)\xi_z v & \begin{array}{l} \xi_t + \xi_z(2-\gamma)w \\ +(\xi_x u + \xi_y v + \xi_z w) \end{array} & (\gamma-1)\xi_z \\ \hdashline
\begin{array}{l} (\xi_x u + \xi_y v + \xi_z w)\,[-\gamma e_t \\ +\ (\gamma-1)(u^2+v^2+w^2)] \end{array} & \begin{array}{l} \xi_x\left[\gamma e_t - \frac{1}{2}(\gamma-1)\cdot\right. \\ \left.(u^2+v^2+w^2)\right] \\ -(\gamma-1)\,[\xi_x u + \xi_y v \\ +\xi_z w]\,u \end{array} & \begin{array}{l} \xi_y\left[\gamma e_t - \frac{1}{2}(\gamma-1)\cdot\right. \\ \left.(u^2+v^2+w^2)\right] \\ -(\gamma-1)\,[\xi_x u + \xi_y v \\ +\xi_z w]\,v \end{array} & \begin{array}{l} \xi_z\left[\gamma e_t - \frac{1}{2}(\gamma-1)\cdot\right. \\ \left.(u^2+v^2+w^2)\right] \\ -(\gamma-1)\,[\xi_x u + \xi_y v \\ +\xi_z w]\,w \end{array} & \begin{array}{l} \xi_t + \\ \gamma\,[\xi_x u + \xi_y v + \xi_z w] \end{array}
\end{array}\right] \tag{11-87}$$

$$
\frac{\partial \bar{F}}{\partial \bar{Q}} =
\left[\begin{array}{c:c:c:c:c}
\eta_t & \eta_x & \eta_y & \eta_z & 0 \\
\hdashline
\begin{array}{l} -\,u(\eta_x u + \eta_y v + \eta_z w) \\ +\,\eta_x \left[\frac{1}{2}(\gamma - 1)(u^2 + v^2 + w^2)\right] \end{array} &
\begin{array}{l} \eta_t + \eta_x(2-\gamma)u \\ +\,(\eta_x u + \eta_y v + \eta_z w) \end{array} &
\eta_y u - (\gamma - 1)\eta_x v &
\eta_z u - (\gamma - 1)\eta_x w &
(\gamma - 1)\eta_x \\
\hdashline
\begin{array}{l} -\,v(\eta_x u + \eta_y v + \eta_z w) \\ +\,\eta_y \left[\frac{1}{2}(\gamma - 1)(u^2 + v^2 + w^2)\right] \end{array} &
\eta_x v - (\gamma - 1)\eta_y u &
\begin{array}{l} \eta_t + \eta_y(2-\gamma)v \\ +\,(\eta_x u + \eta_y v + \eta_z w) \end{array} &
\eta_z v - (\gamma - 1)\eta_y w &
(\gamma - 1)\eta_y \\
\hdashline
\begin{array}{l} -\,w(\eta_x u + \eta_y v + \eta_z w) \\ +\,\eta_z \left[\frac{1}{2}(\gamma - 1)(u^2 + v^2 + w^2)\right] \end{array} &
\eta_x w - (\gamma - 1)\eta_z u &
\eta_y w - (\gamma - 1)\eta_z v &
\begin{array}{l} \eta_t + \eta_z(2-\gamma)w \\ +\,(\eta_x u + \eta_y v + \eta_z w) \end{array} &
(\gamma - 1)\eta_z \\
\hdashline
\begin{array}{l} (\eta_x u + \eta_y v + \eta_z w)\,[-\gamma e_t \\ +\,(\gamma - 1)(u^2 + v^2 + w^2)] \end{array} &
\begin{array}{l} \eta_x \left[\gamma e_t - \frac{1}{2}(\gamma - 1)\cdot \right. \\ \left. (u^2 + v^2 + w^2)\right] \\ -\,(\gamma - 1)\,[\eta_x u + \eta_y v \\ +\,\eta_z w]\,u \end{array} &
\begin{array}{l} \eta_y \left[\gamma e_t - \frac{1}{2}(\gamma - 1)\cdot \right. \\ \left. (u^2 + v^2 + w^2)\right] \\ -\,(\gamma - 1)\,[\eta_x u + \eta_y v \\ +\,\eta_z w]\,v \end{array} &
\begin{array}{l} \eta_z \left[\gamma e_t - \frac{1}{2}(\gamma - 1)\cdot \right. \\ \left. (u^2 + v^2 + w^2)\right] \\ -\,(\gamma - 1)\,[\eta_x u + \eta_y v \\ +\,\eta_z w]\,w \end{array} &
\begin{array}{l} \eta_t + \\ \gamma\,[\eta_x u + \eta_y v + \eta_z w] \end{array}
\end{array}\right]
\qquad (11\text{-}88)
$$

$$
\frac{\partial \bar{G}}{\partial \bar{Q}} = \left[\begin{array}{c:c:c:c:c}
\zeta_t & \zeta_x & \zeta_y & \zeta_z & 0 \\ \hdashline
\begin{array}{l} -\,u(\zeta_x u + \zeta_y v + \zeta_z w) \\ +\,\zeta_x \left[\frac{1}{2}(\gamma-1)(u^2+v^2+w^2)\right] \end{array} & \begin{array}{l} \zeta_t + \zeta_x(2-\gamma)u \\ +\,(\zeta_x u + \zeta_y v + \zeta_z w) \end{array} & \zeta_y u - (\gamma-1)\zeta_x v & \zeta_z u - (\gamma-1)\zeta_x w & (\gamma-1)\zeta_x \\ \hdashline
\begin{array}{l} -\,v(\zeta_x u + \zeta_y v + \zeta_z w) \\ +\,\zeta_y \left[\frac{1}{2}(\gamma-1)(u^2+v^2+w^2)\right] \end{array} & \zeta_x v - (\gamma-1)\zeta_y u & \begin{array}{l} \zeta_t + \zeta_y(2-\gamma)v \\ +\,(\zeta_x u + \zeta_y v + \zeta_z w) \end{array} & \zeta_z v - (\gamma-1)\zeta_y w & (\gamma-1)\zeta_y \\ \hdashline
\begin{array}{l} -\,w(\zeta_x u + \zeta_y v + \zeta_z w) \\ +\,\zeta_z \left[\frac{1}{2}(\gamma-1)(u^2+v^2+w^2)\right] \end{array} & \zeta_x w - (\gamma-1)\zeta_z u & \zeta_y w - (\gamma-1)\zeta_z v & \begin{array}{l} \zeta_t + \zeta_z(2-\gamma)w \\ +\,(\zeta_x u + \zeta_y v + \zeta_z w) \end{array} & (\gamma-1)\zeta_z \\ \hdashline
\begin{array}{l} (\zeta_x u + \zeta_y v + \zeta_z w)\,[-\gamma e_t \\ +\;(\gamma-1)(u^2+v^2+w^2)] \end{array} & \begin{array}{l} \zeta_x \Big[\gamma e_t - \frac{1}{2}(\gamma-1)\cdot \\ \;(u^2+v^2+w^2)\Big] \\ -\,(\gamma-1)\,[\zeta_x u + \zeta_y v \\ +\,\zeta_z w]\,u \end{array} & \begin{array}{l} \zeta_y \Big[\gamma e_t - \frac{1}{2}(\gamma-1)\cdot \\ \;(u^2+v^2+w^2)\Big] \\ -\,(\gamma-1)\,[\zeta_x u + \zeta_y v \\ +\,\zeta_z w]\,v \end{array} & \begin{array}{l} \zeta_z \Big[\gamma e_t - \frac{1}{2}(\gamma-1)\cdot \\ \;(u^2+v^2+w^2)\Big] \\ -\,(\gamma-1)\,[\zeta_x u + \zeta_y v \\ +\,\zeta_z w]\,w \end{array} & \begin{array}{l} \zeta_t + \\ \gamma\,[\zeta_x u + \zeta_y v + \zeta_z w] \end{array}
\end{array}\right] \tag{11-89}
$$

Some of the terms appearing in the Jacobian matrices may be defined such that they are expressed in a compact form. For example,

$$U = \xi_t + \xi_x u + \xi_y v + \xi_z w \tag{11-90a}$$

$$V = \eta_t + \eta_x u + \eta_y v + \eta_z w \tag{11-90b}$$

$$W = \zeta_t + \zeta_x u + \zeta_y v + \zeta_z w \tag{11-90c}$$

are known as the contravariant velocity components. They represent velocity components which are perpendicular to planes of constant ξ, η, and ζ, respectively. Furthermore, define

$$q^2 = u^2 + v^2 + w^2 \tag{11-91}$$

Before attempting to determine the viscous Jacobians, the viscous flux vectors must be rearranged by substituting expressions for shear stress and heat conduction terms. For example, consider the second component of $\bar{E}_v$, which from Equation (11-65) is given by

$$\bar{E}_{v_2} = \frac{1}{J}\left(\xi_x E_{v_2} + \xi_y F_{v_2} + \xi_z G_{v_2}\right) \tag{11-92}$$

From Equations (11-52), (11-54) and (11-56), the following is observed:

$$E_{v_2} = \tau_{xx} \quad , \quad F_{v_2} = \tau_{xy} \quad , \quad \text{and} \;\; G_{v_2} = \tau_{xz}$$

Hence, Equation (11-92) may be expressed as

$$\bar{E}_{v_2} = \frac{1}{J}\left(\xi_x \tau_{xx} + \xi_y \tau_{xy} + \xi_z \tau_{xz}\right) \tag{11-93}$$

Now, using the expressions for the shear stresses given by Equations (11-68), (11-71) and (11-72), Equation (11-93) is rearranged as

$$\begin{aligned}\bar{E}_{v_2} = \frac{1}{Re_\infty J}\Bigg\{ & \xi_x\left[\frac{4}{3}\mu(\xi_x u_\xi + \eta_x u_\eta + \zeta_x u_\zeta) - \frac{2}{3}\mu(\xi_y v_\xi + \eta_y v_\eta\right. \\ & \left. + \;\zeta_y v_\zeta) - \frac{2}{3}\mu(\xi_z w_\xi + \eta_z w_\eta + \zeta_z w_\zeta)\right] + \xi_y\Big[\mu(\xi_y u_\xi + \eta_y u_\eta + \\ & + \;\zeta_y u_\zeta + \xi_x v_\xi + \eta_x v_\eta + \zeta_x v_\zeta)\Big] + \xi_z\Big[\mu(\xi_z u_\xi + \eta_z u_\eta + \\ & + \;\zeta_z u_\zeta + \xi_x w_\xi + \eta_x w_\eta + \zeta_x w_\zeta)\Big]\Bigg\}\end{aligned}$$

which is factored as

$$\bar{E}_{v_2} = \frac{\mu}{Re_\infty J}\Big\{(\frac{4}{3}\xi_x^2 + \xi_y^2 + \xi_z^2)u_\xi + (\frac{1}{3}\xi_x\xi_y)v_\xi +$$

$$(\frac{1}{3}\xi_x\xi_z)w_\xi + (\frac{4}{3}\xi_x\eta_x + \xi_y\eta_y + \xi_z\eta_z)u_\eta + (\xi_y\eta_x - \frac{2}{3}\xi_x\eta_y)v_\eta +$$

$$(\xi_z\eta_x - \frac{2}{3}\xi_x\eta_z)w_\eta + (\frac{4}{3}\xi_x\zeta_x + \xi_y\zeta_y + \xi_z\zeta_z)u_\zeta +$$

$$(\xi_y\zeta_x - \frac{2}{3}\xi_x\zeta_y)v_\zeta + (\xi_z\zeta_x - \frac{2}{3}\xi_x\zeta_z)w_\zeta\Big\}$$

All the components of flux vector $\bar{E}_v$ are reformulated accordingly. In addition, viscous flux vectors $\bar{F}_v$ and $\bar{G}_v$ are modified in a similar fashion, resulting in the following form of the viscous flux vectors:

$$\bar{E}_v = \frac{\mu}{Re_\infty J}\left[\begin{array}{l}
0 \\
\hline
a_1 u_\xi + a_5 v_\xi + a_7 w_\xi + d_1 u_\eta + d_7 v_\eta + d_9 w_\eta + e_1 u_\zeta + e_7 v_\zeta + e_9 w_\zeta \\
\hline
a_5 u_\xi + a_2 v_\xi + a_6 w_\xi + d_5 u_\eta + d_2 v_\eta + d_{10} w_\eta + e_5 u_\zeta + e_2 v_\zeta + e_{10} w_\zeta \\
\hline
a_7 u_\xi + a_6 v_\xi + a_3 w_\xi + d_6 u_\eta + d_8 v_\eta + d_3 w_\eta + e_6 u_\zeta + e_8 v_\zeta + e_3 w_\zeta \\
\hline
\frac{1}{2}a_1(u^2)_\xi + \frac{1}{2}a_2(v^2)_\xi + \frac{1}{2}a_3(w^2)_\xi + a_5(uv)_\xi + a_6(vw)_\xi + a_7(uw)_\xi \\
+\ \frac{1}{Pr(\gamma-1)M_\infty^2}a_4 T_\xi + \frac{1}{2}d_1(u^2)_\eta + \frac{1}{2}d_2(v^2)_\eta + \frac{1}{2}d_3(w^2)_\eta + d_5 v u_\eta \\
+\ d_6 w u_\eta + d_7 u v_\eta + d_8 w v_\eta + d_9 u w_\eta + d_{10} v w_\eta + \frac{1}{Pr(\gamma-1)M_\infty^2} d_4 T_\eta \\
+\ \frac{1}{2}e_1(u^2)_\zeta + \frac{1}{2}e_2(v^2)_\zeta + \frac{1}{2}e_3(w^2)_\zeta + e_5 v u_\zeta + e_6 w u_\zeta + e_7 u v_\zeta \\
+\ e_8 w v_\zeta + e_9 u w_\zeta + e_{10} v w_\zeta + \frac{1}{Pr(\gamma-1)M_\infty^2} e_4 T_\zeta
\end{array}\right] \tag{11-94}$$

where, for a perfect gas,

$$\frac{1}{Pr(\gamma-1)M_\infty^2}T = \frac{\gamma}{Pr}\left[e_t - \frac{1}{2}(u^2+v^2+w^2)\right] \tag{11-95}$$

$$\bar{F}_v = \frac{\mu}{Re_\infty J} \left[\begin{array}{l} 0 \\ \hline d_1 u_\xi + d_5 v_\xi + d_6 w_\xi + b_1 u_\eta + b_5 v_\eta + b_7 w_\eta + f_1 u_\zeta + f_7 v_\zeta + f_9 w_\zeta \\ \hline d_7 u_\xi + d_2 v_\xi + d_8 w_\xi + b_5 u_\eta + b_2 v_\eta + b_6 w_\eta + f_5 u_\zeta + f_2 v_\zeta + f_{10} w_\zeta \\ \hline d_9 u_\xi + d_{10} v_\xi + d_3 w_\xi + b_7 u_\eta + b_6 v_\eta + b_3 w_\eta + f_6 u_\zeta + f_8 v_\zeta + f_3 w_\zeta \\ \hline \frac{1}{2} d_1 (u^2)_\xi + \frac{1}{2} d_2 (v^2)_\xi + \frac{1}{2} d_3 (w^2)_\xi + d_5 u v_\xi + d_6 u w_\xi + d_7 v u_\xi \\ + \; d_8 v w_\xi + d_9 w u_\xi + d_{10} w v_\xi + \frac{1}{Pr(\gamma-1)M_\infty^2} d_4 T_\xi \\ + \; \frac{1}{2} b_1 (u^2)_\eta + \frac{1}{2} b_2 (v^2)_\eta + \frac{1}{2} b_3 (w^2)_\eta + b_5 (uv)_\eta + b_6 (vw)_\eta + b_7 (uw)_\eta \\ + \; \frac{1}{Pr(\gamma-1)M_\infty^2} b_4 T_\eta + \frac{1}{2} f_1 (u^2)_\zeta + \frac{1}{2} f_2 (v^2)_\zeta + \frac{1}{2} f_3 (w^2)_\zeta \\ + \; f_5 v u_\zeta + f_6 w u_\zeta + f_7 u v_\zeta + f_8 w v_\zeta + f_9 u w_\zeta + f_{10} v w_\zeta \\ + \; \frac{1}{Pr(\gamma-1)M_\infty^2} f_4 T_\zeta \end{array} \right] \tag{11-96}$$

$$\bar{G}_v = \frac{\mu}{Re_\infty J}\left[\begin{array}{l} 0 \\ \hline e_1 u_\xi + e_5 v_\xi + e_6 w_\xi + f_1 u_\eta + f_5 v_\eta + f_6 w_\eta + c_1 u_\zeta + c_5 v_\zeta + c_7 w_\zeta \\ \hline e_7 u_\xi + e_2 v_\xi + e_8 w_\xi + f_7 u_\eta + f_2 v_\eta + f_8 w_\eta + c_5 u_\zeta + c_2 v_\zeta + c_6 w_\zeta \\ \hline e_9 u_\xi + e_{10} v_\xi + e_3 w_\xi + f_9 u_\eta + f_{10} v_\eta + f_3 w_\eta + c_7 u_\zeta + c_6 v_\zeta + c_3 w_\zeta \\ \hline \frac{1}{2} e_1 (u^2)_\xi + \frac{1}{2} e_2 (v^2)_\xi + \frac{1}{2} e_3 (w^2)_\xi + e_5 u v_\xi + e_6 u w_\xi + e_7 v u_\xi \\ +\ e_8 v w_\xi + e_9 w u_\xi + e_{10} w v_\xi + \frac{1}{Pr(\gamma-1)M_\infty^2} e_4 T_\xi + \frac{1}{2} f_1 (u^2)_\eta \\ +\ \frac{1}{2} f_2 (v^2)_\eta + \frac{1}{2} f_3 (w^2)_\eta + f_5 u v_\eta + f_6 u w_\eta + f_7 v u_\eta + f_8 v w_\eta \\ +\ f_9 w u_\eta + f_{10} w v_\eta + \frac{1}{Pr(\gamma-1)M_\infty^2} f_4 T_\eta + \frac{1}{2} c_1 (u^2)_\zeta + \frac{1}{2} c_2 (v^2)_\zeta \\ +\ \frac{1}{2} c_3 (w^2)_\zeta + c_5 (uv)_\zeta + c_6 (vw)_\zeta + c_7 (uw)_\zeta + \frac{1}{Pr(\gamma-1)M_\infty^2} c_4 T_\zeta \end{array}\right] \tag{11-97}$$

where

$$a_1 = \tfrac{4}{3}\xi_x^2 + \xi_y^2 + \xi_z^2 \quad (11\text{-}98)$$

$$a_2 = \xi_x^2 + \tfrac{4}{3}\xi_y^2 + \xi_z^2 \quad (11\text{-}99)$$

$$a_3 = \xi_x^2 + \xi_y^2 + \tfrac{4}{3}\xi_z^2 \quad (11\text{-}100)$$

$$a_4 = \xi_x^2 + \xi_y^2 + \xi_z^2 \quad (11\text{-}101)$$

$$a_5 = \tfrac{1}{3}\xi_x\xi_y \quad (11\text{-}102)$$

$$a_6 = \tfrac{1}{3}\xi_y\xi_z \quad (11\text{-}103)$$

$$a_7 = \tfrac{1}{3}\xi_x\xi_z \quad (11\text{-}104)$$

$$b_1 = \tfrac{4}{3}\eta_x^2 + \eta_y^2 + \eta_z^2 \quad (11\text{-}105)$$

$$b_2 = \eta_x^2 + \tfrac{4}{3}\eta_y^2 + \eta_z^2 \quad (11\text{-}106)$$

$$b_3 = \eta_x^2 + \eta_y^2 + \tfrac{4}{3}\eta_z^2 \quad (11\text{-}107)$$

$$b_4 = \eta_x^2 + \eta_y^2 + \eta_z^2 \quad (11\text{-}108)$$

$$b_5 = \tfrac{1}{3}\eta_x\eta_y \quad (11\text{-}109)$$

$$b_6 = \tfrac{1}{3}\eta_y\eta_z \quad (11\text{-}110)$$

$$b_7 = \tfrac{1}{3}\eta_x\eta_z \quad (11\text{-}111)$$

$$c_1 = \tfrac{4}{3}\zeta_x^2 + \zeta_y^2 + \zeta_z^2 \quad (11\text{-}112)$$

$$c_2 = \zeta_x^2 + \tfrac{4}{3}\zeta_y^2 + \zeta_z^2 \quad (11\text{-}113)$$

$$c_3 = \zeta_x^2 + \zeta_y^2 + \tfrac{4}{3}\zeta_z^2 \quad (11\text{-}114)$$

$$c_4 = \zeta_x^2 + \zeta_y^2 + \zeta_z^2 \quad (11\text{-}115)$$

$$c_5 = \tfrac{1}{3}\zeta_x\zeta_y \quad (11\text{-}116)$$

$$c_6 = \tfrac{1}{3}\zeta_y\zeta_z \quad (11\text{-}117)$$

$$c_7 = \tfrac{1}{3}\zeta_x\zeta_z \quad (11\text{-}118)$$

$$d_1 = \tfrac{4}{3}\xi_x\eta_x + \xi_y\eta_y + \xi_z\eta_z \quad (11\text{-}119)$$

$$d_2 = \xi_x\eta_x + \tfrac{4}{3}\xi_y\eta_y + \xi_z\eta_z \quad (11\text{-}120)$$

$$d_3 = \xi_x\eta_x + \xi_y\eta_y + \tfrac{4}{3}\xi_z\eta_z \quad (11\text{-}121)$$

$$d_4 = \xi_x\eta_x + \xi_y\eta_y + \xi_z\eta_z \quad (11\text{-}122)$$

$$d_5 = \xi_x\eta_y - \tfrac{2}{3}\xi_y\eta_x \quad (11\text{-}123)$$

$$d_6 = \xi_x\eta_z - \tfrac{2}{3}\xi_z\eta_x \quad (11\text{-}124)$$

$$d_7 = \xi_y\eta_x - \tfrac{2}{3}\xi_x\eta_y \quad (11\text{-}125)$$

$$d_8 = \xi_y\eta_z - \tfrac{2}{3}\xi_z\eta_y \quad (11\text{-}126)$$

$$d_9 = \xi_z\eta_x - \tfrac{2}{3}\xi_x\eta_z \quad (11\text{-}127)$$

$$d_{10} = \xi_z\eta_y - \tfrac{2}{3}\xi_y\eta_z \quad (11\text{-}128)$$

$$e_1 = \tfrac{4}{3}\xi_x\zeta_x + \xi_y\zeta_y + \xi_z\zeta_z \qquad (11\text{-}129)$$

$$e_2 = \xi_x\zeta_x + \tfrac{4}{3}\xi_y\zeta_y + \xi_z\zeta_z \qquad (11\text{-}130)$$

$$e_3 = \xi_x\zeta_x + \xi_y\zeta_y + \tfrac{4}{3}\xi_z\zeta_z \qquad (11\text{-}131)$$

$$e_4 = \xi_x\zeta_x + \xi_y\zeta_y + \xi_z\zeta_z \qquad (11\text{-}132)$$

$$e_5 = \xi_x\zeta_y - \tfrac{2}{3}\xi_y\zeta_x \qquad (11\text{-}133)$$

$$e_6 = \xi_x\zeta_z - \tfrac{2}{3}\xi_z\zeta_x \qquad (11\text{-}134)$$

$$e_7 = \xi_y\zeta_x - \tfrac{2}{3}\xi_x\zeta_y \qquad (11\text{-}135)$$

$$e_8 = \xi_y\zeta_z - \tfrac{2}{3}\xi_z\zeta_y \qquad (11\text{-}136)$$

$$e_9 = \xi_z\zeta_x - \tfrac{2}{3}\xi_x\zeta_z \qquad (11\text{-}137)$$

$$e_{10} = \xi_z\zeta_y - \tfrac{2}{3}\xi_y\zeta_z \qquad (11\text{-}138)$$

$$f_1 = \tfrac{4}{3}\eta_x\zeta_x + \eta_y\zeta_y + \eta_z\zeta_z \qquad (11\text{-}139)$$

$$f_2 = \eta_x\zeta_x + \tfrac{4}{3}\eta_y\zeta_y + \eta_z\zeta_z \qquad (11\text{-}140)$$

$$f_3 = \eta_x\zeta_x + \eta_y\zeta_y + \tfrac{4}{3}\eta_z\zeta_z \qquad (11\text{-}141)$$

$$f_4 = \eta_x\zeta_x + \eta_y\zeta_y + \eta_z\zeta_z \qquad (11\text{-}142)$$

$$f_5 = \eta_x\zeta_y - \tfrac{2}{3}\eta_y\zeta_x \qquad (11\text{-}143)$$

$$f_6 = \eta_x\zeta_z - \tfrac{2}{3}\eta_z\zeta_x \qquad (11\text{-}144)$$

$$f_7 = \eta_y\zeta_x - \tfrac{2}{3}\eta_x\zeta_y \qquad (11\text{-}145)$$

$$f_8 = \eta_y\zeta_z - \tfrac{2}{3}\eta_z\zeta_y \qquad (11\text{-}146)$$

$$f_9 = \eta_z\zeta_x - \tfrac{2}{3}\eta_x\zeta_z \qquad (11\text{-}147)$$

$$f_{10} = \eta_z\zeta_y - \tfrac{2}{3}\eta_y\zeta_z \qquad (11\text{-}148)$$

The derivation of viscous Jacobian matrices must be handled with special care, due to the fact that the components of the viscous flux vectors involve gradients of the dependent variable and the Jacobian of transformation, which is itself a function of the independent variables. Thus, J remains embedded inside the viscous Jacobians, unlike the inviscid Jacobians where the Js were cancelled. In order to generalize the mathematical procedure, it is recognized that all of the terms in the viscous flux vector may be expressed as

$$\bar{E}_t = (\text{Factor})\frac{1}{J}(\phi_\psi) \; , \qquad (11\text{-}149)$$

where ϕ represents the dependent variable such as u, or a combination of dependent variables such as u^2 and ψ represent an independent variable such as ξ, η, or ζ. For example, the first term of the second component of $\bar{E}_v$ (from Equation (11-94)) is

$$\bar{E}_{v2,\text{1st term}} = \frac{\mu}{Re_\infty}\frac{1}{J}a_1(u_\xi)$$

which may be expressed as

$$\bar{E}_{v2,1T} = (\frac{\mu}{Re_\infty}a_1)\frac{1}{J}(u_\xi)$$

or, in the general formulation of (11-149),

$$\bar{E}_{v2,1T} = (\text{Factor})\frac{1}{J}\phi_{\psi}$$

To obtain the viscous Jacobian $\partial\bar{E}_v/\partial\bar{Q}$, it is necessary to compute each of the terms in $\partial(\bar{E}_{v1}, \bar{E}_{v2}, \bar{E}_{v3}, \bar{E}_{v4}, \bar{E}_{v5})/\partial(\bar{Q}_1, \bar{Q}_2, \bar{Q}_3, \bar{Q}_4, \bar{Q}_5)$. In a similar fashion, the Jacobians associated with flux vectors $\bar{F}_v$ and $\bar{G}_v$ may be determined. In order to perform this differentiation, the general expression (11-149) will be used to illustrate the procedure. Consider the determination of

$$\begin{aligned}\frac{\partial\bar{E}_t}{\partial\bar{Q}} &= \frac{\partial}{\partial\bar{Q}}\left[(\text{Factor})\frac{1}{J}(\phi_{\psi})\right] = (\text{Factor})(\frac{1}{J})\frac{\partial}{\partial\bar{Q}}(\phi_{\psi}) \\ &= (\text{Factor})(\frac{1}{J})\frac{\partial}{\partial\bar{Q}}\frac{\partial\phi}{\partial\psi}\end{aligned}$$

At this point the order of differentiation is interchanged to yield:

$$\frac{\partial\bar{E}_t}{\partial\bar{Q}} = (\text{Factor})(\frac{1}{J})\frac{\partial}{\partial\psi}(\frac{\partial\phi}{\partial\bar{Q}}) \tag{11-150}$$

Recall that $\bar{Q} = Q/J$. Hence, after substituting into (11-150), one obtains

$$\frac{\partial\bar{E}_t}{\partial\bar{Q}} = (\text{Factor})\left[\frac{1}{J}\frac{\partial}{\partial\psi}(J\frac{\partial\phi}{\partial Q})\right] = (\text{Factor})\left[\frac{1}{J}(J\frac{\partial\phi}{\partial Q})_{\psi}\right] \tag{11-151}$$

This formulation is used to obtain the viscous Jacobians given by Equations (11-152), (11-153), and (11-154):

$$\frac{\partial\bar{E}_v}{\partial\bar{Q}} = \frac{\mu}{Re_{\infty}J}\begin{bmatrix} 0 & 0 & 0 & 0 & 0 \\ EQ_{2,1} & EQ_{2,2} & EQ_{2,3} & EQ_{2,4} & 0 \\ EQ_{3,1} & EQ_{3,2} & EQ_{3,3} & EQ_{3,4} & 0 \\ EQ_{4,1} & EQ_{4,2} & EQ_{4,3} & EQ_{4,4} & 0 \\ EQ_{5,1} & EQ_{5,2} & EQ_{5,3} & EQ_{5,4} & EQ_{5,5} \end{bmatrix} \tag{11-152}$$

$$\frac{\partial \bar{F}_v}{\partial \bar{Q}} = \frac{\mu}{Re_\infty J} \begin{bmatrix} 0 & 0 & 0 & 0 & 0 \\ FQ_{2,1} & FQ_{2,2} & FQ_{2,3} & FQ_{2,4} & 0 \\ FQ_{3,1} & FQ_{3,2} & FQ_{3,3} & FQ_{3,4} & 0 \\ FQ_{4,1} & FQ_{4,2} & FQ_{4,3} & FQ_{4,4} & 0 \\ FQ_{5,1} & FQ_{5,2} & FQ_{5,3} & FQ_{5,4} & FQ_{5,5} \end{bmatrix} \tag{11-153}$$

$$\frac{\partial \bar{G}_v}{\partial \bar{Q}} = \frac{\mu}{Re_\infty J} \begin{bmatrix} 0 & 0 & 0 & 0 & 0 \\ GQ_{2,1} & GQ_{2,2} & GQ_{2,3} & GQ_{2,4} & 0 \\ GQ_{3,1} & GQ_{3,2} & GQ_{3,3} & GQ_{3,4} & 0 \\ GQ_{4,1} & GQ_{4,2} & GQ_{4,3} & GQ_{4,4} & 0 \\ GQ_{5,1} & GQ_{5,2} & GQ_{5,3} & GQ_{5,4} & GQ_{5,5} \end{bmatrix} \tag{11-154}$$

where

$$\begin{aligned} EQ_{2,1} &= -\left[a_1(J\frac{u}{\rho})_\xi + a_5(J\frac{v}{\rho})_\xi + a_7(J\frac{w}{\rho})_\xi\right] \\ &\quad - \left[d_1(J\frac{u}{\rho})_\eta + d_7(J\frac{v}{\rho})_\eta + d_9(J\frac{w}{\rho})_\eta\right] \\ &\quad - \left[e_1(J\frac{u}{\rho})_\zeta + e_7(J\frac{v}{\rho})_\zeta + e_9(J\frac{w}{\rho})_\zeta\right] \end{aligned}$$

$$EQ_{2,2} = a_1(\frac{J}{\rho})_\xi + d_1(\frac{J}{\rho})_\eta + e_1(\frac{J}{\rho})_\zeta$$

$$EQ_{2,3} = a_5(\frac{J}{\rho})_\xi + d_7(\frac{J}{\rho})_\eta + e_7(\frac{J}{\rho})_\zeta$$

$$EQ_{2,4} = a_7(\frac{J}{\rho})_\xi + d_9(\frac{J}{\rho})_\eta + e_9(\frac{J}{\rho})_\zeta$$

$$\begin{aligned} EQ_{3,1} = & - \left[a_5(J\frac{u}{\rho})_\xi + a_2(J\frac{v}{\rho})_\xi + a_6(J\frac{w}{\rho})_\xi\right] \\ & - \left[d_5(J\frac{u}{\rho})_\eta + d_2(J\frac{v}{\rho})_\eta + d_{10}(J\frac{w}{\rho})_\eta\right] \\ & - \left[e_5(J\frac{u}{\rho})_\zeta + e_2(J\frac{v}{\rho})_\zeta + e_{10}(J\frac{w}{\rho})_\zeta\right] \end{aligned}$$

$$EQ_{3,2} = a_5(\frac{J}{\rho})_\xi + d_5(\frac{J}{\rho})_\eta + e_5(\frac{J}{\rho})_\zeta$$

$$EQ_{3,3} = a_2(\frac{J}{\rho})_\xi + d_2(\frac{J}{\rho})_\eta + e_2(\frac{J}{\rho})_\zeta$$

$$EQ_{3,4} = a_6(\frac{J}{\rho})_\xi + d_{10}(\frac{J}{\rho})_\eta + e_{10}(\frac{J}{\rho})_\zeta$$

$$\begin{aligned} EQ_{4,1} = & - \left[a_7(J\frac{u}{\rho})_\xi + a_6(J\frac{v}{\rho})_\xi + a_3(J\frac{w}{\rho})_\xi\right] \\ & - \left[d_6(J\frac{u}{\rho})_\eta + d_8(J\frac{v}{\rho})_\eta + d_3(J\frac{w}{\rho})_\eta\right] \\ & - \left[e_6(J\frac{u}{\rho})_\zeta + e_8(J\frac{v}{\rho})_\zeta + e_3(J\frac{w}{\rho})_\zeta\right] \end{aligned}$$

$$EQ_{4,2} = a_7(\frac{J}{\rho})_\xi + d_6(\frac{J}{\rho})_\eta + e_6(\frac{J}{\rho})_\zeta$$

$$EQ_{4,3} = a_6(\frac{J}{\rho})_\xi + d_8(\frac{J}{\rho})_\eta + e_8(\frac{J}{\rho})_\zeta$$

$$EQ_{4,4} = a_3(\frac{J}{\rho})_\xi + d_3(\frac{J}{\rho})_\eta + e_3(\frac{J}{\rho})_\zeta$$

$$\begin{aligned}
EQ_{5,1} = & -\Big\{(a_1 - \frac{\gamma}{Pr}a_4)(J\frac{u^2}{\rho})_\xi + (a_2 - \frac{\gamma}{Pr}a_4)(J\frac{v^2}{\rho})_\xi + \\
& (a_3 - \frac{\gamma}{Pr}a_4)(J\frac{w^2}{\rho})_\xi + \frac{\gamma}{Pr}a_4(J\frac{e_t}{\rho})_\xi + 2a_5(J\frac{uv}{\rho})_\xi + 2a_6(J\frac{vw}{\rho})_\xi \\
& + 2a_7(J\frac{uw}{\rho})_\xi\Big\} - \Big\{(d_1 - \frac{\gamma}{Pr}d_4)(J\frac{u^2}{\rho})_\eta + (d_2 - \frac{\gamma}{Pr}d_4)(J\frac{v^2}{\rho})_\eta \\
& + (d_3 - \frac{\gamma}{Pr}a_4)(\gamma\frac{w^2}{\rho})_\eta + \frac{\gamma}{Pr}d_4(J\frac{e_t}{\rho})_\eta + (d_5v + d_6w)(J\frac{u}{\rho})_\eta \\
& + (d_7u + d_8w)(J\frac{v}{\rho})_\eta + (d_9u + d_{10}v)(J\frac{w}{\rho})_\eta + \frac{J}{\rho}(d_5v + d_6w)u_\eta \\
& + \frac{J}{\rho}(d_7u + d_8w)v_\eta + \frac{J}{\rho}(d_9u + d_{10}v)w_\eta\Big\} - \Big\{(e_1 - \frac{\gamma e_4}{Pr})(J\frac{u^2}{\rho})_\zeta \\
& + (e_2 - \frac{\gamma}{Pr}e_4)(J\frac{v^2}{\rho})_\zeta + (e_3 - \frac{\gamma}{Pr}e_4)(J\frac{w^2}{\rho})_\zeta + \frac{\gamma}{Pr}e_4(J\frac{e_t}{\rho})_\zeta \\
& + (e_5v + e_6w)(J\frac{u}{\rho})_\zeta + (e_7u + e_8w)(J\frac{v}{\rho})_\zeta + (e_9u + e_{10}v)(J\frac{w}{\rho})_\zeta \\
& + \frac{J}{\rho}(e_5v + e_6w)u_\zeta + \frac{J}{\rho}(e_7u + e_8w)v_\zeta + \frac{J}{\rho}(e_9u + e_{10}v)w_\zeta\Big\}
\end{aligned}$$

$$
\begin{aligned}
EQ_{5,2} &= \left[(a_1 - \frac{\gamma}{Pr}a_4)(J\frac{u}{\rho})_\xi + a_5(J\frac{v}{\rho})_\xi + a_7(J\frac{w}{\rho})_\xi\right] \\
&+ \left[(d_1 - \frac{\gamma}{Pr}d_4)(J\frac{u}{\rho})_\eta + (d_5 v + d_6 w)(\frac{J}{\rho})_\eta + d_7(\frac{J}{\rho})v_\eta + d_9(\frac{J}{\rho})w_\eta\right] \\
&+ \left[(e_1 - \frac{\gamma}{Pr}e_4)(J\frac{u}{\rho})_\zeta + (e_5 v + e_6 w)(\frac{J}{\rho})_\zeta + e_7(\frac{J}{\rho})v_\zeta + e_9(\frac{J}{\rho})w_\zeta\right]
\end{aligned}
$$

$$
\begin{aligned}
EQ_{5,3} &= \left[(a_2 - \frac{\gamma}{Pr}a_4)(J\frac{v}{\rho})_\xi + a_5(J\frac{u}{\rho})_\xi + a_6(J\frac{w}{\rho})_\xi\right] \\
&+ \left[(d_2 - \frac{\gamma}{Pr}d_4)(J\frac{v}{\rho})_\eta + (d_7 u + d_{10} w)(\frac{J}{\rho})_\eta + d_5(\frac{J}{\rho})u_\eta + d_{10}(\frac{J}{\rho})w_\eta\right] \\
&+ \left[(e_2 - \frac{\gamma}{Pr}e_4)(J\frac{v}{\rho})_\eta + (e_7 u + e_8 w)(\frac{J}{\rho})_\zeta + e_5(\frac{J}{\rho})u_\zeta + e_{10}(\frac{J}{\rho})w_\zeta\right]
\end{aligned}
$$

$$
\begin{aligned}
EQ_{5,4} &= \left[(a_3 - \frac{\gamma}{Pr}a_4)(J\frac{w}{\rho})_\xi + a_7(J\frac{u}{\rho})_\xi + a_6(J\frac{v}{\rho})_\xi\right] \\
&+ \left[(d_3 - \frac{\gamma}{Pr}d_4)(J\frac{w}{\rho})_\eta + (d_9 u + d_{10} v)(\frac{J}{\rho})_\eta + d_6(\frac{J}{\rho})u_\eta + d_8(\frac{J}{\rho})v_\eta\right] \\
&+ \left[(e_3 - \frac{\gamma}{Pr}e_4)(J\frac{w}{\rho})_\zeta + (e_9 u + e_{10} v)(\frac{J}{\rho})_\zeta + e_6(\frac{J}{\rho})u_\zeta + e_8(\frac{J}{\rho})v_\zeta\right]
\end{aligned}
$$

$$
EQ_{5,5} = \frac{\gamma}{Pr}\left[a_4(\frac{J}{\rho})_\xi + d_4(\frac{J}{\rho})_\eta + e_4(\frac{J}{\rho})_\zeta\right]
$$

$$
\begin{aligned}
FQ_{2,1} &= -\left[d_1(J\frac{u}{\rho})_\xi + d_5(J\frac{v}{\rho})_\xi + d_6(J\frac{w}{\rho})_\xi\right] \\
&- \left[b_1(J\frac{u}{\rho})_\eta + b_5(J\frac{v}{\rho})_\eta + b_7(J\frac{w}{\rho})_\eta\right]
\end{aligned}
$$

$$- \left[f_1(J\frac{u}{\rho})_\zeta + f_7(J\frac{v}{\rho})_\zeta + f_9(J\frac{w}{\rho})_\zeta \right]$$

$$FQ_{2,2} = d_1(\frac{J}{\rho})_\xi + b_1(\frac{J}{\rho})_\eta + f_1(\frac{J}{\rho})_\zeta$$

$$FQ_{2,3} = d_5(\frac{J}{\rho})_\xi + b_5(\frac{J}{\rho})_\eta + f_7(\frac{J}{\rho})_\zeta$$

$$FQ_{2,4} = d_6(\frac{J}{\rho})_\xi + b_7(\frac{J}{\rho})_\eta + f_9(\frac{J}{\rho})_\zeta$$

$$\begin{aligned} FQ_{3,1} = & - \left[d_7(J\frac{u}{\rho})_\xi + d_2(J\frac{v}{\rho})_\xi + d_8(J\frac{w}{\rho})_\xi \right] \\ & - \left[b_5(J\frac{u}{\rho})_\eta + b_2(J\frac{v}{\rho})_\eta + b_6(J\frac{w}{\rho})_\eta \right] \\ & - \left[f_5(J\frac{u}{\rho})_\zeta + f_2(J\frac{v}{\rho})_\zeta + f_{10}(J\frac{w}{\rho})_\zeta \right] \end{aligned}$$

$$FQ_{3,2} = d_7(\frac{J}{\rho})_\xi + b_5(\frac{J}{\rho})_\eta + f_5(\frac{J}{\rho})_\zeta$$

$$FQ_{3,3} = d_2(\frac{J}{\rho})_\xi + b_2(\frac{J}{\rho})_\eta + f_2(\frac{J}{\rho})_\zeta$$

$$FQ_{3,4} = d_8(\frac{J}{\rho})_\xi + b_6(\frac{J}{\rho})_\eta + f_{10}(\frac{J}{\rho})_\zeta$$

$$FQ_{4,1} = -\left[d_9(J\frac{u}{\rho})_\xi + d_{10}(J\frac{v}{\rho})_\xi + d_3(J\frac{w}{\rho})_\xi\right]$$

$$-\left[b_7(J\frac{u}{\rho})_\eta + b_6(J\frac{v}{\rho})_\eta + b_3(J\frac{w}{\rho})_\eta\right]$$

$$-\left[f_6(J\frac{u}{\rho})_\zeta + f_8(J\frac{v}{\rho})_\zeta + f_3(J\frac{w}{\rho})_\zeta\right]$$

$$FQ_{4,2} = d_9(\frac{J}{\rho})_\xi + b_7(\frac{J}{\rho})_\eta + f_6(\frac{J}{\rho})_\zeta$$

$$FQ_{4,3} = d_{10}(\frac{J}{\rho})_\xi + b_6(\frac{J}{\rho})_\eta + f_8(\frac{J}{\rho})_\zeta$$

$$FQ_{4,4} = d_3(\frac{J}{\rho})_\xi + b_3(\frac{J}{\rho})_\eta + f_3(\frac{J}{\rho})_\zeta$$

$$FQ_{5,1} = -\left\{(d_1 - \frac{\gamma}{Pr}d_4)(J\frac{u^2}{\rho})_\xi + (d_2 - \frac{\gamma}{Pr}d_4)(J\frac{v^2}{\rho})_\xi\right.$$

$$+ (d_3 - \frac{\gamma}{Pr}d_4)(J\frac{w^2}{\rho})_\xi + \frac{\gamma}{Pr}d_4(J\frac{e_t}{\rho})_\xi + (d_7 v + d_9 w)(J\frac{u}{\rho})_\xi$$

$$+ (d_5 u + d_{10} w)(J\frac{v}{\rho})_\xi + (d_6 u + d_8 v)(J\frac{w}{\rho})_\xi + \frac{J}{\rho}(d_7 v + d_9 w)u_\xi$$

$$\left. + \frac{J}{\rho}(d_5 u + d_{10} w)v_\xi + \frac{J}{\rho}(d_6 u + d_8 v)w_\xi\right\} - \left\{ (b_1 - \frac{\gamma}{Pr}b_4)(J\frac{u^2}{\rho})_\eta\right.$$

$$+ (b_2 - \frac{\gamma}{Pr}b_4)(J\frac{v^2}{\rho})_\eta + (b_3 - \frac{\gamma}{Pr}b_4)(J\frac{w^2}{\rho})_\eta + \frac{\gamma}{Pr}b_4(J\frac{e_t}{\rho})_\eta$$

$$\left. + 2b_5(J\frac{uv}{\rho})_\eta + 2b_6(J\frac{vw}{\rho})_\eta + 2b_7(J\frac{uw}{\rho})_\eta\right\} - \left\{(f_1 - \frac{\gamma}{Pr}f_4)(J\frac{u^2}{\rho})_\zeta + \right.$$

$$+ (f_2 - \frac{\gamma}{Pr} f_4)(J\frac{v^2}{\rho})_\zeta + (f_3 - \frac{\gamma}{Pr} f_4)(J\frac{w^2}{\rho})_\zeta + \frac{\gamma}{Pr} f_4 (J\frac{e_t}{\rho})_\zeta$$

$$+ (f_5 v + f_6 w)(J\frac{u}{\rho})_\zeta + (f_7 u + f_8 w)(J\frac{v}{\rho})_\zeta + (f_9 u + f_{10} v)(J\frac{w}{\rho})_\zeta$$

$$+ \frac{J}{\rho}(f_5 v + f_6 w)u_\zeta + \frac{J}{\rho}(f_7 u + f_8 w)v_\zeta + \frac{J}{\rho}(f_9 u + f_{10} v)w_\zeta \Big\}$$

$$FQ_{5,2} = \left[(d_1 - \frac{\gamma}{Pr} d_4)(J\frac{u}{\rho})_\xi + (d_7 v + d_9 w)(\frac{J}{\rho})_\xi + d_5(\frac{J}{\rho})v_\xi + d_6(\frac{J}{\rho})w_\xi\right]$$

$$+ \left[(b_1 - \frac{\gamma}{Pr} b_4)(J\frac{u}{\rho})_\eta + b_5(J\frac{v}{\rho})_\eta + b_7(J\frac{w}{\rho})_\eta\right] + \left[(f_1 - \frac{\gamma}{Pr} f_4)(J\frac{u}{\rho})_\zeta\right.$$

$$\left. + (f_5 v + f_6 w)(\frac{J}{\rho})_\xi + f_7(\frac{J}{\rho})v_\zeta + f_9(\frac{J}{\rho})w_\zeta\right]$$

$$FQ_{5,3} = \left[(d_2 - \frac{\gamma}{Pr} d_4)(J\frac{v}{\rho})_\xi + (d_5 u + d_{10} w)(\frac{J}{\rho})_\xi + d_7(\frac{J}{\rho})u_\xi + d_8(\frac{J}{\rho})w_\xi\right]$$

$$+ \left[(b_2 - \frac{\gamma}{Pr} b_4)(J\frac{v}{\rho})_\eta + b_5(J\frac{u}{\rho})_\eta + b_6(J\frac{w}{\rho})_\eta\right]$$

$$+ \left[(f_2 - \frac{\gamma}{Pr} f_4)(J\frac{v}{\rho})_\zeta + (f_7 u + f_8 w)(\frac{J}{\rho})_\zeta + f_5(\frac{J}{\rho})u_\zeta + f_{10}(\frac{J}{\rho})w_\zeta\right]$$

$$FQ_{5,4} = \left[(d_3 - \frac{\gamma}{Pr} d_4)(J\frac{w}{\rho})_\xi + (d_6 u + d_8 v)(\frac{J}{\rho})_\xi + d_9(\frac{J}{\rho})u_\xi\right.$$

$$\left. + d_{10}(\frac{J}{\rho})v_\xi\right] + \left[(b_3 - \frac{\gamma}{Pr} b_4)(J\frac{w}{\rho})_\eta + b_7(J\frac{u}{\rho})_\eta + b_6(J\frac{v}{\rho})_\eta\right]$$

$$+ \left[(f_3 - \frac{\gamma}{Pr} f_4)(J\frac{w}{\rho})_\zeta + (f_9 u + f_{10} v)(\frac{J}{\rho})_\zeta + f_6(\frac{J}{\rho})u_\zeta + f_8(\frac{J}{\rho})v_\zeta\right]$$

$$FQ_{5,5} = \frac{\gamma}{Pr}\left[d_4(\frac{J}{\rho})_\xi + b_4(\frac{J}{\rho})_\eta + f_4(\frac{J}{\rho})_\zeta\right]$$

$$\begin{aligned} GQ_{2,1} &= -\left[e_1(J\frac{u}{\rho})_\xi + e_5(J\frac{v}{\rho})_\xi + e_6(J\frac{w}{\rho})_\xi\right] \\ &\quad -\left[f_1(J\frac{u}{\rho})_\eta + f_5(J\frac{v}{\rho})_\eta + f_6(J\frac{w}{\rho})_\eta\right] \\ &\quad -\left[c_1(J\frac{u}{\rho})_\zeta + c_5(J\frac{v}{\rho})_\zeta + c_7(J\frac{w}{\rho})_\zeta\right] \end{aligned}$$

$$GQ_{2,2} = e_1(\frac{J}{\rho})_\xi + f_1(\frac{J}{\rho})_\eta + c_1(\frac{J}{\rho})_\zeta$$

$$GQ_{2,3} = e_5(\frac{J}{\rho})_\xi + f_5(\frac{J}{\rho})_\eta + c_5(\frac{J}{\rho})_\zeta$$

$$GQ_{2,4} = e_6(\frac{J}{\rho})_\xi + f_6(\frac{J}{\rho})_\eta + c_7(\frac{J}{\rho})_\zeta$$

$$\begin{aligned} GQ_{3,1} &= -\left[e_7(J\frac{u}{\rho})_\xi + e_2(J\frac{v}{\rho})_\xi + e_8(J\frac{w}{\rho})_\xi\right] \\ &\quad -\left[f_7(J\frac{u}{\rho})_\eta + f_2(J\frac{v}{\rho})_\eta + f_8(J\frac{w}{\rho})_\eta\right] \\ &\quad -\left[c_5(J\frac{u}{\rho})_\zeta + c_2(J\frac{v}{\rho})_\zeta + c_6(J\frac{w}{\rho})_\zeta\right] \end{aligned}$$

$$GQ_{3,2} = e_7(\frac{J}{\rho})_\xi + f_7(\frac{J}{\rho})_\eta + c_5(\frac{J}{\rho})_\zeta$$

$$GQ_{3,3} = e_2(\frac{J}{\rho})_\xi + f_2(\frac{J}{\rho})_\eta + c_2(\frac{J}{\rho})_\zeta$$

$$GQ_{3,4} = e_8(\frac{J}{\rho})_\xi + f_8(\frac{J}{\rho})_\eta + c_6(\frac{J}{\rho})_\zeta$$

$$\begin{aligned} GQ_{4,1} = &- \left[e_9(J\frac{u}{\rho})_\xi + e_{10}(J\frac{v}{\rho})_\xi + e_3(J\frac{w}{\rho})_\xi\right] \\ &- \left[f_9(J\frac{u}{\rho})_\eta + f_{10}(J\frac{v}{\rho})_\eta + f_3(J\frac{w}{\rho})_\eta\right] \\ &- \left[c_7(J\frac{u}{\rho})_\zeta + c_6(J\frac{v}{\rho})_\zeta + c_3(J\frac{w}{\rho})_\zeta\right] \end{aligned}$$

$$GQ_{4,2} = e_9(\frac{J}{\rho})_\xi + f_9(\frac{J}{\rho})_\eta + c_7(\frac{J}{\rho})_\zeta$$

$$GQ_{4,3} = e_{10}(\frac{J}{\rho})_\xi + f_{10}(\frac{J}{\rho})_\eta + c_6(\frac{J}{\rho})_\zeta$$

$$GQ_{4,4} = e_3(\frac{J}{\rho})_\xi + f_3(\frac{J}{\rho})_\eta + c_3(\frac{J}{\rho})_\zeta$$

$$\begin{aligned} GQ_{5,1} = &- \Big\{(e_1 - \frac{\gamma}{Pr}e_4)(J\frac{u^2}{\rho})_\xi + (e_2 - \frac{\gamma}{Pr}e_4)(J\frac{v^2}{\rho})_\xi \\ &+ (e_3 - \frac{\gamma}{Pr}e_4)(J\frac{w^2}{\rho})_\xi + \frac{\gamma}{Pr}e_4(J\frac{e_t}{\rho})_\xi + (e_7 v + e_9 w)(J\frac{u}{\rho})_\xi \\ &+ (e_5 u + e_{10} w)(J\frac{v}{\rho})_\xi + (e_6 u + e_8 v)(J\frac{w}{\rho})_\xi + \frac{J}{\rho}(e_7 v + e_9 w)u_\xi \\ &+ \frac{J}{\rho}(e_5 u + e_{10} w)v_\xi + \frac{J}{\rho}(e_6 u + e_8 v)w_\xi\Big\} - \Big\{(f_1 - \frac{\gamma}{Pr}f_4)(J\frac{u^2}{\rho})_\eta \\ &+ (f_2 - \frac{\gamma}{Pr}f_4)(J\frac{v^2}{\rho})_\eta + (f_3 - \frac{\gamma}{Pr}f_4)(J\frac{w^2}{\rho})_\eta + \frac{\gamma}{Pr}f_4(J\frac{e_t}{\rho})_\eta + \end{aligned}$$

$$+ (f_7 v + f_9 w)(J\frac{u}{\rho})_\eta + (f_5 u + f_{10} w)(J\frac{v}{\rho})_\eta + (f_6 u + f_8 v)(J\frac{w}{\rho})_\eta$$

$$+ \frac{J}{\rho}(f_7 v + f_9 w)u_\eta + \frac{J}{\rho}(f_5 u + f_{10} w)v_\eta + \frac{J}{\rho}(f_6 u + f_8 v)w_\eta \Big\}$$

$$- \Big\{ (c_1 - \frac{\gamma}{Pr}c_4)(J\frac{u^2}{\rho})_\zeta + (c_2 - \frac{\gamma}{Pr}c_4)(J\frac{v^2}{\rho})_\zeta + (c_3 - \frac{\gamma}{Pr}c_4)(J\frac{w^2}{\rho})_\zeta$$

$$+ \frac{\gamma}{Pr}c_4(J\frac{e_t}{\rho})_\zeta + 2c_5(J\frac{uv}{\rho})_\zeta + 2c_6(J\frac{vw}{\rho})_\zeta + 2c_7(J\frac{uw}{\rho})_\zeta \Big\}$$

$$GQ_{5,2} = \Big[(e_1 - \frac{\gamma}{Pr}e_4)(J\frac{u}{\rho})_\xi + (e_7 v + e_9 w)(\frac{J}{\rho})_\xi + e_5(\frac{J}{\rho})v_\xi$$

$$+ e_6(\frac{J}{\rho})w_\xi \Big] + \Big[(f_1 - \frac{\gamma}{Pr}f_4)(J\frac{u}{\rho})_\eta + (f_7 v + f_9 w)(\frac{J}{\rho})_\eta + f_5(\frac{J}{\rho})v_\eta$$

$$+ f_6(\frac{J}{\rho})w_\eta \Big] + \Big[(c_1 - \frac{\gamma}{Pr}c_4)(J\frac{u}{\rho})_\zeta + c_5(J\frac{v}{\rho})_\zeta + c_7(J\frac{w}{\rho})_\zeta \Big]$$

$$GQ_{5,3} = \Big[(e_2 - \frac{\gamma}{Pr}e_4)(J\frac{v}{\rho})_\xi + (e_5 u + e_{10} w)(\frac{J}{\rho})_\xi + e_7(\frac{J}{\rho})u_\xi + e_8(\frac{J}{\rho})w_\xi) \Big]$$

$$+ \Big[(f_2 - \frac{\gamma}{Pr}f_4)(J\frac{v}{\rho})_\eta + (f_5 u + f_{10} w)(\frac{J}{\rho})_\eta + f_7(\frac{J}{\rho})u_\eta + f_8(\frac{J}{\rho})w_\eta \Big]$$

$$+ \Big[(c_2 - \frac{\gamma}{Pr}c_4)(J\frac{v}{\rho})_\zeta + c_5(J\frac{u}{\rho})_\zeta + c_6(J\frac{w}{\rho})_\zeta \Big]$$

$$GQ_{5,4} = \Big[(e_3 - \frac{\gamma}{Pr}e_4)(J\frac{w}{\rho})_\xi + (e_6 u + e_8 v)(\frac{J}{\rho})_\xi + e_9(\frac{J}{\rho})u_\xi + e_{10}(\frac{J}{\rho})v_\xi \Big] +$$

$$+ \left[(f_3 - \frac{\gamma}{Pr} f_4)(J\frac{w}{\rho})_\eta + (f_6 u + f_8 v)(\frac{J}{\rho})_\eta + f_9(\frac{J}{\rho})u_\eta + f_{10}(\frac{J}{\rho})v_\eta \right]$$

$$+ \left[(c_3 - \frac{\gamma}{Pr} c_4)(J\frac{w}{\rho})_\zeta + c_7(J\frac{u}{\rho})_\zeta + c_6(J\frac{v}{\rho})_\zeta \right]$$

$$GQ_{5,5} = \frac{\gamma}{Pr} \left[e_4(\frac{J}{\rho})_\xi + f_4(\frac{J}{\rho})_\eta + c_4(\frac{J}{\rho})_\zeta \right]$$

11.5 Thin-Layer Approximation

The Navier-Stokes equations are reduced rather drastically by assuming thin-layer approximation. Under this assumption the gradients of the viscous stresses in the direction parallel to the surface (ξ and ζ directions) are neglected. The Thin-Layer Navier-Stokes equations are expressed as

$$\frac{\partial \bar{Q}}{\partial \tau} + \frac{\partial \bar{E}}{\partial \xi} + \frac{\partial \bar{F}}{\partial \eta} + \frac{\partial \bar{G}}{\partial \zeta} = \frac{\partial \bar{F}_{vT}}{\partial \eta} \tag{11-155}$$

where flux vectors $\bar{Q}$, $\bar{E}$, $\bar{F}$, and $\bar{G}$ are given by (11-61), (11-62), (11-63), (11-64) and

$$\bar{F}_{vT} = \frac{\mu}{Re_\infty J} \begin{bmatrix} 0 \\ \hline b_1 u_\eta + b_5 v_\eta + b_7 w_\eta \\ \hline b_5 u_\eta + b_2 v_\eta + b_6 w_\eta \\ \hline b_7 u_\eta + b_6 v_\eta + b_3 w_\eta \\ \hline \frac{1}{2} b_1 (u^2)_\eta + \frac{1}{2} b_2 (v^2)_\eta + \frac{1}{2} b_3 (w^2)_\eta + b_5 (uv)_\eta \\ \quad + \ b_6 (vw)_\eta + b_7 (uw)_\eta + \frac{1}{Pr(\gamma-1)M_\infty^2} b_4 T_\eta \end{bmatrix} \tag{11-156}$$

The viscous flux vector $\bar{F}_{vT}$ is obtained from the viscous flux vector $\bar{F}_v$ given as Equation (11-96) by omitting all the mixed partial derivatives. To illustrate this point, consider for example the second component of the flux vector $\bar{F}_v$, where

$$\bar{F}_{v_2} = d_1 u_\xi + d_5 v_\xi + d_6 w_\xi + b_1 u_\eta + b_5 v_\eta + b_7 w_\eta + f_1 u_\zeta + f_7 v_\zeta + f_9 w_\zeta$$

The partial differential equation (11-155) includes the η gradient of the viscous terms. Therefore, this gradient will include terms such as $\partial^2 u/\partial\eta\partial\xi$, $\partial^2 v/\partial\eta\partial\xi$, ... and so on. These and similar mixed partial derivatives are omitted. As a result, only η derivatives of the flow variables remain, which have been redefined as the flux vector $\bar{F}_{vT}$. The reason these terms are dropped will be discussed in Chapter 14 where the numerical solution techniques for thin-layer and Navier-Stokes equations are considered. The elimination of these terms is not necessary; however, this reduction does simplify the equations and reduces computer storage and time requirements.

The inviscid Jacobian matrices are the same as those of the Navier-Stokes equations, and the Jacobian matrix $\partial\bar{F}_{vT}/\partial\bar{Q}$ is

$$
\frac{\partial \bar{F}_{vT}}{\partial \bar{Q}} = \frac{\mu}{Re_\infty J}
\left[
\begin{array}{l:l:l:l:l}
0 & 0 & 0 & 0 & 0 \\
\hdashline
-\left[b_1(J\frac{u}{\rho})_\eta + b_5(J\frac{v}{\rho})_\eta + b_7(J\frac{w}{\rho})_\eta\right] & b_1(\frac{J}{\rho})_\eta & b_5(\frac{J}{\rho})_\eta & b_7(\frac{J}{\rho})_\eta & 0 \\
\hdashline
-\left[b_5(J\frac{u}{\rho})_\eta + b_2(J\frac{v}{\rho})_\eta + b_6(J\frac{w}{\rho})_\eta\right] & b_5(\frac{J}{\rho})_\eta & b_2(\frac{J}{\rho})_\eta & b_6(\frac{J}{\rho})_\eta & 0 \\
\hdashline
-\left[b_7(J\frac{u}{\rho})_\eta + b_6(J\frac{v}{\rho})_\eta + b_3(J\frac{w}{\rho})_\eta\right] & b_7(\frac{J}{\rho})_\eta & b_6(\frac{J}{\rho})_\eta & b_3(\frac{J}{\rho})_\eta & 0 \\
\hdashline
-\left\{(b_1 - \frac{\gamma}{Pr}b_4)(J\frac{u^2}{\rho})_\eta\right. & (b_1 - \frac{\gamma}{Pr}b_4)(J\frac{u}{\rho})_\eta & b_5(J\frac{u}{\rho})_\eta & b_7(J\frac{u}{\rho})_\eta & \frac{\gamma}{Pr}b_4(\frac{J}{\rho})_\eta \\
+(b_2 - \frac{\gamma}{Pr}b_4)(J\frac{v^2}{\rho})_\eta & +b_5(J\frac{v}{\rho})_\eta & +(b_2 - \frac{\gamma}{Pr}b_4)(J\frac{v}{\rho})_\eta & +b_6(J\frac{v}{\rho})_\eta & \\
+(b_3 - \frac{\gamma}{Pr}b_4)(J\frac{w^2}{\rho})_\eta & +b_7(J\frac{w}{\rho})_\eta & +b_6(J\frac{w}{\rho})_\eta & +(b_3 - \frac{\gamma}{Pr}b_4)(J\frac{w}{\rho})_\eta & \\
+\frac{\gamma}{Pr}b_4(J\frac{e_t}{\rho})_\eta & & & & \\
+2b_5(J\frac{uv}{\rho})_\eta + 2b_6(J\frac{vw}{\rho})_\eta & & & & \\
+2b_7(J\frac{uw}{\rho})_\eta & & & &
\end{array}
\right]
$$

11.6 Parabolized Navier-Stokes Equations

For steady supersonic flow fields, PNS equations provide an efficient method of solution. It is assumed that the streamwise gradient of viscous stress is small compared to the gradients of stresses in the η and ζ directions. Therefore, the streamwise gradient of viscous stress is dropped. In addition, the streamwise pressure gradient within the subsonic portion of the boundary layer must be approximated to prevent a departure solution. This approximation may employ any one of the following methods:

a. Neglect the pressure gradient completely.
b. Treat the pressure gradient explicitly.
c. Impose the pressure from the first supersonic point—this method is known as the sublayer approximation.
d. Retain a fraction ω of the pressure gradient.

The PNS approximations will exclude streamwise flow separation; however, cross flow separation is predicted. In the following formulation of the PNS equation, the streamwise pressure gradient within the subsonic portion is approximated by method d. The PNS equations are expressed as

$$\frac{\partial \bar{E}_P}{\partial \xi} + \frac{\partial \bar{E}_{PP}}{\partial \xi} + \frac{\partial \bar{F}}{\partial \eta} + \frac{\partial \bar{G}}{\partial \zeta} = \frac{\partial \bar{F}_{vP}}{\partial \eta} + \frac{\partial \bar{G}_{vP}}{\partial \zeta} \tag{11-157}$$

where

$$\bar{E}_P = \frac{1}{J}\left[\xi_x E_P + \xi_y F_P + \xi_z G_P\right] \tag{11-158}$$

$$\bar{E}_{PP} = \frac{1}{J}\left[\xi_x E_{PP} + \xi_y F_{PP} + \xi_z G_{PP}\right] \tag{11-159}$$

$$\bar{F} = \frac{1}{J}\left[\eta_x E + \eta_y F + \eta_z G\right] \tag{11-160}$$

$$\bar{G} = \frac{1}{J}\left[\zeta_x E + \zeta_y F + \zeta_z G\right] \tag{11-161}$$

$$\bar{F}_{vP} = \frac{1}{J}\left[\eta_x E_v + \eta_y F_v + \eta_z G_v\right] \tag{11-162}$$

$$\bar{G}_{vP} = \frac{1}{J}\left[\zeta_x E_v + \zeta_y F_v + \zeta_z G_v\right] \tag{11-163}$$

and flux vectors are defined as

$$E_P = \begin{bmatrix} \rho u \\ \rho u^2 + \omega p \\ \rho uv \\ \rho uw \\ (\rho e_t + p)u \end{bmatrix} \quad (11\text{-}165) \qquad F_P = \begin{bmatrix} \rho v \\ \rho vu \\ \rho v^2 + \omega p \\ \rho vw \\ (\rho e_t + p)v \end{bmatrix} \quad (11\text{-}166)$$

$$G_P = \begin{bmatrix} \rho w \\ \rho wu \\ \rho wv \\ \rho w^2 + \omega p \\ (\rho e_t + p)w \end{bmatrix} \quad (11\text{-}167) \qquad E_{PP} = \begin{bmatrix} 0 \\ (1-\omega)p \\ 0 \\ 0 \\ 0 \end{bmatrix} \quad (11\text{-}168)$$

$$F_{PP} = \begin{bmatrix} 0 \\ 0 \\ (1-\omega)p \\ 0 \\ 0 \end{bmatrix} \quad (11\text{-}169) \qquad G_{PP} = \begin{bmatrix} 0 \\ 0 \\ 0 \\ (1-\omega)p \\ 0 \end{bmatrix} \quad (11\text{-}170)$$

The flux vectors E, F, and G are the same as E_P, F_P, and G_P with $\omega = 1$. The parameter ω, introduced in Equations (11-165) through (11-170), is determined by stability analysis. The viscous flux vectors $\bar{F}_{vP}$ and $\bar{G}_{vP}$ are

$$\bar{F}_{vP} = \frac{\mu}{Re_\infty J}\begin{bmatrix} 0 \\ \hline b_1 u_\eta + b_5 v_\eta + b_7 w_\eta \\ \hline b_5 u_\eta + b_2 v_\eta + b_6 w_\eta \\ \hline b_7 u_\eta + b_6 v_\eta + b_3 w_\eta \\ \hline \frac{1}{2}b_1(u^2)_\eta + \frac{1}{2}b_2(v^2)_\eta + \frac{1}{2}b_3(w^2)_\eta + b_5(uv)_\eta + b_6(uw)_\eta \\ +b_7(uw)_\eta + \frac{1}{Pr(\gamma-1)M_\infty^2}b_4 T_\eta \end{bmatrix} \tag{11-171}$$

$$\bar{G}_{vP} = \frac{\mu}{Re_\infty J}\begin{bmatrix} 0 \\ \hline c_1 u_\zeta + c_5 v_\zeta + c_7 w_\zeta \\ \hline c_5 u_\zeta + c_2 v_\zeta + c_6 w_\zeta \\ \hline c_7 u_\zeta + c_6 v_\zeta + c_3 w_\zeta \\ \hline \frac{1}{2}c_1(u^2)_\zeta + \frac{1}{2}c_2(v^2)_\zeta + \frac{1}{2}c_3(w^2)_\zeta + c_5(uv)_\zeta + c_6(uw)_\zeta \\ +c_7(uw)_\zeta + \frac{1}{Pr(\gamma-1)M_\infty^2}c_4 T_\zeta \end{bmatrix} \tag{11-172}$$

where

$$b_1 = \tfrac{4}{3}\eta_x^2 + \eta_y^2 + \eta_z^2 \qquad (11\text{-}173)$$

$$b_2 = \eta_x^2 + \tfrac{4}{3}\eta_y^2 + \eta_z^2 \qquad (11\text{-}174)$$

$$b_3 = \eta_x^2 + \eta_y^2 + \tfrac{4}{3}\eta_y^2 \qquad (11\text{-}175)$$

$$b_4 = \eta_x^2 + \eta_y^2 + \eta_z^2 \qquad (11\text{-}176)$$

$$b_5 = \tfrac{1}{3}\eta_x\eta_y \qquad (11\text{-}177)$$

$$b_6 = \tfrac{1}{3}\eta_y\eta_z \qquad (11\text{-}178)$$

$$b_7 = \tfrac{1}{3}\eta_x\eta_z \qquad (11\text{-}179)$$

$$c_1 = \tfrac{4}{3}\zeta_x^2 + \zeta_y^2 + \zeta_z^2 \qquad (11\text{-}180)$$

$$c_2 = \zeta_x^2 + \tfrac{4}{3}\zeta_y^2 + \zeta_z^2 \qquad (11\text{-}181)$$

$$c_3 = \zeta_x^2 + \zeta_y^2 + \tfrac{4}{3}\zeta_z^2 \qquad (11\text{-}182)$$

$$c_4 = \zeta_x^2 + \zeta_y^2 + \zeta_z^2 \qquad (11\text{-}183)$$

$$c_5 = \tfrac{1}{3}\zeta_x\zeta_y \qquad (11\text{-}184)$$

$$c_6 = \tfrac{1}{3}\zeta_y\zeta_z \qquad (11\text{-}185)$$

$$c_7 = \tfrac{1}{3}\zeta_x\zeta_z \qquad (11\text{-}186)$$

Note that the viscous flux vectors $\bar{F}_{vP}$ and $\bar{G}_{vP}$ are obtained from relations (11-96) and (11-97) by omitting all the mixed partial derivatives, as was illustrated previously for the Thin-Layer Navier-Stokes equations.

The inviscid and viscous Jacobian matrices for the PNS equations are as follows:

$$
\frac{\partial \bar{E}_p}{\partial \bar{Q}} =
\left[\begin{array}{c:c:c:c:c}
0 & \xi_x & \xi_y & \xi_z & 0 \\ \hdashline
\begin{array}{l} -u(\xi_x u + \xi_y v + \xi_z w) \\ + \xi_x \left[\frac{1}{2}\omega(\gamma-1)(u^2+v^2+w^2)\right] \end{array} &
\begin{array}{l} \xi_x[1-\omega(\gamma-1)]u \\ + (\xi_x u + \xi_y v + \xi_z w) \end{array} &
\xi_y u - \omega(\gamma-1)\xi_x v &
\xi_z u - \omega(\gamma-1)\xi_x w &
(\gamma-1)\xi_x \\ \hdashline
\begin{array}{l} -v(\xi_x u + \xi_y v + \xi_z w) \\ + \xi_y \left[\frac{1}{2}\omega(\gamma-1)(u^2+v^2+w^2)\right] \end{array} &
\xi_x v - \omega(\gamma-1)\xi_y u &
\begin{array}{l} \xi_y[1-\omega(\gamma-1)]v \\ + (\xi_x u + \xi_y v + \xi_z w) \end{array} &
\xi_z v - \omega(\gamma-1)\xi_y w &
(\gamma-1)\xi_y \\ \hdashline
\begin{array}{l} -w(\xi_x u + \xi_y v + \xi_z w) \\ + \xi_z \left[\frac{1}{2}\omega(\gamma-1)(u^2+v^2+w^2)\right] \end{array} &
\xi_x w - \omega(\gamma-1)\xi_z u &
\xi_y w - \omega(\gamma-1)\xi_z v &
\begin{array}{l} \xi_z[1-\omega(\gamma-1)]w \\ + (\xi_x u + \xi_y v + \xi_z w) \end{array} &
(\gamma-1)\xi_z \\ \hdashline
\begin{array}{l} (\xi_x u + \xi_y v + \xi_z w)[-\gamma e_t \\ + (\gamma-1)(u^2+v^2+w^2)] \end{array} &
\begin{array}{l} \xi_x \left[\gamma e_t - \frac{1}{2}(\gamma-1) \cdot \right. \\ \left. (u^2+v^2+w^2)\right] \\ - (\gamma-1)[\xi_x u + \xi_y v \\ + \xi_z w]\, u \end{array} &
\begin{array}{l} \xi_y \left[\gamma e_t - \frac{1}{2}(\gamma-1) \cdot \right. \\ \left. (u^2+v^2+w^2)\right] \\ - (\gamma-1)[\xi_x u + \xi_y v \\ + \xi_z w]\, v \end{array} &
\begin{array}{l} \xi_z \left[\gamma e_t - \frac{1}{2}(\gamma-1) \cdot \right. \\ \left. (u^2+v^2+w^2)\right] \\ - (\gamma-1)[\xi_x u + \xi_y v \\ + \xi_z w]\, w \end{array} &
\gamma[\xi_x u + \xi_y v + \xi_z w]
\end{array}\right]
\qquad (11\text{-}187)
$$

$$
\frac{\partial \bar{F}}{\partial \bar{Q}} = \left[\begin{array}{c:c:c:c:c}
0 & \eta_x & \eta_y & \eta_z & 0 \\ \hdashline
\begin{array}{l} -u(\eta_x u + \eta_y v + \eta_z w) \\ +\ \eta_x \left[\frac{1}{2}(\gamma-1)(u^2+v^2+w^2)\right] \end{array} & \begin{array}{l} \eta_x(2-\gamma)u \\ +\ (\eta_x u + \eta_y v + \eta_z w) \end{array} & \eta_y u - (\gamma-1)\eta_x v & \eta_z u - (\gamma-1)\eta_x w & (\gamma-1)\eta_x \\ \hdashline
\begin{array}{l} -v(\eta_x u + \eta_y v + \eta_z w) \\ +\ \eta_y \left[\frac{1}{2}(\gamma-1)(u^2+v^2+w^2)\right] \end{array} & \eta_x v - (\gamma-1)\eta_y u & \begin{array}{l} \eta_y(2-\gamma)v \\ +\ (\eta_x u + \eta_y v + \eta_z w) \end{array} & \eta_z v - (\gamma-1)\eta_y w & (\gamma-1)\eta_y \\ \hdashline
\begin{array}{l} -w(\eta_x u + \eta_y v + \eta_z w) \\ +\ \eta_z \left[\frac{1}{2}(\gamma-1)(u^2+v^2+w^2)\right] \end{array} & \eta_x w - (\gamma-1)\eta_z u & \eta_y w - (\gamma-1)\eta_z v & \begin{array}{l} \eta_z(2-\gamma)w \\ +\ (\eta_x u + \eta_y v + \eta_z w) \end{array} & (\gamma-1)\eta_z \\ \hdashline
\begin{array}{l} (\eta_x u + \eta_y v + \eta_z w)\,[-\gamma e_t \\ +\ (\gamma-1)(u^2+v^2+w^2)] \end{array} & \begin{array}{l} \eta_x \left[\gamma e_t - \frac{1}{2}(\gamma-1)\cdot \right. \\ \left. (u^2+v^2+w^2)\right] \\ -\ (\gamma-1)\,[\eta_x u + \eta_y v \\ +\ \eta_z w]\,u \end{array} & \begin{array}{l} \eta_y \left[\gamma e_t - \frac{1}{2}(\gamma-1)\cdot \right. \\ \left. (u^2+v^2+w^2)\right] \\ -\ (\gamma-1)\,[\eta_x u + \eta_y v \\ +\ \eta_z w]\,v \end{array} & \begin{array}{l} \eta_z \left[\gamma e_t - \frac{1}{2}(\gamma-1)\cdot \right. \\ \left. (u^2+v^2+w^2)\right] \\ -\ (\gamma-1)\,[\eta_x u + \eta_y v \\ +\ \eta_z w]\,w \end{array} & \gamma\,[\eta_x u + \eta_y v + \eta_z w]
\end{array} \right] \tag{11-188}
$$

$$
\frac{\partial \bar{G}}{\partial \bar{Q}} = \left[\begin{array}{c:c:c:c:c}
0 & \zeta_x & \zeta_y & \zeta_z & 0 \\
\hdashline
\begin{array}{l} -\, u(\zeta_x u + \zeta_y v + \zeta_z w) \\ +\, \zeta_x \left[\frac{1}{2}(\gamma - 1)(u^2 + v^2 + w^2)\right] \end{array} & \begin{array}{l} \zeta_x(2-\gamma)u \\ +\, (\zeta_x u + \zeta_y v + \zeta_z w) \end{array} & \zeta_y u - (\gamma - 1)\zeta_x v & \zeta_z u - (\gamma - 1)\zeta_x w & (\gamma - 1)\zeta_x \\
\hdashline
\begin{array}{l} -\, v(\zeta_x u + \zeta_y v + \zeta_z w) \\ +\, \zeta_y \left[\frac{1}{2}(\gamma - 1)(u^2 + v^2 + w^2)\right] \end{array} & \zeta_x v - (\gamma - 1)\zeta_y u & \begin{array}{l} \zeta_y(2-\gamma)v \\ +\, (\zeta_x u + \zeta_y v + \zeta_z w) \end{array} & \zeta_z v - (\gamma - 1)\zeta_y w & (\gamma - 1)\zeta_y \\
\hdashline
\begin{array}{l} -\, w(\zeta_x u + \zeta_y v + \zeta_z w) \\ +\, \zeta_z \left[\frac{1}{2}(\gamma - 1)(u^2 + v^2 + w^2)\right] \end{array} & \zeta_x w - (\gamma - 1)\zeta_z u & \zeta_y w - (\gamma - 1)\zeta_z v & \begin{array}{l} \zeta_z(2-\gamma)w \\ +\, (\zeta_x u + \zeta_y v + \zeta_z w) \end{array} & (\gamma - 1)\zeta_z \\
\hdashline
\begin{array}{l} (\zeta_x u + \zeta_y v + \zeta_z w)\,[-\gamma e_t \\ +\, (\gamma - 1)(u^2 + v^2 + w^2)] \end{array} & \begin{array}{l} \zeta_x \left[\gamma e_t - \frac{1}{2}(\gamma - 1)\cdot \right. \\ \left. (u^2 + v^2 + w^2)\right] \\ -\, (\gamma - 1)\,[\zeta_x u + \zeta_y v \\ +\, \zeta_z w]\, u \end{array} & \begin{array}{l} \zeta_y \left[\gamma e_t - \frac{1}{2}(\gamma - 1)\cdot \right. \\ \left. (u^2 + v^2 + w^2)\right] \\ -\, (\gamma - 1)\,[\zeta_x u + \zeta_y v \\ +\, \zeta_z w]\, v \end{array} & \begin{array}{l} \zeta_z \left[\gamma e_t - \frac{1}{2}(\gamma - 1)\cdot \right. \\ \left. (u^2 + v^2 + w^2)\right] \\ -\, (\gamma - 1)\,[\zeta_x u + \zeta_y v \\ +\, \zeta_z w]\, w \end{array} & \gamma\,[\zeta_x u + \zeta_y v + \zeta_z w]
\end{array} \right]
\tag{11-189}
$$

$$\frac{\partial \bar{F}_{vP}}{\partial \bar{Q}} = \frac{\mu}{Re_\infty J}\left[\begin{array}{c:c:c:c:c}
0 & 0 & 0 & 0 & 0 \\ \hdashline
-\left[b_1\left(J\frac{u}{\rho}\right)_\eta + b_5\left(J\frac{v}{\rho}\right)_\eta + b_7\left(J\frac{w}{\rho}\right)_\eta\right] & b_1\left(\frac{J}{\rho}\right)_\eta & b_5\left(\frac{J}{\rho}\right)_\eta & b_7\left(\frac{J}{\rho}\right)_\eta & 0 \\ \hdashline
-\left[b_5\left(J\frac{u}{\rho}\right)_\eta + b_2\left(J\frac{v}{\rho}\right)_\eta + b_6\left(J\frac{w}{\rho}\right)_\eta\right] & b_5\left(\frac{J}{\rho}\right)_\eta & b_2\left(\frac{J}{\rho}\right)_\eta & b_6\left(\frac{J}{\rho}\right)_\eta & 0 \\ \hdashline
-\left[b_7\left(J\frac{u}{\rho}\right)_\eta + b_6\left(J\frac{v}{\rho}\right)_\eta + b_3\left(J\frac{w}{\rho}\right)_\eta\right] & b_7\left(\frac{J}{\rho}\right)_\eta & b_6\left(\frac{J}{\rho}\right)_\eta & b_3\left(\frac{J}{\rho}\right)_\eta & 0 \\ \hdashline
\begin{array}{l} -\left[\left(b_1 - \frac{\gamma}{Pr}b_4\right)\left(J\frac{u^2}{\rho}\right)_\eta\right. \\ +\left(b_2 - \frac{\gamma}{Pr}b_4\right)\left(J\frac{v^2}{\rho}\right)_\eta \\ +\left(b_3 - \frac{\gamma}{Pr}b_4\right)\left(J\frac{w^2}{\rho}\right)_\eta + \frac{\gamma}{Pr}b_4\left(J\frac{e_t}{\rho}\right)_\eta \\ +2b_5\left(J\frac{uv}{\rho}\right)_\eta + 2b_6\left(J\frac{vw}{\rho}\right)_\eta \\ \left. +2b_7\left(J\frac{uw}{\rho}\right)_\eta\right] \end{array} &
\begin{array}{l} \left(b_1 - \frac{\gamma}{Pr}b_4\right)\left(J\frac{u}{\rho}\right)_\eta \\ +b_5\left(J\frac{v}{\rho}\right)_\eta \\ +b_7\left(J\frac{w}{\rho}\right)_\eta \end{array} &
\begin{array}{l} \left(b_2 - \frac{\gamma}{Pr}b_4\right)\left(J\frac{v}{\rho}\right)_\eta \\ +b_5\left(J\frac{u}{\rho}\right)_\eta \\ +b_6\left(J\frac{w}{\rho}\right)_\eta \end{array} &
\begin{array}{l} \left(b_3 - \frac{\gamma}{Pr}b_4\right)\left(J\frac{w}{\rho}\right)_\eta \\ +b_7\left(J\frac{u}{\rho}\right)_\eta \\ +b_6\left(J\frac{v}{\rho}\right)_\eta \end{array} &
\frac{\gamma}{Pr}b_4\left(\frac{J}{\rho}\right)_\eta
\end{array}\right] \tag{11-190}$$

$$\frac{\partial \bar{G}_{vP}}{\partial \bar{Q}} = \frac{\mu}{Re_\infty J}\left[\begin{array}{c:c:c:c:c}
0 & 0 & 0 & 0 & 0 \\ \hdashline
-\left[c_1\left(J\frac{u}{\rho}\right)_\zeta + c_5\left(J\frac{v}{\rho}\right)_\zeta + c_7\left(J\frac{w}{\rho}\right)_\zeta\right] & c_1\left(\frac{J}{\rho}\right)_\zeta & c_5\left(\frac{J}{\rho}\right)_\zeta & c_7\left(\frac{J}{\rho}\right)_\zeta & 0 \\ \hdashline
-\left[c_5\left(J\frac{u}{\rho}\right)_\zeta + c_2\left(J\frac{v}{\rho}\right)_\zeta + c_6\left(J\frac{w}{\rho}\right)_\zeta\right] & c_5\left(\frac{J}{\rho}\right)_\zeta & c_2\left(\frac{J}{\rho}\right)_\zeta & c_6\left(\frac{J}{\rho}\right)_\zeta & 0 \\ \hdashline
-\left[c_7\left(J\frac{u}{\rho}\right)_\zeta + c_6\left(J\frac{v}{\rho}\right)_\zeta + c_3\left(J\frac{w}{\rho}\right)_\zeta\right] & c_7\left(\frac{J}{\rho}\right)_\zeta & c_6\left(\frac{J}{\rho}\right)_\zeta & c_3\left(\frac{J}{\rho}\right)_\zeta & 0 \\ \hdashline
\begin{array}{l}
-\left[\left(c_1 - \frac{\gamma}{Pr}c_4\right)\left(J\frac{u^2}{\rho}\right)_\zeta\right. \\
+\left(c_2 - \frac{\gamma}{Pr}c_4\right)\left(J\frac{v^2}{\rho}\right)_\zeta \\
+\left(c_3 - \frac{\gamma}{Pr}c_4\right)\left(J\frac{w^2}{\rho}\right)_\zeta + \frac{\gamma}{Pr}c_4\left(J\frac{e_t}{\rho}\right)_\zeta \\
+2c_5\left(J\frac{uv}{\rho}\right)_\zeta + 2c_6\left(J\frac{vw}{\rho}\right)_\zeta \\
\left.+2c_7\left(J\frac{uw}{\rho}\right)_\zeta\right]
\end{array} &
\begin{array}{l}
\left(c_1 - \frac{\gamma}{Pr}c_4\right)\left(J\frac{u}{\rho}\right)_\zeta \\
+c_5\left(J\frac{v}{\rho}\right)_\zeta \\
+c_7\left(J\frac{w}{\rho}\right)_\zeta
\end{array} &
\begin{array}{l}
\left(c_2 - \frac{\gamma}{Pr}c_4\right)\left(J\frac{v}{\rho}\right)_\zeta \\
+c_5\left(J\frac{u}{\rho}\right)_\zeta \\
+c_6\left(J\frac{w}{\rho}\right)_\zeta
\end{array} &
\begin{array}{l}
\left(c_3 - \frac{\gamma}{Pr}c_4\right)\left(J\frac{w}{\rho}\right)_\zeta \\
+c_7\left(J\frac{u}{\rho}\right)_\zeta \\
+c_6\left(J\frac{v}{\rho}\right)_\zeta
\end{array} &
\frac{\gamma}{Pr}c_4\left(\frac{J}{\rho}\right)_\zeta
\end{array}\right] \tag{11-191}$$

11.7 Two-Dimensional Planar or Axisymmetric Formulation

For a two-dimensional planar or axisymmetric flow field, the nondimensionalized Navier-Stokes equation may be expressed in a combined form as

$$\frac{\partial Q}{\partial t} + \frac{\partial E}{\partial x} + \frac{\partial F}{\partial y} + \alpha H = \frac{\partial E_v}{\partial x} + \frac{\partial F_v}{\partial y} + \alpha H_v \tag{11-192}$$

where $\alpha = 0$ represents two-dimensional planar flow and $\alpha = 1$ represents two-dimensional axisymmetric flow. The flux vectors in Equation (11-192) are

$$Q = \begin{bmatrix} \rho \\ \rho u \\ \rho v \\ \rho e_t \end{bmatrix} \quad \text{(11-193a)} \qquad E = \begin{bmatrix} \rho u \\ \rho u^2 + p \\ \rho uv \\ (\rho e_t + p)u \end{bmatrix} \quad \text{(11-193b)}$$

$$F = \begin{bmatrix} \rho v \\ \rho uv \\ \rho v^2 + p \\ (\rho e_t + p)v \end{bmatrix} \quad \text{(11-193c)} \qquad H = \frac{1}{y}\begin{bmatrix} \rho v \\ \rho uv \\ \rho v^2 \\ (\rho e_t + p)v \end{bmatrix} \quad \text{(11-193d)}$$

$$E_v = \begin{bmatrix} 0 \\ \tau_{xxp} \\ \tau_{xy} \\ u\tau_{xxp} + v\tau_{xy} - q_x \end{bmatrix} \quad \text{(11-193e)} \qquad F_v = \begin{bmatrix} 0 \\ \tau_{xy} \\ \tau_{yyp} \\ u\tau_{xy} + v\tau_{yyp} - q_y \end{bmatrix} \quad \text{(11-193f)}$$

$$H_v = \frac{1}{y}\begin{bmatrix} 0 \\ \hline \tau_{xy} - \frac{2}{3}\frac{y}{Re_\infty}\frac{\partial}{\partial x}(\mu\frac{v}{y}) \\ \hline \tau_{yyp} - \tau_{\theta\theta} - \frac{2}{3}\frac{\mu}{Re_\infty}(\frac{v}{y}) - \frac{y}{Re_\infty}\frac{2}{3}\frac{\partial}{\partial y}(\mu\frac{v}{y}) \\ \hline u\tau_{xy} + v\tau_{yyp} - q_y - \frac{2}{3}\frac{\mu}{Re_\infty}\frac{v^2}{y} - \frac{y}{Re_\infty}\frac{\partial}{\partial y}(\frac{2}{3}\mu\frac{v^2}{y}) \\ -\frac{y}{Re_\infty}\frac{\partial}{\partial x}(\frac{2}{3}\mu\frac{uv}{y}) \end{bmatrix} \tag{11-193g}$$

where

$$\tau_{xxp} = \frac{\mu}{Re_\infty}\left(\frac{4}{3}\frac{\partial u}{\partial x} - \frac{2}{3}\frac{\partial v}{\partial y}\right) \tag{11-194}$$

$$\tau_{yyp} = \frac{\mu}{Re_\infty}\left(\frac{4}{3}\frac{\partial v}{\partial y} - \frac{2}{3}\frac{\partial u}{\partial x}\right) \tag{11-195}$$

$$\tau_{xy} = \frac{\mu}{Re_\infty}\left(\frac{\partial u}{\partial y} + \frac{\partial v}{\partial x}\right) \tag{11-196}$$

$$\tau_{\theta\theta} = \frac{\mu}{Re_\infty}\left[-\frac{2}{3}(\frac{\partial u}{\partial x} + \frac{\partial v}{\partial y}) + \frac{4}{3}\frac{v}{y}\right] \tag{11-197}$$

$$q_x = -\frac{\mu}{Re_\infty Pr(\gamma - 1)M_\infty^2}\frac{\partial T}{\partial x} \tag{11-198}$$

$$q_y = -\frac{\mu}{Re_\infty Pr(\gamma - 1)M_\infty^2}\frac{\partial T}{\partial y} \tag{11-199}$$

The Navier-Stokes equations given by Equation (11-192) are transformed to the computational space by relations (11-6) and (11-7), resulting in

$$\frac{\partial \bar{Q}}{\partial \tau} + \frac{\partial \bar{E}}{\partial \xi} + \frac{\partial \bar{F}}{\partial \eta} + \alpha \bar{H} = \frac{\partial \bar{E}_v}{\partial \xi} + \frac{\partial \bar{F}_v}{\partial \eta} + \alpha \bar{H}_v \tag{11-200}$$

where

$$\bar{Q} = \frac{Q}{J} \tag{11-201}$$

$$\bar{E} = \frac{1}{J}[\xi_t Q + \xi_x E + \xi_y F] \tag{11-202}$$

$$\bar{F} = \frac{1}{J}[\eta_t Q + \eta_x E + \eta_y F] \tag{11-203}$$

$$\bar{H} = \frac{H}{J} \tag{11-204}$$

$$\bar{E}_v = \frac{1}{J}[\xi_x E_v + \xi_y F_v] \tag{11-205}$$

$$\bar{F}_v = \frac{1}{J}[\eta_x E_v + \eta_y F_v] \tag{11-206}$$

$$\bar{H}_v = \frac{H_v}{J} \tag{11-207}$$

Substitution of viscous shear stresses into expressions (11-205) and (11-206) provides

$$\bar{E}_v = \frac{\mu}{Re_\infty J}\left[\begin{array}{l} 0 \\ \hline a_1 u_\xi + a_3 v_\xi + c_1 u_\eta + c_3 v_\eta \\ \hline a_3 u_\xi + a_2 v_\xi + c_4 u_\eta + c_2 v_\eta \\ \hline \frac{1}{2}a_1(u^2)_\xi + \frac{1}{2}a_2(v^2)_\xi + a_3(uv)_\xi + \frac{1}{Pr(\gamma-1)M_\infty^2}a_4 T_\xi \\ +\frac{1}{2}c_1(u^2)_\eta + \frac{1}{2}c_2(v^2)_\eta + c_3 u v_\eta + c_4 v u_\eta + \frac{1}{Pr(\gamma-1)M_\infty^2}c_5 T_\eta \end{array}\right] \tag{11-208}$$

and

$$\bar{F}_v = \frac{\mu}{Re_\infty J}\left[\begin{array}{l} 0 \\ \hline c_1 u_\xi + c_4 v_\xi + b_1 u_\eta + b_3 v_\eta \\ \hline c_3 u_\xi + c_2 v_\xi + b_3 u_\eta + b_2 v_\eta \\ \hline \frac{1}{2}c_1(u^2)_\xi + \frac{1}{2}c_2(v^2)_\xi + c_3 v u_\xi + c_4 u v_\xi + \frac{1}{Pr(\gamma-1)M_\infty^2}c_5 T_\xi \\ +\frac{1}{2}b_1(u^2)_\eta + \frac{1}{2}b_2(v^2)_\eta + b_3(uv)_\eta + \frac{1}{Pr(\gamma-1)M_\infty^2}b_4 T_\eta \end{array}\right] \tag{11-209}$$

where

$a_1 = \frac{4}{3}\xi_x^2 + \xi_y^2$ (11-210a) $a_2 = \xi_x^2 + \frac{4}{3}\xi_y^2$ (11-210b)

$a_3 = \frac{1}{3}\xi_x\xi_y$ (11-210c) $a_4 = \xi_x^2 + \xi_y^2$ (11-210d)

$b_1 = \frac{4}{3}\eta_x^2 + \eta_y^2$ (11-211a) $b_2 = \eta_x^2 + \frac{4}{3}\eta_y^2$ (11-211b)

$b_3 = \frac{1}{3}\eta_x\eta_y$ (11-211c) $b_4 = \eta_x^2 + \eta_y^2$ (11-211d)

$c_1 = \frac{4}{3}\eta_x\xi_x + \xi_y\eta_y$ (11-212a) $c_2 = \xi_x\eta_x + \frac{4}{3}\xi_y\eta_y$ (11-212b)

$c_3 = \eta_x\xi_y - \frac{2}{3}\xi_x\eta_y$ (11-212c) $c_4 = \xi_x\eta_y - \frac{2}{3}\xi_y\eta_x$ (11-212d)

$c_5 = \xi_x\eta_x + \xi_y\eta_y$ (11-212e)

The inviscid and viscous Jacobian matrices are:

$$\frac{\partial \bar{E}}{\partial \bar{Q}} = \begin{bmatrix} \xi_t & \xi_x & \xi_y & 0 \\ -u(\xi_x u + \xi_y v) + \xi_x[\frac{1}{2}(\gamma - 1)(u^2 + v^2)] & \xi_t + \xi_x(2 - \gamma)u + \xi_x u + \xi_y v & -(\gamma - 1)\xi_x v + \xi_y u & (\gamma - 1)\xi_x \\ -v(\xi_x u + \xi_y v) + \xi_y[\frac{1}{2}(\gamma - 1)(u^2 + v^2)] & -(\gamma - 1)\xi_y u + \xi_x v & \xi_t + \xi_y(2 - \gamma)v + \xi_x u + \xi_y v & (\gamma - 1)\xi_y \\ (\xi_x u + \xi_y v)[-\gamma e_t + (\gamma - 1)(u^2 + v^2)] & \xi_x[\gamma e_t - \frac{1}{2}(\gamma - 1)(u^2 + v^2)] - (\gamma - 1)(\xi_x u + \xi_y v)u & \xi_y[\gamma e_t - \frac{1}{2}(\gamma - 1)(u^2 + v^2)] - (\gamma - 1)(\xi_x u + \xi_y v)v & \xi_t + \gamma(\xi_x u + \xi_y v) \end{bmatrix} \tag{11-213}$$

$$\frac{\partial \bar{F}}{\partial \bar{Q}} = \begin{bmatrix} \eta_t & \eta_x & \eta_y & 0 \\ \begin{array}{l} -u(\eta_x u + \eta_y v) + \\ \eta_x[\frac{1}{2}(\gamma - 1)(u^2 + v^2)] \end{array} & \begin{array}{l} \eta_t + \eta_x(2 - \gamma)u \\ +\eta_x u + \eta_y v \end{array} & -(\gamma - 1)\eta_x v + \eta_y u & (\gamma - 1)\eta_x \\ \begin{array}{l} -v(\eta_x u + \eta_y v) + \\ \eta_y[\frac{1}{2}(\gamma - 1)(u^2 + v^2)] \end{array} & -(\gamma - 1)\eta_y u + \eta_x v & \begin{array}{l} \eta_t + \eta_y(2 - \gamma)v \\ +\eta_x u + \eta_y v \end{array} & (\gamma - 1)\eta_y \\ \begin{array}{l} (\eta_x u + \eta_y v)[-\gamma e_t + \\ (\gamma - 1)(u^2 + v^2)] \end{array} & \begin{array}{l} \eta_x[\gamma e_t - \frac{1}{2}(\gamma - 1)(u^2 + v^2)] \\ -(\gamma - 1)(\eta_x u + \eta_y v)u \end{array} & \begin{array}{l} \eta_y[\gamma e_t - \frac{1}{2}(\gamma - 1)(u^2 + v^2)] \\ -(\gamma - 1)(\eta_x u + \eta_y v)v \end{array} & \eta_t + \gamma(\eta_x u + \eta_y v) \end{bmatrix}$$

(11-214)

$$C = \frac{\partial \bar{H}}{\partial \bar{Q}} = \frac{1}{y} \left[\begin{array}{c|c|c|c} 0 & 0 & 1 & 0 \\ \hline -uv & v & u & 0 \\ \hline -v^2 & 0 & 2v & 0 \\ \hline v[-\gamma e_t + (\gamma - 1)(u^2 + v^2)] & -(\gamma - 1)uv & \gamma e_t - \left(\dfrac{\gamma - 1}{2}\right)(3v^2 + u^2) & \gamma v \end{array} \right] \tag{11-215}$$

$$
\frac{\partial \bar{E}_v}{\partial \bar{Q}} = \frac{\mu}{Re_\infty J}
\left[\begin{array}{l:l:l:l}
0 & 0 & 0 & 0 \\ \hdashline
\begin{array}{l}
-\left[a_1\left(J\frac{u}{\rho}\right)_\xi + a_3\left(J\frac{v}{\rho}\right)_\xi\right] \\
-\left[c_1\left(J\frac{u}{\rho}\right)_\eta + c_3\left(J\frac{v}{\rho}\right)_\eta\right]
\end{array}
&
\begin{array}{l}
a_1\left(\frac{J}{\rho}\right)_\xi \\
+c_1\left(\frac{J}{\rho}\right)_\eta
\end{array}
&
\begin{array}{l}
a_3\left(\frac{J}{\rho}\right)_\xi \\
+c_3\left(\frac{J}{\rho}\right)_\eta
\end{array}
& 0 \\ \hdashline
\begin{array}{l}
-\left[a_2\left(J\frac{v}{\rho}\right)_\xi + a_3\left(J\frac{u}{\rho}\right)_\xi\right] \\
-\left[c_2\left(J\frac{v}{\rho}\right)_\eta + c_4\left(J\frac{u}{\rho}\right)_\eta\right]
\end{array}
&
\begin{array}{l}
a_3\left(\frac{J}{\rho}\right)_\xi \\
+c_4\left(\frac{J}{\rho}\right)_\eta
\end{array}
&
\begin{array}{l}
a_2\left(\frac{J}{\rho}\right)_\xi \\
+c_2\left(\frac{J}{\rho}\right)_\eta
\end{array}
& 0 \\ \hdashline
\begin{array}{l}
-\left\{\left(a_1 - \frac{\gamma}{Pr}a_4\right)\left(J\frac{u^2}{\rho}\right)_\xi\right. \\
+\left(a_2 - \frac{\gamma}{Pr}a_4\right)\left(J\frac{v^2}{\rho}\right)_\xi + 2a_3\left(J\frac{uv}{\rho}\right)_\xi \\
\left.+\frac{\gamma}{Pr}a_4\left(J\frac{e_t}{\rho}\right)_\xi\right\} \\
-\left\{\left(c_1 - \frac{\gamma}{Pr}c_5\right)\left(J\frac{u^2}{\rho}\right)_\eta\right. \\
+\left(c_2 - \frac{\gamma}{Pr}c_5\right)\left(J\frac{v^2}{\rho}\right)_\eta \\
+c_3 u\left(J\frac{v}{\rho}\right)_\eta + c_4 v\left(J\frac{u}{\rho}\right)_\eta \\
+c_3\left(J\frac{u}{\rho}\right)v_\eta + c_4\left(J\frac{v}{\rho}\right)u_\eta \\
\left.+\frac{\gamma}{Pr}c_5\left(J\frac{e_t}{\rho}\right)_\eta\right\}
\end{array}
&
\begin{array}{l}
\left[a_3\left(J\frac{v}{\rho}\right)_\xi\right. \\
\left.+\left(a_1 - \frac{\gamma}{Pr}a_4\right)\left(J\frac{u}{\rho}\right)_\xi\right] \\
+\left[c_4 v\left(\frac{J}{\rho}\right)_\eta + c_3\frac{J}{\rho}(v_\eta)\right. \\
\left.+\left(c_1 - \frac{\gamma}{Pr}c_5\right)\left(J\frac{u}{\rho}\right)_\eta\right]
\end{array}
&
\begin{array}{l}
\left[a_3\left(J\frac{u}{\rho}\right)_\xi\right. \\
\left.+\left(a_2 - \frac{\gamma}{Pr}a_4\right)\left(J\frac{v}{\rho}\right)_\xi\right] \\
+\left[c_3 u\left(\frac{J}{\rho}\right)_\eta + c_4\frac{J}{\rho}(u_\eta)\right. \\
\left.+\left(c_2 - \frac{\gamma}{Pr}c_5\right)\left(J\frac{v}{\rho}\right)_\eta\right]
\end{array}
&
\begin{array}{l}
\frac{\gamma}{Pr}a_4\left(\frac{J}{\rho}\right)_\xi \\
+\frac{\gamma}{Pr}c_5\left(\frac{J}{\rho}\right)_\eta
\end{array}
\end{array}\right]
\tag{11-216}
$$

$$\frac{\partial \bar{F}_v}{\partial \bar{Q}} = \frac{\mu}{Re_\infty J}\left[\begin{array}{l:l:l:l}
0 & 0 & 0 & 0 \\ \hdashline
\begin{array}{l} -\left[c_1\left(J\frac{u}{\rho}\right)_\xi + c_4\left(J\frac{v}{\rho}\right)_\xi\right] \\ -\left[b_1\left(J\frac{u}{\rho}\right)_\eta + b_3\left(J\frac{v}{\rho}\right)_\eta\right] \end{array} & \begin{array}{l} c_1\left(\frac{J}{\rho}\right)_\xi \\ +b_1\left(\frac{J}{\rho}\right)_\eta \end{array} & \begin{array}{l} c_4\left(\frac{J}{\rho}\right)_\xi \\ +b_3\left(\frac{J}{\rho}\right)_\eta \end{array} & 0 \\ \hdashline
\begin{array}{l} -\left[c_3\left(J\frac{u}{\rho}\right)_\xi + c_2\left(J\frac{v}{\rho}\right)_\xi\right] \\ -\left[b_3\left(J\frac{u}{\rho}\right)_\eta + b_2\left(J\frac{v}{\rho}\right)_\eta\right] \end{array} & \begin{array}{l} c_3\left(\frac{J}{\rho}\right)_\xi \\ +b_3\left(\frac{J}{\rho}\right)_\eta \end{array} & \begin{array}{l} c_2\left(\frac{J}{\rho}\right)_\xi \\ +b_2\left(\frac{J}{\rho}\right)_\eta \end{array} & 0 \\ \hdashline
\begin{array}{l} -\left\{\left(c_1 - \frac{\gamma}{Pr}c_5\right)\left(J\frac{u^2}{\rho}\right)_\xi \right. \\ +\left(c_2 - \frac{\gamma}{Pr}c_5\right)\left(J\frac{v^2}{\rho}\right)_\xi + c_3 v\left(J\frac{u}{\rho}\right)_\xi \\ +c_4 u\left(J\frac{v}{\rho}\right)_\xi + c_3\left(J\frac{v}{\rho}\right)u_\xi + c_4\left(J\frac{u}{\rho}\right)v_\xi \\ \left. +\frac{\gamma}{Pr}c_5\left(J\frac{e_t}{\rho}\right)_\xi\right\} \\ -\left\{\left(b_1 - \frac{\gamma}{Pr}b_4\right)\left(J\frac{u^2}{\rho}\right)_\eta \right. \\ +\left(b_2 - \frac{\gamma}{Pr}b_4\right)\left(J\frac{v^2}{\rho}\right)_\eta + 2b_3\left(J\frac{uv}{\rho}\right)_\eta \\ \left. +\frac{\gamma}{Pr}b_4\left(J\frac{e_t}{\rho}\right)_\eta\right\} \end{array} & \begin{array}{l} \left[c_3 v\left(\frac{J}{\rho}\right)_\xi + c_4\frac{J}{\rho}v_\xi \right. \\ \left. +\left(c_1 - \frac{\gamma}{Pr}c_5\right)\left(J\frac{u}{\rho}\right)_\xi\right] \\ +\left[b_3\left(J\frac{v}{\rho}\right)_\eta \right. \\ \left. +\left(b_1 - \frac{\gamma}{Pr}b_4\right)\left(J\frac{u}{\rho}\right)_\eta\right] \end{array} & \begin{array}{l} \left[c_3\frac{J}{\rho}u_\xi + c_4 u\left(\frac{J}{\rho}\right)_\xi \right. \\ \left. +\left(c_2 - \frac{\gamma}{Pr}c_5\right)\left(J\frac{v}{\rho}\right)_\xi\right] \\ +\left[b_3\left(J\frac{u}{\rho}\right)_\eta \right. \\ \left. +\left(b_2 - \frac{\gamma}{Pr}b_4\right)\left(J\frac{v}{\rho}\right)_\eta\right] \end{array} & \begin{array}{l} \frac{\gamma}{Pr}c_5\left(\frac{J}{\rho}\right)_\xi \\ +\frac{\gamma}{Pr}b_4\left(\frac{J}{\rho}\right)_\eta \end{array}
\end{array}\right] \tag{11-217}$$

$$C_v = \frac{1}{Re_\infty Jy} \begin{bmatrix} 0 & 0 & 0 & 0 \\ HQ_{2,1} & HQ_{2,2} & HQ_{2,3} & 0 \\ HQ_{3,1} & HQ_{3,2} & HQ_{3,3} & 0 \\ HQ_{4,1} & HQ_{4,2} & HQ_{4,3} & HQ_{4,4} \end{bmatrix} \tag{11-218}$$

where

$$HQ_{2,1} = -\mu\Big[\xi_x(J\frac{v}{\rho})_\xi + \xi_y(J\frac{u}{\rho})_\xi\Big] + \frac{2}{3}y\xi_x\Big[(\frac{\mu}{y})(J\frac{v}{\rho})\Big]_\xi$$

$$- \mu\Big[\eta_x(J\frac{v}{\rho})_\eta + \eta_y(J\frac{u}{\rho})_\eta\Big] + \frac{2}{3}y\eta_x\Big[(\frac{\mu}{y})(J\frac{v}{\rho})_\eta\Big]$$

$$HQ_{2,2} = \Big[\mu\xi_y(\frac{J}{\rho})_\xi + \mu\eta_y(\frac{J}{\rho})_\eta\Big]$$

$$HQ_{2,3} = \mu\xi_x(\frac{J}{\rho})_\xi - \frac{2}{3}y\xi_x\Big[(\frac{J}{\rho})(\frac{\mu}{y})\Big]_\xi + \mu\eta_x(\frac{J}{\rho})_\eta - \frac{2}{3}y\eta_x\Big[(\frac{J}{\rho})(\frac{\mu}{y})\Big]_\eta$$

$$HQ_{3,1} = 2\mu\Big[-\xi_y(J\frac{v}{\rho})_\xi\Big] + \frac{2}{3}y\xi_y\Big[(\frac{\mu}{y})(J\frac{v}{\rho})\Big]_\xi + 2\mu\Big[-\eta_y(J\frac{v}{\rho})_\eta + \frac{J}{y}(\frac{v}{\rho})\Big]$$

$$+ \frac{2}{3}y\eta_y\Big[(\frac{\mu}{y})(J\frac{v}{\rho})\Big]_\eta$$

$$HQ_{3,2} = 0$$

$$HQ_{3,3} = 2\mu\xi_y(\frac{J}{\rho})_\xi - \frac{2}{3}y\xi_y\Big[(\frac{\mu}{y})(\frac{J}{\rho})\Big]_\xi + 2\mu\Big[\eta_y(\frac{J}{\rho})_\eta - \frac{1}{y}\frac{J}{\rho}\Big] - \frac{2}{3}y\eta_y\Big[(\frac{\mu}{y})(\frac{J}{\rho})\Big]_\eta$$

$$HQ_{4,1} = -\mu(J\frac{u}{\rho})\Big[\xi_x v_\xi + \xi_y u_\xi\Big] - \mu u\Big[\xi_x(J\frac{v}{\rho})_\xi + \xi_y(J\frac{u}{\rho})_\xi\Big] - \frac{2}{3}\mu(J\frac{v}{\rho})\Big[2\xi_y v_\xi - \xi_x u_\xi\Big]$$

$$+ \frac{2}{3}\mu v\Big[-2\xi_y(J\frac{v}{\rho})_\xi + \xi_x(J\frac{u}{\rho})_\xi\Big] + \frac{4}{3}y\xi_y\Big[(\frac{\mu}{y})(J\frac{v^2}{\rho})\Big]_\xi + \frac{4}{3}y\xi_x\Big[(\frac{\mu}{y})(J\frac{uv}{\rho})\Big]_\xi$$

$$+ \frac{\gamma\mu}{Pr}\xi_y\Big[-J\frac{e_t}{\rho} + (\frac{J}{\rho}u^2 + \frac{J}{\rho}v^2)\Big]_\xi - \mu(J\frac{u}{\rho})\Big[\eta_x v_\eta + \eta_y u_\eta\Big]$$

$$- \mu u\Big[\eta_x(J\frac{v}{\rho})_\eta + \eta_y(J\frac{u}{\rho})_\eta\Big] - \frac{2}{3}\mu(J\frac{v}{\rho})\Big[2\eta_y v_\eta - \eta_x u_\eta\Big]$$

$$+ \frac{2}{3}\mu v\Big[-2\eta_y(J\frac{v}{\rho})_\eta + \eta_x(J\frac{u}{\rho})_\eta\Big] + \frac{4}{3}\frac{\mu}{y}(J\frac{v^2}{\rho}) + \frac{4}{3}y\eta_y\Big[(\frac{\mu}{y})(J\frac{v^2}{\rho})\Big]_\eta$$

$$+ \frac{4}{3}y\eta_x\Big[(\frac{\mu}{y})(J\frac{uv}{\rho})\Big]_\eta + \frac{\gamma\mu}{Pr}\eta_y\Big[-J\frac{e_t}{\rho} + (\frac{J}{\rho}u^2 + \frac{J}{\rho}v^2)\Big]_\eta$$

$$HQ_{4,2} = \mu(\frac{J}{\rho})(\xi_x v_\xi + \xi_y u_\xi) + \mu u\xi_y(\frac{J}{\rho})_\xi - \frac{2}{3}\mu v\xi_x(\frac{J}{\rho})_\xi - \frac{2}{3}y\xi_x\Big[(\frac{\mu}{y})(J\frac{v}{\rho})\Big]_\xi$$

$$+ \mu(\frac{J}{\rho})(\eta_x v_\eta + \eta_y u_\eta) + \mu u\eta_y(\frac{J}{\rho})_\eta - \frac{2}{3}\mu v\eta_x(\frac{J}{\rho})_\eta - \frac{\gamma\mu}{Pr}\xi_y(J\frac{u}{\rho})_\xi$$

$$- \frac{2}{3}y\eta_x\Big[(\frac{\mu}{y})(J\frac{v}{\rho})\Big]_\eta - \frac{\gamma\mu}{Pr}\eta_y(J\frac{u}{\rho})_\eta$$

$$HQ_{4,3} = \mu u\xi_x(\frac{J}{\rho})_\xi + \frac{2}{3}\mu\frac{J}{\rho}(2\xi_y v_\xi - \xi_x u_\xi) + \frac{4}{3}\mu v\xi_y(\frac{J}{\rho})_\xi$$

$$- \frac{4}{3}y\xi_y\Big[(\frac{\mu}{\rho})(J\frac{v}{\rho})\Big]_\xi - \frac{2}{3}y\xi_x\Big[(\frac{\mu}{y})(J\frac{u}{\rho})\Big]_\xi - \frac{\gamma\mu}{Pr}\xi_y(J\frac{v}{\rho})_\xi + \mu u\eta_x(\frac{J}{\rho})_\eta$$

$$+ \frac{2}{3}\mu\frac{J}{\rho}(2\eta_y v_\eta - \eta_x u_\eta) + \frac{4}{3}\mu v\eta_y(\frac{J}{\rho})_\eta - \frac{4}{3}\frac{\mu}{y}(J\frac{v}{\rho}) - \frac{4}{3}y\eta_y\Big[(\frac{\mu}{y})(J\frac{v}{\rho})\Big]_\eta +$$

$$-\frac{2}{3}y\eta_x\left[(\frac{\mu}{y})(J\frac{u}{\rho})\right]_\eta - \frac{\gamma\mu}{Pr}y_\eta(J\frac{v}{\rho})_\eta$$

$$HQ_{4,4} = \frac{\gamma\mu}{Pr}\left[\eta_y(\frac{J}{\rho})_\eta + \xi_y(\frac{J}{\rho})_\xi\right]$$

The Thin-Layer Navier-Stokes equation for two-dimensional planar or axisymmetric flow is reduced to

$$\frac{\partial \bar{Q}}{\partial \tau} + \frac{\partial \bar{E}}{\partial \xi} + \frac{\partial \bar{F}}{\partial \eta} + \alpha\bar{H} = \frac{\partial \bar{F}_{vT}}{\partial \eta} + \alpha\bar{H}_v \tag{11-219}$$

where flux vectors $\bar{Q}$, $\bar{E}$, and $\bar{F}$ are given by (11-201), (11-202) and (11-203), and

$$\bar{F}_{vT} = \frac{\mu}{Re_\infty J}\begin{bmatrix} 0 \\ \hline b_1u_\eta + b_3v_\eta \\ \hline b_3u_\eta + b_2v_\eta \\ \hline \frac{1}{2}b_1(u^2)_\eta + \frac{1}{2}b_2(v^2)_\eta + b_3(uv)_\eta + \frac{1}{Pr(\gamma-1)M_\infty^2}b_4T_\eta \end{bmatrix} \tag{11-220}$$

The Jacobian matrices $\partial\bar{E}/\partial\bar{Q}$ and $\partial\bar{F}/\partial\bar{Q}$ are given by (11-213) and (11-214), and

$$\frac{\partial \bar{F}_{vT}}{\partial \bar{Q}} = \frac{\mu}{Re_\infty J}\left[\begin{array}{c|c|c|c} 0 & 0 & 0 & 0 \\ \hline -\left[b_1\left(J\frac{u}{\rho}\right)_\eta + b_3\left(J\frac{v}{\rho}\right)_\eta\right] & b_1\left(\frac{J}{\rho}\right)_\eta & b_3\left(\frac{J}{\rho}\right)_\eta & 0 \\ \hline -\left[b_3\left(J\frac{u}{\rho}\right)_\eta + b_2\left(J\frac{v}{\rho}\right)_\eta\right] & b_3\left(\frac{J}{\rho}\right)_\eta & b_2\left(\frac{J}{\rho}\right)_\eta & 0 \\ \hline \begin{array}{l} -\left\{\left(b_1 - \frac{\gamma}{Pr}b_4\right)\left(J\frac{u^2}{\rho}\right)_\eta \right. \\ +\left(b_2 - \frac{\gamma}{Pr}b_4\right)\left(J\frac{v^2}{\rho}\right)_\eta \\ +\frac{\gamma}{Pr}b_4\left(J\frac{e_t}{\rho}\right)_\eta \\ \left. +2b_3\left(J\frac{uv}{\rho}\right)_\eta\right\} \end{array} & \begin{array}{l} \left(b_1 - \frac{\gamma}{Pr}b_4\right)\left(J\frac{u}{\rho}\right)_\eta \\ +b_3\left(J\frac{v}{\rho}\right)_\eta \end{array} & \begin{array}{l} b_3\left(J\frac{u}{\rho}\right)_\eta \\ +\left(b_2 - \frac{\gamma}{Pr}b_4\right)\left(J\frac{v}{\rho}\right)_\eta \end{array} & \frac{\gamma}{Pr}b_4\left(\frac{J}{\rho}\right)_\eta \end{array}\right] \quad (11\text{-}221)$$

The Parabolized Navier-Stokes equations under the assumptions stated previously are formulated as

$$\frac{\partial \bar{E}_P}{\partial \xi} + \frac{\partial \bar{E}_{PP}}{\partial \xi} + \frac{\partial \bar{F}}{\partial \eta} + \alpha \bar{H} = \frac{\partial \bar{F}_{vP}}{\partial \eta} + \alpha \bar{H}_v \tag{11-222}$$

where

$$\bar{E}_P = \frac{1}{J}[\xi_x E_P + \xi_y F_P] \tag{11-223}$$

$$\bar{E}_{PP} = \frac{1}{J}[\xi_x E_{PP} + \xi_y F_{PP}] \tag{11-224}$$

and

$$E_P = \begin{bmatrix} \rho u \\ \rho u^2 + \omega p \\ \rho uv \\ (\rho e_t + p)u \end{bmatrix} \tag{11-225a}$$

$$F_P = \begin{bmatrix} \rho v \\ \rho uv \\ \rho v^2 + \omega p \\ (\rho e_t + p)v \end{bmatrix} \tag{11-225b}$$

$$E_{PP} = \begin{bmatrix} 0 \\ (1-\omega)p \\ 0 \\ 0 \end{bmatrix} \tag{11-226a}$$

$$F_{PP} = \begin{bmatrix} 0 \\ 0 \\ (1-\omega)p \\ 0 \end{bmatrix} \tag{11-226b}$$

and the viscous flux vector $\bar{F}_{vP}$ is

$$\bar{F}_{vP} = \frac{\mu}{Re_\infty J} \begin{bmatrix} 0 \\ b_1 u_\eta + b_3 v_\eta \\ b_3 u_\eta + b_2 v_\eta \\ \frac{1}{2} b_1 (u^2)_\eta + \frac{1}{2} b_2 (v^2)_\eta + b_3 (uv)_\eta + \frac{1}{Pr(\gamma-1)M_\infty^2} b_4 T_\eta \end{bmatrix} \tag{11-227}$$

The inviscid and viscous Jacobian matrices are:

$$\frac{\partial \bar{E}_P}{\partial \bar{Q}} = \left[\begin{array}{c|c|c|c} 0 & \xi_x & \xi_y & 0 \\ \hline -u\left(\xi_x u + \xi_y v\right) + \xi_x \left[\frac{1}{2}\omega\left(\gamma - 1\right)\left(u^2 + v^2\right)\right] & \begin{array}{c}\xi_x\left[1 - \omega\left(\gamma - 1\right)\right]u \\ +\xi_x u + \xi_y v\end{array} & \begin{array}{c}-\omega\left(\gamma - 1\right)\xi_x v \\ +\xi_y u\end{array} & \omega\left(\gamma - 1\right)\xi_x \\ \hline -v\left(\xi_x u + \xi_y v\right) + \xi_y \left[\frac{1}{2}\omega\left(\gamma - 1\right)\left(u^2 + v^2\right)\right] & \xi_x v - \omega\left(\gamma - 1\right)\xi_y u & \begin{array}{c}\xi_y\left[1 - \omega\left(\gamma - 1\right)\right]v \\ +\xi_x u + \xi_y v\end{array} & \omega\left(\gamma - 1\right)\xi_y \\ \hline \left(\xi_x u + \xi_y v\right)\left[-\gamma e_t + \left(\gamma - 1\right)\left(u^2 + v^2\right)\right] & \begin{array}{c}\xi_x\left[\gamma e_t - \frac{1}{2}\left(\gamma - 1\right)\cdot\right. \\ \left.\left(u^2 + v^2\right)\right] \\ -\left(\gamma - 1\right)\left[\xi_x u + \xi_y v\right]u\end{array} & \begin{array}{c}\xi_y\left[\gamma e_t - \frac{1}{2}\left(\gamma - 1\right)\cdot\right. \\ \left.\left(u^2 + v^2\right)\right] \\ -\left(\gamma - 1\right)\left[\xi_x u + \xi_y v\right]v\end{array} & \gamma\left[\xi_x u + \xi_y v\right] \end{array}\right] \tag{11-228}$$

$$\frac{\partial \bar{F}}{\partial \bar{Q}} = \left[\begin{array}{c:c:c:c} 0 & \eta_x & \eta_y & 0 \\ \hdashline -u(\eta_x u + \eta_y v) + & \eta_x(2-\gamma)u & -(\gamma-1)\eta_x v + \eta_y u & (\gamma-1)\eta_x \\ \eta_x[\frac{1}{2}(\gamma-1)(u^2+v^2)] & +\eta_x u + \eta_y v & & \\ \hdashline -v(\eta_x u + \eta_y v) + & -(\gamma-1)\eta_y u + \eta_x v & \eta_y(2-\gamma)v & (\gamma-1)\eta_y \\ \eta_y[\frac{1}{2}(\gamma-1)(u^2+v^2)] & & +\eta_x u + \eta_y v & \\ \hdashline (\eta_x u + \eta_y v)[-\gamma e_t + & \eta_x[\gamma e_t - \frac{1}{2}(\gamma-1)(u^2+v^2)] & \eta_y[\gamma e_t - \frac{1}{2}(\gamma-1)(u^2+v^2)] & \gamma(\eta_x u + \eta_y v) \\ (\gamma-1)(u^2+v^2)] & -(\gamma-1)(\eta_x u + \eta_y v)u & -(\gamma-1)(\eta_x u + \eta_y v)v & \end{array}\right] \tag{11-229}$$

$$\frac{\partial \bar{F}_{vP}}{\partial \bar{Q}} = \frac{\mu}{Re_\infty J} \left[\begin{array}{c:c:c:c} 0 & 0 & 0 & 0 \\ \hdashline -\left[b_1\left(J\frac{u}{\rho}\right)_\eta + b_3\left(J\frac{v}{\rho}\right)_\eta\right] & b_1\left(\frac{J}{\rho}\right)_\eta & b_3\left(\frac{J}{\rho}\right)_\eta & 0 \\ \hdashline -\left[b_3\left(J\frac{u}{\rho}\right)_\eta + b_2\left(J\frac{v}{\rho}\right)_\eta\right] & b_3\left(\frac{J}{\rho}\right)_\eta & b_2\left(\frac{J}{\rho}\right)_\eta & 0 \\ \hdashline \begin{array}{l} -\left\{\left(b_1 - \frac{\gamma}{Pr}b_4\right)\left(J\frac{u^2}{\rho}\right)_\eta \right. \\ +\left(b_2 - \frac{\gamma}{Pr}b_4\right)\left(J\frac{v^2}{\rho}\right)_\eta \\ +\frac{\gamma}{Pr}b_4\left(J\frac{e_t}{\rho}\right)_\eta \\ \left. +2b_3\left(J\frac{uv}{\rho}\right)_\eta\right\} \end{array} & \begin{array}{l} \left(b_1 - \frac{\gamma}{Pr}b_4\right)\left(J\frac{u}{\rho}\right)_\eta \\ +b_3\left(J\frac{v}{\rho}\right)_\eta \end{array} & \begin{array}{l} b_3\left(J\frac{u}{\rho}\right)_\eta \\ +\left(b_2 - \frac{\gamma}{Pr}b_4\right)\left(J\frac{v}{\rho}\right)_\eta \end{array} & \frac{\gamma}{Pr}b_4\left(\frac{J}{\rho}\right)_\eta \end{array} \right] \tag{11-230}$$

11.8 Incompressible Navier-Stokes Equations

As introduced in Chapter 8, a common scheme to solve numerically the incompressible Navier-Stokes equations is the modification of the continuity equation to include an artificial compressibility term. If τ is used to denote this artificial compressibility and β is used to represent its inverse, then the incompressible Navier-Stokes equations in dimensional form are given by

$$\frac{\partial Q}{\partial t} + \frac{\partial E}{\partial x} + \frac{\partial F}{\partial y} + \frac{\partial G}{\partial z} = \frac{\partial E_v}{\partial x} + \frac{\partial F_v}{\partial y} + \frac{\partial G_v}{\partial z} \tag{11-231}$$

where

$$Q = \begin{bmatrix} p \\ u \\ v \\ w \end{bmatrix} \tag{11-232}$$

$$E = \begin{bmatrix} \beta u \\ u^2 + p \\ uv \\ uw \end{bmatrix} \quad (11\text{-}233) \qquad E_v = \begin{bmatrix} 0 \\ \tau_{xx} \\ \tau_{xy} \\ \tau_{xz} \end{bmatrix} \quad (11\text{-}234)$$

$$F = \begin{bmatrix} \beta v \\ vu \\ v^2 + p \\ vw \end{bmatrix} \quad (11\text{-}235) \qquad F_v = \begin{bmatrix} 0 \\ \tau_{yx} \\ \tau_{yy} \\ \tau_{yz} \end{bmatrix} \quad (11\text{-}236)$$

$$G = \begin{bmatrix} \beta w \\ wu \\ wv \\ w^2 + p \end{bmatrix} \quad (11\text{-}237) \qquad G_v = \begin{bmatrix} 0 \\ \tau_{zx} \\ \tau_{zy} \\ \tau_{zz} \end{bmatrix} \quad (11\text{-}238)$$

and

$$\begin{aligned} \tau_{xx} &= 2\nu \frac{\partial u}{\partial x} \\ \tau_{xy} &= \tau_{yx} = \nu \left(\frac{\partial u}{\partial y} + \frac{\partial v}{\partial x} \right) \\ \tau_{yy} &= 2\nu \frac{\partial u}{\partial y} \end{aligned}$$

$$\tau_{xz} = \tau_{zx} = \nu \left(\frac{\partial w}{\partial x} + \frac{\partial u}{\partial z} \right)$$

$$\tau_{zz} = 2\nu \frac{\partial w}{\partial z}$$

$$\tau_{yz} = \tau_{zy} = \nu \left(\frac{\partial w}{\partial y} + \frac{\partial v}{\partial z} \right)$$

Equation (11-231) may be expressed in a nondimensional form if the variables are nondimensionalized according to nondimensional terms defined previously in Section 8.2.1. The nondimensional form of the incompressible Navier-Stokes equation in a flux vector form is:

$$\frac{\partial Q^*}{\partial t^*} + \frac{\partial E^*}{\partial x^*} + \frac{\partial F^*}{\partial y^*} + \frac{\partial G^*}{\partial z^*} = \frac{\partial E_v^*}{\partial x^*} + \frac{\partial F_v^*}{\partial y^*} + \frac{\partial G_v^*}{\partial z^*} \tag{11-239}$$

where

$$Q^* = \begin{bmatrix} p^* \\ u^* \\ v^* \\ w^* \end{bmatrix}$$

$$E^* = \begin{bmatrix} \beta^* u^* \\ {u^*}^2 + p^* \\ u^* v^* \\ u^* w^* \end{bmatrix} \tag{11-240}$$

$$E_v^* = \begin{bmatrix} 0 \\ \tau_{xx}^* \\ \tau_{xy}^* \\ \tau_{xz}^* \end{bmatrix} \tag{11-241}$$

$$F^* = \begin{bmatrix} \beta^* v^* \\ v^* u^* \\ v^* + p^* \\ v^* w^* \end{bmatrix} \tag{11-242}$$

$$F_v^* = \begin{bmatrix} 0 \\ \tau_{yx}^* \\ \tau_{yy}^* \\ \tau_{yz}^* \end{bmatrix} \tag{11-243}$$

$$G^* = \begin{bmatrix} \beta^* w^* \\ w^* u^* \\ w^* v^* \\ {w^*}^2 + p^* \end{bmatrix} \tag{11-244}$$

$$G_v^* = \begin{bmatrix} 0 \\ \tau_{zx}^* \\ \tau_{zy}^* \\ \tau_{zz}^* \end{bmatrix} \tag{11-245}$$

where

$$\tau_{xx}^* = \frac{2}{Re_\infty} \frac{\partial u^*}{\partial x^*} \tag{11-246}$$

$$\tau^*_{xy} = \tau^*_{yx} = \frac{1}{Re_\infty}\left(\frac{\partial u^*}{\partial y^*} + \frac{\partial v^*}{\partial x^*}\right) \tag{11-247}$$

$$\tau^*_{yy} = \frac{2}{Re_\infty}\frac{\partial v^*}{\partial y^*} \tag{11-248}$$

$$\tau^*_{xz} = \tau^*_{zx} = \frac{1}{Re_\infty}\left(\frac{\partial w^*}{\partial x^*} + \frac{\partial u^*}{\partial z^*}\right) \tag{11-249}$$

$$\tau^*_{zz} = \frac{2}{Re_\infty}\frac{\partial w^*}{\partial z^*} \tag{11-250}$$

$$\tau^*_{yz} = \tau^*_{zy} = \frac{1}{Re_\infty}\left(\frac{\partial w^*}{\partial y^*} + \frac{\partial v^*}{\partial z^*}\right) \tag{11-251}$$

and

$$Re_\infty = \frac{\rho_\infty u_\infty L}{\mu_\infty}$$

The asterisk which is used to denote the nondimensional quantities will be dropped for convenience. Therefore, all the expressions to follow are in nondimensional form unless specified otherwise. Note that the flux vector formulations in either dimensional or nondimensional forms are similar. Therefore, the transformed formulation applies to either one. However, appropriate nondimensional terms and expressions for shear stresses must be utilized.

Following the procedure of Section 11-4, the nondimensional incompressible Navier-Stokes equations in the computational space are expressed as

$$\frac{\partial \bar{Q}}{\partial t} + \frac{\partial \bar{E}}{\partial \xi} + \frac{\partial \bar{F}}{\partial \eta} + \frac{\partial \bar{G}}{\partial \zeta} = \frac{\partial \bar{E}_v}{\partial \xi} + \frac{\partial \bar{F}_v}{\partial \eta} + \frac{\partial \bar{G}_v}{\partial \zeta} \tag{11-252}$$

where

$$\bar{Q} = \frac{Q}{J}$$

and

$$\bar{E} = \frac{1}{J}\left(\xi_x E + \xi_y F + \xi_z G\right) \tag{11-253}$$

$$\bar{F} = \frac{1}{J}\left(\eta_x E + \eta_y F + \eta_z G\right) \tag{11-254}$$

$$\bar{G} = \frac{1}{J}\left(\zeta_x E + \zeta_y F + \zeta_z G\right) \tag{11-255}$$

$$\bar{E}_v = \frac{1}{J}\left(\xi_x E_v + \xi_y F_v + \xi_z G_v\right) \tag{11-256}$$

$$\bar{F}_v = \frac{1}{J}\left(\eta_x E_v + \eta_y F_v + \eta_z G_v\right) \tag{11-257}$$

$$\bar{G}_v = \frac{1}{J}\left(\zeta_x E_v + \zeta_y F_v + \zeta_z G_v\right) \tag{11-258}$$

The shear stresses given by Equations (11-246) through (11-251) expressed in the computational space are as follow:

$$\tau_{xx} = \frac{2}{Re_\infty}\left(\xi_x u_\xi + \eta_x u_\eta + \zeta_x u_\zeta\right) \tag{11-259}$$

$$\tau_{yy} = \frac{2}{Re_\infty}\left(\xi_y v_\xi + \eta_y v_\eta + \zeta_x v_\zeta\right) \tag{11-260}$$

$$\tau_{zz} = \frac{2}{Re_\infty}\left(\xi_z w_\xi + \eta_z w_\eta + \zeta_z w_\zeta\right) \tag{11-261}$$

$$\tau_{xy} = \tau_{yx} = \frac{1}{Re_\infty}\left(\xi_y u_\xi + \eta_z u_\eta + \zeta_y u_\zeta + \xi_x v_\xi + \eta_x v_\eta + \zeta_x v_\zeta\right) \tag{11-262}$$

$$\tau_{xz} = \tau_{zx} = \frac{1}{Re_\infty}\left(\xi_z u_\xi + \eta_z u_\eta + \zeta_z u_\zeta + \xi_x w_\xi + \eta_x w_\eta + \zeta_x w_\zeta\right) \tag{11-263}$$

$$\tau_{yz} = \tau_{zy} = \frac{1}{Re_\infty}\left(\xi_y w_\xi + \eta_y w_\eta + \zeta_y w_\zeta + \xi_z v_\xi + \eta_y v_\eta + \zeta_z v_\zeta\right) \tag{11-264}$$

The expressions for the shear stresses given by Equations (11-259) through (11-264) are substituted into the viscous flux vectors $\bar{E}_v$, $\bar{F}_v$, and $\bar{G}_v$ to provide the following:

$$\bar{E}_v = \frac{1}{JRe_\infty}\begin{bmatrix} 0 \\ a_1 u_\xi + b_1 u_\eta - c_1 v_\eta + c_2 w_\eta + b_2 u_\zeta - c_4 v_\zeta + c_5 w_\zeta \\ a_1 v_\xi + c_1 u_\eta + b_1 v_\eta - c_3 w_\eta + c_4 u_\zeta + b_2 v_\zeta - c_6 w_\zeta \\ a_1 w_\xi - c_2 u_\eta + c_3 v_\eta + b_1 w_\eta - c_5 u_\zeta + c_6 v_\zeta + b_2 w_\zeta \end{bmatrix} \tag{11-265}$$

$$\bar{F}_v = \frac{1}{JRe_\infty}\begin{bmatrix} 0 \\ a_2 u_\eta + b_1 u_\xi + c_1 v_\xi - c_2 w_\xi + b_2 u_\zeta - c_7 v_\zeta + c_8 w_\zeta \\ a_2 v_\eta - c_1 u_\xi + b_1 v_\xi + c_3 w_\xi + c_7 u_\zeta + b_3 v_\zeta - c_9 w_\zeta \\ a_2 w_\eta + c_2 u_\xi - c_3 v_\xi + b_1 w_\xi - c_8 u_\zeta + c_9 v_\zeta + b_3 w_\zeta \end{bmatrix} \tag{11-266}$$

$$\bar{G}_v = \frac{1}{JRe_\infty}\begin{bmatrix} 0 \\ a_3 u_\zeta + b_2 u_\xi + c_4 v_\xi - c_5 w_\xi + b_3 u_\eta + c_7 v_\eta - c_8 w_\eta \\ a_3 v_\zeta - c_4 u_\xi + b_2 v_\xi + c_6 w_\xi - c_7 u_\eta + b_3 v_\eta + c_9 w_\eta \\ a_3 w_\zeta + c_5 u_\xi - c_6 v_\xi + b_2 w_\xi + c_8 u_\eta - c_9 v_\eta + b_3 w_\eta \end{bmatrix} \tag{11-267}$$

where

$$a_1 = \xi_x^2 + \xi_y^2 + \xi_z^2 \tag{11-268}$$

$$a_2 = \eta_x^2 + \eta_y^2 + \eta_z^2 \tag{11-269}$$

$$a_3 = \zeta_x^2 + \zeta_y^2 + \zeta_z^2 \tag{11-270}$$

$$b_1 = \xi_x\eta_x + \xi_y\eta_y + \xi_z\eta_z \tag{11-271}$$

$$b_2 = \xi_x\zeta_x + \xi_y\zeta_y + \xi_z\zeta_z \tag{11-272}$$

$$b_3 = \zeta_x\eta_x + \zeta_y\eta_y + \zeta_z\eta_z \tag{11-273}$$

$c_1 = \xi_x\eta_y - \eta_x\xi_y$ (11-274) $c_2 = \eta_x\xi_z - \xi_x\eta_z$ (11-275) $c_3 = \xi_y\eta_z - \eta_y\xi_z$ (11-276)

$c_4 = \xi_x\zeta_y - \zeta_x\xi_y$ (11-277) $c_5 = \zeta_x\xi_z - \xi_x\zeta_z$ (11-278) $c_6 = \xi_y\zeta_z - \zeta_y\xi_z$ (11-279)

$c_7 = \eta_x\zeta_y - \zeta_x\eta_y$ (11-280) $c_8 = \zeta_x\eta_z - \eta_x\zeta_z$ (11-281) $c_9 = \eta_y\zeta_z - \zeta_y\eta_z$ (11-282)

11.8.1 Inviscid and Viscous Jacobian Matrices

The inviscid Jacobian matrices are determined according to

$$A = \frac{\partial \bar{E}}{\partial \bar{Q}}$$

$$B = \frac{\partial \bar{F}}{\partial \bar{Q}}$$

and

$$C = \frac{\partial \bar{G}}{\partial \bar{Q}}$$

Using the contravariant velocity components defined by

$$U = \xi_x u + \xi_y v + \xi_z w \tag{11-283}$$

$$V = \eta_x u + \eta_y v + \eta_z w \tag{11-284}$$

and

$$W = \zeta_x u + \zeta_y v + \zeta_z w \tag{11-285}$$

the resulting inviscid Jacobian matrices are:

$$A = \begin{bmatrix} 0 & \xi_x\beta & \xi_y\beta & \xi_z\beta \\ \xi_x & U + \xi_x u & \xi_y u & \xi_z u \\ \xi_y & \xi_x v & U + \xi_y v & \xi_z v \\ \xi_z & \xi_x w & \xi_y w & U + \xi_z w \end{bmatrix} \tag{11-286}$$

$$B = \begin{bmatrix} 0 & \eta_x\beta & \eta_y\beta & \eta_z\beta \\ \eta_x & V+\eta_x u & \eta_y u & \eta_z u \\ \eta_y & \eta_x v & V+\eta_y v & \eta_z v \\ \eta_z & \eta_x w & \eta_y w & V+\eta_z w \end{bmatrix} \tag{11-287}$$

$$C = \begin{bmatrix} 0 & \zeta_x\beta & \zeta_y\beta & \zeta_z\beta \\ \zeta_x & W+\zeta_x u & \zeta_y u & \zeta_z u \\ \zeta_y & \zeta_x v & W+\zeta_y v & \zeta_z v \\ \zeta_z & \zeta_x w & \zeta_y w & W+\zeta_z w \end{bmatrix} \tag{11-288}$$

Following the procedure outlined in Section 11.4.2 and given by the general expression (11-151), the viscous Jacobian matrices are determined and are provided by the following:

$$A_v = \frac{\partial \bar{E}_v}{\partial \bar{Q}} = \frac{1}{Re_\infty J}\begin{bmatrix} 0 & 0 & 0 & 0 \\ 0 & A_1 & -B_1 & C_1 \\ 0 & B_1 & A_1 & -D_1 \\ 0 & -C_1 & D_1 & A_1 \end{bmatrix} \tag{11-289}$$

where

$$A_1 = a_1(J)_\xi + b_1(J)_\eta + b_2(J)_\zeta \tag{11-290}$$

$$B_1 = c_1(J)_\eta + c_4(J)_\zeta \tag{11-291}$$

$$C_1 = c_2(J)_\eta + c_5(J)_\zeta \tag{11-292}$$

$$D_1 = c_3(J)_\eta + c_6(J)_\zeta \tag{11-293}$$

$$B_v = \frac{\partial \bar{F}_v}{\partial \bar{Q}} = \frac{1}{Re_\infty J}\begin{bmatrix} 0 & 0 & 0 & 0 \\ 0 & A_2 & -B_2 & C_2 \\ 0 & B_2 & A_2 & -D_2 \\ 0 & -C_2 & D_2 & A_2 \end{bmatrix} \tag{11-294}$$

where

$$A_2 = a_2(J)_\eta + b_1(J)_\xi + b_3(J)_\zeta \tag{11-295}$$

$$B_2 = -c_1(J)_\xi + c_7(J)_\zeta \tag{11-296}$$

$$C_2 = -c_2(J)_\xi + c_8(J)_\zeta \tag{11-297}$$

$$D_2 = -c_3(J)_\xi + c_9(J)_\zeta \tag{11-298}$$

$$C_v = \frac{\partial \bar{G}_v}{\partial \bar{Q}} = \frac{1}{Re_\infty J}\begin{bmatrix} 0 & 0 & 0 & 0 \\ 0 & A_3 & -B_3 & C_3 \\ 0 & B_3 & A_3 & -D_3 \\ 0 & -C_3 & D_3 & A_3 \end{bmatrix} \tag{11-299}$$

where

$$A_3 = a_3(J)_\zeta + b_2(J)_\xi + b_3(J)_\eta \tag{11-300}$$

$$B_3 = -c_4(J)_\xi - c_7(J)_\eta \tag{11-301}$$

$$C_3 = -c_5(J)_\xi - c_8(J)_\eta \tag{11-302}$$

$$D_3 = -c_6(J)_\xi - c_9(J)_\eta \tag{11-303}$$

11.8.2 Two-Dimensional Incompressible Navier-Stokes Equations

The nondimensional, incompressible Navier-Stokes equations in the computational domain ξ,η formulated in a flux vector form are expressed as

$$\frac{\partial \bar{Q}}{\partial t} + \frac{\partial \bar{E}}{\partial \xi} + \frac{\partial \bar{F}}{\partial \eta} = \frac{\partial \bar{E}_v}{\partial \xi} + \frac{\partial \bar{F}_v}{\partial \eta} \tag{11-304}$$

where the flux vectors are defined by the following:

$$\bar{Q} = \frac{Q}{J} \tag{11-305}$$

$$\bar{E} = \frac{1}{J}(\xi_x E + \xi_y F) \tag{11-306}$$

$$\bar{F} = \frac{1}{J}(\eta_x E + \eta_y F) \tag{11-307}$$

$$\bar{E}_v = \frac{1}{J}(\xi_x E_v + \xi_y F_v) \tag{11-308}$$

$$\bar{F}_v = \frac{1}{J}(\eta_x E_v + \eta_y F_v) \tag{11-309}$$

and

$$Q = \begin{bmatrix} p \\ u \\ v \end{bmatrix} \tag{11-310}$$

$$E = \begin{bmatrix} \beta u \\ u^2 + p \\ uv \end{bmatrix} \tag{11-311}$$

$$F = \begin{bmatrix} \beta v \\ vu \\ v^2 + p \end{bmatrix} \tag{11-312}$$

$$E_v = \begin{bmatrix} 0 \\ \tau_{xx} \\ \tau_{xy} \end{bmatrix} \quad (11\text{-}313) \qquad F_v = \begin{bmatrix} 0 \\ \tau_{yx} \\ \tau_{yy} \end{bmatrix} \quad (11\text{-}314)$$

The shear stresses in the physical space are

$$\tau_{xx} = \frac{2}{Re_\infty}\left(\frac{\partial u}{\partial x}\right) \quad (11\text{-}315)$$

$$\tau_{xy} = \tau_{yx} = \frac{1}{Re_\infty}\left(\frac{\partial u}{\partial y} + \frac{\partial v}{\partial x}\right) \quad (11\text{-}316)$$

$$\tau_{yy} = \frac{2}{Re_\infty}\left(\frac{\partial v}{\partial y}\right) \quad (11\text{-}317)$$

The shear stresses in the computational space are

$$\tau_{xx} = \frac{2}{Re_\infty}\left(\xi_x u_\xi + \eta_x u_\eta\right) \quad (11\text{-}318)$$

$$\tau_{xy} = \tau_{yx} = \frac{1}{Re_\infty}\left(\xi_y u_\xi + \eta_y u_\eta + \xi_x v_\xi + \eta_x v_\eta\right) \quad (11\text{-}319)$$

$$\tau_{yy} = \frac{2}{Re_\infty}\left(\xi_y v_\xi + \eta_y v_\eta\right) \quad (11\text{-}320)$$

Upon substitution of shear stresses into the viscous flux vectors $\bar{E}_v$ and $\bar{F}_v$ given by Equations (11-308) and (11-309) and utilizing continuity, one obtains

$$\bar{E}_v = \frac{1}{JRe_\infty}\begin{bmatrix} 0 \\ au_\xi + bu_\eta - cv_\eta \\ av_\xi + cu_\eta + bv_\eta \end{bmatrix} \quad (11\text{-}321)$$

and

$$\bar{F}_v = \frac{1}{JRe_\infty}\begin{bmatrix} 0 \\ du_\eta + bu_\xi + cv_\xi \\ dv_\eta - cu_\xi + bv_\xi \end{bmatrix} \quad (11\text{-}322)$$

where

$$a = \xi_x^2 + \xi_y^2 \quad (11\text{-}323)$$

$$b = \xi_x\eta_x + \xi_y\eta_y \quad (11\text{-}324)$$

$$c = \xi_x\eta_y - \xi_y\eta_x \quad (11\text{-}325)$$

$$d = \eta_x^2 + \eta_y^2 \quad (11\text{-}326)$$

The inviscid and viscous Jacobian matrices are determined to be as follows

$$A = \frac{\partial \bar{E}}{\partial \bar{Q}} = \begin{bmatrix} 0 & \xi_x \beta & \xi_y \beta \\ \xi_x & U + \xi_x u & \xi_y u \\ \xi_y & \xi_x u & U + \xi_y v \end{bmatrix} \tag{11-327}$$

where the contravariant velocity U is

$$U = \xi_x u + \xi_y v \tag{11-328}$$

$$B = \frac{\partial \bar{F}}{\partial \bar{Q}} = \begin{bmatrix} 0 & \eta_x \beta & \eta_y \beta \\ \eta_x & V + \eta_x u & \eta_y u \\ \eta_y & \eta_x v & V + \eta_y v \end{bmatrix} \tag{11-329}$$

The contravariant velocity V is defined as

$$V = \eta_x u + \eta_y v \tag{11-330}$$

$$A_v = \frac{\partial \bar{E}_v}{\partial \bar{Q}} = \frac{1}{JRe_\infty} \begin{bmatrix} 0 & 0 & 0 \\ 0 & a(J)_\xi + b(J)_\eta & -c(J)_\eta \\ 0 & c(J)_\eta & a(J)_\xi + b(J)_\eta \end{bmatrix} \tag{11-331}$$

and

$$B_v = \frac{1}{JRe_\infty} \begin{bmatrix} 0 & 0 & 0 \\ 0 & b(J)_\xi + d(J)_\eta & c(J)_\xi \\ 0 & -c(J)_\xi & b(J)_\xi + d(J)_\eta \end{bmatrix} \tag{11-332}$$

11.9 Problems

11.1 Nondimensionalize the energy equation given by

$$\frac{\partial}{\partial t}(\rho e_t) + \frac{\partial}{\partial x}(\rho u e_t + pu) + \frac{\partial}{\partial y}(\rho v e_t + pv) + \frac{\partial}{\partial z}(\rho w e_t + pw) =$$

$$-\frac{\partial q_x}{\partial x} - \frac{\partial q_y}{\partial y} - \frac{\partial q_z}{\partial z} + \frac{\partial}{\partial x}\left[u\tau_{xx} + v\tau_{xy} + w\tau_{xz}\right] +$$

$$\frac{\partial}{\partial y}\left[u\tau_{xy} + v\tau_{yy} + w\tau_{yz}\right] + \frac{\partial}{\partial z}\left[u\tau_{xz} + v\tau_{yz} + w\tau_{zz}\right]$$

where

$$q_x = -k\frac{\partial T}{\partial x} \;, \qquad q_y = -k\frac{\partial T}{\partial y} \;, \qquad q_z = -k\frac{\partial T}{\partial z}$$

$$\tau_{xx} = 2\mu\frac{\partial u}{\partial x} + \lambda\nabla\cdot\vec{V}$$

$$\tau_{xy} = \mu\left(\frac{\partial u}{\partial y} + \frac{\partial v}{\partial x}\right)$$

$$\tau_{yy} = 2\mu\frac{\partial v}{\partial y} + \lambda\nabla\cdot\vec{V}$$

$$\tau_{xz} = \mu\left(\frac{\partial u}{\partial z} + \frac{\partial w}{\partial x}\right)$$

$$\tau_{zz} = 2\mu\frac{\partial w}{\partial z} + \lambda\nabla\cdot\vec{V}$$

$$\tau_{yz} = \mu\left(\frac{\partial v}{\partial z} + \frac{\partial w}{\partial y}\right)$$

Use the nondimensional variables defined in Section 11.3.

11.2 What are the physical implications of Stokes hypothesis?

11.3 Determine the following components of the inviscid Jacobian matrix $\partial\bar{F}/\partial\bar{Q}$, (a) $\partial\bar{F}_2/\partial\bar{Q}_2$, (b) $\partial\bar{F}_3/\partial\bar{Q}_2$, and (c) $\partial\bar{F}_5/\partial\bar{Q}_4$.

11.4 Start with

$$(\bar{F}_v)_3 = \frac{1}{J}\Big[\eta_x (E_v)_3 + \eta_y (F_v)_3 + \eta_z (G_v)_3\Big]$$

and determine the third component of viscous flux vector $\bar{F}_v$ given by relation (11-96).

11.5 Determine the following components of the viscous Jacobian matrix $\partial\bar{F}_v/\partial\bar{Q}$, (a) $\partial\bar{F}v_2/\partial\bar{Q}_1$, (b) $\partial\bar{F}v_2/\partial\bar{Q}_2$, and (c) $\partial\bar{F}v_5/\partial\bar{Q}_4$.

11.6 What are the assumptions used in the PNS equations? What are the physical implications of these assumptions?

11.7 What are the assumptions used in the TLNS equations? What are the physical implications of these assumptions?

11.8 Show that with a generalized coordinate transformation, the equation

$$\frac{\partial u}{\partial x} + \frac{\partial v}{\partial y} = 0$$

may be expressed as

$$\frac{\partial \bar{U}}{\partial \xi} + \frac{\partial \bar{V}}{\partial \eta} = 0$$

where

$$\bar{U} = \frac{U}{J} \quad , \quad \bar{V} = \frac{V}{J}$$

and

$$U = u\xi_x + v\xi_y \quad , \quad V = u\eta_x + v\eta_y$$

11.9 Consider the system of partial differential equations given by

$$\frac{\partial u}{\partial x} + \frac{\partial v}{\partial y} = 0$$

$$u\frac{\partial u}{\partial x} + v\frac{\partial u}{\partial y} + \frac{1}{\rho}\frac{\partial p}{\partial x} = 0$$

$$u\frac{\partial v}{\partial x} + v\frac{\partial v}{\partial y} + \frac{1}{\rho}\frac{\partial p}{\partial y} = 0$$

where ρ, the density, is assumed to be a constant.

(a) Write the system of equations in a vector form, where the unknown vector Q is

$$Q = \begin{bmatrix} u \\ v \\ p \end{bmatrix}$$

(b) Recast the system of equations to a conservative form. Subsequently, write a vector form of the system and determine the Jacobian matrices A and B, where

$$A = \frac{\partial E}{\partial Q} \quad \text{and} \quad B = \frac{\partial F}{\partial Q}$$

Define the flux vectors as

$$E = \begin{bmatrix} u \\ u^2 + \dfrac{p}{\rho} \\ uv \end{bmatrix} \quad \text{and} \quad F = \begin{bmatrix} v \\ uv \\ v^2 + \dfrac{p}{\rho} \end{bmatrix}$$

(c) Transform the vector form of the system obtained in (b) to a (ξ,η) coordinate system.

11.10 The x-component of the momentum equation for an incompressible boundary layer with zero pressure gradient is

$$u\frac{\partial u}{\partial x} + v\frac{\partial u}{\partial y} = \nu\frac{\partial^2 u}{\partial y^2}$$

Transform the equation into generalized coordinate system.

11.11 Transform the x-component of the momentum equation into the generalized coordinate system. Assume a steady, incompressible, zero pressure gradient flow, so that the governing equation is written as

$$u\frac{\partial u}{\partial x} + v\frac{\partial u}{\partial y} = \nu\left(\frac{\partial^2 u}{\partial x^2} + \frac{\partial^2 u}{\partial y^2}\right)$$

11.12 Consider the generalized coordinate system in Problem 11.11 to be body fitted such that the ξ coordinate is along the body and the η coordinate is perpendicular to it. Use the boundary layer assumptions to reduce the transformed equation. Compare the result to the transformed equation of Problem 11.10.

Chapter 12
Euler Equations

12.1 Introductory Remarks

Reduced forms of the equations of fluid motion are used for many practical applications. Obviously, the assumptions involved in the reduction of the equations of motion must represent the physics of a particular problem. A commonly used method is the decoupling of the equations of motion for viscous and inviscid regions. This approach is suitable for problems where the viscous/inviscid interaction is weak. For high Reynolds number flows, viscous effects are confined to the vicinity of the surface, where large velocity gradients exist. This region is known as the boundary layer. Outside the boundary layer, the velocity gradients are negligible resulting in zero shear stresses. This region is called the inviscid region.

In order to solve the boundary layer equations, the flow properties at the boundary layer edge (usually defined at a location of $u/u_e \cong 0.99$, where u_e is the velocity at the edge of the boundary layer) are required. One method of providing this information is to solve the inviscid region initially and impose the result on the boundary layer. An iterative procedure between the inviscid flowfield and the boundary layer may be used in order to include boundary layer displacement effect. This chapter will investigate the solution procedures for the inviscid flow region. The governing equations are known as the Euler equations.

Several solution schemes will be investigated in detail in this chapter. This is important not only because we are seeking the solution of the Euler equation, but also because the Euler equation form the left-hand side of the Navier-Stokes equation. In fact, the solution of the Navier-Stokes equation is not much more difficult than the Euler equation. Practically, there is only one method of approximating the viscous terms which form the right-hand side of the Navier-Stokes equation. Typically, that approximation is accomplished by the use of second-order central difference formulation. However, there is a wide range of formulations to approximate the convective term, each with its own advantages and disadvantages.

The numerical schemes and associated issues which are presented in this chapter can be directly extended to the Navier-Stokes equation. An issue of importance is, of course, the application of boundary conditions which are different for the Euler and Navier-Stokes equations. The Euler equation requires slip condition at the surface, whereas the Navier-Stokes equation requires the no-slip condition at the surface. The application of appropriate boundary conditions will be addressed in each chapter.

12.2 Euler Equations

Recall the equations of fluid motion in a flux vector form, given by Equation (11-49), which is repeated here for convenience

$$\frac{\partial Q}{\partial t} + \frac{\partial E}{\partial x} + \frac{\partial F}{\partial y} + \frac{\partial G}{\partial z} = \frac{\partial E_v}{\partial x} + \frac{\partial F_v}{\partial y} + \frac{\partial G_v}{\partial z} \tag{12-1}$$

where all quantities have been nondimensionalized. For an inviscid flow, the viscous forces are negligible and, therefore, Equation (12-1) is reduced to

$$\frac{\partial Q}{\partial t} + \frac{\partial E}{\partial x} + \frac{\partial F}{\partial y} + \frac{\partial G}{\partial z} = 0 \tag{12-2}$$

where

$$Q = \begin{bmatrix} \rho \\ \rho u \\ \rho v \\ \rho w \\ \rho e_t \end{bmatrix}, \quad E = \begin{bmatrix} \rho u \\ \rho u^2 + p \\ \rho u v \\ \rho u w \\ (\rho e_t + p)u \end{bmatrix}, \quad F = \begin{bmatrix} \rho v \\ \rho v u \\ \rho v^2 + p \\ \rho v w \\ (\rho e_t + p)v \end{bmatrix}, \quad G = \begin{bmatrix} \rho w \\ \rho w u \\ \rho w v \\ \rho w^2 + p \\ (\rho e_t + p)w \end{bmatrix}$$

Due to advantages discussed previously in Chapter 9, Equation (12-2) is transformed to a computational domain where grid point spacing is uniform and the domain is rectangular. The transformed Euler equation is given by

$$\frac{\partial \bar{Q}}{\partial \tau} + \frac{\partial \bar{E}}{\partial \xi} + \frac{\partial \bar{F}}{\partial \eta} + \frac{\partial \bar{G}}{\partial \zeta} = 0 \tag{12-3}$$

where

$$\bar{Q} = \frac{Q}{J} \tag{12-4}$$

$$\bar{E} = \frac{1}{J}(\xi_t Q + \xi_x E + \xi_y F + \xi_z G) \tag{12-5}$$

$$\bar{F} = \frac{1}{J}(\eta_t Q + \eta_x E + \eta_y F + \eta_z G) \tag{12-6}$$

$$\bar{G} = \frac{1}{J}(\zeta_t Q + \zeta_x E + \zeta_y F + \zeta_z G) \tag{12-7}$$

To investigate several solution procedures, we will consider as a first step a model system of equations, namely, the quasi one-dimensional Euler equations. With this simple system of equations, the effect of time and spatial stepsizes, boundary and initial conditions, convergence, and stability will be explored. Subsequently, numerical procedures are extended to two-dimensional problems.

12.3 Quasi One-Dimensional Euler Equations

The Euler equations for a quasi one-dimensional flow may be expressed as:

continuity,

$$\frac{\partial}{\partial t}(\rho S) + \frac{\partial}{\partial x}(\rho u S) = 0 \tag{12-8}$$

momentum,

$$\frac{\partial}{\partial t}(\rho u S) + \frac{\partial}{\partial x}[(\rho u^2 + p)S] - p\frac{dS}{dx} = 0 \tag{12-9}$$

energy,

$$\frac{\partial}{\partial t}(\rho e_t S) + \frac{\partial}{\partial x}[(\rho e_t + p)uS] = 0 \tag{12-10}$$

where S is the cross-sectional area assumed independent of time, i.e., $S = S(x)$, and

$$e_t = e + \frac{1}{2}u^2$$

Equations (12-8) through (12-10) are expressed in a flux vector form in a similar fashion as the previous equations. Hence,

$$\frac{\partial}{\partial t}(SQ) + \frac{\partial E}{\partial x} - H = 0 \tag{12-11}$$

where

$$Q = \begin{bmatrix} \rho \\ \rho u \\ \rho e_t \end{bmatrix}, \quad E = S\begin{bmatrix} \rho u \\ \rho u^2 + p \\ (\rho e_t + p)u \end{bmatrix}, \quad \text{and} \quad H = \frac{dS}{dx}\begin{bmatrix} 0 \\ p \\ 0 \end{bmatrix}$$

12.3.1 Numerical Issues

Before proceeding to explore specific numerical schemes, several issues are addressed. First is the choice between an implicit scheme or an explicit scheme. Some of the advantages and disadvantages of each category were explored previously; however, at this point, consider the difference between the two schemes with regard to linearization.

An explicit scheme can be formulated when the time derivative is approximated by a forward difference approximation. For simplicity, a first-order approximation is used and the explicit formulation is written as

$$S\frac{Q^{n+1}-Q^{n}}{\Delta t}+\left(\frac{\partial E}{\partial x}\right)^{n}-H^{n}=0 \tag{12-12}$$

Now, consider an implicit algorithm for Equation (12-11). The time derivative is approximated by a first-order backward difference approximation to provide

$$S\frac{Q^{n+1}-Q^{n}}{\Delta t}+\left(\frac{\partial E}{\partial x}\right)^{n+1}-H^{n+1}=0 \tag{12-13}$$

Since formulation (12-13) is implicit, the second and third terms have been expressed at the $n+1$ time level.

The change in flow properties per time step will be defined as

$$\Delta Q = Q^{n+1}-Q^{n}$$

Typically, the FDE is formulated in terms of ΔQ, which is referred to as the delta formulation.

Observe that the nonlinear term given by the flux vector E in Equation (12-11) is evaluated at the known time level in the formulation (12-12), and, therefore, it does not require linearization. However, in the implicit formulation (12-13), a linearization procedure must be considered. The linearization process introduces additional approximations and associated errors into the equations. Thus, explicit schemes have an advantage over implicit schemes with regard to linearization, namely, they do not require linearization, and, therefore, associated errors are not introduced into the equations.

Since both explicit and implicit schemes will be considered in this chapter, linearization of implicit formulation is addressed next. However, since a linearization procedure was explored previously in Chapter 11, only a brief review is presented here.

Consider a Taylor series expansion about time level n as follows:

$$E^{n+1}=E^{n}+\frac{\partial E}{\partial t}\Delta t+O(\Delta t)^{2} \tag{12-14}$$

Since $E = f(Q, S)$, the chain rule of differentiation yields

$$\frac{\partial E}{\partial t} = \frac{\partial E}{\partial Q}\frac{\partial Q}{\partial t} + \frac{\partial E}{\partial S}\frac{\partial S}{\partial t}$$

from which (note that S was assumed a function of x only and, therefore, the second term is omitted)

$$\frac{\partial E}{\partial t} = \frac{\partial E}{\partial Q}\frac{\partial Q}{\partial t} \cong \frac{\partial E}{\partial Q}\frac{Q^{n+1} - Q^n}{\Delta t} = \frac{\partial E}{\partial Q}\frac{\Delta Q}{\Delta t} \tag{12-15}$$

Substitute (12-15) into Equation (12-14) to obtain

$$E^{n+1} = E^n + \frac{\partial E}{\partial Q}\frac{\Delta Q}{\Delta t}\Delta t + O(\Delta t)^2$$

or

$$E^{n+1} = E^n + \frac{\partial E}{\partial Q}\Delta Q + O(\Delta t)^2 \tag{12-16}$$

In a similar fashion the following may be derived:

$$H^{n+1} = H^n + \frac{\partial H}{\partial Q}\Delta Q + O(\Delta t)^2 \tag{12-17}$$

Recall that the terms such as $\partial E/\partial Q$ and $\partial H/\partial Q$ were defined as the flux Jacobian matrices. Derivation of Jacobian matrices was fully discussed in Chapter 11 and, therefore, only the results are given below. The Jacobian matrix $\partial E/\partial Q$ will be denoted by A and is

$$A = \frac{\partial E}{\partial Q} = S \begin{bmatrix} 0 & 1 & 0 \\ \left(\frac{\gamma-3}{2}\right)u^2 & -(\gamma-3)u & (\gamma-1) \\ -\gamma u e_t + (\gamma-1)u^3 & \gamma e_t - \frac{3(\gamma-1)}{2}u^2 & \gamma u \end{bmatrix} \tag{12-18}$$

The Jacobian matrix $\partial H/\partial Q$ is denoted by B and is

$$B = \frac{\partial H}{\partial Q} = (\gamma - 1)\frac{dS}{dx} \begin{bmatrix} 0 & 0 & 0 \\ \frac{1}{2}u^2 & -u & 1 \\ 0 & 0 & 0 \end{bmatrix} \tag{12-19}$$

The total energy e_t in the Jacobian matrix A may be expressed in terms of the speed of sound. We will assume a perfect gas and, therefore,

$$p = \rho e(\gamma - 1) \tag{12-20}$$

and since

$$e = e_t - \frac{1}{2}u^2$$

thus,

$$p = \rho(\gamma - 1)(e_t - \frac{1}{2}u^2)$$

from which

$$e_t = \frac{p}{\rho(\gamma - 1)} + \frac{1}{2}u^2 \tag{12-21}$$

Speed of sound, denoted by a, is given by

$$a^2 = \gamma\frac{p}{\rho} \tag{12-22}$$

After substitution of (12-22) into (12-21) the following is obtained:

$$e_t = \frac{a^2}{\gamma(\gamma - 1)} + \frac{1}{2}u^2 \tag{12-23}$$

Equation (12-23) is used to rearrange the Jacobian matrix as (note that only the third row is affected)

$$A = S\begin{bmatrix} 0 & 1 & 0 \\ \left(\frac{\gamma - 3}{2}\right)u^2 & -(\gamma - 3)u & (\gamma - 1) \\ -\frac{ua^2}{\gamma - 1} + \left(\frac{1}{2}\gamma - 1\right)u^3 & \frac{a^2}{\gamma - 1} + \left(\frac{3}{2} - \gamma\right)u^2 & \gamma u \end{bmatrix} \tag{12-24}$$

The second issue to address is to identify the properties of the system of PDE's under consideration.

The first order hyperbolic equation (12-11) has the property that the flux vector E is a homogeneous function of degree one in Q; i.e., for any value of α, $E(\alpha Q) = \alpha E(Q)$. This property is referred to as the homogeneous property. In general, Euler equations possess this property. Recall that for a system of equations to be classified as hyperbolic, the Jacobian matrix A (for our model equation) must possess real eigenvalues. The eigenvalues of A represent the characteristic direction of the hyperbolic system and thus provide the direction of the propagation of information. Additional materials with regard to characteristics are provided in Appendix G. Now consider the following statement. If a matrix has real eigenvalues and associated eigenvectors, it may be diagonalized, i.e., a similarity transformation exists such that

$$A = XDX^{-1}$$

where D is a diagonal matrix with its elements being the eigenvalues of A, and X is the eigenvector matrix.

These mathematical observations are important in the development of the numerical schemes to solve the model equation (12-11) and also in the specification of the boundary conditions. The knowledge gained is easily extended to two- and three-dimensional problems.

Before proceeding further, the following observations are re-emphasized about the Euler equations:

(1) The flux vectors are a homogeneous function of degree one;

(2) For our hyperbolic system, the eigenvalues are real and, in general, consist of mixed positive and negative eigenvalues;

(3) The signs of the eigenvalues indicate the direction of data propagation;

(4) The flux Jacobian matrices may be diagonalized.

The third issue to consider is the approximation of the convective term in Equation (12-11). Recall that several schemes were investigated for the solution of simple hyperbolic equations in Chapter 6. These schemes can be extended to the Euler equation, that is, Equation (12-11). That includes TVD schemes, Runge-Kutta schemes, and upwind schemes, to name a few. Furthermore, recall that, in general, the convective terms can be approximated either by a central difference approximation such as the Beam and Warming implicit scheme and the Runge-Kutta scheme, or by a one-sided difference approximations such as the upwind schemes. Methods which employ central difference approximation of convective terms typically require the addition of numerical viscosity in the form of damping terms or TVD. Both categories of schemes will be explored in this chapter.

Upwind schemes take advantage of the physics of the problem in the development of a numerical scheme. That is, the finite difference approximation is consistent with the direction of signal propagation. To proceed with the development of upwind type numerical schemes, the flux vector splitting scheme introduced by Steger and Warming [12-1] is explored first. Subsequently, the flux vector splitting scheme of van Leer [12-2] will be introduced in Section 12.3.2.2.

The following discussion is pertinent specifically to the Steger and Warming flux vector splitting. However, similar concepts are used in any flux vector splitting scheme. Now return back to the flux matrix A, given by (12-24), and determine its eigenvalues. For this purpose any program capable of symbolic manipulation such as MACSYMA [12-3] or Maple [12-4] can be used. The resulting eigenvalues are

$$\lambda_1 = u \tag{12-25}$$

$$\lambda_2 = u + a \tag{12-26}$$

$$\lambda_3 = u - a \tag{12-27}$$

and the associated eigenvectors (also determined by MACSYMA, Maple, etc.) are

$$X_1 = \begin{bmatrix} 1 \\ u \\ \frac{1}{2}u^2 \end{bmatrix}, \quad X_2 = \begin{bmatrix} 1 \\ u + a \\ \frac{1}{2}u^2 + ua + \frac{a^2}{\gamma - 1} \end{bmatrix}, \quad X_3 = \begin{bmatrix} 1 \\ u - a \\ \frac{1}{2}u^2 - ua + \frac{a^2}{\gamma - 1} \end{bmatrix}$$

Since the flux Jacobian matrix A possesses a complete set of eigenvalues and eigenvectors, a similarity transformation exists such that

$$A = XDX^{-1}$$

where

$$D = S\begin{bmatrix} u & 0 & 0 \\ 0 & u + a & 0 \\ 0 & 0 & u - a \end{bmatrix} \tag{12-28}$$

and

$$X = \begin{bmatrix} 1 & \alpha & \alpha \\ u & \alpha(u + a) & \alpha(u - a) \\ \frac{1}{2}u^2 & \alpha(\frac{1}{2}u^2 + ua + \frac{a^2}{\gamma - 1}) & \alpha(\frac{1}{2}u^2 - ua + \frac{a^2}{\gamma - 1}) \end{bmatrix} \tag{12-29}$$

where $\alpha = \rho / \left(a\sqrt{2}\right)$. The inverse of the eigenvector matrix X is

$$X^{-1} = \begin{bmatrix} 1 - \frac{u^2}{2}\frac{\gamma - 1}{a^2} & (\gamma - 1)\frac{u}{a^2} & -\frac{(\gamma - 1)}{a^2} \\ \beta[(\gamma - 1)\frac{u^2}{2} - ua] & \beta[a - (\gamma - 1)u] & \beta(\gamma - 1) \\ \beta[(\gamma - 1)\frac{u^2}{2} + ua] & -\beta[a + (\gamma - 1)u] & \beta(\gamma - 1) \end{bmatrix} \tag{12-30}$$

where $\beta = 1 / \left(\rho a\sqrt{2}\right)$.

Note that the flux vector E equals AQ. Also, recall that the flux vector E possesses the homogenous property; therefore, it may be split into subvectors such that each subvector is associated with positive or negative eigenvalues of the flux matrix Jacobian. Thus, the eigenvalues may be grouped as positive or negative. For example, for a subsonic flow, two of the eigenvalues, namely u and $u + a$, are positive, whereas the third eigenvalue, $u - a$, is negative. Therefore, the Jacobian matrix A is split according to

$$A = A^+ + A^-$$

where

$$A^+ = XD^+X^{-1}$$

and

$$A^- = XD^-X^{-1}$$

The elements of the diagonal matrices D^+ and D^- are the positive and negative eigenvalues, i.e.,

$$D = S\begin{bmatrix} u & 0 & 0 \\ 0 & u+a & 0 \\ 0 & 0 & u-a \end{bmatrix} = S\begin{bmatrix} u & 0 & 0 \\ 0 & u+a & 0 \\ 0 & 0 & 0 \end{bmatrix} + S\begin{bmatrix} 0 & 0 & 0 \\ 0 & 0 & 0 \\ 0 & 0 & u-a \end{bmatrix}$$

Now, the flux vector E may be split according to

$$E^+ = A^+Q \tag{12-31}$$

and

$$E^- = A^-Q \tag{12-32}$$

Note that for a supersonic flow, all three eigenvalues are positive and, therefore,

$$A^+ = A$$

and

$$A^- = 0$$

The flux Jacobian matrices A^+ and A^- (for the subsonic flow) are easily determined by MACSYMA, Maple, etc. The result is:

$$A^+ = S\begin{bmatrix} -\dfrac{(\gamma-1)u^3+a(3-\gamma)u^2-2a^2u}{4a^2} & \dfrac{(\gamma-1)u^2+a(2-\gamma)u+a^2}{2a^2} & -\dfrac{(\gamma-1)u-\gamma a+a}{2a^2} \\ -\dfrac{(\gamma-1)u^4+2a(2-\gamma)u^3+a^2(1-\gamma)u^2+2a^3u}{4a^2} & \dfrac{(\gamma-1)u^3+a(3-2\gamma)u^2+a^2(3-\gamma)u+a^3}{2a^2} & -\dfrac{(\gamma-1)u^2+2a(1-\gamma)u-\gamma a^2+a^2}{2a^2} \\ \begin{array}{l} -\Big[(\gamma^2-2\gamma+1)u^5+(-3\gamma^2+8\gamma-5)au^4 \\ -2a^2(\gamma^2-2\gamma+1)u^3+2a^3(\gamma+1)u^2 \\ +4a^4u\Big] \,/\, \Big[8a^2(\gamma-1)\Big] \end{array} & \begin{array}{l} \Big[(\gamma^2-2\gamma+1)u^4+(-3\gamma^2+7\gamma-4)au^3 \\ -a^2(2\gamma^2-5\gamma+3)u^2+2a^3(u+a)\Big] \\ /\, \Big[4a^2(\gamma-1)\Big] \end{array} & -\dfrac{(\gamma-1)u^3+3a(1-\gamma)u^2-2\gamma a^2u-2a^3}{4a^2} \end{bmatrix} \tag{12-33}$$

$$A^- = S\begin{bmatrix} \dfrac{(u-a)[(\gamma-1)u^2+2au]}{4a^2} & -\dfrac{(u-a)[(\gamma-1)u+a]}{2a^2} & \dfrac{(\gamma-1)(u-a)}{2a^2} \\ \dfrac{(u-a)^2[(\gamma-1)u^2+2au]}{4a^2} & -\dfrac{(u-a)^2[(\gamma-1)u+a]}{2a^2} & \dfrac{(\gamma-1)(u-a)^2}{2a^2} \\ \dfrac{(u-a)\left[\frac{1}{2}u^2-au+\frac{a^2}{\gamma-1}\right]\ [(\gamma-1)u^2+2au]}{4a^2} & -\dfrac{(u-a)\left[\frac{1}{2}u^2-au+\frac{a^2}{\gamma-1}\right][(\gamma-1)u+a]}{2a^2} & \dfrac{(\gamma-1)(u-a)\left[\frac{1}{2}u^2-au+\frac{a^2}{\gamma-1}\right]}{2a^2} \end{bmatrix} \tag{12-34}$$

The flux vector E for a subsonic flow is split according to

$$E^+ = S\frac{\rho}{2\gamma}\begin{bmatrix} 2\gamma u + a - u \\ 2(\gamma-1)u^2 + (u+a)^2 \\ (\gamma-1)u^3 + \left[(u+a)^3\right]/2 + \left[(3-\gamma)(u+a)a^2\right]/\left[2(\gamma-1)\right] \end{bmatrix} \tag{12-35}$$

and

$$E^- = S\frac{\rho}{2\gamma}\begin{bmatrix} u - a \\ (u-a)^2 \\ \left[(u-a)^3\right]/2 + \left[(3-\gamma)(u-a)a^2\right]/\left[2(\gamma-1)\right] \end{bmatrix} \tag{12-36}$$

At this point, pause a moment to determine the reason for all the mathematical manipulations considered so far. Recall that the objective is to develop efficient and stable numerical schemes to solve a system of hyperbolic PDEs, for the time being the model equation (12-11). To investigate the stability requirement of the equation, a linear stability analysis is employed (Reference [12-1]). The results indicate that if one-sided differencing is used for the spatial derivatives, it must be a forward differencing for the terms associated with the negative eigenvalues and a backward differencing for the terms associated with the positive eigenvalues. This requirement is used to write the FDEs where one-sided differences are used. A second consideration, a very important one, is the specification of the inflow and outflow boundary conditions based on the eigenvalues. This point will be explored after the examination of the FDEs.

12.3.2 Explicit Formulations

A first-order forward difference approximation in time introduced in Equation (12-11) would provide the following explicit formulation.

$$S\frac{Q^{n+1} - Q^n}{\Delta t} + \frac{\partial}{\partial x}(E^n) - H^n = 0 \tag{12-37}$$

Now several schemes are available to approximate the spatial term. The resulting finite difference equations are as follows.

12.3.2.1 Steger and Warming Flux Vector Splitting: The flux vector E is split into positive and negative components, and Equation (12-37) is written as

$$S\frac{\Delta Q}{\Delta t} + \frac{\partial}{\partial x}(E^+ + E^-)^n - H^n = 0 \tag{12-38}$$

where $\Delta Q = Q^{n+1} - Q^n$.

Now the spatial derivative can be replaced either by a first-order or a second-order one-sided approximation. According to the previous discussion, a backward difference approximation is used for the positive terms and a forward difference approximation must be used for the negative terms. Thus, a first-order approximation in space will provide the following FDE

$$\Delta Q = -\frac{1}{S}\left(\frac{\Delta t}{\Delta x}\right)\left[E_i^+ - E_{i-1}^+ + E_{i+1}^- - E_i^-\right] + \frac{1}{S}(\Delta t)H_i \qquad (12\text{-}39)$$

Observe that all the terms on the right-hand side are at time level n, that is, at a known time level, and that the superscript n is dropped for convenience.

When a second-order approximation is used, the FDE becomes

$$\Delta Q = -\frac{1}{S}\left(\frac{\Delta t}{2\Delta x}\right)\left[E_{i-2}^+ - 4E_{i-1}^+ + 3E_i^+ - 3E_i^- + 4E_{i+1}^- - E_{i+2}^-\right] + \frac{1}{S}(\Delta t)H_i \quad (12\text{-}40)$$

Note that, when Equation (12-40) is computed at points $i = 2$ and $i = IMM1 = IMAX - 1$ (assuming that points $i = 1$ and $i = IMAX$ are the boundary points), a difficulty is encountered, that is, at points $i-2$ and $i+2$. The required points ($i = 0$ and $i = IMP1 = IMAX + 1$) do not exist! There are several methods to overcome this problem. One simple method is to switch from the second-order scheme to the first-order scheme. That is, at points 2 and $IMM1$, Equation (12-39) is used. Another approach would be to employ the concept of fictitious points similar to that discussed in Appendix B.

The flux vector E in the Steger and Warming formulation is given by relations (12-35) and (12-36) for a subsonic flow and for a supersonic flow by $E^+ = E$ and $E^- = 0$.

Equations (12-39) and (12-40) will be referred to as first-order and second-order explicit, respectively. In fact, observe that both are first-order accurate in time; however, the spatial order of accuracy will be used to distinguish between the two.

12.3.2.2 Van Leer Flux Vector Splitting: Recall that, in the Steger and Warming flux vector splitting scheme, the splitting of flux vector was based on the eigenvalues of the Jacobian matrix $A = \partial E/\partial Q$, and that the flux vector E is split into E^+ and E^-.

Similarly, in the Van Leer flux vector splitting scheme, the flux vector E is expressed as $E = E^+ + E^-$ where the following constraints must hold:

(a) All the eigenvalues of $\partial E^+/\partial Q$ are ≥ 0, and

(b) All the eigenvalues of $\partial E^-/\partial Q$ are ≤ 0.

In addition, the following restrictions are imposed.

(1) The flux vectors E^+ and E^- must be continuous and

$$E^+ = E \quad \text{for} \quad M \geq 1 \ , \quad \text{and}$$

$$E^- = E \quad \text{for} \quad M \leq -1$$

(2) The Jacobian matrices $\partial E^+/\partial Q$ and $\partial E^-/\partial Q$ must be continuous.

(3) One of the eigenvalues of $\partial E^+/\partial Q$ (or $\partial E^-/\partial Q$) must vanish for $|M| < 1$.

(4) The components of the flux vector E^+ (or E^-) must be symmetric to each other with respect to M, as is the case with E. Mathematically, if $E_i(M) = \pm E_i(-M)$, then $E_i^+(M) = \pm E_i^-(-M)$.

(5) The flux vectors E, E^+, and E^- must be polynomials of the lowest degree in M.

To proceed with mathematical details, first the components of flux vector E are written in terms of M. Therefore,

$$E_1 = S\rho u = S\rho a M \tag{12-41}$$

Recall that the speed of sound is given by

$$a^2 = \gamma \frac{p}{\rho}$$

or

$$p = a^2 \frac{\rho}{\gamma}$$

Now

$$E_2 = S(\rho u^2 + p) = S\left(\rho a^2 M^2 + \frac{\rho a^2}{\gamma}\right) = S\left[\rho a^2 \left(M^2 + \frac{1}{\gamma}\right)\right] \tag{12-42}$$

The total energy given by (12-21) is written in terms of M as

$$e_t = \frac{p}{\rho(\gamma - 1)} + \frac{1}{2}u^2 = \frac{a^2}{\gamma(\gamma - 1)} + \frac{1}{2}a^2 M^2$$

Thus,

$$E_3 = S(\rho e_t + p)u = S\left[\rho \frac{a^2}{\gamma(\gamma - 1)} + \frac{1}{2}\rho a^2 M^2 + \rho \frac{a^2}{\gamma}\right] aM$$

or

$$E_3 = S(\rho a^3 M)\left(\frac{1}{2}M^2 + \frac{1}{\gamma - 1}\right) \tag{12-43}$$

Therefore, the flux vector E can be expressed as

$$E = S\begin{bmatrix} \rho a M \\ \rho a^2\left(M^2 + \dfrac{1}{\gamma}\right) \\ \rho a^3 M\left(\dfrac{1}{2}M^2 + \dfrac{1}{\gamma - 1}\right) \end{bmatrix}$$

Now restrictions identified earlier are imposed. Requirements (1) and (2) indicate that E^+ and $\partial E^+/\partial M$ must vanish as M is decreased toward -1. Similarly, E^- and $\partial E^-/\partial M$ must vanish as M is increased toward 1. Furthermore, with the requirement imposed by restriction (5), the flux vectors E^+ and E^- must include a factor of $(1+M)^2$ and $(1-M)^2$, respectively. Thus,

$$E_1 = S(\rho a M) = S(\rho a)\left[\frac{1}{4}(1+M)^2 - \frac{1}{4}(1-M)^2\right] = E_1^+ + E_1^-$$

or

$$E_1^+ = S\left(\frac{1}{4}\rho a\right)(1+M)^2$$

and

$$E_1^- = S\left(-\frac{1}{4}\rho a\right)(1-M)^2$$

for $-1 < M < 1$.

Now, the momentum flux E_2 must include a cubic polynomial in M and, therefore, it is expressed as

$$E_2 = S(\rho a^2)\left(M^2 + \frac{1}{\gamma}\right) = S(\rho a^2)\left[\frac{1}{4}(1+M)^2\left(\frac{\gamma-1}{\gamma}M + \frac{2}{\gamma}\right) + \frac{1}{4}(1-M)^2\left(-\frac{\gamma-1}{\gamma}M + \frac{2}{\gamma}\right)\right]$$

or

$$E_2^+ = S(\rho a^2)\frac{1}{4}(1+M)^2\left(\frac{\gamma-1}{\gamma}M + \frac{2}{\gamma}\right)$$

and

$$E_2^- = S(\rho a^2)\frac{1}{4}(1-M)^2\left(-\frac{\gamma-1}{\gamma}M + \frac{2}{\gamma}\right)$$

for $-1 < M < 1$.

The energy flux E_3 can be split based on E_2 and E_1 according to

$$E_3^+ = S\frac{\gamma^2}{2(\gamma^2-1)}\frac{(E_2^+)^2}{(E_1^+)}$$

and

$$E_3^- = S\frac{\gamma^2}{2(\gamma^2-1)}\frac{(E_2^-)^2}{(E_1^-)}$$

for $-1 < M < 1$.

Finally, the flux vectors E^+ and E^- are written as

$$E^+ = \begin{bmatrix} E_1^+ \\ E_2^+ \\ E_3^+ \end{bmatrix} = S\begin{bmatrix} \frac{1}{4}\rho a(1+M)^2 \\ \frac{1}{4}\rho a^2(1+M)^2\left(\frac{\gamma-1}{\gamma}M+\frac{2}{\gamma}\right) \\ \frac{1}{4}\rho a^3(1+M)^2\frac{\gamma^2}{2(\gamma^2-1)}\left(\frac{\gamma-1}{\gamma}M+\frac{2}{\gamma}\right)^2 \end{bmatrix}$$

or

$$E^+ = E_1^+\begin{bmatrix} 1 \\ a\left(\frac{\gamma-1}{\gamma}M+\frac{2}{\gamma}\right) \\ a^2\frac{\gamma^2}{2(\gamma^2-1)}\left(\frac{\gamma-1}{\gamma}M+\frac{2}{\gamma}\right)^2 \end{bmatrix} \tag{12-44}$$

and

$$E^- = \begin{bmatrix} E_1^- \\ E_2^- \\ E_3^- \end{bmatrix} = S\begin{bmatrix} -\frac{1}{4}\rho a(1-M)^2 \\ \frac{1}{4}\rho a^2(1-M)^2\left(-\frac{\gamma-1}{\gamma}M+\frac{2}{\gamma}\right) \\ \frac{1}{4}\rho a^3(1-M)^2\frac{\gamma^2}{2(\gamma^2-1)}\left(-\frac{\gamma-1}{\gamma}M+\frac{2}{\gamma}\right)^2 \end{bmatrix}$$

or

$$E^- = E_1^-\begin{bmatrix} 1 \\ -a\left(-\frac{\gamma-1}{\gamma}M+\frac{2}{\gamma}\right) \\ -a^2\frac{\gamma^2}{2(\gamma^2-1)}\left(-\frac{\gamma-1}{\gamma}M+\frac{2}{\gamma}\right)^2 \end{bmatrix} \tag{12-45}$$

for $-1 < M < 1$.

Recall that, if $M \geq 1$, then $E^+ = E$ and $E^- = 0$, and if $M \leq -1$, then $E^+ = 0$ and $E^- = E$.

Now the finite difference equations (12-39) or (12-40), with the appropriate flux vector splitting as described above, can be used for solution.

12.3.2.3 Modified Runge-Kutta Formulation: The modified fourth-order Runge-Kutta scheme introduced in Section 6.6.8 can be extended to Equation (12-12) to provide the following FDE

$$Q_i^{(1)} = Q_i^n \tag{12-46}$$

$$Q_i^{(2)} = Q_i^n - \frac{\Delta t}{4}\left(\frac{1}{S}\right)\left[\left(\frac{\partial E}{\partial x}\right)_i^{(1)} - H_i^{(1)}\right] \tag{12-47}$$

$$Q_i^{(3)} = Q_i^n - \frac{\Delta t}{3}\left(\frac{1}{S}\right)\left[\left(\frac{\partial E}{\partial x}\right)_i^{(2)} - H_i^{(2)}\right] \tag{12-48}$$

$$Q_i^{(4)} = Q_i^n - \frac{\Delta t}{2}\left(\frac{1}{S}\right)\left[\left(\frac{\partial E}{\partial x}\right)_i^{(3)} - H_i^{(3)}\right] \tag{12-49}$$

$$Q_i^{n+1} = Q_i^n - \Delta t\left(\frac{1}{S}\right)\left[\left(\frac{\partial E}{\partial x}\right)_i^{(4)} - H_i^{(4)}\right] \tag{12-50}$$

where a second-order central difference approximation is used for thc spatial derivative, that is,

$$\left(\frac{\partial E}{\partial x}\right)_i = \frac{E_{i+1} - E_{i-1}}{2\Delta x} \tag{12-51}$$

12.3.2.4 Second-Order TVD Formulation: Several second-order TVD schemes presented in Section 6.10.4 will be extended to a system of equations and will be applied to the quasi one-dimensional Euler equation in this section. Following Equation (6-123), a finite difference equation for the vector equation given by (12-11) is written as follows

$$Q_i^{n+1} = Q_i^n - \frac{1}{S}\left(\frac{\Delta t}{\Delta x}\right)\left[R_{i+\frac{1}{2}}^n - R_{i-\frac{1}{2}}^n\right] + \frac{\Delta t}{S} H_i^n \tag{12-52}$$

where

$$R_{i+\frac{1}{2}}^n = \frac{1}{2}\left(E_{i+1}^n + E_i^n + X_{i+\frac{1}{2}}^n \Phi_{i+\frac{1}{2}}^n\right) \tag{12-53}$$

and

$$R_{i-\frac{1}{2}}^n = \frac{1}{2}\left(E_i^n + E_{i-1}^n + X_{i-\frac{1}{2}}^n \Phi_{i-\frac{1}{2}}^n\right) \tag{12-54}$$

The eigenvector matrix X is given by (12-29), and the components of the flux limiter vector Φ are provided in the subsequent sections. Note that now the flux limiter Φ is a vector with three components, that is,

$$\Phi = \begin{bmatrix} \phi_1 \\ \phi_2 \\ \phi_3 \end{bmatrix}$$

12.3.2.4.1 Harten-Yee Upwind TVD: The general expression for the components of the flux vector limiter is defined as

$$\phi_{i+\frac{1}{2}} = \sigma(\alpha_{i+\frac{1}{2}})\,(G_{i+1} + G_i) - \psi(\alpha_{i+\frac{1}{2}} + \beta_{i+\frac{1}{2}})\delta_{i+\frac{1}{2}} \tag{12-55}$$

$$\phi_{i-\frac{1}{2}} = \sigma(\alpha_{i-\frac{1}{2}})\,(G_i + G_{i-1}) - \psi(\alpha_{i-\frac{1}{2}} + \beta_{i-\frac{1}{2}})\delta_{i-\frac{1}{2}} \tag{12-56}$$

where α is used to denote $S\lambda$. Recall that the eigenvalues for the one-dimensional problem are $\lambda_1 = u$, $\lambda_2 = u + a$, and $\lambda_3 = u - a$. Now consider, for example, the computation of ϕ_1 for which $\alpha_1 = Su$, and δ_1 is calculated from (12-60) as

$$(\delta_{i+\frac{1}{2}})_1 = (X^{-1}_{i+\frac{1}{2}})_{\text{First row}}(Q_{i+1} - Q_i)$$

Components ϕ_2 and ϕ_3 are determined similarly, and subsequently the flux limiter vector Φ is formed.

The various terms appearing in $\phi_{i+\frac{1}{2}}$ are defined as follows

$$\sigma(\alpha_{i+\frac{1}{2}}) = \frac{1}{2}\psi(\alpha_{i+\frac{1}{2}}) + \frac{\Delta t}{\Delta x}(\alpha_{i+\frac{1}{2}})^2 \tag{12-57}$$

$$\beta_{i+\frac{1}{2}} = \sigma(\alpha_{i+\frac{1}{2}}) \begin{cases} \dfrac{G_{i+1} - G_i}{\delta_{i+\frac{1}{2}}} & \text{for} \quad \delta_{i+\frac{1}{2}} \neq 0 \\ 0 & \text{for} \quad \delta_{i+\frac{1}{2}} = 0 \end{cases} \tag{12-58}$$

$$\psi(y) = \begin{cases} |y| & \text{for} \quad |y| \geq \epsilon \\ \dfrac{y^2 + \epsilon^2}{2\epsilon} & \text{for} \quad |y| < \epsilon \end{cases} \tag{12-59}$$

with $0 \leq \epsilon \leq 0.125$
and

$$\delta_{i+\frac{1}{2}} = \left(X^{-1}_{i+\frac{1}{2}}\right)(Q_{i+1} - Q_i) \tag{12-60}$$

The inverse eigenvector matrix X^{-1} is given by (12-30). The terms for $\phi_{i-\frac{1}{2}}$ are defined in a similar fashion.

Several limiters have been proposed, as follow

$$G_i = \text{minmod}\ (\delta_{i-\frac{1}{2}}\ ,\ \delta_{i+\frac{1}{2}}) \tag{12-61}$$

$$G_i = \frac{\delta_{i+\frac{1}{2}}\delta_{i-\frac{1}{2}} + \left|\delta_{i+\frac{1}{2}}\delta_{i-\frac{1}{2}}\right|}{\delta_{i+\frac{1}{2}} + \delta_{i-\frac{1}{2}}} \tag{12-62}$$

$$G_i = \frac{\delta_{i-\frac{1}{2}}\left[(\delta_{i+\frac{1}{2}})^2 + \omega\right] + \delta_{i+\frac{1}{2}}\left[(\delta_{i-\frac{1}{2}})^2 + \omega\right]}{(\delta_{i+\frac{1}{2}})^2 + (\delta_{i-\frac{1}{2}})^2 + 2\omega} \tag{12-63}$$

where $10^{-7} \leq \omega \leq 10^{-5}$

$$G_i = \text{minmod}\left[2\delta_{i-\frac{1}{2}}\ ,\ 2\delta_{i+\frac{1}{2}}\ ,\ \frac{1}{2}(\delta_{i+\frac{1}{2}} + \delta_{i-\frac{1}{2}})\right] \tag{12-64}$$

$$G_i = S * \max\left[0\ ,\ \min\left(2\left|\delta_{i+\frac{1}{2}}\right|\ ,\ S * \delta_{i-\frac{1}{2}}\right)\ ,\ \min\left(\left|\delta_{i+\frac{1}{2}}\right|\ ,\ 2S * \delta_{i-\frac{1}{2}}\right)\right] \tag{12-65}$$

where, as before,

$$S = Sgn\,(\delta_{i+\frac{1}{2}}) = \frac{ABS(\delta_{i+\frac{1}{2}})}{\delta_{i+\frac{1}{2}}}$$

and

$$\text{minmod}\,(a, b, c, \ldots, n) = S * \max\,[0,\ \min\ (\,|a|,\ S * b,\ S * c,\ \ldots,\ S * n)\,] \tag{12-66}$$

with

$$S = \frac{ABS(a)}{a}$$

12.3.2.4.2 Roe-Sweby Upwind TVD: The general expression for the components of the flux limiter vector is given by

$$\phi_{i+\frac{1}{2}} = \left[\frac{G_i}{2}\left(\left|\alpha_{i+\frac{1}{2}}\right| + \frac{\Delta t}{\Delta x}\alpha^2_{i+\frac{1}{2}}\right) - \left|\alpha_{i+\frac{1}{2}}\right|\right]\delta_{i+\frac{1}{2}} \tag{12-67}$$

$$\phi_{i-\frac{1}{2}} = \left[\frac{G_{i-1}}{2}\left(\left|\alpha_{i-\frac{1}{2}}\right| + \frac{\Delta t}{\Delta x}\alpha^2_{i-\frac{1}{2}}\right) - \left|\alpha_{i-\frac{1}{2}}\right|\right]\delta_{i-\frac{1}{2}}$$

and the following limiters have been proposed.

$$G_i = \text{minmod}\ (1, r) \tag{12-68}$$

$$G_i = \frac{r + |r|}{1 + r} \tag{12-69}$$

$$G_i = \max\ [0\ ,\ \min\,(2r, 1)\ ,\ \min\,(r, 2)\,] \tag{12-70}$$

where

$$r = \begin{cases} \dfrac{X^{-1}_{i+1+\sigma}Q_{i+1+\sigma} - X^{-1}_{i+\sigma}Q_{i+\sigma}}{\delta_{i+\frac{1}{2}}} & \text{for} \quad \delta_{i+\frac{1}{2}} \neq 0 \\ 0 & \text{for} \quad \delta_{i+\frac{1}{2}} = 0 \end{cases} \tag{12-71}$$

and $\sigma = Sgn(\alpha_{i+\frac{1}{2}})$

12.3.2.4.3 Davis-Yee Symmetric TVD: The general expression for the components of the flux limiter vector is defined as

$$\phi_{i+\frac{1}{2}} = -\left[\frac{\Delta t}{\Delta x}(\alpha_{i+\frac{1}{2}})^2 G_{i+\frac{1}{2}} + \psi(\alpha_{i+\frac{1}{2}})\,(\delta_{i+\frac{1}{2}} - G_{i+\frac{1}{2}})\right] \tag{12-72}$$

Again, several limiters have been introduced as follow

$$G_{i+\frac{1}{2}} = \text{minmod}\left[2\delta_{i-\frac{1}{2}}\,,\; 2\delta_{i+\frac{1}{2}}\,,\; 2\delta_{i+\frac{3}{2}}\,,\; \frac{1}{2}\left(\delta_{i-\frac{1}{2}} + \delta_{i+\frac{3}{2}}\right)\right] \tag{12-73}$$

$$G_{i+\frac{1}{2}} = \text{minmod}\left[\delta_{i-\frac{1}{2}}\,,\; \delta_{i+\frac{1}{2}}\,,\; \delta_{i+\frac{3}{2}}\right] \tag{12-74}$$

$$G_{i+\frac{1}{2}} = \text{minmod}\left[\delta_{i+\frac{1}{2}}\,,\; \delta_{i-\frac{1}{2}}\right] + \text{minmod}\left[\delta_{i+\frac{1}{2}}\,,\; \delta_{i+\frac{3}{2}}\right] - \delta_{i+\frac{1}{2}} \tag{12-75}$$

where $\delta_{\frac{1}{2}} = 0$ and $\delta_{IMAX+\frac{3}{2}} = 0$.

12.3.2.5 Modified Runge-Kutta Scheme with TVD: The modified fourth-order Runge-Kutta scheme discussed in Section 12.3.2.3 can be amended by a TVD scheme as a post processor step to provide a mechanism to reduce dispersion error. In this case, after the computation of Equation (12-50), the value of Q is updated according to

$$Q_i^{n+1} = Q_i^{n+1} - \frac{1}{S}\frac{1}{2}\frac{\Delta t}{\Delta x}\left[X^n_{i+\frac{1}{2}}\Phi^n_{i+\frac{1}{2}} - X^n_{i-\frac{1}{2}}\Phi^n_{i-\frac{1}{2}}\right] \tag{12-76}$$

where any one of the flux limiter functions and limiters can be used.

12.3.3 Implicit Formulations

The implicit formulation for the one-dimensional Euler equation is given by

$$S\frac{\Delta Q}{\Delta t} + \left(\frac{\partial E}{\partial x}\right)^{n+1} - H^{n+1} = 0 \tag{12-77}$$

Since Equation (12-77) is nonlinear, the linearization procedure described in Section 12.3.1 is used. Substitution of (12-16) and (12-17) into (12-77) yields

$$S\frac{\Delta Q}{\Delta t} + \frac{\partial}{\partial x}\left(E^n + \frac{\partial E}{\partial Q}\Delta Q\right) - \left(H^n + \frac{\partial H}{\partial Q}\Delta Q\right) = 0 \tag{12-78}$$

This equation may be expressed in terms of the Jacobian matrices A and B as

$$S\frac{\Delta Q}{\Delta t} + \frac{\partial}{\partial x}(A\Delta Q) - B\Delta Q = -\frac{\partial E^n}{\partial x} + H^n \tag{12-79}$$

and is factored as

$$\left[SI + \Delta t\left(\frac{\partial A}{\partial x}\right) - B\Delta t\right]\Delta Q = -\Delta t\left(\frac{\partial E^n}{\partial x} - H^n\right) \tag{12-80}$$

where I is the identity matrix and $(\partial A/\partial x)\Delta Q$ implies $\partial(A\Delta Q)/\partial x$.

Several explicit formulations described in Section 12.3.2 can be extended and applied to Equation (12-80) to provide equivalent implicit formulations.

Typically, the Runge-Kutta scheme is used in its explicit formulation because, first, the stability requirement is larger than the typical CFL number of one, and, second, the implicit formulation would require an excessive amount of computation time. However, the flux vector splitting schemes can be used for implicit formulations. In the following section, the implicit Steger and Warming flux vector splitting formulation is used to illustrate the development of an implicit scheme. Furthermore, the modification of the coefficient matrix due to implementation of the boundary condition is described. Other implicit schemes can be formulated in a similar fashion.

12.3.3.1 Steger and Warming Flux Vector Splitting: The flux vector E and the flux Jacobian matrix A are split according to the previous discussion to provide

$$\left[SI + \Delta t\left\{\frac{\partial}{\partial x}(A^+ + A^-) - B\right\}\right]\Delta Q = -\Delta t\left[\frac{\partial}{\partial x}(E^+ + E^-) - H\right]$$

Note that the superscript n on the right-hand side has been dropped for convenience. Following previous deliberations, a backward difference approximation is used for the positive terms, and a forward difference approximation is used for the negative terms. Hence, when first-order approximations are used, the following finite difference equation is obtained.

$$\left[SI + \frac{\Delta t}{\Delta x}(A_i^+ - A_{i-1}^+ + A_{i+1}^- - A_i^-) - \Delta t\, B_i\right]\Delta Q =$$

$$-\Delta t\left[\frac{1}{\Delta x}(E_i^+ - E_{i-1}^+ + E_{i+1}^- - E_i^-) - H_i\right]$$

This equation is rearranged to provide

$$-\left(\frac{\Delta t}{\Delta x}A_{i-1}^+\right)\Delta Q_{i-1} + \left[SI + \frac{\Delta t}{\Delta x}(A_i^+ - A_i^-) - \Delta t B_i\right]\Delta Q_i + \left(\frac{\Delta t}{\Delta x}A_{i+1}^-\right)\Delta Q_{i+1} =$$

$$-\frac{\Delta t}{\Delta x}(E_i^+ - E_{i-1}^+ + E_{i+1}^- - E_i^-) + \Delta t\, H_i \tag{12-81}$$

In order to write this equation in a manageable fashion, the following are defined:

$$AM = -\frac{\Delta t}{\Delta x} A^+_{i-1}$$

$$AA = \left[SI + \frac{\Delta t}{\Delta x}(A^+_i - A^-_i) - B_i \, \Delta t \right]$$

$$AP = \frac{\Delta t}{\Delta x} A^-_{i+1}$$

$$RHS = -\frac{\Delta t}{\Delta x}(E^+_i - E^+_{i-1} + E^-_{i+1} - E^-_i) + \Delta t \, H_i$$

Thus, Equation (12-81) is expressed as

$$AM_i \Delta Q_{i-1} + AA_i \Delta Q_i + AP_i \Delta Q_{i+1} = RHS_i \tag{12-82}$$

This equation is solved in a computational domain shown in Figure 12-1.

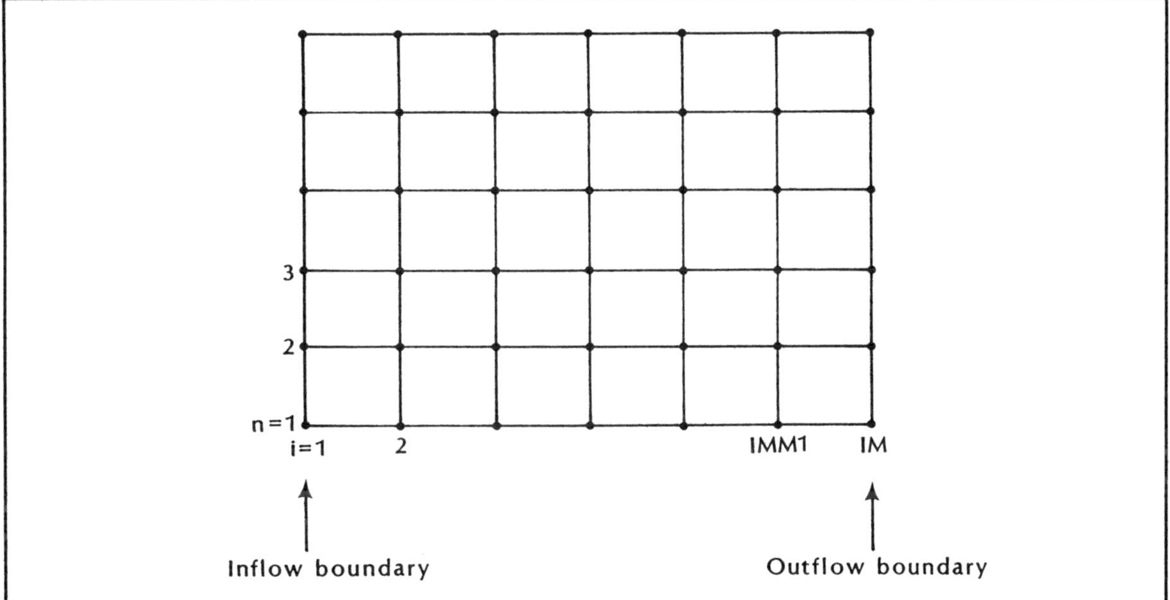

Figure 12-1. The computational domain used for the solution of Equation (12-81).

Once Equation (12-81) is applied to each grid point i, a block tridiagonal system is produced. Note that the elements of the coefficient matrix are themselves matrices and, for our 1-D problem, they are 3×3 matrices. As a result, it is referred to as a block tridiagonal system. To illustrate this point, consider the following set of equations (i.e., Equation (12-82) at various i grid points):

$$i = 2: \quad AM_2 \Delta Q_1 + AA_2 \Delta Q_2 + AP_2 \Delta Q_3 = RHS_2 \tag{12-83a}$$

$$i = 3: \quad AM_3 \Delta Q_2 + AA_3 \Delta Q_3 + AP_3 \Delta Q_4 = RHS_3 \tag{12-83b}$$

$$\vdots$$

$$i = IMM2: \quad AM_{IMM2}\Delta Q_{IMM3} + AA_{IMM2}\Delta Q_{IMM2} +$$

$$+AP_{IMM2}\Delta Q_{IMM1} = RHS_{IMM2} \tag{12-83c}$$

$$i = IMM1: \quad AM_{IMM1}\Delta Q_{IMM2} + AA_{IMM1}\Delta Q_{IMM1} +$$

$$+AP_{IMM1}\Delta Q_{IM} = RHS_{IMM1} \tag{12-83d}$$

Note that in Equation (12-83a), ΔQ_1 is located at the inflow boundary; and in Equation (12-83d), ΔQ_{IM} is located at the outflow boundary. Specifications of inflow and outflow boundary conditions will be discussed shortly. Equations (12-83a) through (12-83d) are written in a matrix form as:

$$\begin{bmatrix} AA_2 & AP_2 & & & \\ AM_3 & AA_3 & AP_3 & & \\ & & & & \\ & & AM_{IMM2} & AA_{IMM2} & AP_{IMM2} \\ & & & AM_{IMM1} & AA_{IMM1} \end{bmatrix} \begin{bmatrix} \Delta Q_2 \\ \Delta Q_3 \\ | \\ \Delta Q_{IMM2} \\ \Delta Q_{IMM1} \end{bmatrix} =$$

$$\begin{bmatrix} RHS_2 - AM_2\Delta Q_1 \\ RHS_3 \\ | \\ RHS_{IMM2} \\ RHS_{IMM1} - AP_{IMM1}\Delta Q_{IM} \end{bmatrix} \tag{12-84}$$

Any standard block tridiagonal solver may be used to solve this system. Note that, for the supersonic region of the flowfield, AP is zero (because A^- is zero) resulting in a lower diagonal banded matrix (bidiagonal system) which is inverted more efficiently than the tridiagonal system. Recall that a block tridiagonal solver is discussed in Appendix E.

12.4 Boundary Conditions

The Euler equations or, generally, any system of PDEs have an unlimited number of solutions. What makes a solution unique is the proper specification of the initial and boundary conditions. Currently there are intensive investigations on the topics related to the specifications of the boundary conditions and its effect on the stability and accuracy of the solution.

For a given PDE, a set of boundary conditions must be specified. They shall be referred to as the "analytical boundary conditions". Once the PDE is approximated by a FDE, it may be higher order than the PDE. Thus the FDE will require additional boundary conditions. These boundary conditions will be referred to as "numerical boundary conditions."

In order to develop proper boundary conditions, the following points must be considered:

(1) The physics of a particular problem must be modeled correctly. For example, for a viscous flow the no slip condition at the surface is specified.

(2) The physical conditions must be represented correctly by mathematical expressions. In some instances, they are specified by numerical approximations. For example, an adiabatic wall boundary condition requires that the temperature gradient must be zero. This condition is approximated by a finite difference relation.

(3) Additional numerical boundary conditions may be required. These boundary conditions are usually specified by extrapolation from the interior solution.

(4) The manner in which boundary conditions are specified must be considered in the overall stability and accuracy of the numerical scheme used to solve the system.

(5) The boundary conditions may be applied explicitly or implicitly. Some investigators have reported minor differences for the computation of the Euler equations when the boundary conditions were applied either implicitly or explicitly, i.e., the stability requirements are similar, see References [12-5] and [12-6].

(6) No difficulties are observed when the exact boundary conditions are overspecified [12-5].

Note: Some of the statements made above are based on limited investigations and, therefore, they should not be generalized.

Now, consider the specification of the boundary conditions for the quasi one-dimensional problem. Recall that the eigenvalues of the flux Jacobian matrix A are u, $u+a$, and $u-a$. As stated previously, these eigenvalues indicate how information is fed into the domain. To illustrate this point, assume a supersonic inflow and a supersonic outflow. For a supersonic flow, all three eigenvalues are positive. At the inflow, three characteristics enter into the domain and, therefore, three analytical boundary conditions may be specified. At the outflow all the characteristics leave the domain and, as a result, no boundary condition can be specified. These situations are illustrated graphically in Figure 12-2.

The values of the dependent variables at the outflow must be evaluated based on the information reaching the outflow from the interior points. Note that these numerical boundary conditions cannot be specified arbitrarily and must be consistent with the direction of propagation of information determined by the sign of the eigenvalues of A. Usually, extrapolation schemes are used for this purpose. The extrapolation procedure may be either explicit or implicit. If an explicit approach is used, the new value at $n+1$ is determined from the value at time level n or time levels n and $n-1$; in either case, these values are known. For an implicit approach, the values are determined at the time level $n+1$ as a part of the solution. Some extrapolation schemes reported in the literature are (extrapolation for the value of a property f at $i = IM$ from interior points at $IMM1$ or $IMM1$ and $IMM2$, where $IMM1 = IM - 1$ and $IMM2 = IM - 2$):

$$f_{IM}^{n+1} = f_{IMM1}^{n+1} \tag{12-85}$$

$$f_{IM}^{n+1} = 2f_{IMM1}^{n+1} - f_{IMM2}^{n+1} \tag{12-86}$$

$$f_{IM}^{n+1} = f_{IMM1}^{n} \tag{12-87}$$

$$f_{IM}^{n+1} = 2f_{IMM1}^{n} - f_{IMM2}^{n-1} \tag{12-88}$$

$$f_{IM}^{n+1} = 2f_{IMM1}^{n} - f_{IMM2}^{n} \tag{12-89}$$

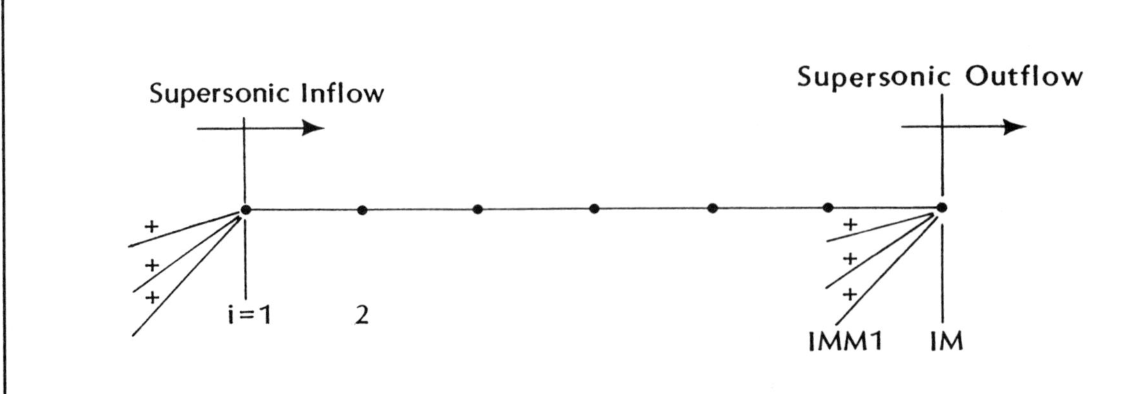

Figure 12-2. Incoming and outgoing characteristics for the supersonic inflow and supersonic outflow conditions.

If the outflow is subsonic, two of the eigenvalues are positive, i.e., outgoing, and one is negative, i.e., incoming from the outside. Therefore, one analytical boundary condition may be specified; and the other two are determined from the interior solution by extrapolation.

A summary of the inflow and outflow boundary conditions for subsonic and supersonic conditions is given in Table 12-1 along with graphical illustrations.

With regard to stability, it must be pointed out that according to the linear stability analysis, the implicit formulation is stable. However, this analysis does not guarantee the unconditional stability of the original nonlinear equations. In practice some stability limits are encountered. That is especially true for highly nonlinear problems, such as domains with shocks. However, the linear stability analysis is valuable in providing some guidelines for the stability requirements of the numerical schemes. In addition, the manner by which boundary conditions are specified will affect the stability of the numerical scheme. Therefore, stability analysis must be extended to include the boundary conditions. Some findings on the effect of boundary condition implementation are reported in Reference [12-5].

	INFLOW		OUTFLOW	
	Subsonic	Supersonic	Subsonic	Supersonic
Number of B.C. to be specified (Analytical B.C.)	2	3	1	0
Number of B.C. by extrapolation (Numerical B.C.)	1	0	2	3
	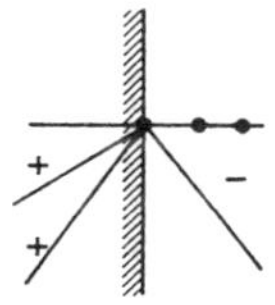 + + −	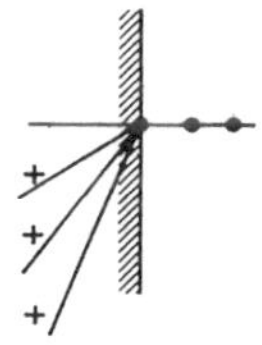 + + +	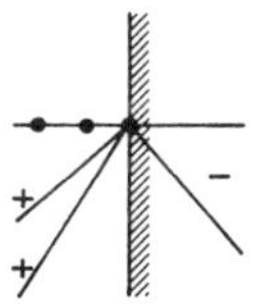 + + −	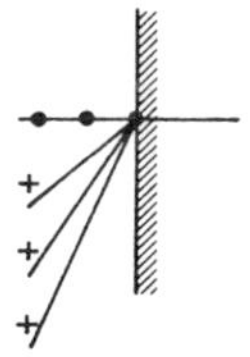 + + +

Table 12-1. Inflow and outflow boundary conditions.

12.5 Application 1: Diverging Nozzle Configuration

Consider a nozzle whose cross-sectional area is defined by

$$S(x) = 1.398 + 0.347\tanh(0.8x - 4) \tag{12-90}$$

Locate the nozzle entrance at $x = 0.0$ and the nozzle exit at $x = 10.0$ ft. Assume air with $\gamma = 1.4$ and $R = 1716$ ft $\text{lb}_f/\text{Slugs}^\circ\text{R}$ enters the nozzle at supersonic speed. The flow leaves the nozzle under two different conditions specified as (1) supersonic and (2) subsonic. We wish to determine the steady-state solution of the quasi one-dimensional model equation within the domain specified in this problem. The physical domain is illustrated in Figure 12-3.

The supersonic flow at the inlet is specified by:

$$M_1 = 1.5$$
$$p_1 = 2000.0 \text{ lb}_f/\text{ft}^2$$
$$T_1 = 520\ ^\circ\text{R}$$

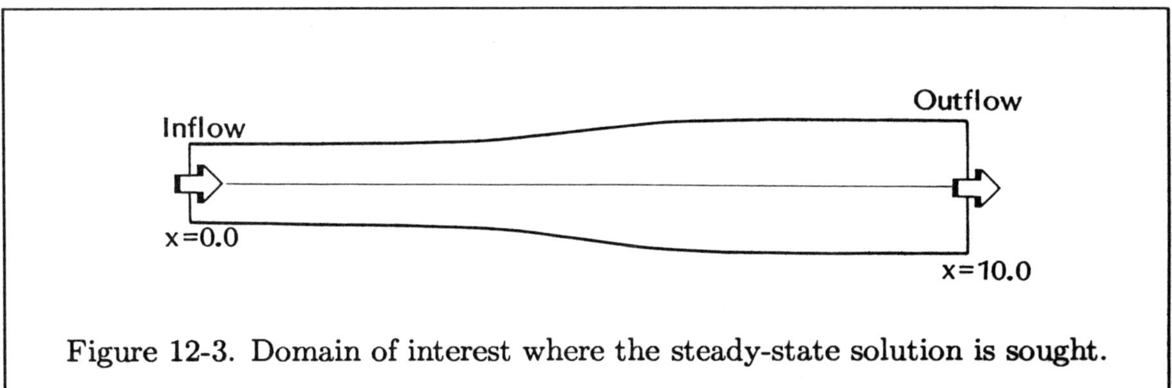

Figure 12-3. Domain of interest where the steady-state solution is sought.

For the first case, where the outflow is supersonic, no analytical boundary conditions can be specified. Recall that our formulation of the Euler equation (quasi one-dimensional) requires ρ, u, and e_t at the inflow. The following relations are used to determine u_1 and e_{t_1} at the inlet:

$$u_1 = M_1\sqrt{\gamma R T_1} \tag{12-91}$$

and

$$e_{t_1} = \frac{RT_1}{(\gamma - 1)} + \frac{1}{2}u_1^2 \tag{12-92}$$

The numerical outflow boundary conditions are evaluated by any of the extrapolation schemes given by (12-85) through (12-89).

To start the solution, a set of initial conditions must be provided. The solution may converge faster for the better initial data. An easy procedure to describe the initial data set is to specify the flow variables everywhere in the domain to be that of the inflow condition.

The graphical illustration of the computational domain including the initial and boundary points is shown in Figure 12-4.

For the second case, the outflow is specified as subsonic and, therefore, one analytical boundary condition must be specified. For our example problem, either a pressure of 4930.07 lb_f/ft^2, or a density of 0.003954 $Slugs/ft^3$, or a subsonic velocity of 572.76 ft/sec at the exit can be specified.

The analytical solution of this problem is easily obtained by using tables in the standard fluid dynamics text [12-7], NACA 1135 Tables [12-8], Tables of [12-9], or relations developed for compressible, one-dimensional flows. With the specified subsonic speed (or pressure) at the exit plane, a normal shock at $x = 5.0$ is expected. The analytical solution will be used for code validation.

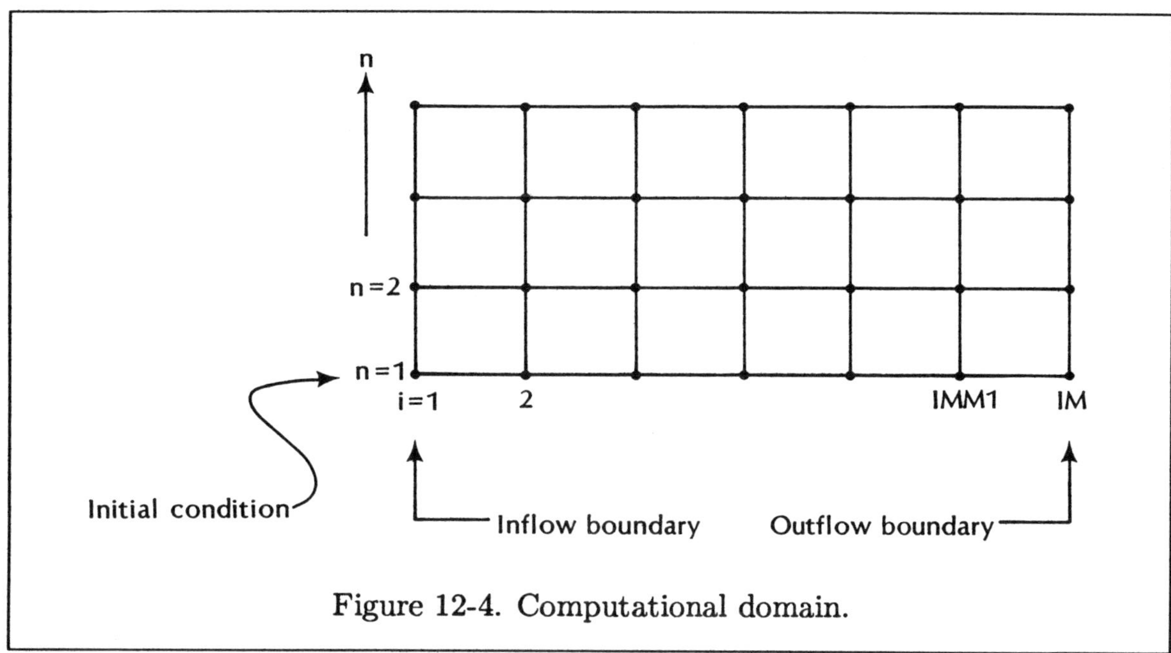

Figure 12-4. Computational domain.

How is it established whether or not a steady-state solution has been reached? Various procedures may be used. Obviously, as the steady-state solution is approached, the variation in the flow properties becomes smaller and smaller. Thus, a convergence criterion is set; and when this condition is satisfied, the steady-state solution has been obtained. The condition in this problem was based on the total variations of pressure within the domain defined as

$$DELP = \sum_{i=1}^{IM} ABS(p_i^{n+1} - p_i^n) \tag{12-93}$$

When $DELP$ was less than $CONV$, which was set to 0.1, the solution was assumed converged, i.e., steady-state.

In this application, solutions by several numerical schemes, the effects of step sizes, stability, and convergence will be investigated.

12.5.1 Supersonic Inflow, Supersonic Outflow

12.5.1.1 Analytical Solution: An expression relating the Mach number distribution to variation in area can be established which is known as the *area-Mach number relation.* The derivation of this expression is given in any standard text such as [12-10, 12-11]. The area-Mach number relation is given by

$$\frac{S}{S^*} = \frac{1}{M}\left[\left(\frac{2}{\gamma+1}\right)\left(1+\frac{\gamma-1}{2}M^2\right)\right]^{\frac{\gamma+1}{2(\gamma-1)}} \tag{12-94}$$

where S^* is a location where the Mach number is sonic.

For a given S/S^*, there are two isentropic solutions: one subsonic and one supersonic. The unique solution is determined by the imposed back pressure. With the area variation provided, Equation (12-94) can be used to compute the Mach number distribution throughout the domain. It is noted that Equation (12-94) cannot be solved directly for M. Given an area ratio S/S^*, an iterative scheme must be employed to determine M. Once the Mach number variation has been computed, the temperature, pressure, and density distributions are determined from the following relations.

$$T = T_t\left(1+\frac{\gamma-1}{2}M^2\right)^{-1} \tag{12-95}$$

$$p = p_t\left(1+\frac{\gamma-1}{2}M^2\right)^{-\frac{\gamma}{\gamma-1}} \tag{12-96}$$

$$\rho = \rho_t\left(1+\frac{\gamma-1}{2}M^2\right)^{-\frac{1}{\gamma-1}} \tag{12-97}$$

The stagnation properties used in Equations (12-95) through (12-97) are computed based on the inlet data as follow.

$$T_t = T\left(1+\frac{\gamma-1}{2}M^2\right) = 520\left[1+0.2(1.5)^2\right] = 754\ °\text{R}$$

$$p_t = p\left(1+\frac{\gamma-1}{2}M^2\right)^{\frac{\gamma}{\gamma-1}} = 2000\left[1+0.2(1.5)^2\right]^{3.5} = 7342.1\ \text{lb}_\text{f}/\text{ft}^2$$

To determine the stagnation density, either the local density at the inlet is computed from the equation of state, that is,

$$\rho_1 = \frac{p_1}{RT_1} = 0.002241\ \text{slugs/ft}^3$$

Subsequently, the stagnation density is determined as

$$\rho_t = \rho\left(1+\frac{\gamma-1}{2}M^2\right)^{\frac{1}{\gamma-1}} = 0.002241\left[1+0.2(1.5)^2\right]^{2.5}$$

$$= \ 0.00567 \text{ slugs/ft}^3$$

or, using the values of the stagnation properties ρ_t can be determined from the equation of state as follows

$$\rho_t = \frac{p_t}{RT_t} = 0.00567 \text{ slugs/ft}^3$$

Now relations (12-95) through (12-97) can be used to determine the temperature, pressure, and density distributions. The analytical solution at spatial increments of 1.0 ft is given in Table 12.2.

x (ft)	S (ft^2)	S/S*	M	p (lb$_f$/ft^2)	T (°R)	ρ (slugs/ft^3)	u (ft/s)
0.000	1.0512	1.1762	1.500	2000.00	520.00	0.002241	1676.55
1.000	1.0522	1.1772	1.502	1995.58	519.67	0.002238	1677.72
2.000	1.0567	1.1822	1.509	1974.24	518.08	0.002221	1683.42
3.000	1.0782	1.2063	1.543	1879.39	510.84	0.002144	1709.05
4.000	1.1676	1.3063	1.666	1565.39	484.84	0.001882	1798.09
5.000	1.3980	1.5641	1.907	1083.61	436.47	0.001447	1952.98
6.000	1.6284	1.8220	2.091	814.840	402.33	0.001180	2055.29
7.000	1.7178	1.9220	2.152	740.130	391.43	0.001102	2086.91
8.000	1.7393	1.9460	2.166	723.940	388.96	0.001085	2093.99
9.000	1.7438	1.9511	2.169	720.520	388.44	0.001081	2095.50
10.000	1.7448	1.9521	2.170	719.900	388.34	0.001080	2095.77

Table 12.2. Summary of the analytical solution for supersonic flow within the nozzle.

12.5.1.2 Numerical Solutions: The numerical solutions are obtained by several schemes described in Section 12.3 and are presented in this section.

The flow properties at the inlet may be determined from the given data to be as:

$$u = 1676.55 \text{ ft}/s$$

$$\rho = 0.002241 \text{ slugs/ft}^3$$

$$e_t = 3636204 \text{ ft}^2/s^2$$

The flow at the inlet is supersonic and, therefore, the flow properties specified above are used as the inflow boundary conditions. The flow at the outlet is specified as supersonic; therefore, no boundary conditions can be specified. To determine the properties at the exit plane, extrapolation is used. For this purpose, the following relations are used:

$$\rho_{IM} = \rho_{IMM1}$$

$$u_{IM} = u_{IMM1}$$

$$e_{t_{IM}} = e_{t_{IMM1}}$$

To discretize the domain, equally spaced spatial grid points are used. For now, take the spatial step Δx to be 0.2 ft and the temporal step as 0.00001 sec. This selection of step sizes provides a Courant-Friedrichs and Lewy (CFL) number of about 0.15 which would satisfy the stability requirement of the explicit schemes. The CFL number is defined here as

$$\text{CFL} = (u + a)_{\max} \frac{\Delta t}{\Delta x}$$

Note that the terms *Courant number* and *CFL number* are identical, and therefore they will be used interchangeably throughout the text.

To start the solution, a set of initial data describing the flow properties must be specified which may be accomplished by various methods. Perhaps the simplest is to set all the flow properties within the domain equal to that of the inflow values. With these specified initial conditions, a steady state solution was obtained after 1354 time steps for the first-order Steger and Warming flux vector explicit scheme and after 1361 time steps for the second-order Steger and Warming flux vector explicit scheme. The convergence criterion specified by CONV was set to 0.1. The pressure and Mach number distributions are compared to the analytical solutions in Figures 12.5 and 12.6, respectively.

To illustrate the intermediate solutions, the pressure distribution at various time levels is shown in Figure 12-7. Note that the time level $n = 1$ is the specified initial condition, and $n = 1361$ is the steady-state solution. The intermediate solutions at time levels of $n = 50$, 100, 200, 400, 600, and 800 are shown in this figure. Obviously, these solutions have no significant value; it is just a means of getting to the steady-state solution. However, note that if correct (from the physical point of view) and accurate initial data is provided, the solution at various time levels will represent the time-dependent solution and is usually referred to as the time accurate solution. Since the steady-state solution is of interest in this example, it will be referred to as the converged solution; and the time levels will be referred to as the iteration levels.

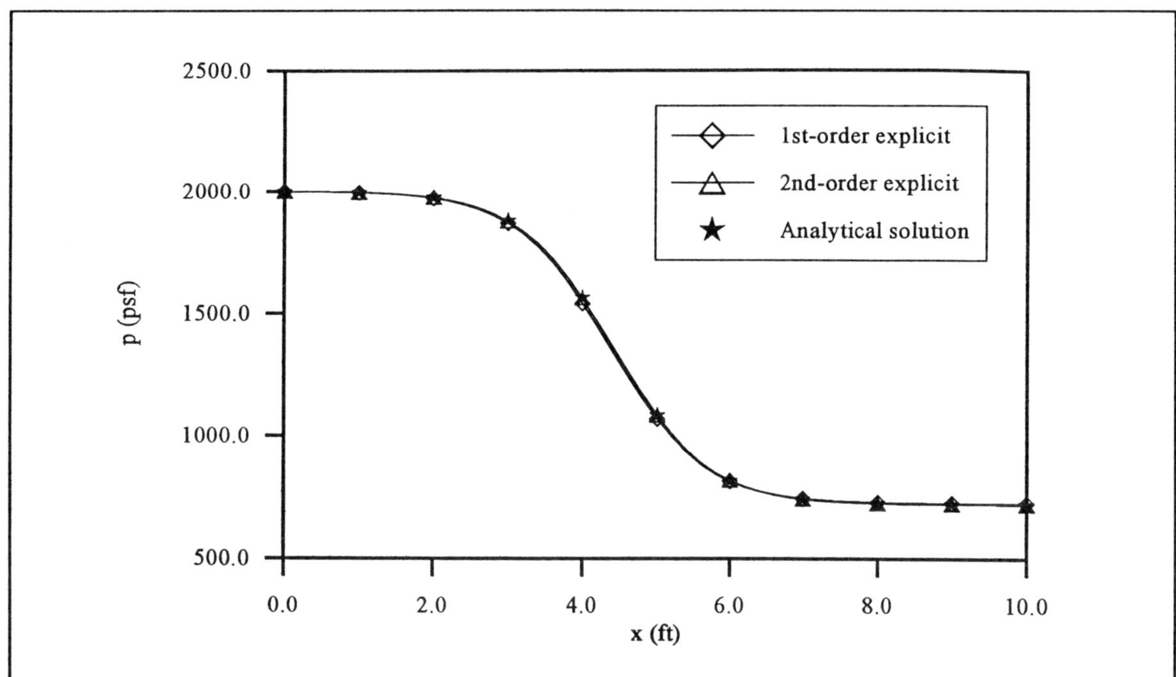

Figure 12-5. Comparison of pressure distributions for the steady-state solution.

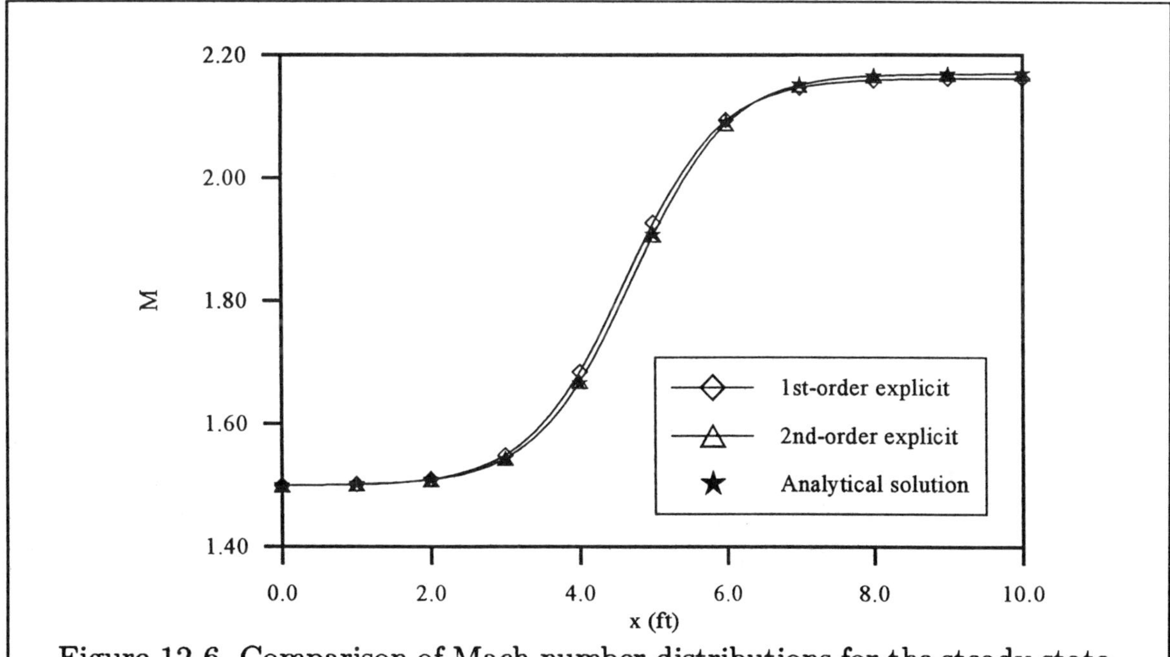

Figure 12-6. Comparison of Mach number distributions for the steady state solution.

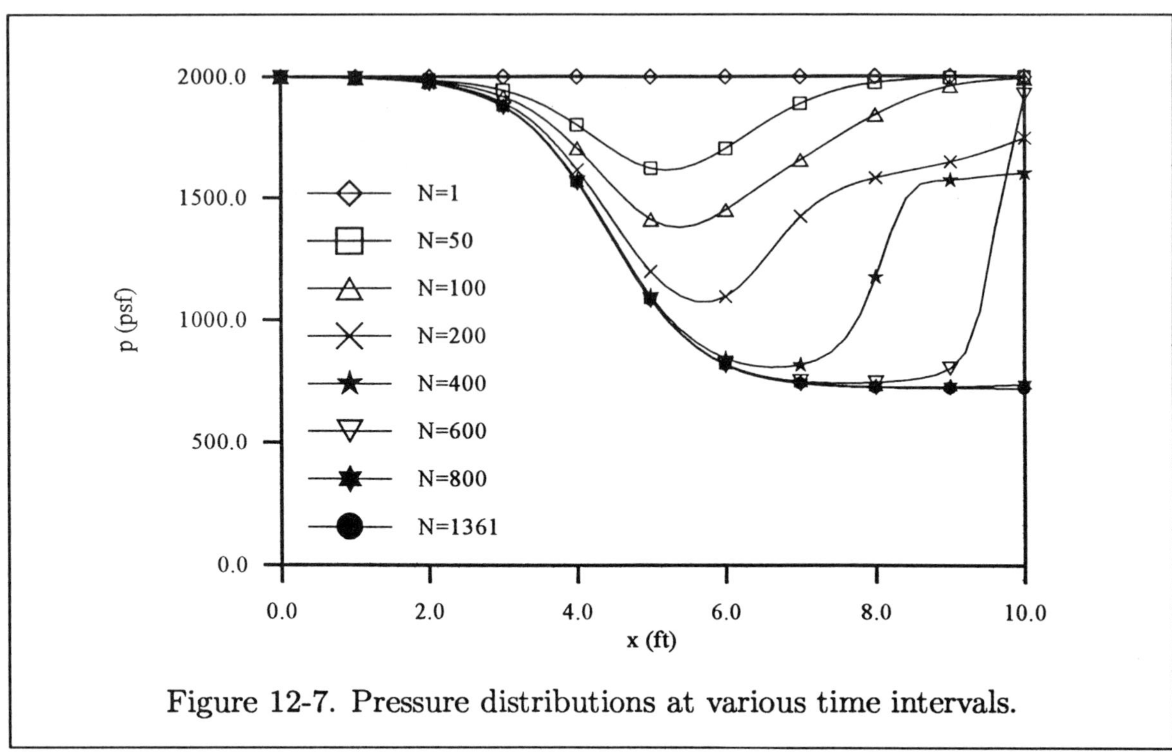

Figure 12-7. Pressure distributions at various time intervals.

To illustrate the effect of stepsizes or the CFL number on convergence, several time steps are used. The results are shown in Figure 12-8. Note that typically drastic changes occur during the first several iterations; thereafter, the solution gradually approaches the steady-state solution. Furthermore, note that log scale is used for the vertical axis which represents the changes in pressure which are used to check the convergence criterion. The numerical solutions by the first and second-order explicit Steger and Warming flux vector splitting schemes are summarized in Tables 12-3 and 12-4, where the spatial and temporal steps were 0.2 ft, and 0.00001 sec, respectively. The solution by the modified Runge-Kutta scheme is given in Table 12-5. Since, in this problem where the flow is supersonic within the domain and a large flow gradient within the domain does not exist, the solution does not encounter any oscillations. Therefore, the addition of a damping term is not necessary, and the solution given in Table 12-5 is obtained without any damping term. A solution by the Davis-Yee TVD scheme with limiter (12-74) is provided in Table 12-6.

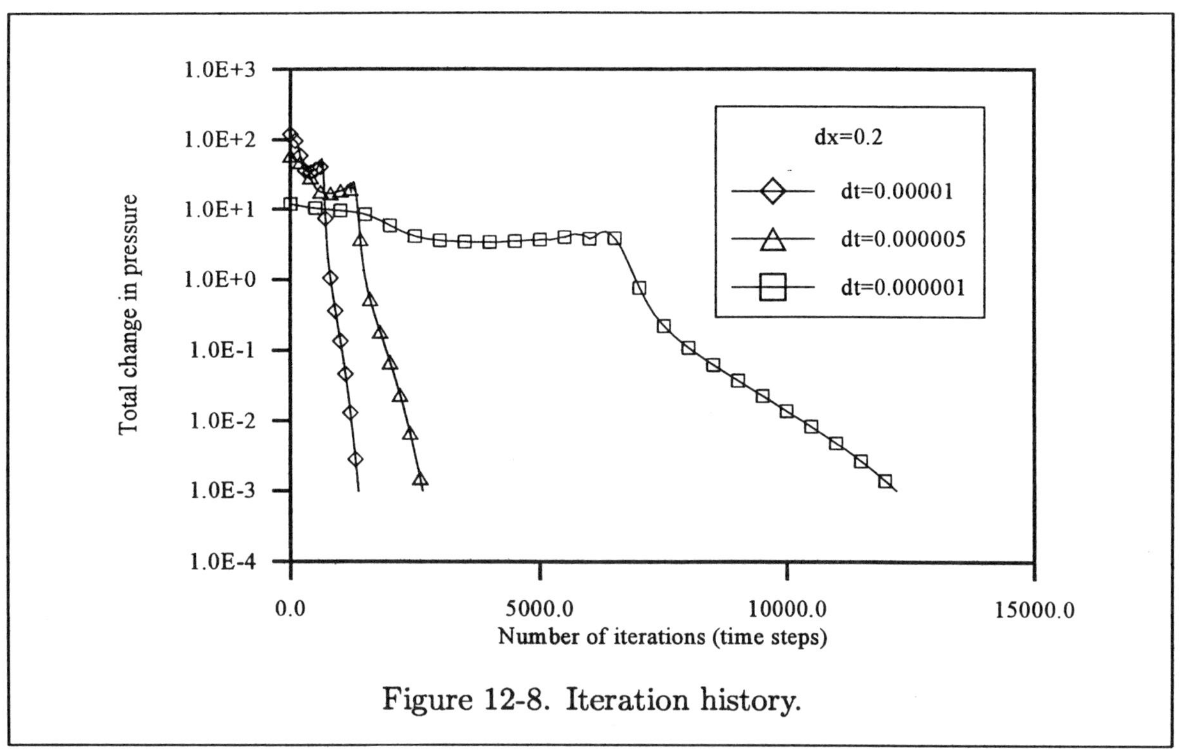

Figure 12-8. Iteration history.

x (ft)	M	p (lb$_f$/ft^2)	ρ (slugs/ft^3)	u (ft/s)
0.00	1.500	2000.00	0.002241	1676.55
0.40	1.500	1998.90	0.002240	1676.86
0.80	1.501	1996.82	0.002239	1677.44
1.20	1.503	1992.89	0.002236	1678.55
1.60	1.505	1985.52	0.002230	1680.64
2.00	1.510	1971.80	0.002219	1684.52
2.40	1.520	1946.65	0.002199	1691.66
2.80	1.537	1901.90	0.002164	1704.44
3.20	1.566	1826.07	0.002103	1726.29
3.60	1.614	1707.48	0.002007	1761.01
4.00	1.684	1542.75	0.001869	1810.52
4.40	1.776	1347.19	0.001698	1871.56
4.80	1.877	1153.87	0.001520	1934.91

x (ft)	M	p (lb$_f$/ft^2)	ρ (slugs/ft^3)	u (ft/s)
5.20	1.971	994.86	0.001366	1989.90
5.60	2.044	883.09	0.001253	2030.49
6.00	2.094	813.10	0.001179	2056.92
6.40	2.124	772.41	0.001136	2072.68
6.80	2.142	749.77	0.001111	2081.59
7.20	2.151	737.49	0.001098	2086.46
7.60	2.156	730.91	0.001090	2089.09
8.00	2.159	727.42	0.001086	2090.48
8.40	2.161	725.58	0.001084	2091.21
8.80	2.161	724.62	0.001083	2091.58
9.20	2.162	724.13	0.001083	2091.77
9.60	2.162	723.90	0.001083	2091.84
10.00	2.162	723.79	0.001082	2091.85

Table 12-3. Solutions by the explicit first-order SWFVS scheme $\Delta x = 0.2$ ft, $\Delta t = 0.00001$ sec.

x (ft)	M	p (lb_f/ft^2)	ρ ($slugs/ft^3$)	u (ft/s)
0.00	1.500	2000.00	0.002241	1676.55
0.40	1.500	1998.94	0.002241	1676.84
0.80	1.501	1997.01	0.002239	1677.36
1.20	1.502	1993.38	0.002236	1678.34
1.60	1.505	1986.57	0.002231	1680.17
2.00	1.509	1973.85	0.002220	1683.60
2.40	1.517	1950.49	0.002202	1689.93
2.80	1.532	1908.66	0.002168	1701.32
3.20	1.559	1837.14	0.002110	1721.02
3.60	1.602	1723.72	0.002016	1752.84
4.00	1.668	1563.24	0.001880	1799.30
4.40	1.756	1368.68	0.001710	1858.42
4.80	1.856	1172.27	0.001530	1922.16

x (ft)	M	p (lb_f/ft^2)	ρ ($slugs/ft^3$)	u (ft/s)
5.20	1.953	1007.74	0.001373	1979.84
5.60	2.032	890.38	0.001257	2024.23
6.00	2.088	816.07	0.001181	2054.19
6.40	2.124	772.52	0.001136	2072.57
6.80	2.145	748.16	0.001110	2083.16
7.20	2.156	734.90	0.001096	2089.04
7.60	2.163	727.78	0.001089	2092.22
8.00	2.166	723.99	0.001085	2093.93
8.40	2.168	721.98	0.001082	2094.83
8.80	2.169	720.92	0.001081	2095.31
9.20	2.169	720.36	0.001081	2095.56
9.60	2.170	720.07	0.001080	2095.69
10.00	2.170	719.90	0.001080	2095.77

Table 12-4. Solutions by the explicit second-order SWFVS scheme
$\Delta t = 0.00001$ sec, $\Delta x = 0.2$ ft.

x (ft)	M	p (lb_f/ft^2)	ρ ($slugs/ft^3$)	u (ft/s)
0.00	1.500	2000.00	0.002241	1676.55
0.40	1.500	1998.76	0.002240	1676.68
0.80	1.501	1997.54	0.002239	1677.53
1.20	1.502	1993.25	0.002236	1677.96
1.60	1.505	1987.19	0.002232	1680.30
2.00	1.508	1974.51	0.002220	1682.95
2.40	1.518	1951.51	0.002204	1689.70
2.80	1.530	1911.20	0.002168	1700.17
3.20	1.558	1839.63	0.002113	1719.59
3.60	1.599	1727.76	0.002017	1751.15
4.00	1.665	1566.70	0.001883	1797.09
4.40	1.754	1370.70	0.001712	1857.48
4.80	1.856	1171.01	0.001528	1922.81

x (ft)	M	p (lb_f/ft^2)	ρ ($slugs/ft^3$)	u (ft/s)
5.20	1.955	1006.99	0.001374	1980.53
5.60	2.040	885.13	0.001253	2028.80
6.00	2.082	821.14	0.001186	2049.51
6.40	2.143	759.78	0.001124	2084.55
6.80	2.122	764.64	0.001127	2067.95
7.20	2.190	712.44	0.001074	2110.56
7.60	2.126	752.49	0.001112	2069.06
8.00	2.206	698.66	0.001060	2118.56
8.40	2.134	745.10	0.001105	2072.91
8.80	2.197	702.68	0.001063	2113.18
9.20	2.152	731.97	0.001092	2084.37
9.60	2.176	716.03	0.001077	2099.58
10.00	2.176	715.66	0.001076	2100.01

Table 12-5. Solutions by the modified Runge-Kutta scheme
$\Delta t = 0.00001$ sec, $\Delta x = 0.2$ ft.

x (ft)	M	p (lb$_f$/ft^2)	ρ (slugs/ft^3)	u (ft/s)
0.00	1.500	2000.00	0.002241	1676.55
0.40	1.500	1999.10	0.002241	1676.78
0.80	1.501	1997.14	0.002239	1677.30
1.20	1.502	1993.43	0.002236	1678.27
1.60	1.505	1986.46	0.002230	1680.11
2.00	1.509	1973.46	0.002220	1683.54
2.40	1.517	1949.59	0.002201	1689.85
2.80	1.532	1906.94	0.002166	1701.22
3.20	1.559	1834.25	0.002107	1720.84
3.60	1.602	1719.50	0.002012	1752.51
4.00	1.667	1558.17	0.001875	1798.74
4.40	1.755	1363.94	0.001705	1857.60
4.80	1.858	1166.13	0.001524	1922.74

x (ft)	M	p (lb$_f$/ft^2)	ρ (slugs/ft^3)	u (ft/s)
5.20	1.958	999.26	0.001365	1982.10
5.60	2.037	883.40	0.001250	2026.38
6.00	2.091	811.81	0.001176	2055.37
6.40	2.125	770.53	0.001133	2072.78
6.80	2.144	747.62	0.001109	2082.72
7.20	2.154	735.30	0.001096	2088.11
7.60	2.160	728.96	0.001089	2090.78
8.00	2.163	725.68	0.001086	2092.12
8.40	2.164	723.70	0.001083	2093.06
8.80	2.166	722.49	0.001082	2093.71
9.20	2.166	722.04	0.001082	2093.88
9.60	2.166	722.01	0.001082	2093.78
10.00	2.165	722.36	0.001082	2093.33

Table 12-6. Solutions by the second-order Davis-Yee TVD scheme with limiter (12-74), $\Delta t = 0.00001$, $\Delta x = 0.2$ ft.

In the applications of various explicit schemes just completed, the selection of stepsizes are limited due to stability requirement of the schemes, namely, that the CFL numbers must be generally less than one. However, implicit schemes are typically less restrictive, and, therefore larger CFL numbers (and corresponding time steps) can be used. Now, consider the application of the implicit formulation given by (12-84). Several combinations of spatial and temporal stepsizes are used to illustrate the effect of CFL number on convergence.

The results are summarized in Figure 12-9 and in Table 12-7, where computation time is included as well. It is clear that as the spatial step Δx is decreased, the number of grid points in the domain is increased, which increases the number of iterations required for a converged solution. It is especially true for lower values of the CFL number. Notice also the reduction of CPU time (Table 12-7) as the iteration is decreased for each converged solution.

For this application where no shock waves appear within the domain, the nonlinearity effect is weak. The linear stability analysis indicates no restriction on the step sizes for the implicit scheme. Indeed, very large time steps corresponding to large CFL numbers may be used to provide a converged solution. However, for problems with shock waves which represent highly nonlinear phenomenon, that is no longer true. A domain with a normal shock will be investigated in Section 12.5.2.

To investigate the effect of the initial condition, consider the following data:

$$\left.\begin{aligned} u &= 1676.55 \text{ ft/sec} \\ \rho &= 0.002241 \text{ slugs/ft}^3 \\ e_t &= 3636204 \text{ ft}^2/\text{sec}^2 \end{aligned}\right\} \qquad 0 \leq x \leq 5$$

$$\left.\begin{aligned} u &= 1441.6 \text{ ft/sec} \\ \rho &= 0.001142 \text{ slugs/ft}^3 \\ e_t &= 1719557 \text{ ft}^2/\text{sec}^2 \end{aligned}\right\} \qquad 5 < x \leq 10$$

When this initial data set was used and the converged solutions were compared to the previous solutions, a negligible effect was indicated.

Δx	Δt	CFL	No. of Iterations	CPU (sec.)
0.1	1×10^{-4}	2.794	138	0.993
0.1	4×10^{-4}	11.177	43	0.314
0.1	1×10^{-3}	27.943	23	0.169
0.1	2×10^{-3}	55.885	15	0.111
0.2	1×10^{-4}	1.397	138	0.485
0.2	4×10^{-4}	5.589	43	0.154
0.2	1×10^{-3}	13.971	22	0.0802
0.2	2×10^{-3}	27.943	15	0.0561
0.2	3×10^{-3}	41.914	12	0.0458
0.4	1×10^{-4}	0.699	142	0.246
0.4	4×10^{-4}	2.794	45	0.0801
0.4	1×10^{-3}	6.986	23	0.0431
0.4	3×10^{-3}	20.957	12	0.0248
0.4	5×10^{-3}	34.928	9	0.0197

Table 12-7: The number of iterations and CPU time required to obtain a steady state solution.

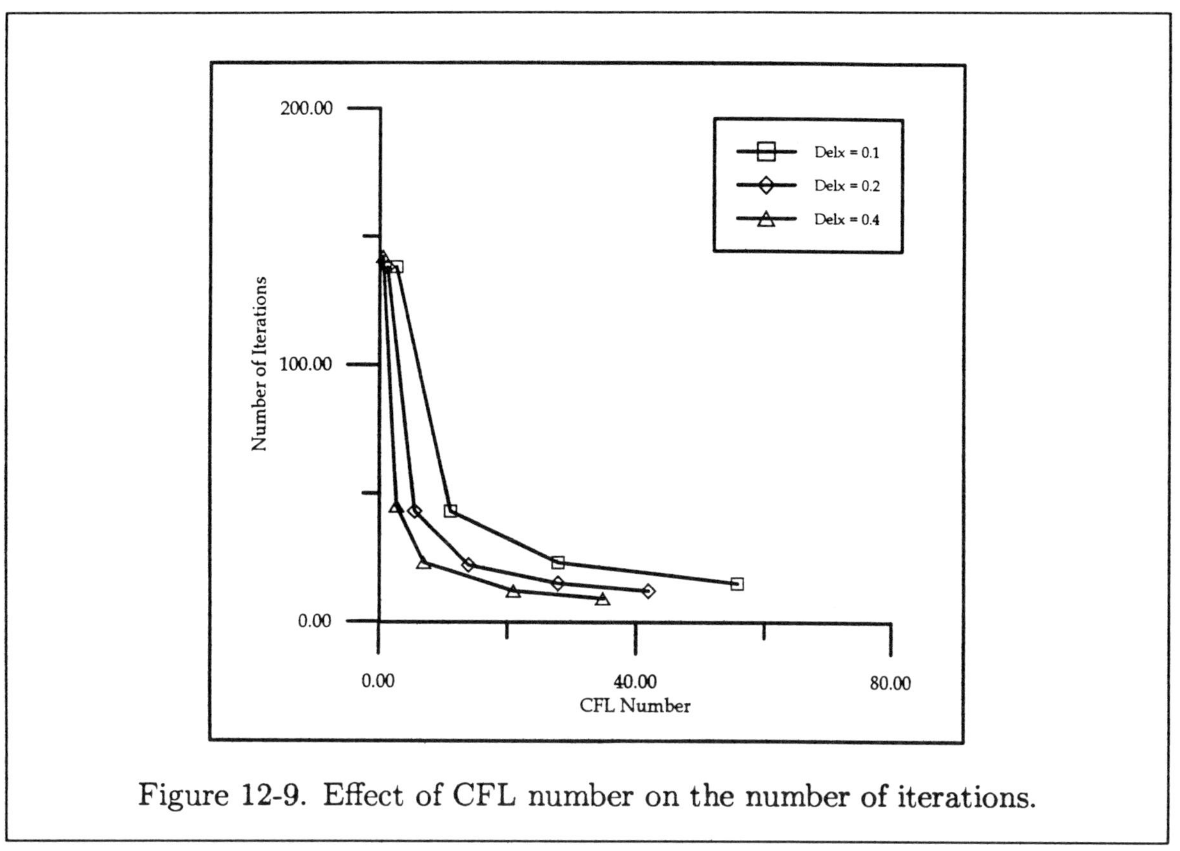

Figure 12-9. Effect of CFL number on the number of iterations.

12.5.2 Supersonic Inflow, Subsonic Outflow

The second part of the proposed problem includes a shock wave within the domain followed by a subsonic flow. Recall that when the flow is subsonic, one of the eigenvalues is negative which dictates that one of the characteristics is providing information into the domain. Thus, one analytical outflow boundary condition may be specified at the exit. The specification may be accomplished by defining ρ, u, or p at the exit. In the following results, the velocity u at the exit is specified such that $u_{IM} = 572.76$ ft/sec. This value of the exit velocity corresponds to a normal shock location of 5.0 ft within the domain.

12.5.2.1 Analytical Solution: The expression for the area-Mach number given by (12-94) can be used to determine the flow properties from the inlet to a location just ahead of the shock, that is, $x = 5.0$ ft. Subsequently, the normal shock relations are used to determine the flow properties after the shock wave. Now again relation (12-94) is used to compute the flow properties from the shock to the exit plane. Recall that relation (12-94) is valid for an isentropic flow and, therefore, it cannot be used across a normal shock wave. The analytical solution at increments of 1.0 ft is given in Table 12.8.

x (ft)	S (ft^2)	S/S*	M	p (lb$_f$/ft^2)	T (°R)	ρ (slugs/ft^3)	u (ft/s)
0.000	1.0512	1.1762	1.500	2000.00	520.00	0.002241	1676.55
1.000	1.0522	1.1772	1.502	1995.58	519.67	0.002238	1677.72
2.000	1.0567	1.1822	1.509	1974.24	518.08	0.002221	1683.42
3.000	1.0782	1.2063	1.543	1879.39	510.84	0.002144	1709.05
4.000	1.1676	1.3063	1.666	1565.39	484.92	0.001882	1798.09
5.000	1.3980	1.5641	1.907	1083.61	436.47	0.001447	1952.98
5.001	1.3983	1.1951	0.594	4417.84	704.27	0.003656	772.93
6.000	1.6284	1.3918	0.475	4808.12	721.51	0.003883	624.74
7.000	1.7178	1.4682	0.442	4904.72	725.62	0.003939	583.87
8.000	1.7393	1.4866	0.435	4925.16	726.48	0.003951	574.93
9.000	1.7438	1.4905	0.434	4929.28	726.66	0.003953	573.11
10.000	1.7448	1.4913	0.433	4930.07	726.69	0.003954	572.76

Table 12.8: Analytical solution with a normal shock wave just downstream of $x = 5$ ft.

12.5.2.2 Numerical Solutions: To start the solution, an initial set of data is required. The following set of data is used for this purpose:

$$u = 1676.55 \text{ ft/sec} \qquad x \leq 2.8$$

$$u = 572.76 \text{ ft/sec} \qquad x > 2.8$$

$$\left.\begin{aligned} p &= 2000.00 \text{ lb}_f/\text{ft}^2 \\ \rho &= 0.002241 \text{ slugs/ft}^3 \end{aligned}\right\} \quad 0.0 \leq x \leq 10.0$$

The solutions by the first-order and second-order explicit Steger and Warming flux vector splitting schemes are compared to the analytical solution in Figure 12-10.

The steady-state solutions are obtained after 8533 time steps by the first-order scheme and after 8877 time steps by the second-order scheme, where the relevant step sizes are $\Delta x = 0.1$ and $\Delta t = 0.00001$ sec.

Observe that, due to the dissipation error of the first-order scheme, shock smearing is much larger compared to the second-order scheme. On the other hand, due to the dispersion error associated with the second-order scheme, some oscillations before and after the shock (discontinuity) are observed. However, shock slope is more accurately predicted. The effect of spatial step on the solution is shown in Figures 12-11 and 12-12 where Δx of 0.2 and 0.4 are used, respectively. As expected, the errors associated with each scheme is increased as the step size Δx is increased from 0.1 to 0.4. The error distributions for the first-order and second-order Steger and Warming flux vector splitting schemes are shown in Figures 12-13 and 12-14, respectively. Note that maximum error occurs in the vicinity of sharp gradients. The convergence histories for the second-order scheme are shown in Figure 12-15. The convergence histories are also shown in Figure 12-16 for several temporal steps and a spatial step of $\Delta x = 0.2$. Note that as the step size is decreased, the accuracy of solution is increased. However, the computation time (indicated by increase in iteration number) is also increased. That is always going to be the case! Thus, a reasonable balance between accuracy and efficiency must be established by the user. The solutions by the explicit first-order and second-order Steger and Warming flux vector splitting are given in Tables 12-9 and 12-10, respectively. The step sizes used are $\Delta x = 0.2$ ft and $\Delta t = 0.00001$ sec. The converged solutions are obtained after 7701 and 8070 time steps (iterations), respectively, for the first-order and second-order schemes.

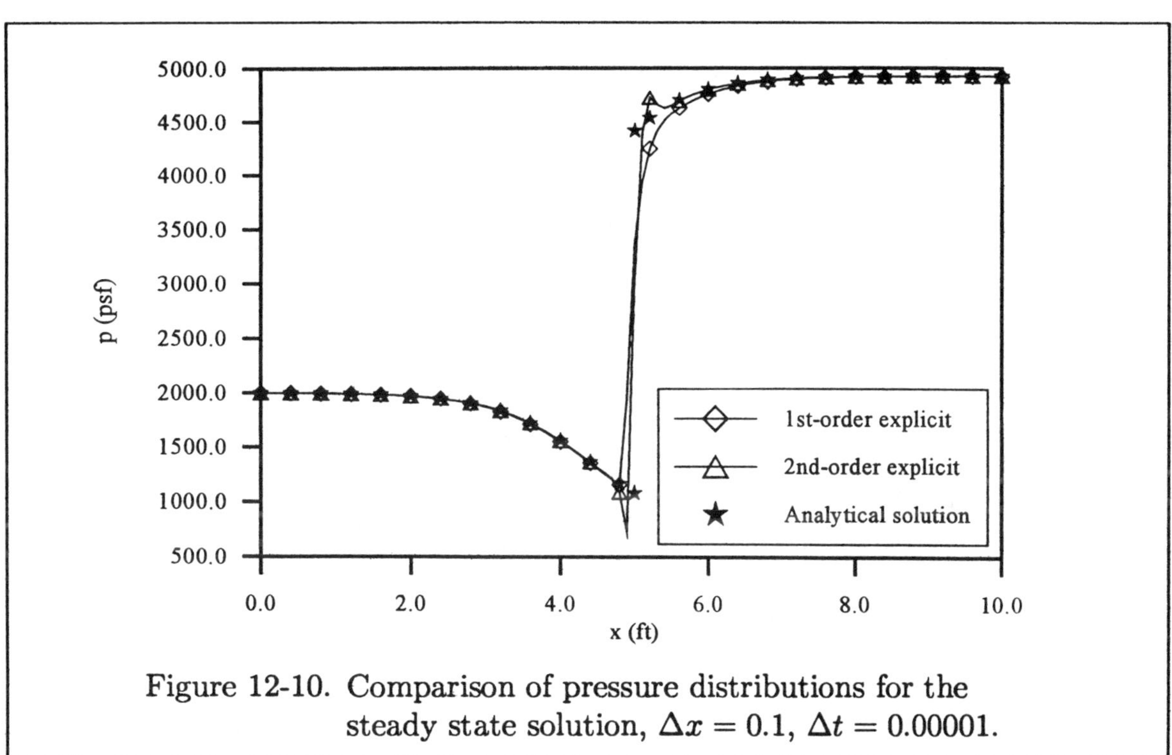

Figure 12-10. Comparison of pressure distributions for the steady state solution, $\Delta x = 0.1$, $\Delta t = 0.00001$.

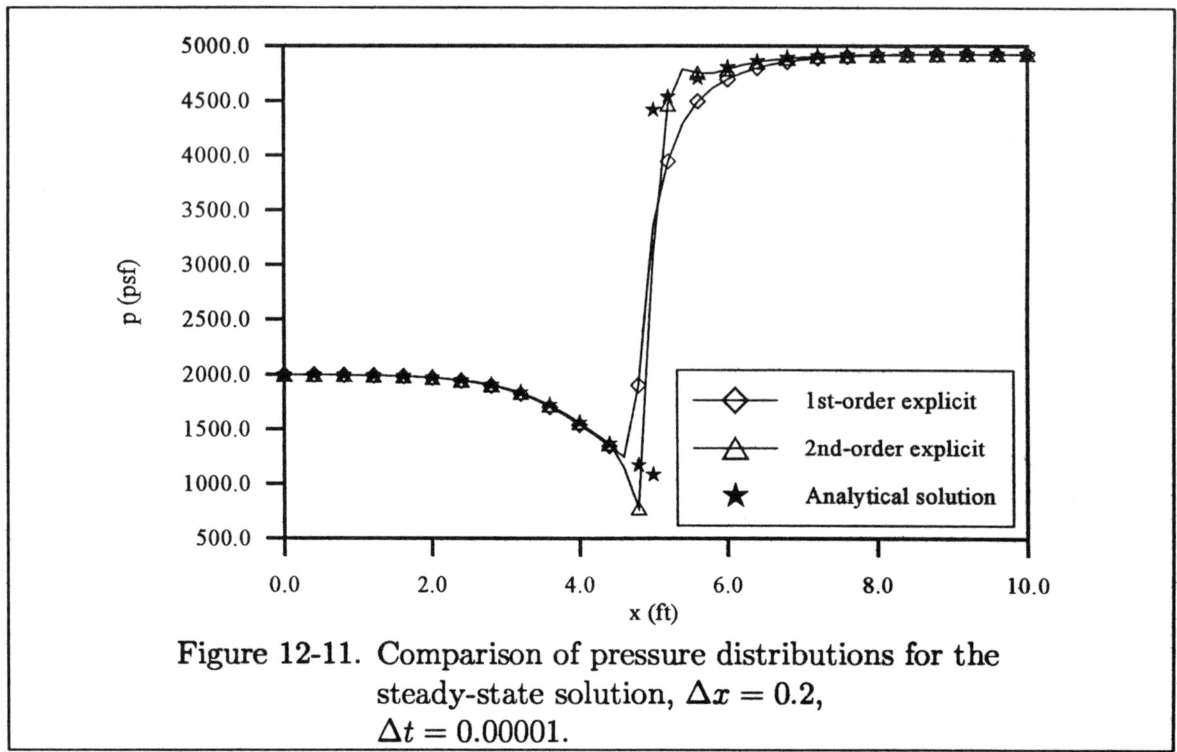

Figure 12-11. Comparison of pressure distributions for the steady-state solution, $\Delta x = 0.2$, $\Delta t = 0.00001$.

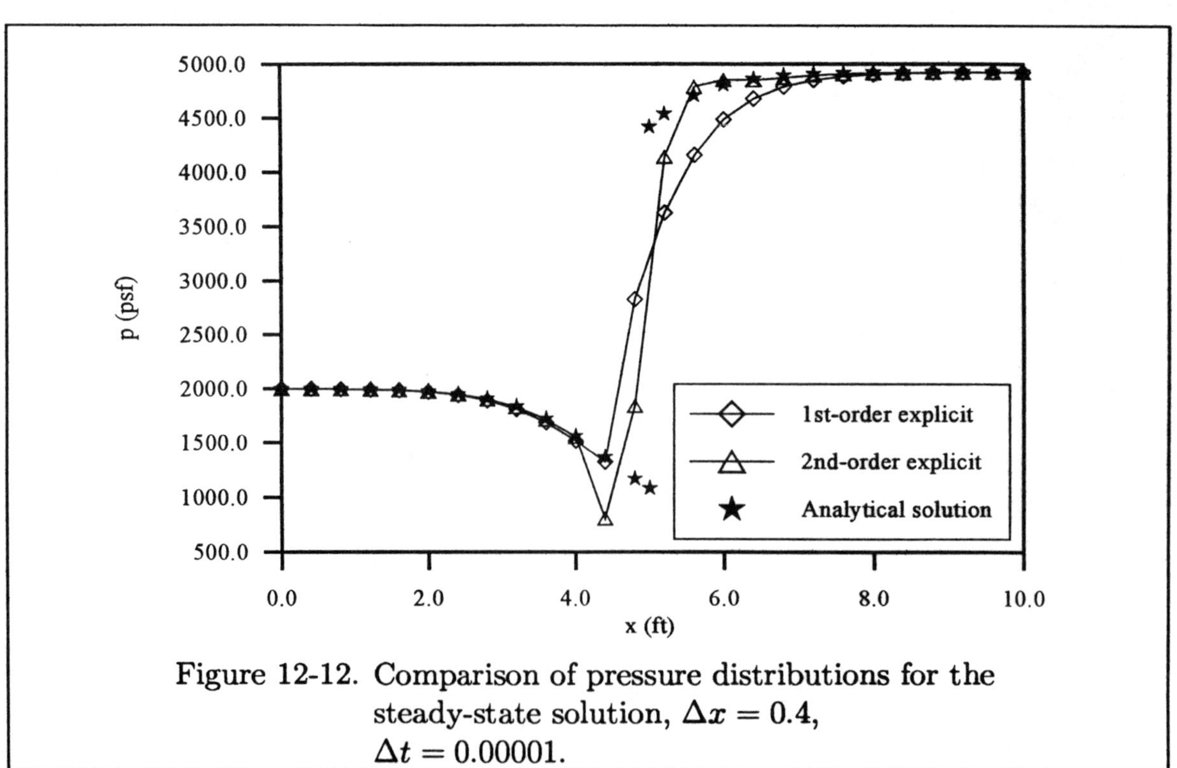

Figure 12-12. Comparison of pressure distributions for the steady-state solution, $\Delta x = 0.4$, $\Delta t = 0.00001$.

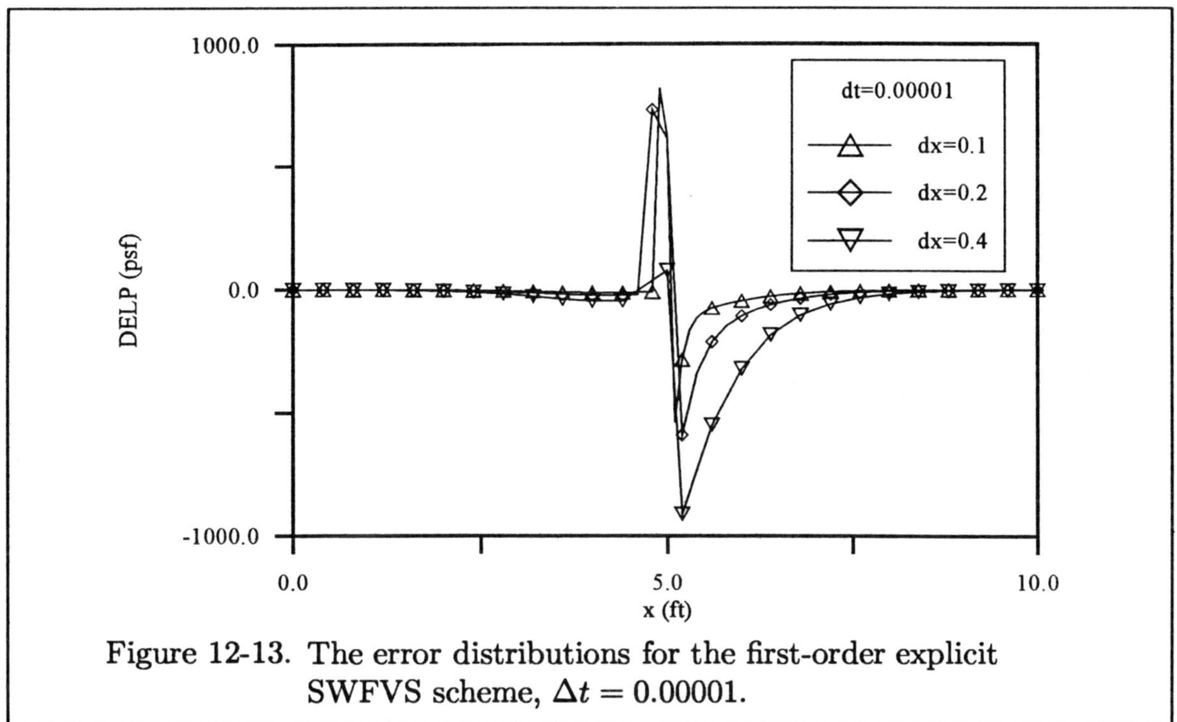

Figure 12-13. The error distributions for the first-order explicit SWFVS scheme, $\Delta t = 0.00001$.

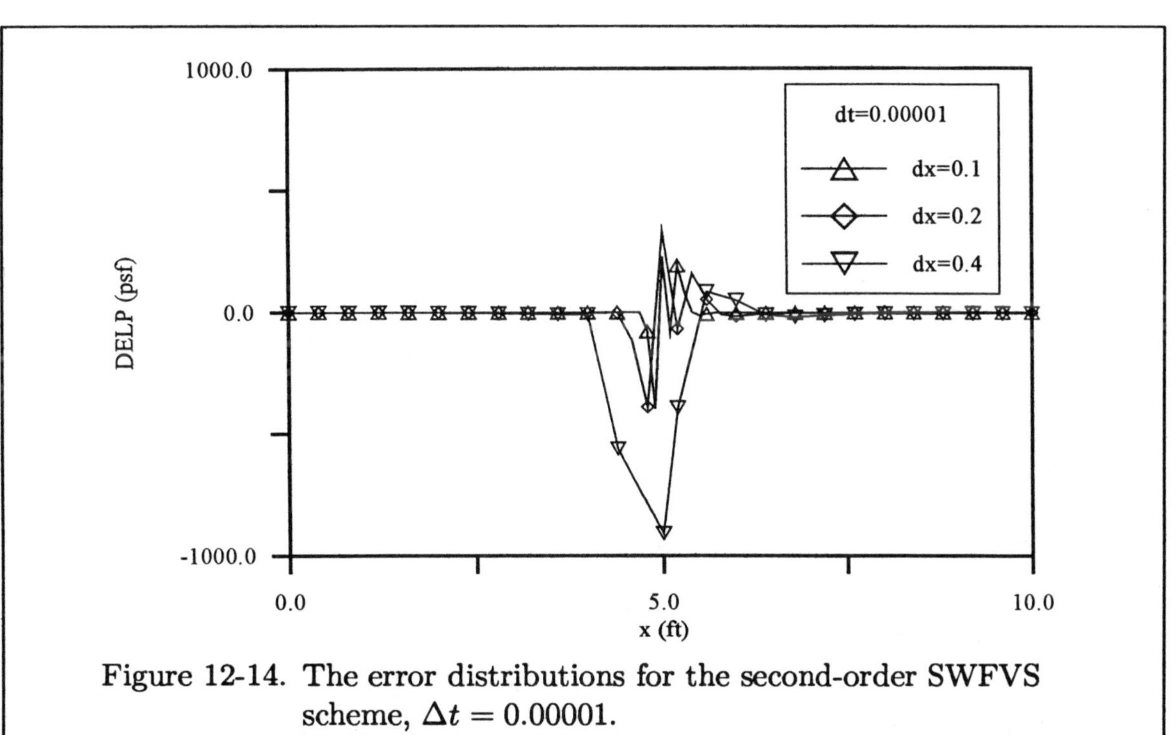

Figure 12-14. The error distributions for the second-order SWFVS scheme, $\Delta t = 0.00001$.

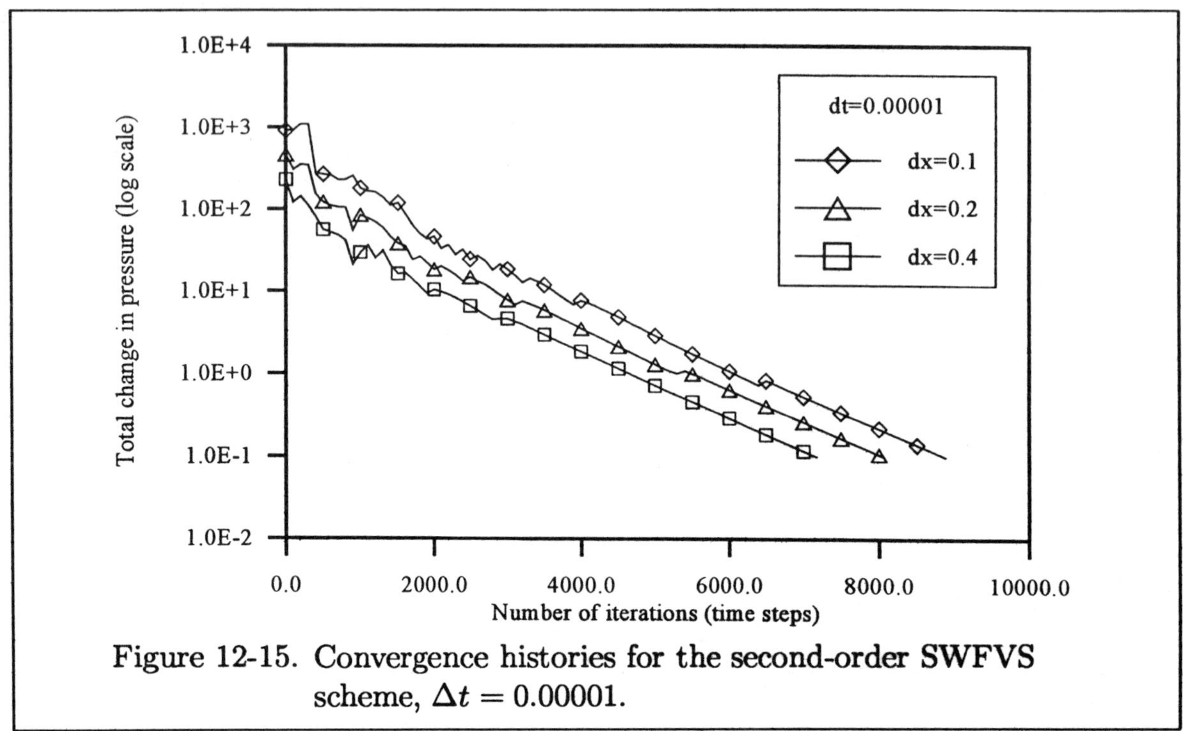

Figure 12-15. Convergence histories for the second-order SWFVS scheme, $\Delta t = 0.00001$.

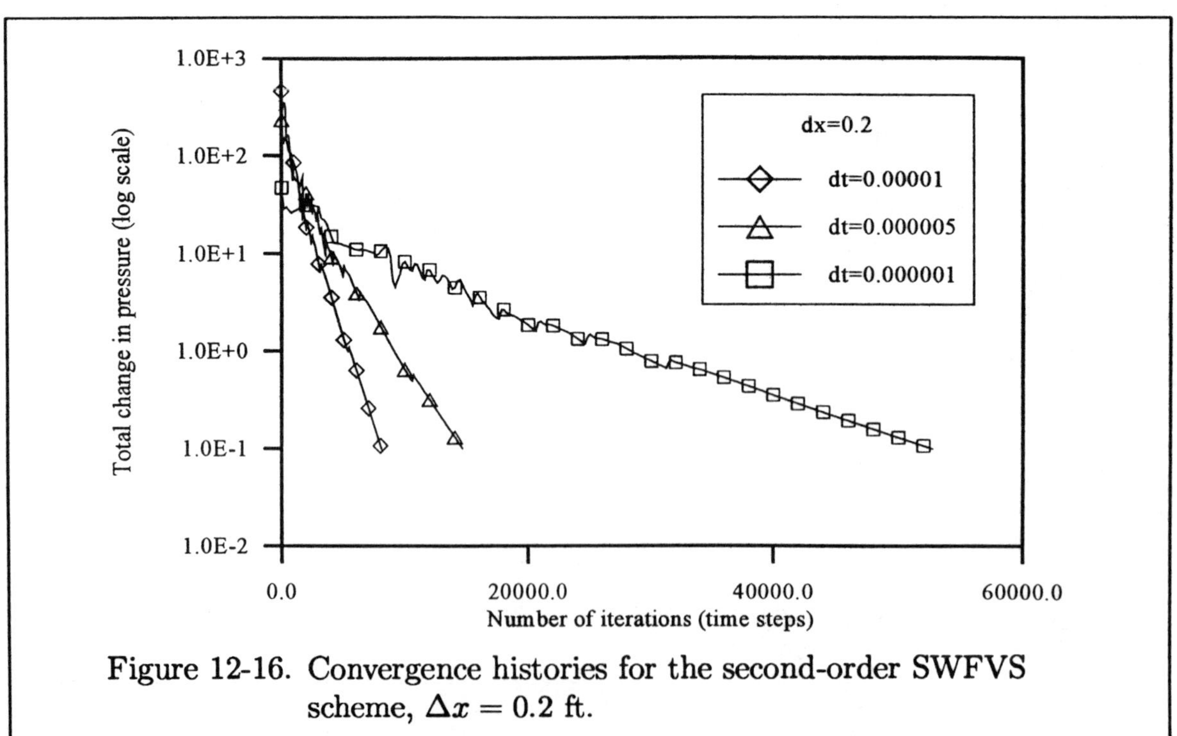

Figure 12-16. Convergence histories for the second-order SWFVS scheme, $\Delta x = 0.2$ ft.

x (ft)	M	p (lb_f/ft^2)	ρ ($slugs/ft^3$)	u (ft/s)	x (ft)	M	p (lb_f/ft^2)	ρ ($slugs/ft^3$)	u (ft/s)
0.00	1.500	2000.00	0.002241	1676.55	5.20	0.676	3948.35	0.003365	866.04
0.40	1.500	1998.90	0.002240	1676.86	5.60	0.547	4494.58	0.003695	713.93
0.80	1.501	1996.82	0.002239	1677.44	6.00	0.495	4699.86	0.003817	649.61
1.20	1.503	1992.89	0.002236	1678.55	6.40	0.467	4802.61	0.003877	615.12
1.60	1.505	1985.52	0.002230	1680.64	6.80	0.452	4859.62	0.003911	595.76
2.00	1.510	1971.80	0.002219	1684.52	7.20	0.443	4891.18	0.003929	585.07
2.40	1.520	1946.65	0.002199	1691.66	7.60	0.439	4908.36	0.003939	579.27
2.80	1.537	1901.90	0.002164	1704.44	8.00	0.436	4917.59	0.003945	576.16
3.20	1.566	1826.07	0.002103	1726.29	8.40	0.435	4922.52	0.003948	574.50
3.60	1.614	1707.48	0.002007	1761.01	8.80	0.434	4925.15	0.003949	573.62
4.00	1.684	1542.75	0.001869	1810.52	9.20	0.434	4926.55	0.003950	573.15
4.40	1.776	1347.19	0.001698	1871.56	9.60	0.434	4927.31	0.003950	572.89
4.80	1.359	1902.99	0.002062	1544.61	10.00	0.433	4927.78	0.003950	572.76

Table 12-9 . Solutions by the explicit first-order SWFVS scheme, $\Delta t = 0.00001$ sec, $\Delta x = 0.2$ ft.

x (ft)	M	p (lb_f/ft^2)	ρ ($slugs/ft^3$)	u (ft/s)	x (ft)	M	p (lb_f/ft^2)	ρ ($slugs/ft^3$)	u (ft/s)
0.00	1.500	2000.00	0.002241	1676.55	5.20	0.568	4470.37	0.003666	741.99
0.40	1.500	1998.94	0.002241	1676.84	5.60	0.498	4759.42	0.003856	654.34
0.80	1.501	1997.01	0.002239	1677.36	6.00	0.476	4790.55	0.003870	627.11
1.20	1.502	1993.38	0.002236	1678.34	6.40	0.456	4856.59	0.003909	601.75
1.60	1.505	1986.57	0.002231	1680.17	6.80	0.446	4889.20	0.003928	588.37
2.00	1.509	1973.85	0.002220	1683.60	7.20	0.440	4907.01	0.003939	581.08
2.40	1.517	1950.49	0.002202	1689.93	7.60	0.437	4916.68	0.003945	577.16
2.80	1.532	1908.66	0.002168	1701.32	8.00	0.435	4921.84	0.003948	575.07
3.20	1.559	1837.14	0.002110	1721.02	8.40	0.434	4924.60	0.003949	573.96
3.60	1.602	1723.72	0.002016	1752.84	8.80	0.434	4926.07	0.003950	573.37
4.00	1.668	1563.24	0.001880	1799.30	9.20	0.434	4926.86	0.003950	573.05
4.40	1.756	1368.68	0.001710	1858.42	9.60	0.434	4927.29	0.003950	572.87
4.80	2.375	780.440	0.001277	2196.91	10.00	0.433	4927.56	0.003951	572.76

Table 12-10 . Solutions by the explicit second-order SWFVS scheme, $\Delta t = 0.00001$ sec, $\Delta x = 0.2$ ft.

The solutions by second-order Harten-Yee upwind TVD scheme with limiters (12-61) through (12-65) are shown in Figure 12-17 where the Mach number distributions are compared to the analytical results. The solution by limiter (12-64) is also provided in Table 12-11.

x (ft)	M	p (lb$_f$/ft^2)	ρ (slugs/ft^3)	u (ft/s)
0.00	1.500	2000.00	0.002241	1676.55
0.40	1.500	1999.19	0.002241	1676.77
0.80	1.501	1997.31	0.002239	1677.27
1.20	1.502	1993.75	0.002236	1678.23
1.60	1.504	1987.06	0.002231	1680.02
2.00	1.509	1974.56	0.002221	1683.38
2.40	1.517	1951.59	0.002202	1689.59
2.80	1.532	1910.43	0.002169	1700.78
3.20	1.557	1839.92	0.002111	1720.20
3.60	1.600	1727.86	0.002019	1751.77
4.00	1.665	1568.95	0.001884	1798.20
4.40	1.753	1376.89	0.001717	1857.19
4.80	1.842	1198.66	0.001559	1911.41
5.20	0.560	4547.50	0.003731	731.19
5.60	0.508	4714.24	0.003828	666.49
6.00	0.475	4813.76	0.003885	625.60
6.40	0.456	4867.82	0.003917	601.70
6.80	0.446	4896.37	0.003935	588.32
7.20	0.440	4911.30	0.003943	581.03
7.60	0.437	4919.06	0.003946	577.13
8.00	0.435	4923.20	0.003947	575.05
8.40	0.434	4925.47	0.003948	573.94
8.80	0.434	4926.68	0.003950	573.34
9.20	0.434	4927.32	0.003952	573.02
9.60	0.434	4927.70	0.003952	572.84
10.00	0.433	4927.86	0.003950	572.76

Table 12-11 . Solutions by the second-order Harten-Yee upwind TVD scheme, $\Delta t = 0.00001$ sec, $\Delta x = 0.2$ ft.

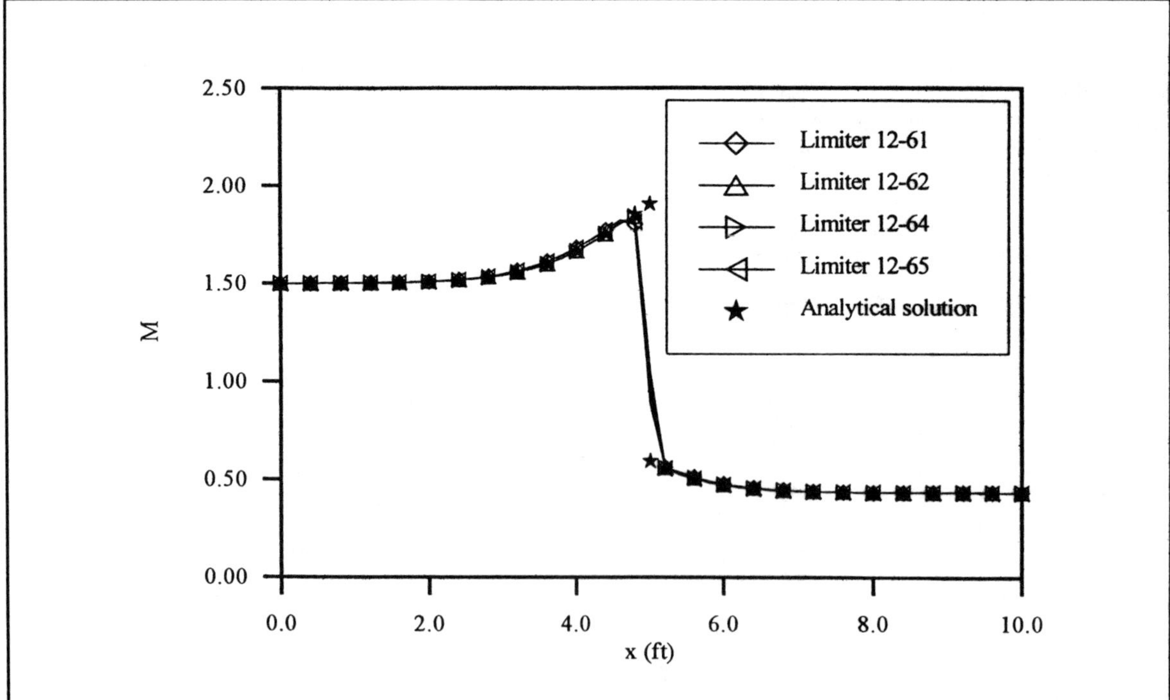

Figure 12-17 . Comparison of Mach number distributions by the second-order Harten-Yee upwind TVD scheme, $\Delta t = 0.00001$ sec, $\Delta x = 0.2$ ft.

x (ft)	M	p (lb_f/ft^2)	ρ (slugs/ft3)	u (ft/s)
0.00	1.500	2000.00	0.002241	1676.55
0.40	1.500	1999.21	0.002241	1676.75
0.80	1.501	1997.35	0.002239	1677.24
1.20	1.502	1993.81	0.002236	1678.17
1.60	1.504	1987.16	0.002231	1679.91
2.00	1.509	1974.74	0.002221	1683.18
2.40	1.517	1951.84	0.002203	1689.24
2.80	1.531	1910.61	0.002169	1700.22
3.20	1.557	1839.88	0.002111	1719.28
3.60	1.600	1724.65	0.002016	1751.41
4.00	1.662	1570.23	0.001885	1794.14
4.40	1.775	1337.77	0.001681	1873.37
4.80	1.832	1199.80	0.001560	1900.67

x (ft)	M	p (lb_f/ft^2)	ρ ($slugs/ft^3$)	u (ft/s)
5.20	0.558	4535.17	0.003724	728.53
5.60	0.506	4708.34	0.003826	664.02
6.00	0.474	4806.73	0.003882	624.43
6.40	0.456	4864.35	0.003915	600.78
6.80	0.445	4894.13	0.003932	587.84
7.20	0.440	4910.19	0.003941	580.76
7.60	0.437	4918.64	0.003946	576.98
8.00	0.435	4923.12	0.003948	574.95
8.40	0.434	4925.50	0.003950	573.88
8.80	0.434	4926.76	0.003950	573.30
9.20	0.434	4927.45	0.003951	573.00
9.60	0.433	4927.78	0.003951	572.82
10.00	0.433	4928.01	0.003951	572.76

Table 12-12 . Solutions by the second-order Davis-Yee symmetric TVD scheme, $\Delta t = 0.00001$ sec, $\Delta x = 0.2$ ft.

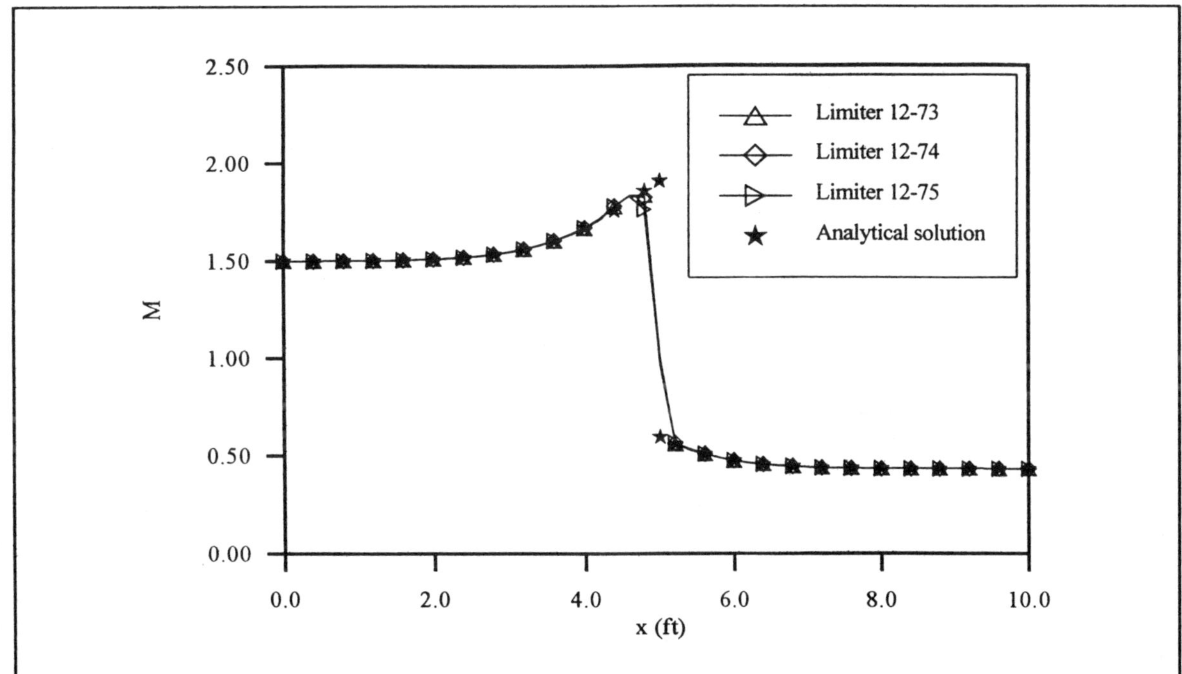

Figure 12-18 . Comparison of Mach number distributions by the second-order Davis-Yee symmetric TVD scheme, $\Delta t = 0.00001$ sec, $\Delta x = 0.2$ ft.

The solutions by the second-order Davis-Yee symmetric TVD scheme are shown in Figure 12-18. Figure 12-18 illustrates the Mach number distributions computed by limiters (12-73) through (12-75). The solution by limiter (12-73) is also provided in Table 12-12.

x (ft)	M	p (lb$_f$/ft^2)	ρ (slugs/ft^3)	u (ft/s)
0.00	1.500	2000.00	0.002241	1676.55
0.40	1.500	1998.82	0.002240	1676.86
0.80	1.501	1996.63	0.002239	1677.45
1.20	1.503	1992.41	0.002235	1678.55
1.60	1.505	1984.71	0.002229	1680.64
2.00	1.510	1969.97	0.002217	1684.52
2.40	1.520	1943.76	0.002196	1691.66
2.80	1.536	1896.98	0.002158	1704.41
3.20	1.566	1817.47	0.002093	1726.17
3.60	1.611	1700.79	0.001996	1759.49
4.00	1.683	1522.19	0.001843	1809.71
4.40	1.753	1408.76	0.001748	1861.65
4.80	1.863	1157.86	0.001531	1917.62
5.20	0.546	4578.15	0.003751	713.73
5.60	0.513	4647.04	0.003790	671.78
6.00	0.477	4773.78	0.003864	627.94
6.40	0.457	4843.96	0.003904	602.96
6.80	0.446	4882.76	0.003926	589.01
7.20	0.440	4903.82	0.003937	581.41
7.60	0.437	4915.08	0.003944	577.33
8.00	0.435	4921.08	0.003947	575.15
8.40	0.434	4924.27	0.003949	573.99
8.80	0.434	4925.97	0.003950	573.37
9.20	0.434	4926.88	0.003950	573.04
9.60	0.434	4927.46	0.003950	572.88
10.00	0.433	4927.66	0.003950	572.76

Table 12-13 . Solutions by the second-order Roe Sweby upwind TVD scheme, $\Delta t = 0.00001$ sec, $\Delta x = 0.2$ ft.

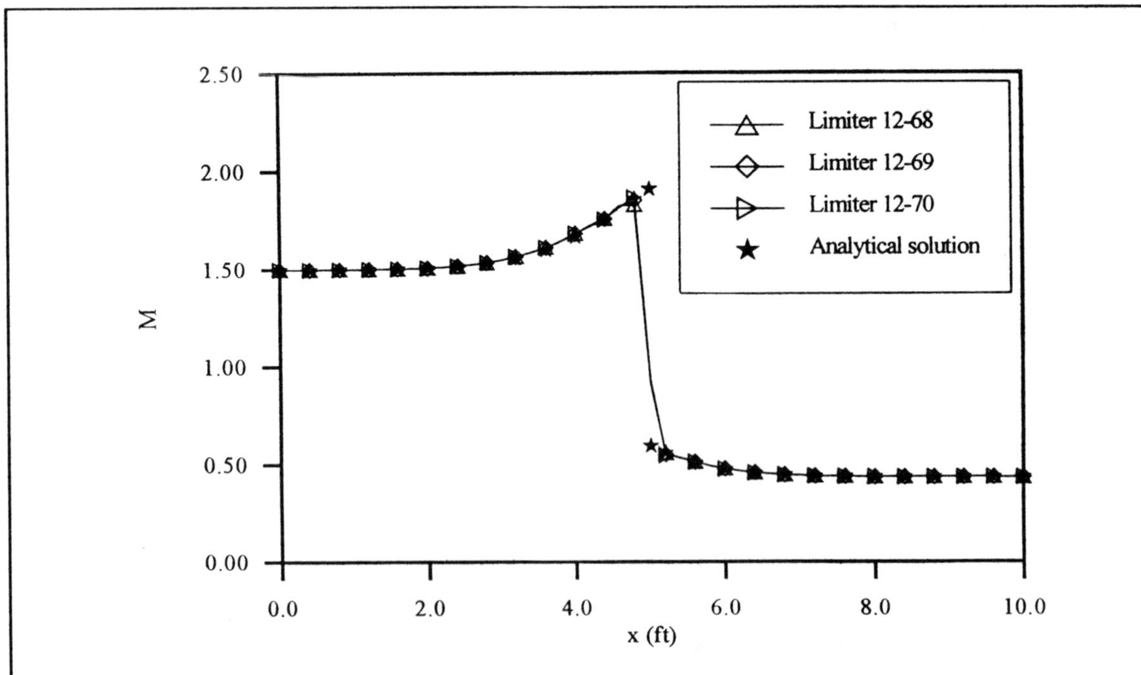

Figure 12-19 . Comparison of Mach number distributions by the second-order Roe Sweby upwind TVD scheme, $\Delta t = 0.00001$ sec, $\Delta x = 0.2$ ft.

The solutions by the Roe-Sweby upwind TVD scheme are shown in Figure 12-19 for limiters (12-68) through (12-70). The solution with limiter (12-70) is also presented in Table 12-13.

Similar to the case of supersonic flow, the implicit formulation of (12-84) is used to investigate the effect of step sizes (CFL number) on the convergence. As in the previous case, the number of iterations required for a converged solution is decreased as the CFL number is increased. However, note that for this case where a normal shock appears within the domain, the number of iterations is larger than for the fully supersonic flow (i.e., no shock wave in the domain). For the CFL number of 2.8 when $\Delta x = 0.1$ and $\Delta t = 0.0001$, this factor is about 9.0. The effect of the spatial step size is shown in Figure 12-20, where the CFL number and the number of iterations for a converged solution are related. When Figure 12-20 is compared with Figure 12-9, it is obvious that the maximum allowable CFL number has decreased for domains with shock waves. This point was mentioned earlier, whereby the high nonlinearity effects due to shock waves were identified as the cause of instability for larger values of the CFL number.

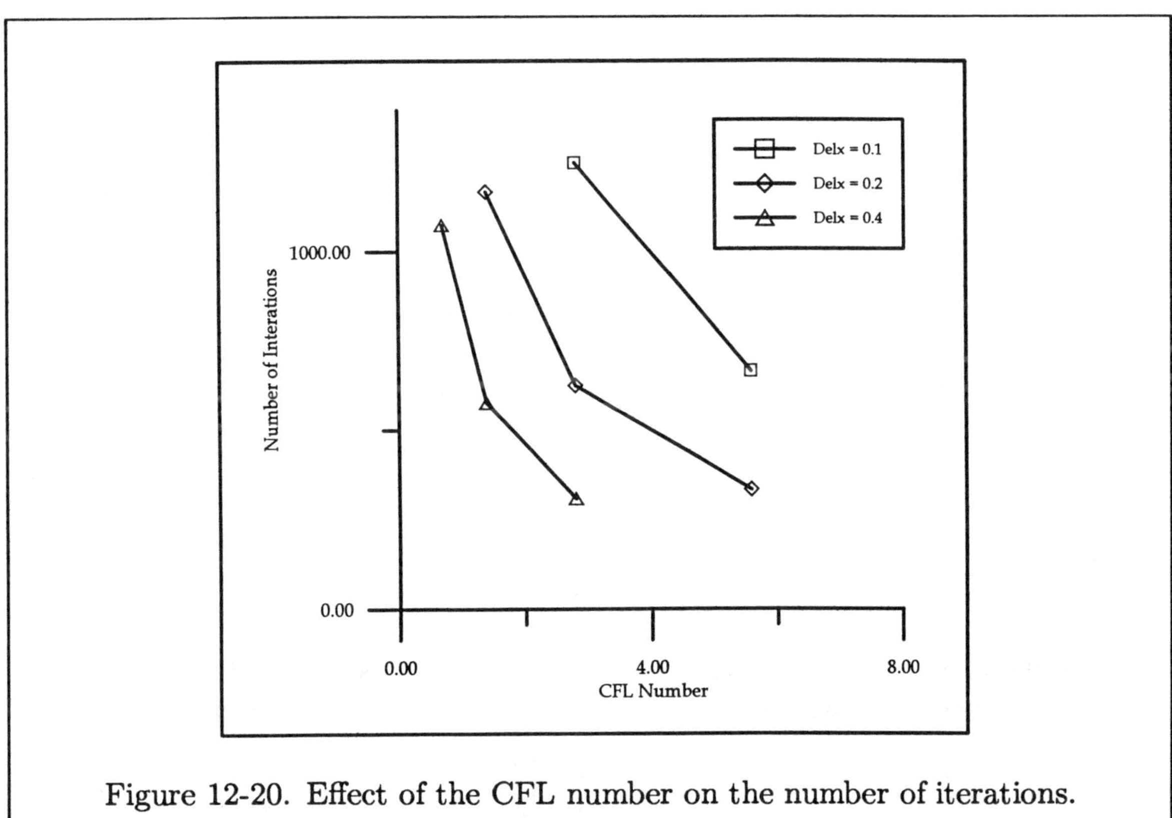

Figure 12-20. Effect of the CFL number on the number of iterations.

12.6 Grid Clustering

In the applications just investigated, equal grid spacing was employed. Shock smearing due to a relatively large spatial grid was clearly illustrated for first-order schemes. To decrease the shock smearing, a finer grid system is required. An efficient scheme to obtain a better flow resolution in the vicinity of high gradients is to use grid clustering. For this purpose, the following coordinate transformation is considered

$$\begin{aligned} \tau &= t \\ \xi &= \xi(x,t) \end{aligned}$$

from which

$$\frac{\partial}{\partial t} = \frac{\partial}{\partial \tau}\frac{\partial \tau}{\partial t} = \frac{\partial}{\partial \tau}$$

and

$$\frac{\partial}{\partial x} = \frac{\partial}{\partial \xi}\frac{\partial \xi}{\partial x} + \frac{\partial}{\partial t}\frac{\partial t}{\partial x} = \xi_x \frac{\partial}{\partial \xi}$$

For the one-dimensional problem, the Jacobian of transformation is defined as

$$J = \frac{1}{x_\xi}$$

Note that for a fixed grid system $\xi_t = 0$ and, therefore, $\xi_x = J$. Now the governing quasi one-dimensional equation is transformed to provide

$$\frac{\partial}{\partial \tau}(SQ) + \xi_x \frac{\partial E}{\partial \xi} - H = 0 \tag{12-98}$$

Once this equation is divided by J and rearranged, the following equation results

$$\frac{\partial}{\partial \tau}(S\bar{Q}) + \frac{\partial \bar{E}}{\partial \xi} - \bar{H} = 0 \tag{12-99}$$

where

$$\bar{Q} = \frac{Q}{J} \ , \quad \bar{E} = E \ , \quad \text{and } \bar{H} = \frac{H}{J}$$

The linearization and numerical scheme is similar to that of the previous problem. Grid point clustering was produced by utilizing the following function

$$x_i = \frac{L}{2}\left\{1 - \beta\left[\left(\frac{\beta+1}{\beta-1}\right)^{1-\xi_i} - 1\right] \Big/ \left[\left(\frac{\beta+1}{\beta-1}\right)^{1-\xi_i} + 1\right]\right\} \tag{12-100}$$

where L is the length of the domain and β is the clustering parameter. The spatial locations of grid points from the entrance to the shock location at $x = 5.0$ ft is determined by Equation (12-100) and subsequently reflected to specify grid points from the shock to the exit plane.

The converged solutions, obtained from Equation (12-99) utilizing the gridpoint clustering relation of (12-100), are compared to the solutions obtained by a constant Δx of 0.2 in Figures 12-21 and 12-22. The clustering parameter β was set to 1.05. The solutions shown in Figure 12-21 are obtained with the explicit first-order SWFVS scheme, whereas the solutions shown in Figure 12-22 are obtained by the explicit second-order SWFVS scheme. By inspection of Figures 12-21 and 12-22, it is evident that the shock smearing has been reduced substantially. However, note that dispersion error indicated by oscillations in the vicinity of shock wave remains, and, in fact, it had increased behind the shockwave. The addition of a fourth-order damping term or TVD will reduce these oscillations.

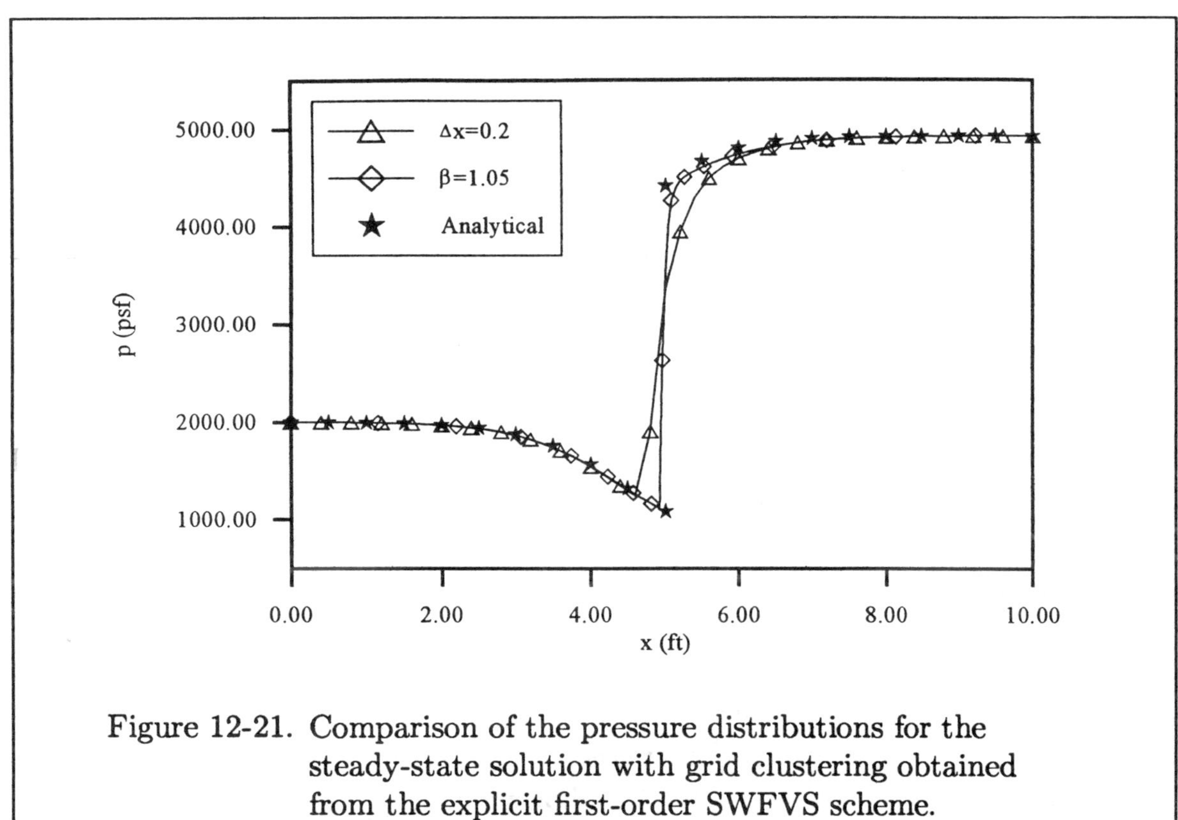

Figure 12-21. Comparison of the pressure distributions for the steady-state solution with grid clustering obtained from the explicit first-order SWFVS scheme.

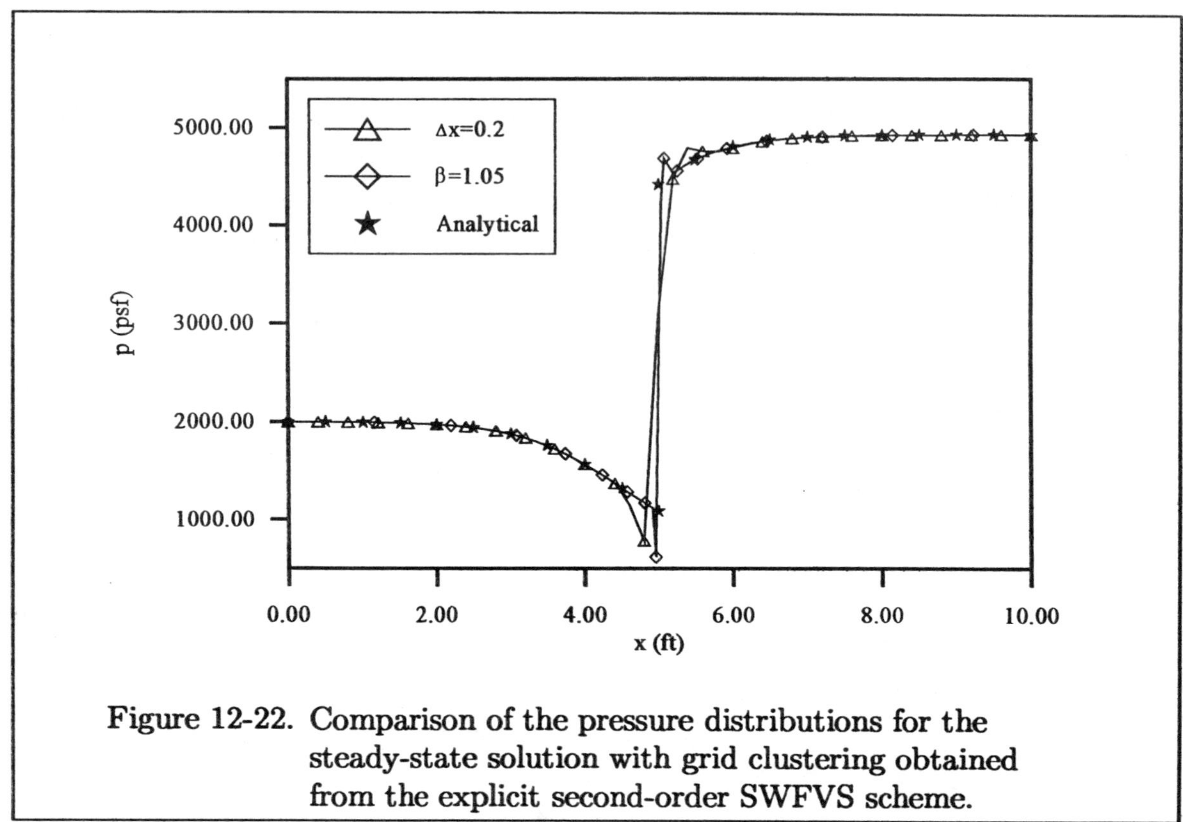

Figure 12-22. Comparison of the pressure distributions for the steady-state solution with grid clustering obtained from the explicit second-order SWFVS scheme.

12.7 Global Time Step and Local Time Step

The solutions in the previous example were obtained by a specified temporal step. A few points, with regard to the problem and the solution schemes, are iterated at this time. Recall that a steady-state solution is to be computed for the given variable area nozzle. The solution was initiated with a specified but arbitrary initial condition. The solution history followed a path to reach a steady-state (or converged) solution. The solution path is, of course, physically meaningless, as discussed previously. Now, if one could introduce a technique by which convergence is accelerated, it would certainly increase the efficiency of the numerical scheme. With these points in mind, consider the following concept.

First, note that an equivalent way of specifying a constant temporal step is to specify the CFL number. For example, one may specify CFL = 0.15, from which a time step is determined according to

$$\Delta t = (CFL)\frac{\Delta x}{(u+a)_{\max}} \tag{12-101}$$

Now one can use this value of Δt and proceed with the solution until a converged solution has been achieved. Obviously this is the general approach that was followed in the previous applications; that is, the solutions were obtained with a constant time step within the domain. Call this constant time step as the global time step, Δt_g. Now consider relation (12-101) and scrutinize it carefully, keeping in mind the discussion just completed about the path to a converged solution. Since the value of $(u + a)$ varies within the domain, the corresponding values of Δt would vary within the domain as well; that is, Δt would have different values at different locations. Call this variable Δt as the local time step designated by Δt_ℓ.

At this point, consider the following concept posed by the question: What if one uses local time steps to advance the solution? This would allow the solution at each point to advance through a local time which may be greater than Δt_g. Therefore, the solution moves faster toward convergence in regions where Δt_ℓ is larger. This has an overall effect of speeding the convergence. This concept is known as "local time stepping." The effect of local time stepping is more dominant in situations when gridpoint spacing varies from location to location, that is, when gridpoint clustering is used. In that situation, the time step (for the one-dimensional problem) is determined from

$$\Delta t_\ell = (CFL) \frac{\Delta \xi}{[\xi_x(u + a)]} \tag{12-102}$$

Note that now, in regions where the grid is coarse, Δt_ℓ would be greater than the Δt_ℓ in regions where the grid is fine for comparable velocities.

The example of Section 12.5 is used to illustrate the effect of local time stepping on convergence. The numerical scheme employed is the explicit first-order Steger and Warming flux vector splitting. The computations will be considered for a specified equal grid spacing as well as with grid point clustering. For the first investigation, an equal grid spacing of 0.2 ft and a CFL number of 0.15 is used. The initial and boundary conditions are the same as those specified in Section 12.5. Two solutions are obtained, one with a global time step and one with a local time step.

Flow	Global time step	Local time step
Fully supersonic	1379	1317
Subsonic/supersonic	7650	5692

Table 12-14: Number of iterations (time steps) required for the converged solution.

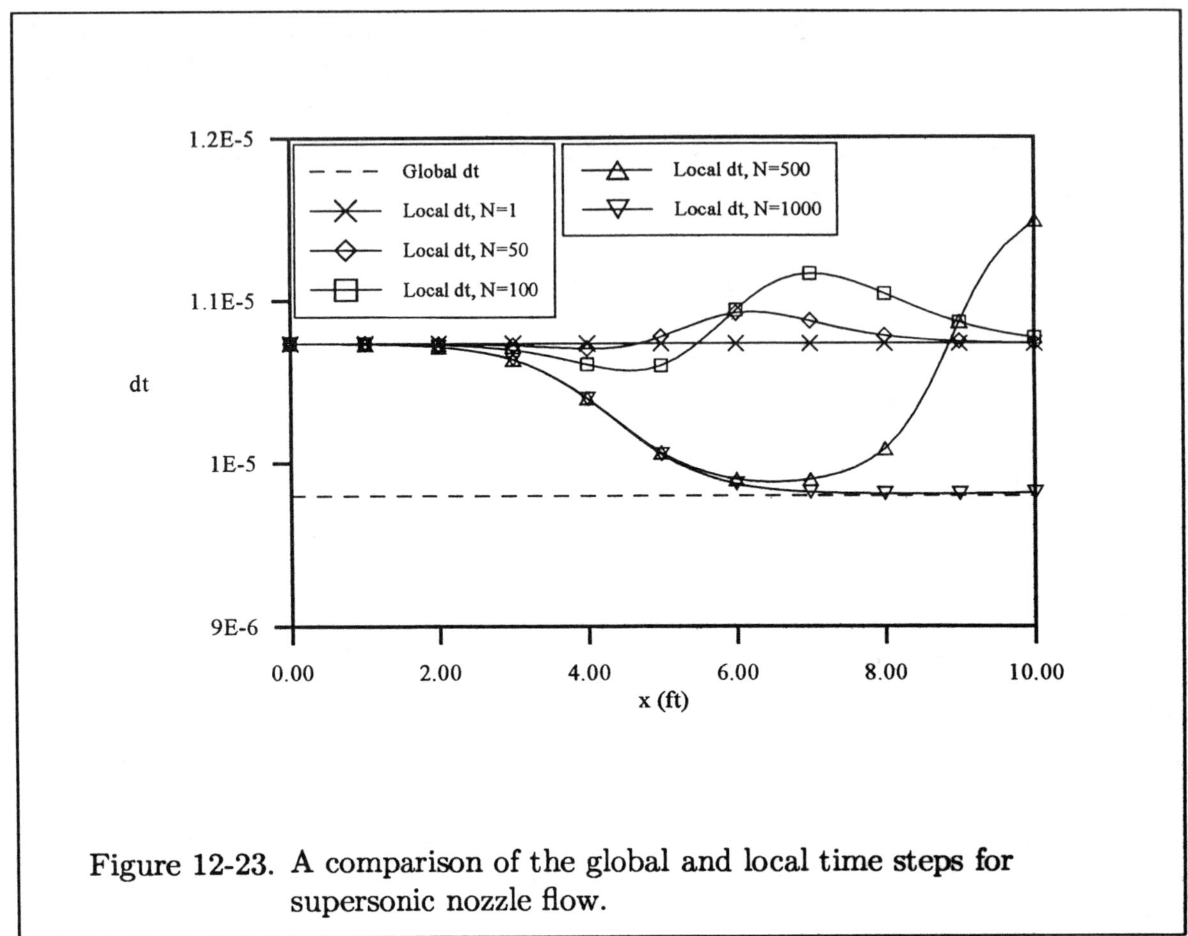

Figure 12-23. A comparison of the global and local time steps for supersonic nozzle flow.

A comparison of the global and local time steps at several iteration levels are shown in Figure 12-23 for the fully supersonic flow of Section 12.5.1 and in Figure 12-24 for the supersonic/subsonic flow of Section 12.5.2. The number of iterations (time steps) required for converged (steady-state) solution is shown in Table 12-14 for each case. The convergence criterion is set at 0.1 based on the total variation in the pressure, as before. The convergence histories for fully supersonic and supersonic/subsonic cases are shown in Figures 12-25 and 12-26, respectively. Similar computations are performed for the second investigation where grid point clustering is used. The results will be illustrated for the supersonic/subsonic problem of Section 12.5.2 with a grid clustering function of Section 12.6 and with $\beta = 1.05$. The global and local time steps at several iteration levels are shown in Figure 12-27. Observe that, due to a large variation of spatial grid spacing, the variation in local time step is much larger than that for equivalent constant spatial grid, illustrated in Figure 12-24. The convergence histories are shown in Figure 12-28. The converged (steady-state) solution with a global time step of 1.9617×10^{-6} corresponding to the CFL number of 0.15 was obtained after 35,438 time steps. When local time stepping was used, the converged solution was obtained after 12,058 time steps.

In the example just completed, the local time step was evaluated at each time level. However, determination of a local time step at each time level is not necessary. In fact, a local time step can be calculated and used for the next few hundred (e.g., 500) time levels and recomputed for the next several hundred time levels and so on.

In closing this section, it is noted that local time stepping accelerates the solution toward steady state with very little modification/addition to a computer code. The total saving in computation time could be substantial (e.g., as high as 50–70%). However, recall that this approach can be used only for steady-state solutions, and it cannot be used for time accurate solutions.

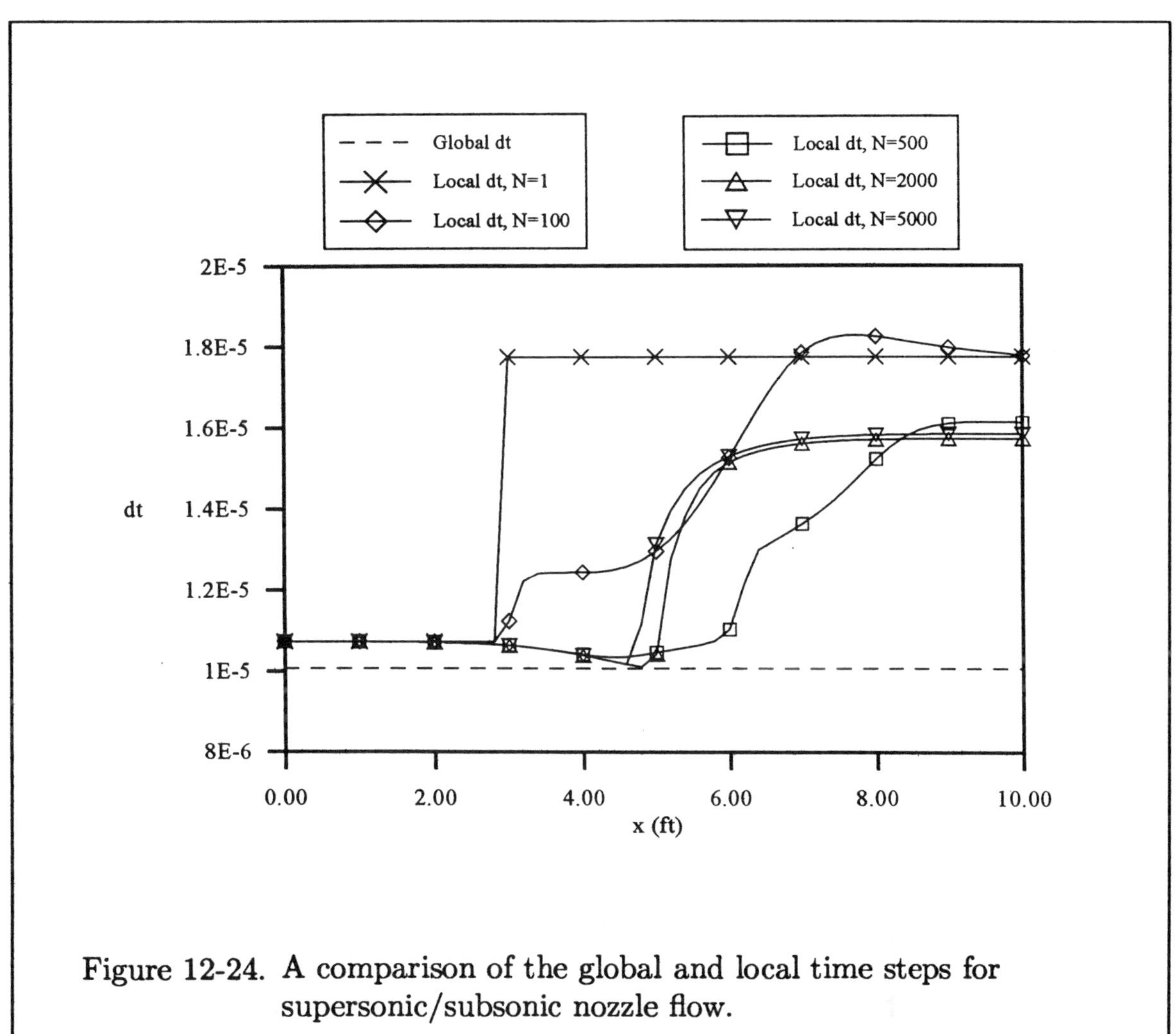

Figure 12-24. A comparison of the global and local time steps for supersonic/subsonic nozzle flow.

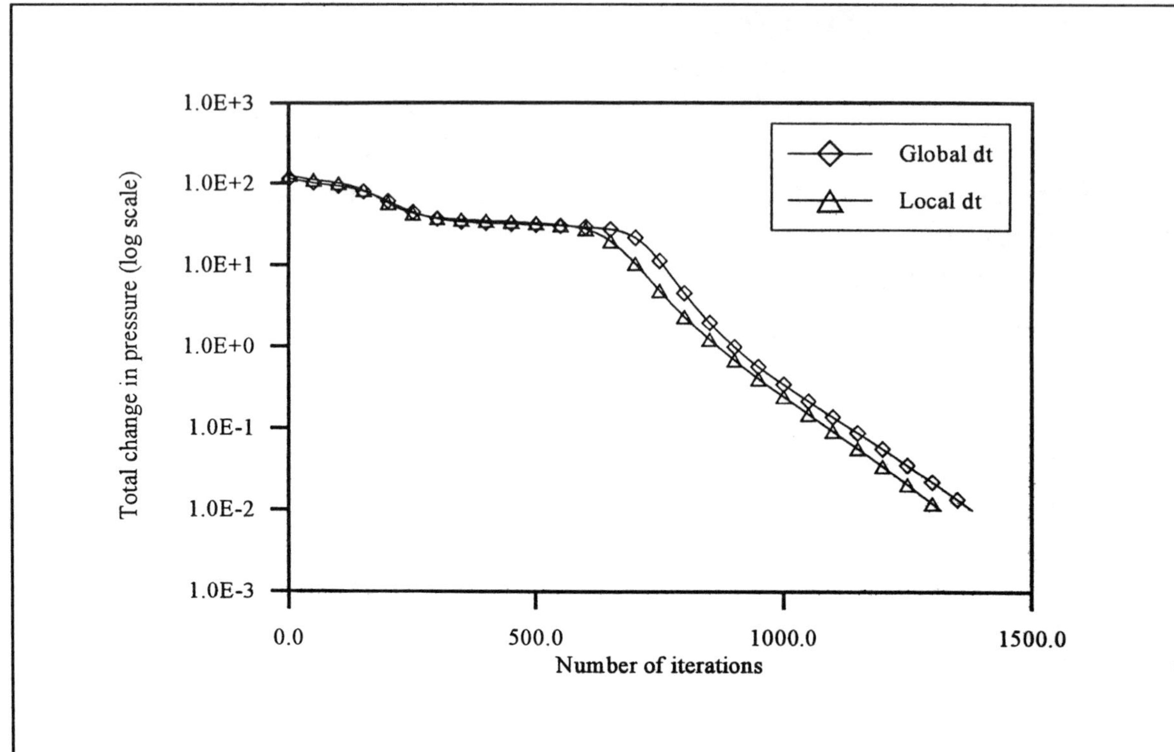

Figure 12-25. Convergence histories for the explicit first-order SWFVS scheme for supersonic nozzle flow.

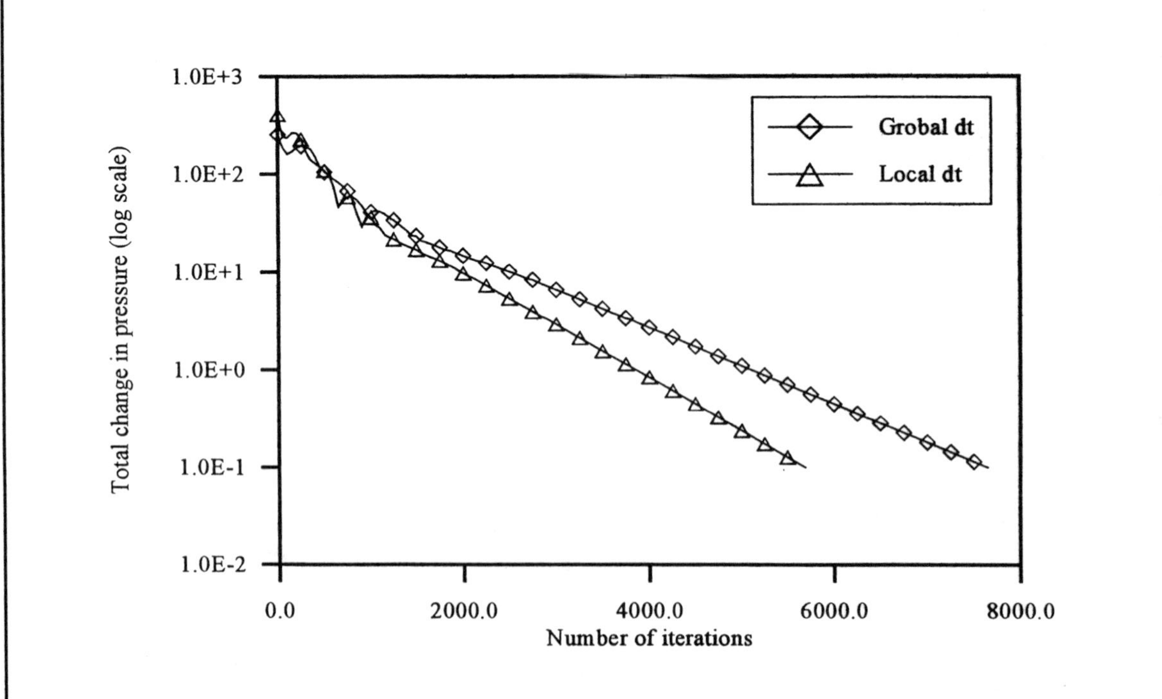

Figure 12-26. Convergence histories for the explicit first-order SWFVS scheme for supersonic/subsonic nozzle flow.

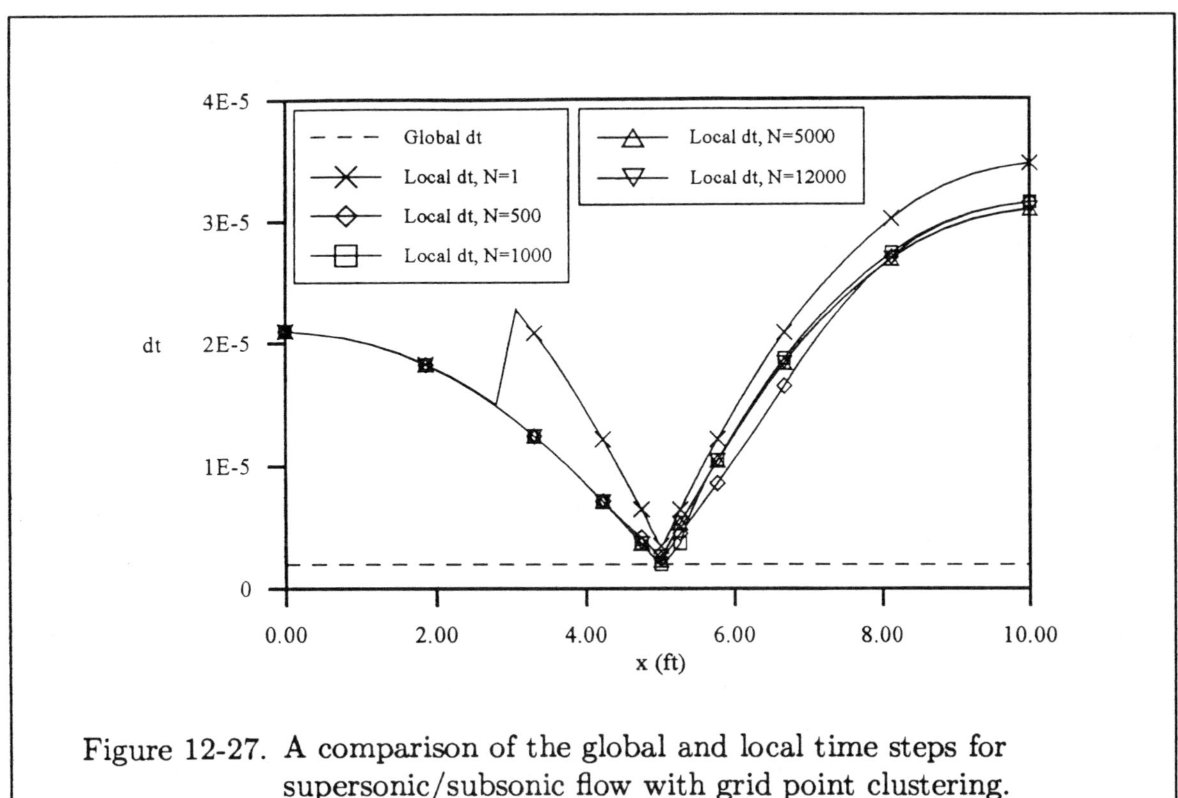

Figure 12-27. A comparison of the global and local time steps for supersonic/subsonic flow with grid point clustering.

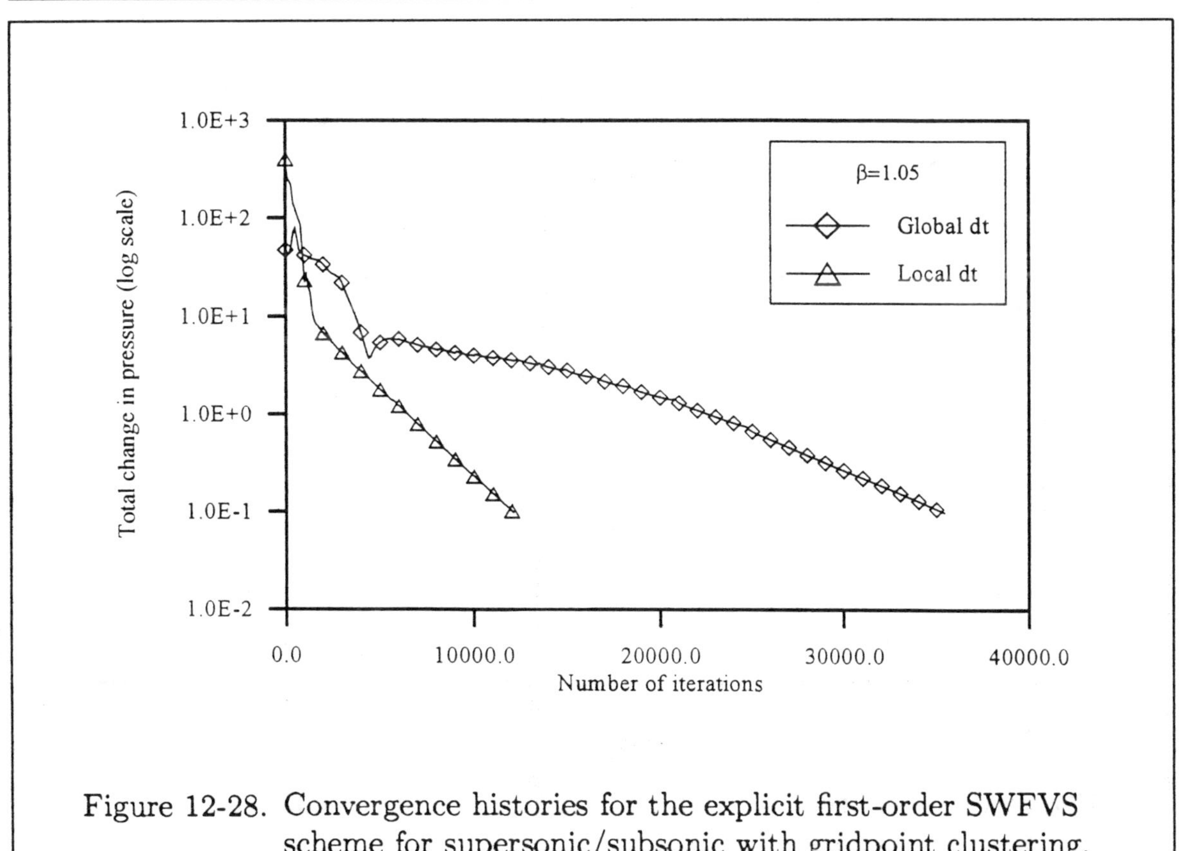

Figure 12-28. Convergence histories for the explicit first-order SWFVS scheme for supersonic/subsonic with gridpoint clustering.

12.8 Application 2: Shock Tube or Riemann Problem

A second example for the solution of the one-dimensional Euler equation is proposed by investigations of the shock tube problem. The shock tube problem is a good case to study because it involves severe situations in a flow involving shock wave, contact surface, and expansion waves.

A description of the shock tube, analytical solution, and numerical solution is to follow.

12.8.1 Problem Description

A shock tube is a device which is used in the experimental investigation of several physical phenomena such as chemical reaction kinetics, shock structure, and aerothermodynamics of supersonic/hypersonic vehicles.

A shock tube is a relatively long and a constant area tube which is divided into two sections by a diaphragm, as shown in Figure 12-29. The section including the high pressure gas is called the *driver* section and will be denoted as ***region 4***, whereas the section with low pressure is called the ***driven*** section (or the ***expansion***

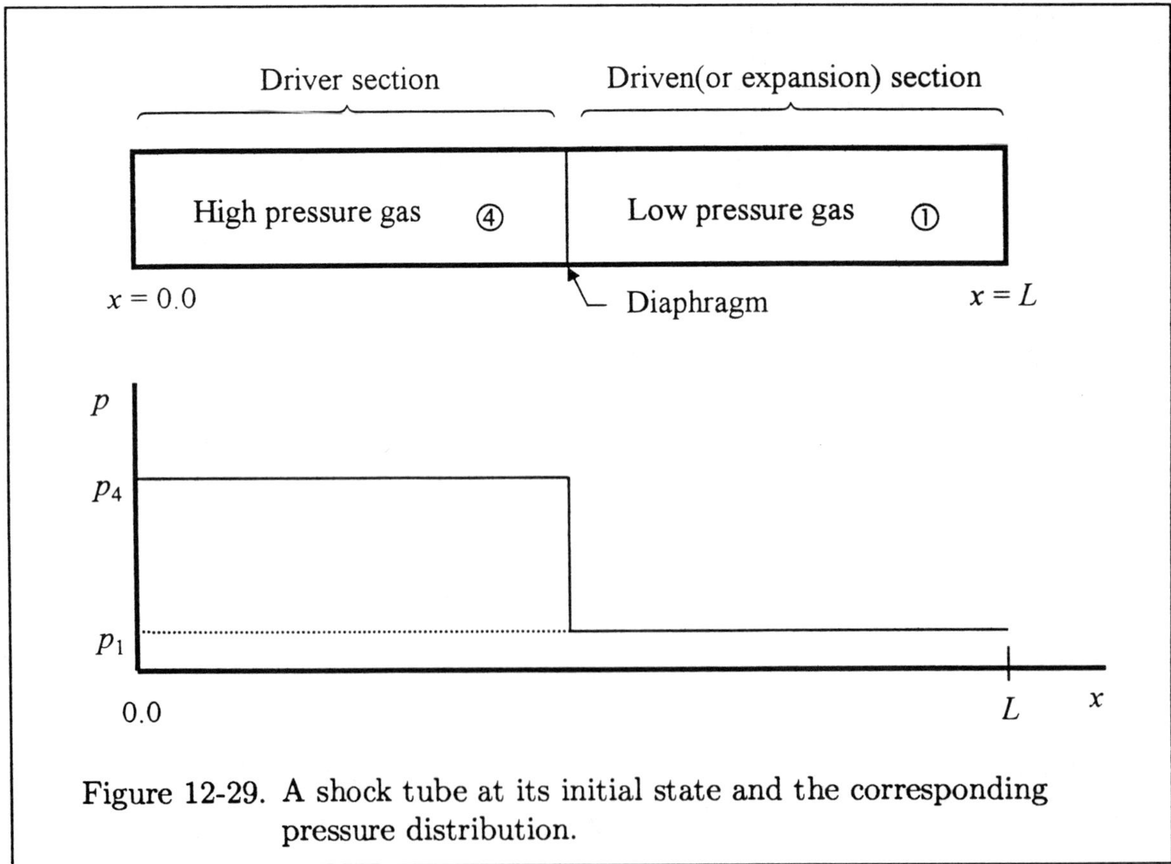

Figure 12-29. A shock tube at its initial state and the corresponding pressure distribution.

section), and will be denoted as *region 1.* The two gases in sections 1 and 4 may be different, and their corresponding ratio of specific heats are given by γ_1 and γ_4, respectively.

Complete specification of driver and expansion gases is sufficient to provide the solution for the shock tube problem which is also known as *the Riemann problem.* Thus, at the initial state, the following data is provided: p_4, T_4, γ_4, and p_1,T_1,γ_1. The operation of a shock tube is initiated by the rupture of the diaphragm, and the resulting flowfield is briefly described in the following.

Once the diaphragm is ruptured, a normal shock propagates into the low pressure region 1, and a series of expansion waves propagate into the high pressure region 4. As the shock wave moves to the right over the gas, it accelerates the gas in the positive x-direction with a velocity of V_2. At the same time, an expansion wave moving to the left accelerates the gas, also in the positive x-direction, with a velocity of V_3. The leading expansion wave moves with a velocity of a_4, whereas the trailing wave moves with a velocity of a_3. Now recall that a shock wave is a nonisentropic process, whereas an expansion wave is an isentropic process. Therefore the entropies behind the shock and expansion waves would be different. Thus two distinct regions are identified separately by what is known as *constant surface.* These regions are commonly identified as regions 2 and 3. A contact surface is defined as a discontinuity in entropy and temperature, whereas pressure and velocity are continuous, that is, $s_2 \neq s_3$, $T_2 \neq T_3$, and $p_2 = p_3$, and $V_2 = V_3$.

The flow pattern and the corresponding pressure, temperature, and velocity distributions are illustrated in Figure 12-30.

12.8.2 Analytical Solution

The analytical solution of the shock tube problem can be achieved by the following relations. It is assumed that the flow is inviscid and one-dimensional and that the gas is a perfect gas. The derivation of these relations can be found in References [12-11, 12-12]. A pressure ratio can be established by the following relation.

$$\frac{p_4}{p_1} = \frac{p_2}{p_1}\left\{1 - \frac{(\gamma_4 - 1)\left(\frac{a_1}{a_4}\right)\left(\frac{p_2}{p_1} - 1\right)}{\left[4\gamma_1^2 + 2\gamma_1(\gamma_1 + 1)\left(\frac{p_2}{p_1} - 1\right)\right]^{\frac{1}{2}}}\right\}^{-\frac{2\gamma_4}{\gamma_4 - 1}} \tag{12-103}$$

Note that Equation (12-103) must be solved by an iterative scheme for p_2/p_1 which defines the shock strength. Equation (12-103) can be rearranged in terms of the shock Mach number M_s, as follows.

$$\frac{p_4}{p_1} = \frac{\gamma_1 - 1}{\gamma_1 + 1}\left[\frac{2\gamma_1}{\gamma_1 - 1}M_s^2 - 1\right]\left[1 - \frac{\frac{\gamma_4 - 1}{\gamma_4 + 1}\left(\frac{a_1}{a_4}\right)(M_s^2 - 1)}{M_s}\right]^{-\frac{2\gamma_4}{\gamma_4 - 1}} \tag{12-104}$$

Again, Equation (12-104) must be solved by an iterative scheme. Once p_2/p_1 is determined, the temperature, density, shock velocity, and the velocity behind the wave V_2 can be determined from the following relations.

$$\frac{T_2}{T_1} = \frac{p_2}{p_1}\frac{\frac{\gamma_1 + 1}{\gamma_1 - 1} + \frac{p_2}{p_1}}{1 + \frac{\gamma_1 + 1}{\gamma_1 - 1}\left(\frac{p_2}{p_1}\right)} \tag{12-105}$$

$$\frac{\rho_2}{\rho_1} = \frac{1 + \frac{\gamma_1 + 1}{\gamma_1 - 1}\left(\frac{p_2}{p_1}\right)}{\frac{\gamma_1 + 1}{\gamma_1 - 1} + \frac{p_2}{p_1}} \tag{12-106}$$

$$V_s = a_1\left[1 + \frac{\gamma_1 + 1}{2\gamma_1}\left(\frac{p_2}{p_1} - 1\right)\right]^{\frac{1}{2}} \tag{12-107}$$

$$M_s = \frac{V_s}{a_1} \tag{12-108}$$

and

$$V_2 = \frac{a_1}{\gamma_1}\left(\frac{p_2}{p_1} - 1\right)\left[\frac{\frac{2\gamma_1}{\gamma_1 + 1}}{\frac{\gamma_1 - 1}{\gamma_1 + 1} + \frac{p_2}{p_1}}\right]^{\frac{1}{2}} \tag{12-109}$$

The flow properties in region 3 are determined as follow. First, the strength of the expansion wave is calculated from

$$\frac{p_3}{p_4} = \frac{p_3}{p_1}\frac{p_1}{p_4} = \frac{p_2}{p_1}\frac{p_1}{p_4} \tag{12-110}$$

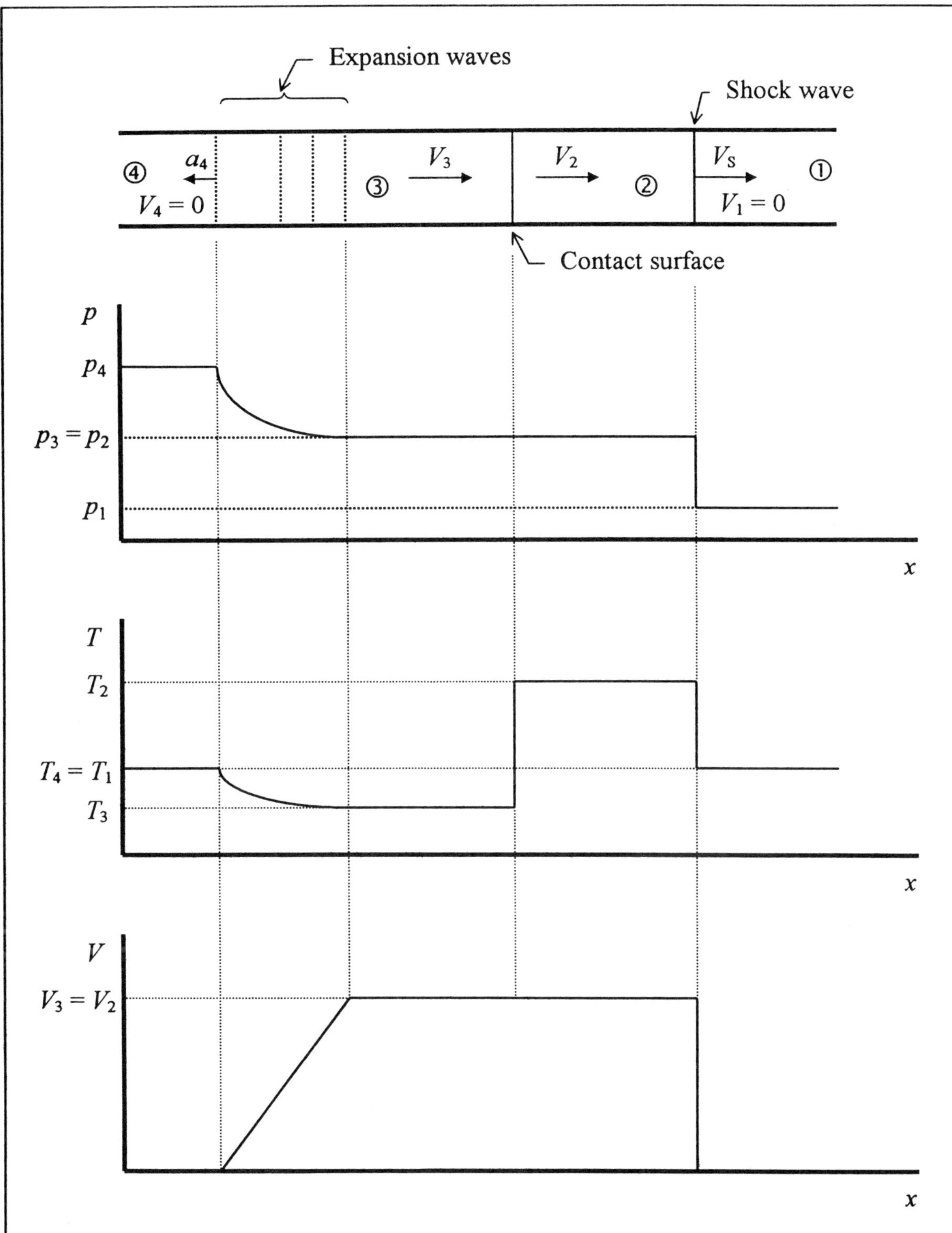

Figure 12-30. Flow pattern in the shock tube and the corresponding pressure, temperature, and velocity distributions.

Since the expansion wave is an isentropic process, the flow properties across the wave are determined from the isentropic relations.

$$\frac{T_3}{T_4} = \left(\frac{p_3}{p_4}\right)^{\frac{\gamma_4 - 1}{\gamma_4}} \tag{12-111}$$

$$\frac{\rho_3}{\rho_4} = \left(\frac{p_3}{p_4}\right)^{\frac{1}{\gamma_4}} \tag{12-112}$$

The speed of sound and the Mach number in region 3 are determined from

$$\frac{a_4}{a_3} = \left(\frac{p_4}{p_3}\right)^{\frac{\gamma_4 - 1}{2\gamma_4}} \tag{12-113}$$

and

$$M_3 = \frac{2}{\gamma_4 - 1}\left[\left(\frac{p_4}{p_3}\right)^{\frac{\gamma_4 - 1}{2\gamma_4}} - 1\right] \tag{12-114}$$

Subsequently, the velocity is given by

$$V_3 = M_3 a_3 \tag{12-115}$$

To calculate the flow variation within the expansion wave, recall that the equation for C^- characteristic is given by Equation (A-30b).

$$\frac{dx}{dt} = V - a \tag{12-116}$$

which can be integrated to yield

$$x = (V - a)t \tag{12-117}$$

Furthermore, since [from Equation (A-35a)]

$$V + \frac{2a}{\gamma - 1} = \text{constant}$$

The following relation can be developed

$$\frac{a}{a_4} = 1 - \frac{\gamma - 1}{2}\left(\frac{V}{a_4}\right) \tag{12-118}$$

Now, from Equations (12-117) and (12-118), one obtains

$$V = \frac{2}{\gamma + 1}\left(a_4 + \frac{x}{t}\right) \tag{12-119}$$

Assuming a perfect gas where $a = (\gamma RT)^{1/2}$, then

$$\frac{T}{T_4} = \left[1 - \frac{\gamma - 1}{2}\left(\frac{V}{a_4}\right)\right]^2 \tag{12-120}$$

and using isentropic relations

$$\frac{\rho}{\rho_4} = \left[1 - \frac{\gamma - 1}{2}\left(\frac{V}{a_4}\right)\right]^{\frac{2}{\gamma - 1}} \tag{12-121}$$

and

$$\frac{p}{p_4} = \left[1 - \frac{\gamma - 1}{2}\left(\frac{V}{a_4}\right)\right]^{\frac{2\gamma}{\gamma - 1}} \tag{12-122}$$

The analytical solution described in this section can be used for validation of the numerical solution as well as for accuracy analysis.

12.8.3 Numerical Solution

The shock tube problem is governed by the one-dimensional Euler equation given by Equation (12-11), which is written as

$$\frac{\partial Q}{\partial t} + \frac{\partial E}{\partial x} = 0 \tag{12-123}$$

for a constant area problem.

The numerical solution is obtained by the modified Runge-Kutta scheme with TVD, as given by Equation (12-76). The shock tube is 10 units long, and the diaphragm is located at 5. The following initial conditions are specified.

$$\rho_4 = 1.0 \qquad \rho_1 = 0.125$$

$$V_4 = 0.0 \qquad V_1 = 0.0$$

$$p_4 = 1.0 \qquad p_1 = 0.1$$

$$\gamma_4 = \gamma = 1.4 \qquad \gamma_1 = \gamma = 1.4$$

A simple zero-order extrapolation is used at the boundaries, that is

$$Q_1^{n+1} = Q_2^{n+1}$$

and

$$Q_{IM}^{n+1} = Q_{IM-1}^{n+1}$$

The solutions obtained with three different TVD models are compared to the analytical solution in Figure 12-31. The spatial step is $\Delta x = 1.0$, and the solution corresponds to a time level of 120.

The specific limiters used are (12-61), (12-68), and (12-73) for *H-Y*, *R-S*, and *D-Y* models, respectively. Solutions are also obtained with finer grid where $\Delta x = 0.25$, as illustrated in Figure 12-32. Solutions are also obtained with the Harten-Yee upwind TVD model with 5 different limiters given in Section 12.3.2.4.1 and are illustrated in Figure 12-33. Several observations can be made at this point. The *R-S* model produces more dissipation than others, and, therefore, shock and contact surface smearing appear in the solution. The *H-Y* model computed the shock and the contact surface more accurately than the other two models. As expected, a reduction in grid spacing increases the accuracy of solution, as seen by comparison of Figures 12-31 and 12-32. Finally, the accuracy of solution depends on the limiter used, as shown in Figure 12-33.

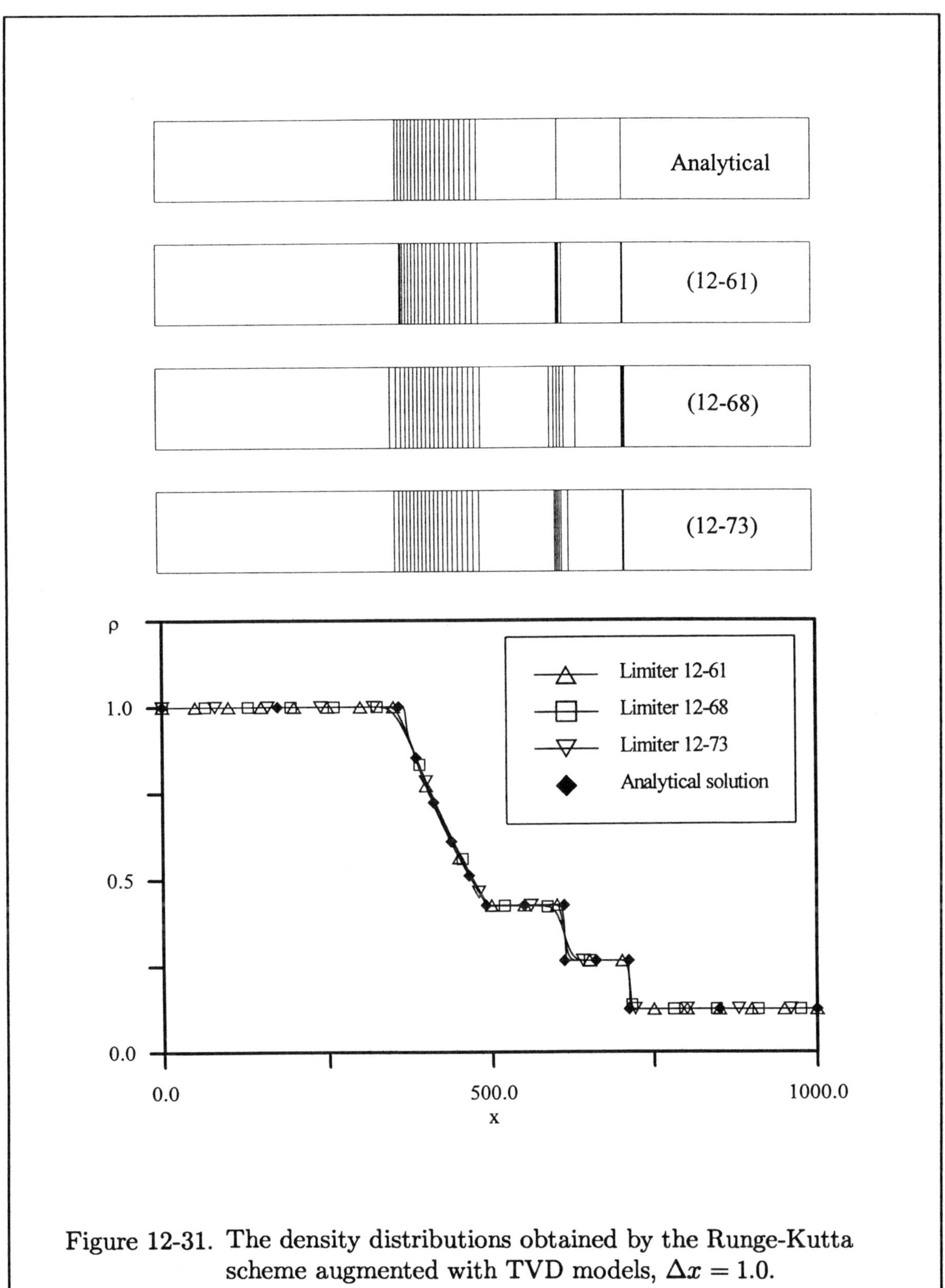

Figure 12-31. The density distributions obtained by the Runge-Kutta scheme augmented with TVD models, $\Delta x = 1.0$.

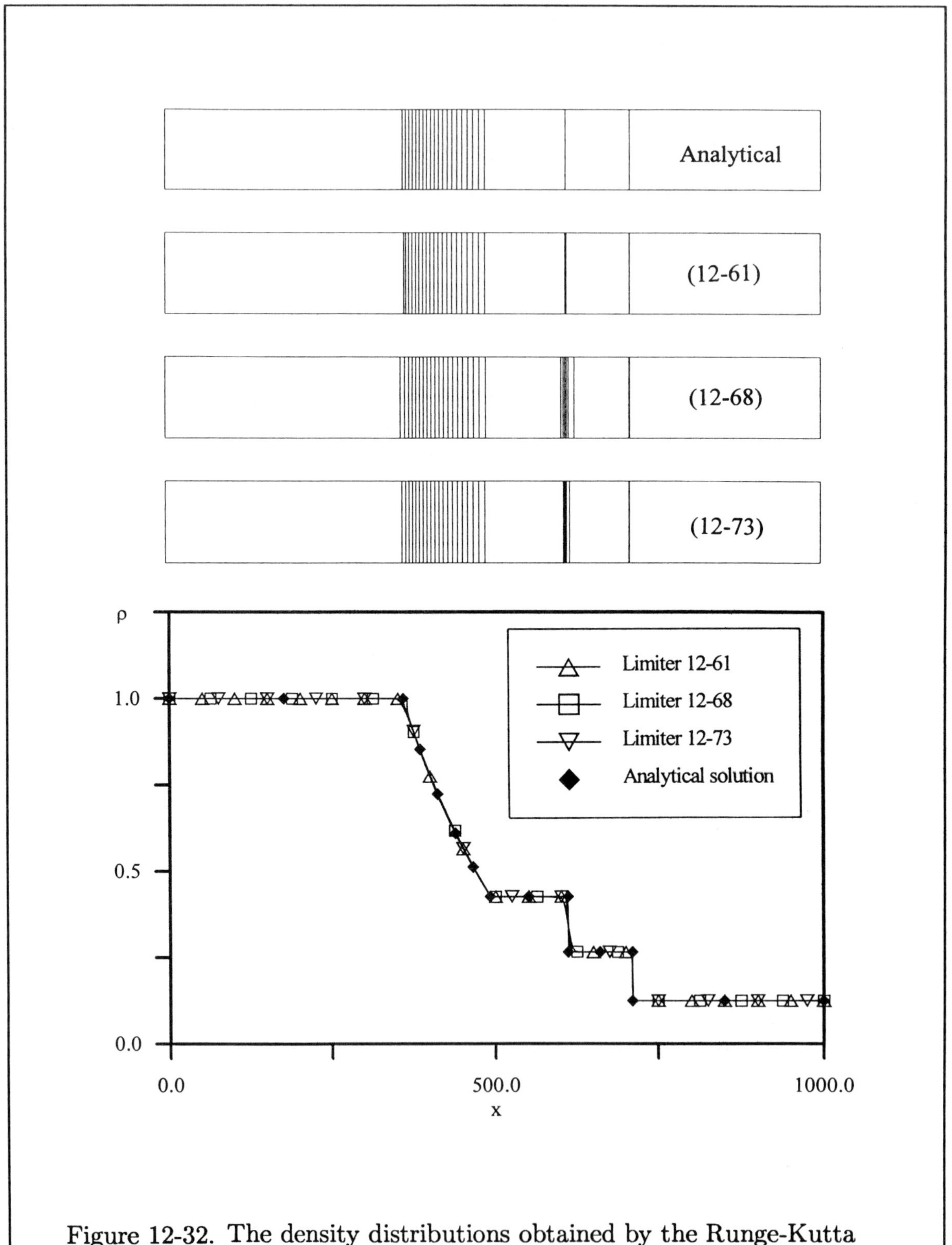

Figure 12-32. The density distributions obtained by the Runge-Kutta scheme augmented with TVD models, $\Delta x = 0.25$.

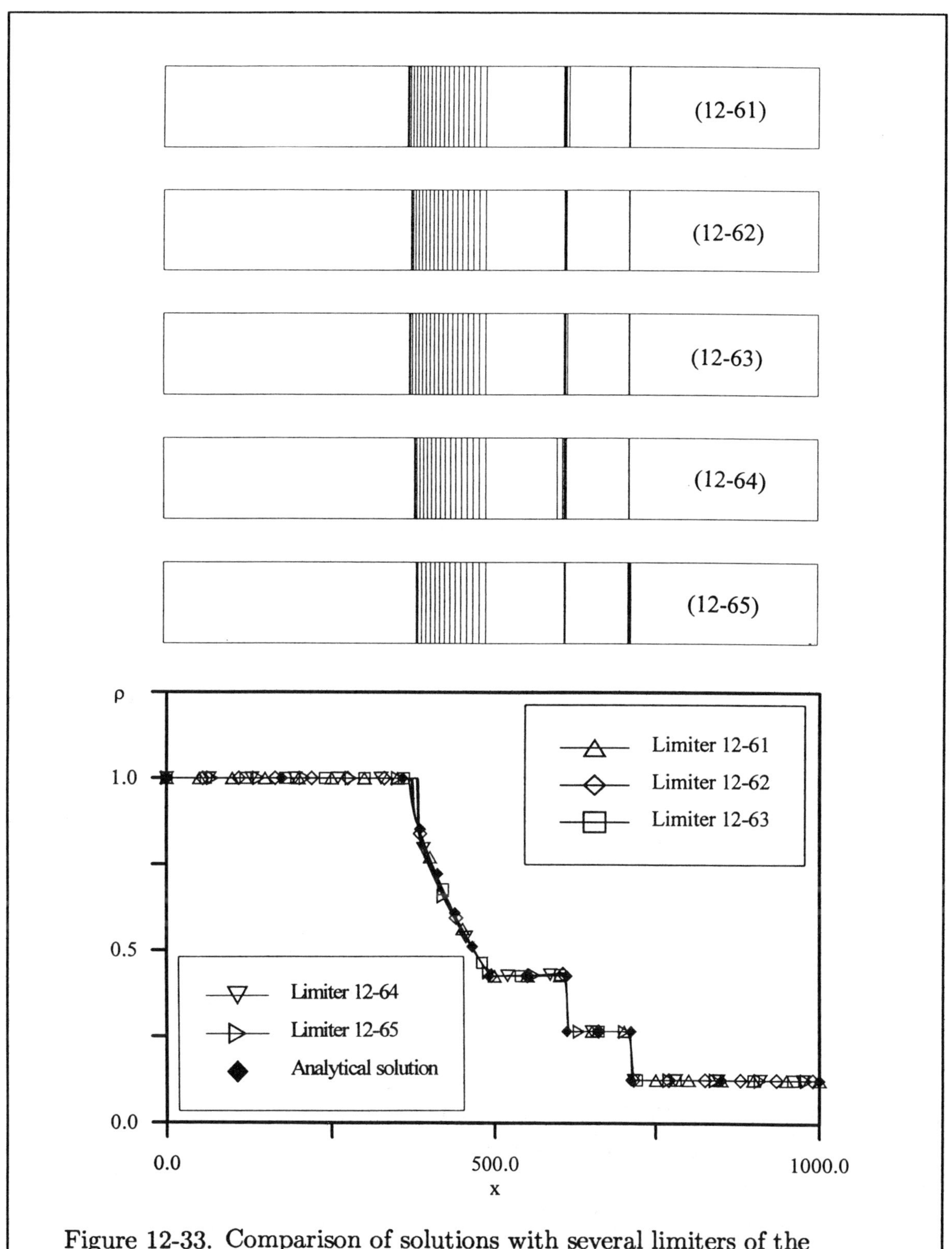

Figure 12-33. Comparison of solutions with several limiters of the Harten-Yee upwind TVD model.

12.9 Two-Dimensional Planar and Axisymmetric Euler Equations

The procedure for the solution of the quasi one-dimensional Euler equation, along with some guidelines on stability, implementation of boundary conditions, and specification of the initial condition, was established in the preceding sections. In this section, two-dimensional planar and axisymmetric Euler equations will be considered.

The governing equation of motion for the two-dimensional/axisymmetric inviscid flow is:

$$\frac{\partial Q}{\partial t} + \frac{\partial E}{\partial x} + \frac{\partial F}{\partial y} + \alpha H = 0 \tag{12-124}$$

where

$$Q = \begin{bmatrix} \rho \\ \rho u \\ \rho v \\ \rho e_t \end{bmatrix} \quad \text{(12-125a)} \qquad E = \begin{bmatrix} \rho u \\ \rho u^2 + p \\ \rho u v \\ (\rho e_t + p)\, u \end{bmatrix} \quad \text{(12-125b)}$$

$$F = \begin{bmatrix} \rho v \\ \rho v u \\ \rho v^2 + p \\ (\rho e_t + p) v \end{bmatrix} \quad \text{(12-125c)} \qquad H = \frac{1}{y} \begin{bmatrix} \rho v \\ \rho v u \\ \rho v^2 \\ (\rho e_t + p) v \end{bmatrix} \quad \text{(12-125d)}$$

and

$$\alpha = \begin{cases} 0 & \text{for a two-dimensional planar flow} \\ 1 & \text{for a two-dimensional axisymmetric flow} \end{cases}$$

The equations of motion were transformed from physical space to computational space in Chapter 11. With the generalized coordinate transformation, the Euler equation in the computational space become

$$\frac{\partial \bar{Q}}{\partial \tau} + \frac{\partial \bar{E}}{\partial \xi} + \frac{\partial \bar{F}}{\partial \eta} + \alpha \bar{H} = 0 \tag{12-126}$$

where

$$\bar{Q} = \frac{Q}{J} \tag{12-127}$$

$$\bar{E} = \frac{1}{J}\left[\xi_t Q + \xi_x E + \xi_y F\right] \tag{12-128}$$

$$\bar{F} = \frac{1}{J}\left[\eta_t Q + \eta_x E + \eta_y F\right] \tag{12-129}$$

$$\bar{H} = \frac{H}{J} \tag{12-130}$$

12.9.1 Numerical Considerations

Issues which were addressed in Section 12.3.1 with regard to the one-dimensional Euler equation are just as valid for the two- and three-dimensional Euler equations. However, it is beneficial to revisit these issues and to investigate the mathematical details for the two-dimensional equations which are addressed in this section. Subsequently, the extension from two-dimensions to three-dimensions is rather straightforward. The three-dimensional equations and the associated Jacobian matrices are addressed in Chapter 11, and a discussion of an implicit solution procedure for the three-dimensional applications is provided in Chapter 14.

First, the issue of linearization of the Euler equation which will be required for implicit schemes is explored. A first-order backward approximation for the Euler equation yields

$$\frac{\bar{Q}^{n+1} - \bar{Q}^n}{\Delta\tau} + \left(\frac{\partial \bar{E}}{\partial \xi}\right)^{n+1} + \left(\frac{\partial \bar{F}}{\partial \eta}\right)^{n+1} + \alpha(\bar{H})^{n+1} = 0 \tag{12-131}$$

Since Equation (12-131) is nonlinear, the following approximations are used to linearize the equation

$$\bar{E}^{n+1} = \bar{E}^n + \left(\frac{\partial \bar{E}}{\partial \bar{Q}}\right)\Delta\bar{Q} + O(\Delta\tau)^2 \tag{12-132}$$

$$\bar{F}^{n+1} = \bar{F}^n + \left(\frac{\partial \bar{F}}{\partial \bar{Q}}\right)\Delta\bar{Q} + O(\Delta\tau)^2 \tag{12-133}$$

$$\bar{H}^{n+1} = \bar{H}^n + \left(\frac{\partial \bar{H}}{\partial \bar{Q}}\right)\Delta\bar{Q} + O(\Delta\tau)^2 \tag{12-134}$$

where $\partial\bar{E}/\partial\bar{Q}$, $\partial\bar{F}/\partial\bar{Q}$ and $\partial\bar{H}/\partial\bar{Q}$ are the flux Jacobian matrices defined as $\partial\bar{E}/\partial\bar{Q} = A$, $\partial\bar{F}/\partial\bar{Q} = B$, and $\partial\bar{H}/\partial\bar{Q} = C$. These matrices are given in Chapter 11 by Equations (11-213), (11-214) and (11-215).

Some of the numerical schemes to be investigated shortly require the modification of the elements of the Jacobian matrices. For this purpose, the elements of the flux Jacobian matrices are rearranged such that e_t is expressed in terms of the speed of sound, a, and the velocity components, u and v. This relation, derived as Equation (12-23) for the one-dimensional case, is extended to the two-dimensional problem as

$$e_t = \frac{a^2}{\gamma(\gamma-1)} + \frac{1}{2}(u^2+v^2) \tag{12-135}$$

The flux Jacobian matrices A and B are now modified using Equation (12-135).

$$A = \frac{\partial \bar{E}}{\partial \bar{Q}} = \left[\begin{array}{c|c|c|c} \xi_t & \xi_x & \xi_y & 0 \\ \hline -u(u\xi_x + v\xi_y) & \xi_t + \xi_x(3-\gamma)u & -(\gamma-1)\xi_x v & (\gamma-1)\xi_x \\ +\xi_x\left[\frac{1}{2}(\gamma-1)(u^2+v^2)\right] & +\xi_y v & +\xi_y u & \\ \hline -v(u\xi_x + v\xi_y) & \xi_x v - (\gamma-1)\xi_y u & \xi_t + \xi_x u & (\gamma-1)\xi_y \\ +\xi_y\left[\frac{1}{2}(\gamma-1)(u^2+v^2)\right] & & +\xi_y(3-\gamma)v & \\ \hline (\xi_x u + \xi_y v)\cdot & \xi_x\left[\frac{a^2}{\gamma-1} + \frac{1}{2}(u^2+v^2)\right] & \xi_y\left[\frac{a^2}{\gamma-1} + \frac{1}{2}(u^2+v^2)\right] & \xi_t + \\ \left[-\frac{a^2}{\gamma-1} + (\frac{1}{2}\gamma - 1)(u^2+v^2)\right] & -(\gamma-1)(\xi_x u + \xi_y v)u & -(\gamma-1)(\xi_x u + \xi_y v)v & \gamma(\xi_x u + \xi_y v) \end{array}\right] \tag{12-136}$$

$$B = \frac{\partial \bar{F}}{\partial \bar{Q}} = \left[\begin{array}{c:c:c:c} \eta_t & \eta_x & \eta_y & 0 \\ \hdashline -u(u\eta_x + v\eta_y) & \eta_t + \eta_x(3-\gamma)u & -(\gamma-1)\eta_x v & (\gamma-1)\eta_x \\ +\eta_x \left[\frac{1}{2}(\gamma-1)(u^2+v^2)\right] & +\eta_y v & +\eta_y u & \\ \hdashline -v(\eta_x u + \eta_y v) & \eta_x v - (\gamma-1)\eta_y u & \eta_t + \eta_x u & (\gamma-1)\eta_y \\ +\eta_y \left[\frac{1}{2}(\gamma-1)(u^2+v^2)\right] & & +\eta_y(3-\gamma)v & \\ \hdashline (u\eta_x + v\eta_y)\cdot & \eta_x \left[\frac{a^2}{\gamma-1} + \frac{1}{2}(u^2+v^2)\right] & \eta_y \left[\frac{a^2}{\gamma-1} + \frac{1}{2}(u^2+v^2)\right] & \eta_t + \\ \left[-\frac{a^2}{\gamma-1} + (\frac{\gamma}{2}-1)(u^2+v^2)\right] & -(\gamma-1)(\eta_x u + \eta_y v)u & -(\gamma-1)(\eta_x u + \eta_y v)v & \gamma(\eta_x u + \eta_y v) \end{array} \right] \tag{12-137}$$

The eigenvalues of matrix A are determined to be

$$\lambda_{\xi_1} = \xi_t + \xi_x u + \xi_y v \tag{12-138}$$

$$\lambda_{\xi_2} = \xi_t + \xi_x u + \xi_y v \tag{12-139}$$

$$\lambda_{\xi_3} = \xi_t + \xi_x u + \xi_y v + a\sqrt{\xi_x^2 + \xi_y^2} \tag{12-140}$$

$$\lambda_{\xi_4} = \xi_t + \xi_x u + \xi_y v - a\sqrt{\xi_x^2 + \xi_y^2} \tag{12-141}$$

The eigenvector matrix as determined by MACSYMA is

$$X_A = \left[\begin{array}{c|c|c|c}
0 & 1 & 1 & 1 \\ \hline
-\dfrac{\xi_y}{\xi_x} & \dfrac{u\xi_x + v\xi_y}{\xi_x} & u + \dfrac{a\xi_x}{\sqrt{\xi_x^2 + \xi_y^2}} & u - \dfrac{a\xi_x}{\sqrt{\xi_x^2 + \xi_y^2}} \\ \hline
1 & 0 & v + \dfrac{a\xi_y}{\sqrt{\xi_x^2 + \xi_y^2}} & v - \dfrac{a\xi_y}{\sqrt{\xi_x^2 + \xi_y^2}} \\ \hline
\dfrac{v\xi_x - u\xi_y}{\xi_x} & \dfrac{2uv\xi_y + (u^2 - v^2)\xi_x}{2\xi_x} & \dfrac{u^2 + v^2}{2} + \dfrac{a^2}{\gamma - 1} + \dfrac{a(u\xi_x + v\xi_y)}{\sqrt{\xi_x^2 + \xi_y^2}} & \dfrac{u^2 + v^2}{2} + \dfrac{a^2}{\gamma - 1} - \dfrac{a(u\xi_x + v\xi_y)}{\sqrt{\xi_x^2 + \xi_y^2}}
\end{array}\right] \tag{12-142}$$

and the inverse of the eigenvector matrix X_A is

$$X_A^{-1} = \left[\begin{array}{c|c|c|c}
\frac{(1-\gamma)v(u^2+v^2)}{2a^2} + \frac{\xi_y(u\xi_x+v\xi_y)}{\xi_x^2+\xi_y^2} & \frac{(\gamma-1)uv}{a^2} - \frac{\xi_x\xi_y}{\xi_x^2+\xi_y^2} & \frac{(\gamma-1)v^2}{a^2} + \frac{\xi_x^2}{\xi_x^2+\xi_y^2} & -\frac{(\gamma-1)v}{a^2} \\
\hline
1+\frac{(1-\gamma)(u^2+v^2)}{2a^2} & \frac{(\gamma-1)u}{a^2} & \frac{(\gamma-1)v}{a^2} & -\frac{\gamma-1}{a^2} \\
\hline
\frac{(\gamma-1)(u^2+v^2)}{4a^2} - \frac{u\xi_x+v\xi_y}{2a\sqrt{\xi_x^2+\xi_y^2}} & \frac{(1-\gamma)u}{2a^2} + \frac{\xi_x}{2a\sqrt{\xi_x^2+\xi_y^2}} & \frac{(1-\gamma)v}{2a^2} + \frac{\xi_y}{2a\sqrt{\xi_x^2+\xi_y^2}} & \frac{\gamma-1}{2a^2} \\
\hline
\frac{(\gamma-1)(u^2+v^2)}{4a^2} + \frac{u\xi_x+v\xi_y}{2a\sqrt{\xi_x^2+\xi_y^2}} & \frac{(1-\gamma)u}{2a^2} - \frac{\xi_x}{2a\sqrt{\xi_x^2+\xi_y^2}} & \frac{(1-\gamma)v}{2a^2} - \frac{\xi_y}{2a\sqrt{\xi_x^2+\xi_y^2}} & \frac{\gamma-1}{2a^2}
\end{array}\right] \tag{12-143}$$

Similarly, the eigenvalues of B are determined as

$$\lambda_{\eta_1} = \eta_t + \eta_x u + \eta_y v \tag{12-144}$$

$$\lambda_{\eta_2} = \eta_t + \eta_x u + \eta_y v \tag{12-145}$$

$$\lambda_{\eta_3} = \eta_t + \eta_x u + \eta_y v + a\sqrt{\eta_x^2+\eta_y^2} \tag{12-146}$$

$$\lambda_{\eta_4} = \eta_t + \eta_x u + \eta_y v - a\sqrt{\eta_x^2+\eta_y^2} \tag{12-147}$$

with the associated eigenvector matrix X_B of

$$X_B = \begin{bmatrix} 0 & 1 & 1 & 1 \\ -\dfrac{\eta_y}{\eta_x} & \dfrac{u\eta_x + v\eta_y}{\eta_x} & u + \dfrac{a\eta_x}{\sqrt{\eta_x^2+\eta_y^2}} & u - \dfrac{a\eta_x}{\sqrt{\eta_x^2+\eta_y^2}} \\ 1 & 0 & v + \dfrac{a\eta_y}{\sqrt{\eta_x^2+\eta_y^2}} & v - \dfrac{a\eta_y}{\sqrt{\eta_x^2+\eta_y^2}} \\ \dfrac{v\eta_x - u\eta_y}{\eta_x} & \dfrac{2uv\eta_y + (u^2-v^2)\eta_x}{2\eta_x} & \dfrac{u^2+v^2}{2} + \dfrac{a^2}{\gamma-1} + \dfrac{a(u\eta_x + v\eta_y)}{\sqrt{\eta_x^2+\eta_y^2}} & \dfrac{u^2+v^2}{2} + \dfrac{a^2}{\gamma-1} - \dfrac{a(u\eta_x + v\eta_y)}{\sqrt{\eta_x^2+\eta_y^2}} \end{bmatrix} \tag{12-148}$$

and the inverse

$$X_B^{-1} = \begin{bmatrix} \dfrac{(1-\gamma)v(u^2+v^2)}{2a^2} + \dfrac{\eta_y(u\eta_x + v\eta_y)}{\eta_x^2+\eta_y^2} & \dfrac{(\gamma-1)uv}{a^2} - \dfrac{\eta_x\eta_y}{\eta_x^2+\eta_y^2} & \dfrac{(\gamma-1)v^2}{a^2} + \dfrac{\eta_x^2}{\eta_x^2+\eta_y^2} & -\dfrac{(\gamma-1)v}{a^2} \\ 1 + \dfrac{(1-\gamma)(u^2+v^2)}{2a^2} & \dfrac{(\gamma-1)u}{a^2} & \dfrac{(\gamma-1)v}{a^2} & -\dfrac{\gamma-1}{a^2} \\ \dfrac{(\gamma-1)(u^2+v^2)}{4a^2} - \dfrac{u\eta_x + v\eta_y}{2a\sqrt{\eta_x^2+\eta_y^2}} & \dfrac{(1-\gamma)u}{2a^2} + \dfrac{\eta_x}{2a\sqrt{\eta_x^2+\eta_y^2}} & \dfrac{(1-\gamma)v}{2a^2} + \dfrac{\eta_y}{2a\sqrt{\eta_x^2+\eta_y^2}} & \dfrac{\gamma-1}{2a^2} \\ \dfrac{(\gamma-1)(u^2+v^2)}{4a^2} + \dfrac{u\eta_x + v\eta_y}{2a\sqrt{\eta_x^2+\eta_y^2}} & \dfrac{(1-\gamma)u}{2a^2} - \dfrac{\eta_x}{2a\sqrt{\eta_x^2+\eta_y^2}} & \dfrac{(1-\gamma)v}{2a^2} - \dfrac{\eta_y}{2a\sqrt{\eta_x^2+\eta_y^2}} & \dfrac{\gamma-1}{2a^2} \end{bmatrix} \tag{12-149}$$

The following splitting procedure corresponds to that of the Steger-Warming scheme. The matrices A and B may be split according to

$$A = A^+ + A^-$$

and

$$B = B^+ + B^-$$

where

$$A^+ = X_A D_A^+ X_A^{-1} \tag{12-150}$$

$$A^- = X_A D_A^- X_A^{-1} \tag{12-151}$$

$$B^+ = X_B D_B^+ X_B^{-1} \tag{12-152}$$

and

$$B^- = X_B D_B^- X_B^{-1} \tag{12-153}$$

As before, matrix D_A^+ is a diagonal matrix whose elements are the positive eigenvalues of A; and D_A^- is a diagonal matrix whose elements are the negative eigenvalues of A. The same analogy is extended to D_B^+ and D_B^-.

It is important at this time to recognize a distinct difference between the two-dimensional problem being considered here and the one-dimensional problem examined previously. Recall that for the one-dimensional problem where no coordinate transformation was used, the metrics of transformation do not appear in the eigenvalues. Indeed, the eigenvalues are explicitly in terms of the flow velocity and the speed of sound, namely u, $u+a$, and $u-a$. Hence, for the one-dimensional problem, eigenvalues are all positive for supersonic flow; whereas for a subsonic flow, one is negative $[u-a]$ and two are positive. In the two-dimensional problem with coordinate transformation, that is no longer true; i.e., the eigenvalues not only depend on the velocity components and the local speed of sound, they also depend on the metrics of transformations as seen by relation (12-138) for example. Thus, the fact that the flow is supersonic in a region does not guarantee that the eigenvalues are all positive. Hence, the expressions for the eigenvalues are used at each point to determine whether they are positive or negative and thus split accordingly. For the one-dimensional flow investigated previously, the flux vector splitting was based on whether the flow is subsonic or supersonic. Also note that if an ideal gas model is not employed, the speed of sound used to modify Equation (12-18) is no longer valid. In that case, the eigenvalues of (12-18) are determined without the modification. Therefore, the sign of the eigenvalue is checked directly at each point rather than whether the flow is subsonic or supersonic.

Returning to the flux vector splitting, $\bar{E}$ and $\bar{F}$ vectors are split according to

$$\bar{E} = \bar{E}^+ + \bar{E}^-$$

and

$$\bar{F} = \bar{F}^+ + \bar{F}^-$$

where

$$\bar{E}^+ = A^+\bar{Q}$$

$$\bar{E}^- = A^-\bar{Q}$$

$$\bar{F}^+ = B^+\bar{Q}$$

and

$$\bar{F}^- = B^-\bar{Q}$$

Several numerical schemes introduced in Section 12.3 for the solution of the one-dimensional Euler equation are extended to two-dimensional applications in the following sections.

12.9.2 Explicit Formulations

A first-order approximation in time provides the following explicit formulation for the Euler equation.

$$\frac{\bar{Q}^{n+1} - \bar{Q}^n}{\Delta\tau} + \frac{\partial}{\partial\xi}(\bar{E}^n) + \frac{\partial}{\partial\eta}(\bar{F}^n) + \alpha\bar{H}^n = 0 \tag{12-154}$$

The convective terms in Equation (12-154) can be approximated by either central difference or one-sided finite difference expressions. And of course, within each category, a different order of accuracy can be introduced. Typically, for general CFD applications, second-order accuracy would be adequate.

Based on error analysis and experience gained by exploring various solution schemes for the Burgers equations as well as for the one-dimensional Euler equation, it is evident that, if a second-order accuracy for the convective terms is used, then the addition of numerical viscosity in the form of damping terms or TVD may be required. In this section, several explicit formulations are explored.

12.9.2.1 Steger and Warming Flux Vector Splitting: The procedure for the flux vector splitting was provided in the previous section. As a consequence, the explicit formulation is expressed as

$$\frac{\Delta\bar{Q}}{\Delta\tau} + \frac{\partial}{\partial\xi}\left(\bar{E}^+ + \bar{E}^-\right) + \frac{\partial}{\partial\eta}\left(\bar{F}^+ + \bar{F}^-\right) + \alpha\bar{H} = 0 \tag{12-155}$$

Again, based on our previous discussions, the positive and negative terms are approximated by backward and forward difference expressions, respectively. A first-order approximation yields

$$\Delta \bar{Q} = -\frac{\Delta\tau}{\Delta\xi}\left[\bar{E}^+_{i,j} - \bar{E}^+_{i-1,j} + \bar{E}^-_{i+1,j} - \bar{E}^-_{i,j}\right]$$

$$- \frac{\Delta\tau}{\Delta\eta}\left[\bar{F}^+_{i,j} - \bar{F}^+_{i,j-1} + \bar{F}^-_{i,j+1} - \bar{F}^-_{i,j}\right] - \Delta\tau(\alpha\bar{H}_{i,j}) \tag{12-156}$$

and a second-order approximation yields

$$\Delta \bar{Q} = -\frac{\Delta\tau}{2\Delta\xi}\left[\bar{E}^+_{i-2,j} - 4\bar{E}^+_{i-1,j} + 3\bar{E}^+_{i,j} - 3\bar{E}^-_{i,j} + 4\bar{E}^-_{i+1,j} - \bar{E}^-_{i+2,j}\right]$$

$$- \frac{\Delta\tau}{2\Delta\eta}\left[\bar{F}^+_{i,j-2} - 4\bar{F}^+_{i,j-1} + 3\bar{F}^+_{i,j} - 3\bar{F}^-_{i,j} + 4\bar{F}^-_{i,j+1} - \bar{F}^-_{i,j+2}\right]$$

$$- \Delta\tau(\alpha\bar{H}_{i,j}) \tag{12-157}$$

Similar to the one-dimensional problem, when Equation (12-157) is used, a difficulty is encountered when the equation is applied along the lines $i = 2$, $i = IMAX - 1$, $j = 2$, and $j = JMAX - 1$. To resolve this difficulty, either one could switch to the first-order schemes at these points or one could introduce fictitious points along $i = 0$, $i = IMAX + 1$, $j = 0$, and $j = JMAX + 1$.

12.9.2.1.1 Matrix Manipulations: An issue which needs further exploration is the calculation of the matrices A^+, A^-, B^+, and B^-. In general, they are calculated within the computer code using the expressions (12-150) through (12-153) by matrix manipulations. In practice, only two sets of multiplication are carried out. For example, expression (12-150) is used to evaluate A^+ and, subsequently, A^- is determined by

$$A^- = A - A^+$$

Similar operations are performed to obtain B^+ and B^-.

A second approach is to carry out the matrix operations external to the computer program and subsequently utilize the results. Recall that this procedure was used for the quasi one-dimensional problem, i.e., the exact values of A^+ and A^- were determined (given by (12-33) and (12-34)).

Since all of the matrices used on the right-hand sides of the expressions (12-150) through (12-153) are known, the operations can be performed for the four combinations of the eigenvalues. Redefine the eigenvalues of A given by (12-138) through (12-141) as follows

$$\lambda_1 = \xi_t + \xi_x u + \xi_y v$$

$$\lambda_2 = \lambda_1$$

$$\lambda_3 = \lambda_1 + a\sqrt{\xi_x^2 + \xi_y^2}$$

$$\lambda_4 = \lambda_1 - a\sqrt{\xi_x^2 + \xi_y^2}$$

where the speed of sound, a, can be written as

$$a = \sqrt{\gamma(\gamma - 1)e}$$

Now, the following four cases are considered.

(1) λ_1 is positive, whereas λ_4 could be either positive or negative; thus,

(a) $\lambda_4 > 0$; therefore, all the eigenvalues of A are positive and, as described previously,
$A^+ = A$ [given by (12-136)]
$A^- = 0$
and subsequently
$\bar{E}^+ = \bar{E}$
$\bar{E}^- = 0$

(b) $\lambda_4 < 0$; therefore, either one of the matrices A^+ or A^- must be evaluated. Due to simplicity of operations, A^- is considered, for which

$$A^- = [X_A]\begin{bmatrix} 0 & 0 & 0 & 0 \\ 0 & 0 & 0 & 0 \\ 0 & 0 & 0 & 0 \\ 0 & 0 & 0 & \lambda_4 \end{bmatrix}[X_A]^{-1} = \lambda_4\,[X_A]\begin{bmatrix} 0 & 0 & 0 & 0 \\ 0 & 0 & 0 & 0 \\ 0 & 0 & 0 & 0 \\ 0 & 0 & 0 & 1 \end{bmatrix}\left[X_A^{-1}\right]$$

or

$$A^- = \lambda_4 \begin{bmatrix} \alpha_1 & \alpha_2 & \alpha_3 & \alpha_4 \\ \alpha_1\beta_1 & \alpha_2\beta_1 & \alpha_3\beta_1 & \alpha_4\beta_1 \\ \alpha_1\beta_2 & \alpha_2\beta_2 & \alpha_3\beta_2 & \alpha_4\beta_2 \\ \alpha_1\beta_3 & \alpha_2\beta_3 & \alpha_3\beta_3 & \alpha_4\beta_3 \end{bmatrix} \tag{12-158}$$

Where

$$\alpha_1 = \frac{2a(u\xi_x + v\xi_y)\sqrt{\xi_x^2+\xi_y^2} + (\gamma-1)\,(u^2+v^2)\,(\xi_x^2+\xi_y^2)}{4a^2(\xi_x^2+\xi_y^2)}$$

$$\alpha_2 = -\left[\frac{a\xi_x\sqrt{\xi_x^2+\xi_y^2} + (\gamma-1)u(\xi_x^2+\xi_y^2)}{2a^2(\xi_x^2+\xi_y^2)}\right]$$

$$\alpha_3 = -\frac{\left[a\xi_y\sqrt{\xi_x^2+\xi_y^2} + (\gamma-1)v(\xi_x^2+\xi_y^2)\right]}{2a^2(\xi_x^2+\xi_y^2)}$$

$$\alpha_4 = \frac{(\gamma-1)}{2a^2}$$

$$\beta_1 = u - \frac{a\xi_x}{\sqrt{\xi_x^2+\xi_y^2}}$$

$$\beta_2 = v - \frac{a\xi_y}{\sqrt{\xi_x^2+\xi_y^2}}$$

$$\beta_3 = \left[-\frac{a(u\xi_x+v\xi_y)}{\sqrt{\xi_x^2+\xi_y^2}} + \frac{u^2+v^2}{2} + \frac{a^2}{\gamma-1}\right]$$

Finally, $A^+ = A - A^-$

The corresponding flux vector $\bar{E}$ is decomposed as follows:

$$\bar{E}^- = \lambda_4\frac{\rho}{2\gamma J}\begin{bmatrix} 1 \\ u - \dfrac{a\xi_x}{\sqrt{\xi_x^2+\xi_y^2}} \\ v - \dfrac{a\xi_y}{\sqrt{\xi_x^2+\xi_y^2}} \\ -\dfrac{a(\xi_x u+\xi_y v)}{\sqrt{\xi_x^2+\xi_y^2}} + \dfrac{u^2+v^2}{2} + \dfrac{a^2}{\gamma-1} \end{bmatrix} \tag{12-159}$$

and

$$\bar{E}^+ = \bar{E} - \bar{E}^-$$

(2) λ_1 is negative and either,

(a) $\lambda_3 < 0$; thus, all the eigenvalues of matrix A are negative. Consequently,
$A^+ = 0$
$A^- = A$ (given by (12-136))

The corresponding flux vector $\bar{E}$ is decomposed as
$\bar{E}^+ = 0$, and
$\bar{E}^- = \bar{E}$

(b) $\lambda_3 > 0$, for which A^+ is evaluated as follows

$$A^+ = [X_A] \begin{bmatrix} 0 & 0 & 0 & 0 \\ 0 & 0 & 0 & 0 \\ 0 & 0 & \lambda_3 & 0 \\ 0 & 0 & 0 & 0 \end{bmatrix} \left[X_A^{-1}\right] = \lambda_3 [X_A] \begin{bmatrix} 0 & 0 & 0 & 0 \\ 0 & 0 & 0 & 0 \\ 0 & 0 & 1 & 0 \\ 0 & 0 & 0 & 0 \end{bmatrix} \left[X_A^{-1}\right]$$

or

$$A^+ = \lambda_3 \begin{bmatrix} \alpha_5 & \alpha_6 & \alpha_7 & \alpha_8 \\ \alpha_5\beta_4 & \alpha_6\beta_4 & \alpha_7\beta_4 & \alpha_8\beta_4 \\ \alpha_5\beta_5 & \alpha_6\beta_5 & \alpha_7\beta_5 & \alpha_8\beta_5 \\ \alpha_5\beta_6 & \alpha_6\beta_6 & \alpha_7\beta_6 & \alpha_8\beta_6 \end{bmatrix} \tag{12-160}$$

where

$$\alpha_5 = \frac{-\left[2a(u\xi_x + v\xi_y)\sqrt{\xi_x^2 + \xi_y^2}\right] + \left[(\gamma - 1)\,(u^2 + v^2)\,(\xi_x^2 + \xi_y^2)\right]}{4a^2(\xi_x^2 + \xi_y^2)}$$

$$\alpha_6 = \frac{a\xi_x\sqrt{\xi_x^2 + \xi_y^2} - (\gamma - 1)u(\xi_x^2 + \xi_y^2)}{2a^2(\xi_x^2 + \xi_y^2)}$$

$$\alpha_7 = \frac{a\xi_y\sqrt{\xi_x^2 + \xi_y^2} - (\gamma - 1)v(\xi_x^2 + \xi_y^2)}{2a^2(\xi_x^2 + \xi_y^2)}$$

$$\alpha_8 = \frac{(\gamma - 1)}{2a^2}$$

$$\beta_4 = u + \frac{a\xi_x}{\sqrt{\xi_x^2 + \xi_y^2}}$$

$$\beta_5 = v + \frac{a\xi_y}{\sqrt{\xi_x^2 + \xi_y^2}}$$

$$\beta_6 = \frac{a(u\xi_x + v\xi_y)}{\sqrt{\xi_x^2 + \xi_y^2}} + \frac{u^2 + v^2}{2} + \frac{a^2}{\gamma - 1}$$

and

$$A^- = A - A^+$$

The decomposition of flux vector $\bar{E}$ provides

$$\bar{E}^+ = \lambda_3 \frac{\rho}{2\gamma J} \begin{bmatrix} 1 \\ u + \dfrac{a\xi_x}{\sqrt{\xi_x^2 + \xi_y^2}} \\ v + \dfrac{a\xi_y}{\sqrt{\xi_x^2 + \xi_y^2}} \\ \dfrac{a(\xi_x u + \xi_y v)}{\sqrt{\xi_x^2 + \xi_y^2}} + \dfrac{u^2 + v^2}{2} + \dfrac{a^2}{\gamma - 1} \end{bmatrix} \tag{12-161}$$

and

$$\bar{E}^- = \bar{E} - \bar{E}^+$$

Similar matrix manipulations are used to determine the Jacobian matrices B^+ and B^- and the flux vectors $\bar{F}^+$ and $\bar{F}^-$. The results are identical to that of A^+, A^-, $\bar{E}^+$ and $\bar{E}^-$, except the metrics ξ_x and ξ_y are replaced by η_x and η_y.

12.9.2.1.2 Existence of Zero Metrics Within the Domain: A difficulty may arise for problems for which the metric ξ_x becomes zero. To overcome this difficulty, the metric ξ_x in the original equation, i.e., (12-126), is set to zero. Subsequently, a new Jacobian matrix A_0 is determined as

$$A_0 = \begin{bmatrix} \xi_t & 0 & \xi_y & 0 \\ -\xi_y uv & \xi_t + \xi_y v & \xi_y u & 0 \\ \xi_y \left[(\gamma - 1)u^2 + (\gamma - 3)v^2\right]/2 & \xi_y u(1 - \gamma) & \xi_t + \xi_y(3 - \gamma)v & \xi_y(\gamma - 1) \\ \xi_y v \left[-a^2/(\gamma - 1) + (\gamma - 2)(u^2 + v^2)/2\right] & \xi_y(1 - \gamma)uv & \xi_y\left[a^2/(\gamma - 1) + u^2/2 + (3/2 - \gamma)v^2\right] & \xi_t + \gamma\xi_y v \end{bmatrix} \tag{12-162}$$

The eigenvalues of A_0 are

$$\lambda_{\xi_{01}} = \xi_t + \xi_y v$$

$$\lambda_{\xi_{02}} = \xi_t + \xi_y v$$

$$\lambda_{\xi_{03}} = \xi_t + (v + a)\xi_y$$

$$\lambda_{\xi_{04}} = \xi_t + (v - a)\xi_y$$

The eigenvector matrix is

$$X_{Ao} = \begin{bmatrix} 0 & 1 & 1 & 1 \\ 1 & 0 & u & u \\ 0 & v & v+a & v-a \\ u & \dfrac{v^2 - u^2}{2} & \dfrac{a^2}{\gamma - 1} + \dfrac{(u^2 + v^2 + 2av)}{2} & \dfrac{a^2}{\gamma - 1} + \dfrac{(u^2 + v^2 - 2av)}{2} \end{bmatrix}$$

(12-163)

where the inverse of the matrix X_{A_0} is

$$X_{A_0}^{-1} = \begin{bmatrix} -\dfrac{(\gamma-1)\,(u^3 + uv^2)}{2a^2} & 1 + \dfrac{(\gamma-1)u^2}{a^2} & \dfrac{(\gamma-1)uv}{a^2} & \dfrac{-(\gamma-1)u}{a^2} \\ 1 - \dfrac{(\gamma-1)\,(u^2+v^2)}{2a^2} & \dfrac{(\gamma-1)u}{a^2} & \dfrac{(\gamma-1)v}{a^2} & \dfrac{-(\gamma-1)}{a^2} \\ \dfrac{(\gamma-1)\,(u^2+v^2) - 2av}{4a^2} & \dfrac{-(\gamma-1)u}{2a^2} & \dfrac{a - (\gamma-1)v}{2a^2} & \dfrac{(\gamma-1)}{2a^2} \\ \dfrac{(\gamma-1)\,(u^2+v^2) + 2av}{4a^2} & \dfrac{-(\gamma-1)u}{2a^2} & -\dfrac{[a + (\gamma-1)v]}{2a^2} & \dfrac{(\gamma-1)}{2a^2} \end{bmatrix}$$

(12-164)

For problems in which η_x becomes zero, the above procedure is repeated. The matrix B_0 has the same form as A_0, except ξ_y is replaced by η_y. The eigenvalues of B_0 are

$$\lambda_{\eta_{01}} = \eta_t + \eta_y v$$

$$\lambda_{\eta_{02}} = \eta_t + \eta_y v$$

$$\lambda_{\eta_{03}} = \eta_t + (v + a)\eta_y$$

$$\lambda_{\eta_{04}} = \eta_t + (v - a)\eta_y$$

The Jacobian matrix B_0 is identical to that of A_0, except ξ_y is replaced by η_y. Since the eigenvectors become independent of the metrics, as seen by the eigenvector matrix X_{A_0}, the eigenvector matrix X_{B_0} and its inverse $X_{B_0}^{-1}$ are identical to that of X_{A_0} and $X_{A_0}^{-1}$.

It should be noted that the formulations above are employed only in regions of the domain where the corresponding metrics become zero. In the remainder of the domain, the original formulations are utilized.

12.9.2.1.3 Eigenvector Matrices: It is beneficial to elaborate on the eigenvector matrices previously derived. For example, consider the eigenvector matrix X_A repeated here for convenience.

$$X_A = \begin{bmatrix} 0 & 1 & 1 & 1 \\ -\dfrac{\xi_y}{\xi_x} & \dfrac{u\xi_x + v\xi_y}{\xi_x} & u + \dfrac{a\xi_x}{\sqrt{\xi_x^2 + \xi_y^2}} & u - \dfrac{a\xi_x}{\sqrt{\xi_x^2 + \xi_y^2}} \\ 1 & 0 & v + \dfrac{a\xi_y}{\sqrt{\xi_x^2 + \xi_y^2}} & v - \dfrac{a\xi_y}{\sqrt{\xi_x^2 + \xi_y^2}} \\ \dfrac{v\xi_x - u\xi_y}{\xi_x} & \dfrac{2uv\xi_y + (u^2 - v^2)\xi_x}{2\xi_x} & \dfrac{u^2 + v^2}{2} + \dfrac{a^2}{\gamma - 1} + \dfrac{a(u\xi_x + v\xi_y)}{\sqrt{\xi_x^2 + \xi_y^2}} & \dfrac{u^2 + v^2}{2} + \dfrac{a^2}{\gamma - 1} - \dfrac{a(u\xi_x + v\xi_y)}{\sqrt{\xi_x^2 + \xi_y^2}} \end{bmatrix}$$

Different forms of the eigenvector matrix X_A will be encountered in literature. That is due to the fact that the constants appearing in the eigenvectors can be set arbitrarily. Therefore, different forms of eigenvector matrices can be obtained. To make this point more clear and to show how the arbitrary constants were selected, consider the general form of the eigenvector associated with λ_{ξ_1} and λ_{ξ_2}. Note that since the eigenvalues are repeated, two arbitrary constants will appear in the general expression for the corresponding eigenvectors. The general expression for the eigenvector is determined to be

$$X_{A_{1,2}} = \begin{bmatrix} \alpha \\ \frac{\xi_y}{\xi_x}(\alpha v - \beta) + \alpha \xi_x u \\ \beta \\ \frac{1}{2\xi_x}\left[(2\alpha u v - 2\beta u)\xi_y + (-\alpha v^2 + 2\beta v + \alpha u^2)\xi_x\right] \end{bmatrix}$$

Various combinations of the constants α and β can be selected arbitrarily to provide the two eigenvectors. For example, the selection of $(\alpha, \beta) = (0, 1)$ and $(\alpha, \beta) = (1, 0)$ provides the first two eigenvectors appearing in matrix X_A.

Since unlimited choices are available for the selection of the arbitrary constants, different forms of the eigenvector matrices can be introduced. However, the final outcome which is to determine, for example A^+, would be the same irregardless of the specific form of X_A used. Obviously, that is due to the fact that for each X_A there is a corresponding inverse matrix, and, once an operation such as (12-150) is carried out, the results would be identical. Clearly, it is advantageous to select suitable constants so as to provide the eigenvector matrix as simple as possible.

12.9.2.2 Van Leer Flux Vector Splitting: The flux vector splitting applied to the convective terms of the Euler equation in the previous section follows the procedure introduced by Steger and Warming [12-1]. A disadvantage of this scheme is that the split fluxes are not continuously differentiable at the sonic point. Therefore, a non-smooth solution in the vicinity of sonic points may appear within the domain. Some simple procedure can be used to overcome this problem. For example, one may exclude sonic points or one may add an arbitrary small number to the eigenvalues, thus preventing them from ever being zero.

A flux splitting scheme which offers continuous differentiability is due to van Leer [12-2]. This scheme was explored in Section 12.3.2.2 for the one-dimensional Euler equation, and, therefore, only a brief discussion is provided.

The scheme requires the splitting of the flux vectors subject to the following conditions:

(1) $\bar{E}^+ = \bar{E}$ for $M_\xi \geq 1$ and $\bar{E}^- = \bar{E}$ for $M_\xi \leq -1$ where

$$M_\xi = \frac{\xi_x u + \xi_y v}{a\sqrt{\xi_x^2 + \xi_y^2}}$$

(2) For $|M_\xi| < 1$, one of the eigenvalues of A^+ (or A^-) must vanish.

(3) The components of the flux vector $\bar{E}^+$ (or $\bar{E}^-$) must be symmetric to each other with respect to M, as is the case with $\bar{E}$. Mathematically, if $\bar{E}_i(M) = \pm\bar{E}_i(-M)$, then $\bar{E}_i^+(M) = \pm\bar{E}_i^-(-M)$.

(4) The flux vectors $\bar{E}$, $\bar{E}^+$, and $\bar{E}^-$ must be polynomials of lowest possible degree in M.

Following the above set conditions, the split fluxes are constructed to provide the following,

(a) $M_\xi \geq +1$, then $\bar{E}^+ = \bar{E}$, and $\bar{E}^- = 0$ (similar to Steger and Warming scheme where all the eigenvalues are positive).

(b) $M_\xi \leq -1$, then $\bar{E}^- = \bar{E}$, and $\bar{E}^+ = 0$.

(c) $0 < M_\xi < 1$, then

$$\bar{E}^+ = \frac{\sqrt{\xi_x^2+\xi_y^2}}{J}\begin{bmatrix} e_1^+ \\ e_1^+\left[\dfrac{\xi_x}{\gamma\sqrt{\xi_x^2+\xi_y^2}}\left(-\bar{U}+2a\right)+u\right] \\ e_1^+\left[\dfrac{\xi_y}{\gamma\sqrt{\xi_x^2+\xi_y^2}}\left(-\bar{U}+2a\right)+v\right] \\ e_1^+\left[\dfrac{-(\gamma-1)\bar{U}^2+2(\gamma-1)a\bar{U}+2a^2}{(\gamma^2-1)}+\dfrac{u^2+v^2}{2}\right] \end{bmatrix} \qquad (12\text{-}165)$$

where $e_1^+ = \frac{1}{4}\rho a(M_\xi+1)^2$, $\bar{E}^- = \bar{E} - \bar{E}^+$, and

$$\bar{U} = \frac{\xi_x u + \xi_y v}{\sqrt{\xi_x^2+\xi_y^2}}$$

(d) $-1 < M_\xi < 0$, then

$$\bar{E}^- = \frac{\sqrt{\xi_x^2+\xi_y^2}}{J}\begin{bmatrix} e_1^- \\ e_1^-\left[\dfrac{\xi_x}{\gamma\sqrt{\xi_x^2+\xi_y^2}}\left(-\bar{U}-2a\right)+u\right] \\ e_1^-\left[\dfrac{\xi_y}{\gamma\sqrt{\xi_x^2+\xi_y^2}}\left(-\bar{U}-2a\right)+v\right] \\ e_1^-\left[\dfrac{-(\gamma-1)\bar{U}^2-2(\gamma-1)a\bar{U}+2a^2}{(\gamma^2-1)}+\dfrac{u^2+v^2}{2}\right] \end{bmatrix} \qquad (12\text{-}166)$$

where $e_1^- = -\frac{1}{4}\rho a(M_\xi - 1)^2$ and $\bar{E}^+ = \bar{E} - \bar{E}^-$.
The flux vectors $\bar{F}^-$ and $\bar{F}^+$ follow the same form as $\bar{E}^-$ and $\bar{E}^+$, where all the "ξ" are replaced by "η."

12.9.2.3 Modified Runge-Kutta Formulation: The modified fourth-order Runge-Kutta scheme introduced in Section 6.6.8 and implemented to write a finite difference equation for the one-dimensional Euler equation in Section 12.3.2.3 is extended to the two-dimensional Euler equation to provide the following finite difference equation.

$$\bar{Q}_{i,j}^{(1)} = \bar{Q}_{i,j}^{n} \tag{12-167}$$

$$\bar{Q}_{i,j}^{(2)} = \bar{Q}_{i,j}^{n} - \frac{\Delta\tau}{4}\left[\left(\frac{\partial\bar{E}}{\partial\xi}\right)_{i,j}^{(1)} + \left(\frac{\partial\bar{F}}{\partial\eta}\right)_{i,j}^{(1)} + \alpha\bar{H}_{i,j}^{(1)}\right] \tag{12-168}$$

$$\bar{Q}_{i,j}^{(3)} = \bar{Q}_{i,j}^{n} - \frac{\Delta\tau}{3}\left[\left(\frac{\partial\bar{E}}{\partial\xi}\right)_{i,j}^{(2)} + \left(\frac{\partial\bar{F}}{\partial\eta}\right)_{i,j}^{(2)} + \alpha\bar{H}_{i,j}^{(2)}\right] \tag{12-169}$$

$$\bar{Q}_{i,j}^{(4)} = \bar{Q}_{i,j}^{n} - \frac{\Delta\tau}{2}\left[\left(\frac{\partial\bar{E}}{\partial\xi}\right)_{i,j}^{(3)} + \left(\frac{\partial\bar{F}}{\partial\eta}\right)_{i,j}^{(3)} + \alpha\bar{H}_{i,j}^{(3)}\right] \tag{12-170}$$

$$\bar{Q}_{i,j}^{n+1} = \bar{Q}_{i,j}^{n} - \Delta\tau\left[\left(\frac{\partial\bar{E}}{\partial\xi}\right)_{i,j}^{(4)} + \left(\frac{\partial\bar{F}}{\partial\eta}\right)_{i,j}^{(4)} + \alpha\bar{H}_{i,j}^{(4)}\right] \tag{12-171}$$

The convective terms in Equations (12-168) through (12-171) are typically approximated by second-order central difference expressions, that is,

$$\left(\frac{\partial\bar{E}}{\partial\xi}\right)_{i,j} + \left(\frac{\partial\bar{F}}{\partial\eta}\right)_{i,j} = \frac{\bar{E}_{i+1,j} - \bar{E}_{i-1,j}}{2\Delta\xi} + \frac{\bar{F}_{i,j+1} - \bar{F}_{i,j-1}}{2\Delta\eta}$$

The addition of a damping term may be required to reduce the dispersion error. In that case, the damping term is added after the final stage, as follows

$$\bar{Q}_{i,j}^{n+1} = \bar{Q}_{i,j}^{n+1} + D_\xi + D_\eta$$

The damping terms could be specified as a combination of both second-order and fourth-order terms. Another scheme by which dispersion error which appears as oscillations in the vicinity of large gradient can be reduced is the use of TVD schemes which is addressed next.

12.9.2.4 Second-Order TVD Formulation: The second-order TVD scheme explored in Section 12.3.2.4 for the quasi one-dimensional Euler equation is extended to the two-dimensional Euler equation in computational space. The finite difference equation is written as

$$\bar{Q}_i^{n+1} = \bar{Q}_i^n - \frac{\Delta\tau}{\Delta\xi}\left[(R_\xi)^n_{i+\frac{1}{2},j} - (R_\xi)^n_{i-\frac{1}{2},j}\right] - \frac{\Delta\tau}{\Delta\eta}\left[(R_\eta)^n_{i,j+\frac{1}{2}} - (R_\eta)^n_{i,j-\frac{1}{2}}\right] - \Delta\tau(\alpha H^n_{i,j}) \tag{12-172}$$

where

$$(R_\xi)^n_{i+\frac{1}{2},j} = \frac{1}{2}\left[\bar{E}^n_{i+1,j} + \bar{E}^n_{i,j} + (X_A)^n_{i+\frac{1}{2},j}(\Phi_\xi)^n_{i+\frac{1}{2},j}\right] \tag{12-173}$$

$$(R_\xi)^n_{i-\frac{1}{2},j} = \frac{1}{2}\left[\bar{E}^n_{i,j} + \bar{E}^n_{i-1,j} + (X_A)^n_{i-\frac{1}{2},j}(\Phi_\xi)^n_{i-\frac{1}{2},j}\right] \tag{12-174}$$

$$(R_\eta)^n_{i,j+\frac{1}{2}} = \frac{1}{2}\left[\bar{F}^n_{i,j+1} + \bar{F}^n_{i,j} + (X_B)^n_{i,j+\frac{1}{2}}(\Phi_\eta)^n_{i,j+\frac{1}{2}}\right] \tag{12-175}$$

$$(R_\eta)^n_{i,j-\frac{1}{2}} = \frac{1}{2}\left[\bar{F}^n_{i,j} + \bar{F}^n_{i,j-1} + (X_B)^n_{i,j-\frac{1}{2}}(\Phi_\eta)^n_{i,j-\frac{1}{2}}\right] \tag{12-176}$$

Now, the flux limiters introduced previously for the one-dimensional Euler equation are extended to the two-dimensional equations.

12.9.2.4.1 Harten-Yee Upwind TVD: The general expressions for the component of the flux limiter vectors are defined as

$$(\Phi_\xi)^n_{i+\frac{1}{2},j} = \sigma\left[(\alpha_\xi)_{i+\frac{1}{2},j}\right]\left[(G_\xi)_{i+1,j} + (G_\xi)_{i,j}\right] - \psi\left[(\alpha_\xi)_{i+\frac{1}{2},j} + (\beta_\xi)_{i+\frac{1}{2},j}\right](\delta_\xi)_{i+\frac{1}{2},j} \tag{12-177}$$

$$(\Phi_\xi)^n_{i-\frac{1}{2},j} = \sigma\left[(\alpha_\xi)_{i-\frac{1}{2},j}\right]\left[(G_\xi)_{i,j} + (G_\xi)_{i-1,j}\right] - \psi\left[(\alpha_\xi)_{i-\frac{1}{2},j} + (\beta_\xi)_{i-\frac{1}{2},j}\right](\delta_\xi)_{i-\frac{1}{2},j} \tag{12-178}$$

$$(\Phi_\eta)^n_{i,j+\frac{1}{2}} = \sigma\left[(\alpha_\eta)_{i,j+\frac{1}{2}}\right]\left[(G_\eta)_{i,j+1} + (G_\eta)_{i,j}\right] - \psi\left[(\alpha_\eta)_{i,j+\frac{1}{2}} + (\beta_\eta)_{i,j+\frac{1}{2}}\right](\delta_\eta)_{i,j+\frac{1}{2}} \tag{12-179}$$

$$(\Phi_\eta)^n_{i,j-\frac{1}{2}} = \sigma\left[(\alpha_\eta)_{i,j-\frac{1}{2}}\right]\left[(G_\eta)_{i,j} + (G_\eta)_{i,j-1}\right] - \psi\left[(\alpha_\eta)_{i,j-\frac{1}{2}} + (\beta_\eta)_{i,j-\frac{1}{2}}\right](\delta_\eta)_{i,j-\frac{1}{2}} \tag{12-180}$$

where α_ξ is used to denote the eigenvalues of A given by (12-138) through (12-141), and α_η is used to denote the eigenvalues of B given by (12-144) through (12-147). As discussed in Section 12.3.2.4.1, each component of Φ_ξ or Φ_η is determined, and subsequently the flux limiter vectors Φ_ξ and Φ_η are formed. The terms appearing in the equations above are defined as follow.

$$\sigma(\alpha_\xi) = \frac{1}{2}\psi(\alpha_\xi) + \frac{\Delta t}{\Delta\xi}(\alpha_\xi)^2$$

$$\sigma(\alpha_\eta) = \frac{1}{2}\psi(\alpha_\eta) + \frac{\Delta t}{\Delta \eta}(\alpha_\eta)^2$$

$$(\delta_\xi)_{i+\frac{1}{2},j} = (X_A)^{-1}_{i+\frac{1}{2},j}\left(J^{-1}_{i+\frac{1}{2},j}\right)(Q_{i+1,j} - Q_{i,j}) \quad , \quad J_{i+\frac{1}{2},j} = \frac{1}{2}(J_{i+1,j} + J_{i,j})$$

$$(\delta_\eta)_{i,j+\frac{1}{2}} = (X_B)^{-1}_{i,j+\frac{1}{2}}\left(J^{-1}_{i,j+\frac{1}{2}}\right)(Q_{i,j+1} - Q_{i,j}) \quad , \quad J_{i,j+\frac{1}{2}} = \frac{1}{2}(J_{i,j+1} + J_{i,j})$$

$$(\beta_\xi)_{i+\frac{1}{2},j} = \sigma\left[(\alpha_\xi)_{i+\frac{1}{2},j}\right]\begin{cases} \dfrac{(G_\xi)_{i+1,j} - (G_\xi)_{i,j}}{(\delta_\xi)_{i+\frac{1}{2},j}} & , \text{ for } \quad (\delta_\xi)_{i+\frac{1}{2},j} \neq 0 \\ 0 & , \text{ for } \quad (\delta_\xi)_{i+\frac{1}{2},j} = 0 \end{cases}$$

and

$$(\beta_\eta)_{i,j+\frac{1}{2}} = \sigma\left[(\alpha_\eta)_{i,j+\frac{1}{2}}\right]\begin{cases} \dfrac{(G_\eta)_{i,j+1} - (G_\eta)_{i,j}}{(\delta_\eta)_{i,j+\frac{1}{2}}} & , \text{ for } \quad (\delta_\eta)_{i,j+\frac{1}{2}} \neq 0 \\ 0 & , \text{ for } \quad (\delta_\eta)_{i,j+\frac{1}{2}} = 0 \end{cases}$$

where ψ is defined by (12-59). The limiters G_ξ and G_η are similar to those given by (12-61) through (12-65) for the one-dimensional case. They are extended to two-dimensions and are as follow.

$$(G_\xi)_{i,j} = \text{minmod}\left[(\delta_\xi)_{i-\frac{1}{2},j} \quad , \quad (\delta_\xi)_{i+\frac{1}{2},j}\right] \tag{12-181}$$

$$(G_\xi)_{i,j} = \frac{(\delta_\xi)_{i+\frac{1}{2},j}(\delta_\xi)_{i-\frac{1}{2},j} + \left|(\delta_\xi)_{i+\frac{1}{2},j}(\delta_\xi)_{i-\frac{1}{2},j}\right|}{(\delta_\xi)_{i+\frac{1}{2},j} + (\delta_\xi)_{i-\frac{1}{2},j}} \tag{12-182}$$

$$(G_\xi)_{i,j} = \frac{(\delta_\xi)_{i-\frac{1}{2},j}\left\{\left[(\delta_\xi)_{i+\frac{1}{2},j}\right]^2 + \omega\right\} + (\delta_\xi)_{i+\frac{1}{2},j}\left\{\left[(\delta_\xi)_{i-\frac{1}{2},j}\right]^2 + \omega\right\}}{\left[(\delta_\xi)_{i+\frac{1}{2},j}\right]^2 + \left[(\delta_\xi)_{i-\frac{1}{2},j}\right]^2 + 2\omega} \tag{12-183}$$

$$(G_\xi)_{i,j} = \text{minmod}\left\{2(\delta_\xi)_{i-\frac{1}{2},j} \quad , \quad 2(\delta_\xi)_{i+\frac{1}{2},j} \quad , \quad \frac{1}{2}\left[(\delta_\xi)_{i+\frac{1}{2},j} + (\delta_\xi)_{i-\frac{1}{2},j}\right]\right\} \tag{12-184}$$

$$(G_\xi)_{i,j} = Sgn * \max\Big\{0 \quad , \quad \min\Big[2\left|(\delta_\xi)_{i+\frac{1}{2},j}\right| \quad , \quad Sgn * (\delta_\xi)_{i-\frac{1}{2},j}\Big] \quad ,$$

$$\min\left[\left|(\delta_\xi)_{i+\frac{1}{2},j}\right| \quad , \quad 2Sgn*(\delta_\xi)_{i-\frac{1}{2},j}\right]\Big\} \tag{12-185}$$

where

$$Sgn = Sgn(\delta_\xi)_{i+\frac{1}{2},j} = \frac{ABS\left[(\delta_\xi)_{i+\frac{1}{2},j}\right]}{(\delta_\xi)_{i+\frac{1}{2},j}}$$

$$(G_\eta)_{i,j} = \text{minmod}\left[(\delta_\eta)_{i,j-\frac{1}{2}} \quad , \quad (\delta_\eta)_{i,j+\frac{1}{2}}\right] \tag{12-186}$$

$$(G_\eta)_{i,j} = \frac{(\delta_\eta)_{i,j+\frac{1}{2}}(\delta_\eta)_{i,j-\frac{1}{2}} + \left|(\delta_\eta)_{i,j+\frac{1}{2}}(\delta_\eta)_{i,j-\frac{1}{2}}\right|}{(\delta_\eta)_{i,j+\frac{1}{2}} + (\delta_\eta)_{i,j-\frac{1}{2}}} \tag{12-187}$$

$$(G_\eta)_{i,j} = \frac{(\delta_\eta)_{i,j-\frac{1}{2}}\left\{\left[(\delta_\eta)_{i,j+\frac{1}{2}}\right]^2 + \omega\right\} + (\delta_\eta)_{i,j+\frac{1}{2}}\left\{\left[(\delta_\eta)_{i,j-\frac{1}{2}}\right]^2 + \omega\right\}}{\left[(\delta_\eta)_{i,j+\frac{1}{2}}\right]^2 + \left[(\delta_\eta)_{i,j-\frac{1}{2}}\right]^2 + 2\omega} \tag{12-188}$$

where $10^{-7} \leq \omega \leq 10^{-5}$

$$(G_\eta)_{i,j} = \text{minmod}\left\{2(\delta_\eta)_{i,j-\frac{1}{2}} \quad , \quad 2(\delta_\eta)_{i,j+\frac{1}{2}} \quad , \quad \frac{1}{2}\left[(\delta_\eta)_{i,j+\frac{1}{2}} + (\delta_\eta)_{i,j-\frac{1}{2}}\right]\right\} \tag{12-189}$$

$$(G_\eta)_{i,j} = Sgn*\max\Big\{0 \quad , \quad \min\left[2\left|(\delta_\eta)_{i,j+\frac{1}{2}}\right| \quad , \quad Sgn*(\delta_\eta)_{i,j-\frac{1}{2}}\right] \quad ,$$

$$\min\left[\left|(\delta_\eta)_{i,j+\frac{1}{2}}\right| \quad , \quad 2Sgn*(\delta_\eta)_{i,j-\frac{1}{2}}\right]\Big\} \tag{12-190}$$

where

$$Sgn = Sgn(\delta_\eta)_{i,j+\frac{1}{2}} = \frac{ABS\left[(\delta_\eta)_{i,j+\frac{1}{2}}\right]}{(\delta_\eta)_{i,j+\frac{1}{2}}}$$

12.9.2.4.2 Roe-Sweby Upwind TVD: The general expression for the components of the flux limiter vectors are given by

$$(\Phi_\xi)^n_{i+\frac{1}{2},j} = \left\{\frac{1}{2}(G_\xi)_{i,j}\left[\left|(\alpha_\xi)_{i+\frac{1}{2},j}\right| + \frac{\Delta t}{\Delta\xi}\left((\alpha_\xi)_{i+\frac{1}{2},j}\right)^2\right] - \left|(\alpha_\xi)_{i+\frac{1}{2},j}\right|\right\}(\delta_\xi)_{i+\frac{1}{2},j} \tag{12-191}$$

$$(\Phi_\xi)^n_{i-\frac{1}{2},j} = \left\{ \frac{1}{2}(G_\xi)_{i-1,j} \left[\left|(\alpha_\xi)_{i-\frac{1}{2},j}\right| + \frac{\Delta t}{\Delta \xi}\left((\alpha_\xi)_{i-\frac{1}{2},j}\right)^2 \right] - \left|(\alpha_\xi)_{i-\frac{1}{2},j}\right| \right\} (\delta_\xi)_{i-\frac{1}{2},j} \quad (12\text{-}192)$$

$$(\Phi_\eta)^n_{i,j+\frac{1}{2}} = \left\{ \frac{1}{2}(G_\eta)_{i,j} \left[\left|(\alpha_\eta)_{i,j+\frac{1}{2}}\right| + \frac{\Delta t}{\Delta \eta}\left((\alpha_\eta)_{i,j+\frac{1}{2}}\right)^2 \right] - \left|(\alpha_\eta)_{i,j+\frac{1}{2}}\right| \right\} (\delta_\eta)_{i,j+\frac{1}{2}} \quad (12\text{-}193)$$

$$(\Phi_\eta)^n_{i,j-\frac{1}{2}} = \left\{ \frac{1}{2}(G_\eta)_{i,j-1} \left[\left|(\alpha_\eta)_{i,j-\frac{1}{2}}\right| + \frac{\Delta t}{\Delta \eta}\left((\alpha_\eta)_{i,j-\frac{1}{2}}\right)^2 \right] - \left|(\alpha_\eta)_{i,j-\frac{1}{2}}\right| \right\} (\delta_\eta)_{i,j-\frac{1}{2}} \quad (12\text{-}194)$$

Any one of the following limiters can be used.

$$(G_\xi)_{i,j} = \text{minmod}\,(1, r_\xi) \quad (12\text{-}195)$$

$$(G_\xi)_{i,j} = \frac{r_\xi + |r_\xi|}{1 + r_\xi} \quad (12\text{-}196)$$

$$(G_\xi)_{i,j} = \max\,[0\,,\ \min\,(2r_\xi,\ 1)\ ,\ \min(2,\ r_\xi)] \quad (12\text{-}197)$$

$$(G_\eta)_{i,j} = \text{minmod}\,(1,\ r_\eta) \quad (12\text{-}198)$$

$$(G_\eta)_{i,j} = \frac{r_\eta + |r_\eta|}{1 + r_\eta} \quad (12\text{-}199)$$

$$(G_\eta)_{i,j} = \max\,[0\,,\ \min(2r_\eta,\ 1)\ ,\ \min(2,\ r_\eta)] \quad (12\text{-}200)$$

where

$$r_\xi = \begin{cases} \dfrac{(X_A^{-1})_{i+1+\sigma,j} Q_{i+1+\sigma,j} - (X_\xi^{-1})_{i+\sigma,j} Q_{i+\sigma,j}}{(\delta_\xi)_{i+\frac{1}{2},j} J_{i+\frac{1}{2}+\sigma,j}} & , \text{ for } (\delta_\xi)_{i+\frac{1}{2},j} \neq 0 \\ 0 & , \text{ for } (\delta_\xi)_{i+\frac{1}{2},j} = 0 \end{cases}$$

with

$$J_{i+\frac{1}{2}+\sigma,j} = \frac{1}{2}\left(J_{i+1+\sigma,j} + J_{i+\sigma,j}\right)$$
$$\sigma = Sgn(\alpha_\xi)_{i+\frac{1}{2},j}$$

and

$$r_\eta = \begin{cases} \dfrac{(X_B^{-1})_{i,j+1+\gamma} Q_{i,j+1+\gamma} - (X_B^{-1})_{i,j+\gamma} Q_{i,j+\gamma}}{(\delta_\eta)_{i,j+\frac{1}{2}} J_{i,j+\frac{1}{2}+\gamma}} & , \text{ for } (\delta_\eta)_{i,j+\frac{1}{2}} \neq 0 \\ 0 & , \text{ for } (\delta_\eta)_{i,j+\frac{1}{2}} = 0 \end{cases}$$

with

$$J_{i,j+\frac{1}{2}+\gamma} = \frac{1}{2}(J_{i,j+1+\gamma} + J_{i,j+\gamma})$$

$$\gamma = Sgn(\alpha_\eta)_{i,j+\frac{1}{2}}$$

12.9.2.4.3 Davis-Yee Symmetric TVD: The general expression for the components of the flux limiter vectors are

$$(\Phi_\xi)^n_{i+\frac{1}{2},j} = -\left\{\frac{\Delta t}{\Delta \xi}\left[(\alpha_\xi)_{i+\frac{1}{2},j}\right]^2 (G_\xi)_{i+\frac{1}{2},j} + \psi\left[(\alpha_\xi)_{i+\frac{1}{2},j}\right]\left[(\delta_\xi)_{i+\frac{1}{2},j} - (G_\xi)_{i+\frac{1}{2},j}\right]\right\} \tag{12-201}$$

$$(\Phi_\xi)^n_{i-\frac{1}{2},j} = -\left\{\frac{\Delta t}{\Delta \xi}\left[(\alpha_\xi)_{i-\frac{1}{2},j}\right]^2 (G_\xi)_{i-\frac{1}{2},j} + \psi\left[(\alpha_\xi)_{i-\frac{1}{2},j}\right]\left[(\delta_\xi)_{i-\frac{1}{2},j} - (G_\xi)_{i-\frac{1}{2},j}\right]\right\} \tag{12-202}$$

$$(\Phi_\eta)^n_{i,j+\frac{1}{2}} = -\left\{\frac{\Delta t}{\Delta \eta}\left[(\alpha_\eta)_{i,j+\frac{1}{2}}\right]^2 (G_\eta)_{i,j+\frac{1}{2}} + \psi\left[(\alpha_\eta)_{i,j+\frac{1}{2}}\right]\left[(\delta_\eta)_{i,j+\frac{1}{2}} - (G_\eta)_{i,j+\frac{1}{2}}\right]\right\} \tag{12-203}$$

$$(\Phi_\eta)^n_{i,j-\frac{1}{2}} = -\left\{\frac{\Delta t}{\Delta \eta}\left[(\alpha_\eta)_{i,j-\frac{1}{2}}\right]^2 (G_\eta)_{i,j-\frac{1}{2}} + \psi\left[(\alpha_\eta)_{i,j-\frac{1}{2}}\right]\left[(\delta_\eta)_{i,j-\frac{1}{2}} - (G_\eta)_{i,j-\frac{1}{2}}\right]\right\} \tag{12-204}$$

Any one of the following limiters can be used.

$$(G_\xi)_{i+\frac{1}{2},j} = \text{minmod}\left\{2(\delta_\xi)_{i-\frac{1}{2},j}, 2(\delta_\xi)_{i+\frac{1}{2},j}, 2(\delta_\xi)_{i+\frac{3}{2},j}, \frac{1}{2}\left[(\delta_\xi)_{i-\frac{1}{2},j} + (\delta_\xi)_{i+\frac{3}{2},j}\right]\right\} \tag{12-205}$$

$$(G_\xi)_{i+\frac{1}{2},j} = \text{minmod}\left\{(\delta_\xi)_{i-\frac{1}{2},j}, (\delta_\xi)_{i+\frac{1}{2},j}, (\delta_\xi)_{i+\frac{3}{2},j}\right\} \tag{12-206}$$

$$(G_\xi)_{i+\frac{1}{2},j} = \text{minmod}\left[(\delta_\xi)_{i+\frac{1}{2},j}, (\delta_\xi)_{i-\frac{1}{2},j}\right] + \text{minmod}\left[(\delta_\xi)_{i+\frac{1}{2},j}, (\delta_\xi)_{i+\frac{3}{2},j}\right] - (\delta_\xi)_{i+\frac{1}{2},j} \tag{12-207}$$

$$(G_\eta)_{i,j+\frac{1}{2}} = \text{minmod}\left\{2(\delta_\eta)_{i,j-\frac{1}{2}}, 2(\delta_\eta)_{i,j+\frac{1}{2}}, 2(\delta_\eta)_{i,j+\frac{3}{2}}, \frac{1}{2}\left[(\delta_\eta)_{i,j-\frac{1}{2}} + (\delta_\eta)_{i,j+\frac{3}{2}}\right]\right\} \tag{12-208}$$

$$(G_\eta)_{i,j+\frac{1}{2}} = \text{minmod}\left[(\delta_\eta)_{i,j-\frac{1}{2}}, (\delta_\eta)_{i,j+\frac{1}{2}}, (\delta_\eta)_{i,j+\frac{3}{2}}\right] \tag{12-209}$$

$$(G_\eta)_{i,j+\frac{1}{2}} = \text{minmod}\left[(\delta_\eta)_{i,j+\frac{1}{2}}, (\delta_\eta)_{i,j-\frac{1}{2}}\right] + \text{minmod}\left[(\delta_\eta)_{i,j+\frac{1}{2}}, (\delta_\eta)_{i,j+\frac{3}{2}}\right] - (\delta_\eta)_{i,j+\frac{1}{2}} \tag{12-210}$$

12.9.2.5 Modified Runge-Kutta Scheme with TVD: The modified fourth-order Runge-Kutta scheme given in Section 12.9.2.3 can be amended by a TVD scheme to reduce any dispersion error that may be developed within the domain of solution. The governing equations are given by Equations (12-167) through (12-171). After the computation of Equation (12-171), the value of $\bar{Q}$ is updated according to

$$\bar{Q}_{i,j}^{n+1} = \bar{Q}_{i,j}^{n+1} - \frac{1}{2}\frac{\Delta\tau}{\Delta\xi}\left[(X_A)_{i+\frac{1}{2},j}^{n}(\Phi_\xi)_{i+\frac{1}{2},j}^{n} - (X_A)_{i-\frac{1}{2},j}^{n}(\Phi_\xi)_{i-\frac{1}{2},j}^{n}\right]$$

$$- \frac{1}{2}\frac{\Delta\tau}{\Delta\eta}\left[(X_B)_{i,j+\frac{1}{2}}^{n}(\Phi_\eta)_{i,j+\frac{1}{2}}^{n} - (X_B)_{i,j-\frac{1}{2}}^{n}(\Phi_\eta)_{i,j-\frac{1}{2}}^{n}\right] \quad (12\text{-}211)$$

where any one of the flux limiter functions and limiters discussed in Section 12.9.2.4 can be used.

12.9.3 Boundary Conditions:

Several types of boundaries and associated boundary conditions were identified in Chapter Eight. The boundaries of a domain may include solid surfaces, symmetry line (or surface), inflow, and outflow. In this section, specification of boundary conditions are reviewed.

12.9.3.1 Body Surface: For an inviscid flow, the slip condition at the surface is used. There are several methods by which the velocity at the surface can be specified. One may extrapolate the velocity from the interior points; or, the same value for $\rho \vec{V}$ can be imposed at the surface with the vector rotated such that the velocity is tangent at the surface. Thus,

$$A_i = |\rho \vec{V}|_{i,2} = \sqrt{(\rho u)_{i,2}^2 + (\rho v)_{i,2}^2} \quad (12\text{-}212)$$

$$C_i = (\rho u)_{i,1} \quad (12\text{-}213)$$

$$D_i = (\rho v)_{i,1} \quad (12\text{-}214)$$

These quantities are illustrated in Figure 12-34.

The pressure at the surface may also be specified by several different methods. Perhaps the simplest is to use a zero-order extrapolation. Therefore,

$$p_{i,1} = p_{i,2} \quad (12\text{-}215)$$

Another scheme by which the pressure at the surface is computed is to use the momentum equation. The governing equation is developed in Appendix H.

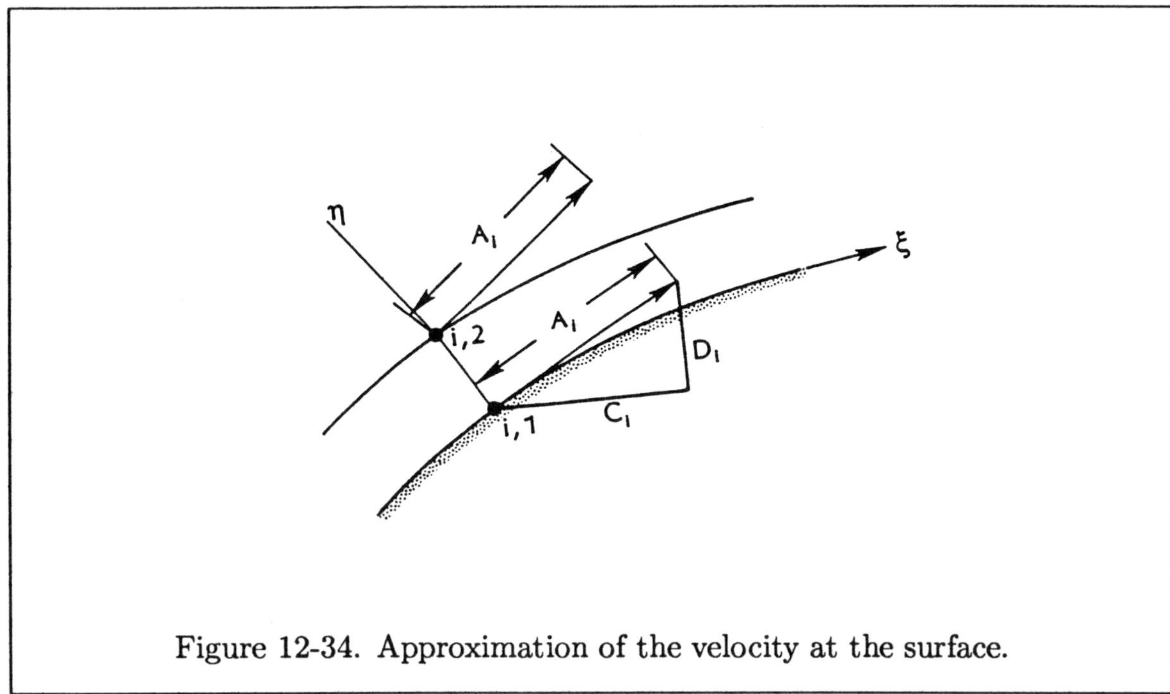

Figure 12-34. Approximation of the velocity at the surface.

An expression for the density at the surface is developed based on the assumption of a constant total enthalpy at the surface. This statement is expressed mathematically as

$$\gamma e + \frac{1}{2}(u^2 + v^2) = (h_t)_{\text{wall}} = \text{constant}$$

or, for a perfect gas

$$\gamma \left[\frac{p_{i,1}}{(\gamma - 1)\rho_{i,1}} \right] + \frac{1}{2} \frac{C_i^2 + D_i^2}{\rho_{i,1}^2} = (h_t)_{\text{wall}} \tag{12-216}$$

Note that $(h_t)_{\text{wall}}$ is known from its value at the freestream. Now Equation (12-216) is rearranged as

$$[2(h_t)_{\text{wall}}(\gamma - 1)]\,\rho_{i,1}^2 - (2\gamma)p_{i,1}\rho_{i,1} + (1 - \gamma)\,(C_i^2 + D_i^2) = 0$$

which may be solved for the density at the wall to yield:

$$\rho_{i,1} = \frac{2\gamma p_{i,1} + \sqrt{4\gamma^2 p_{i,1}^2 + 8(h_t)_{\text{wall}}(\gamma - 1)^2 (C_i^2 + D_i^2)}}{4(\gamma - 1)\,(h_t)_{\text{wall}}} \tag{12-217}$$

Note that the positive sign is used exclusively to prevent negative density values.

Now, with the equations describing the flow properties at the wall, as derived above, an iterative procedure is used for the solution as follows.

(1) Assume $\rho_{i,1}^k = \rho_{i,2}$

(2) Compute u and v from (12-213) and (12-214) as follows.

$$u_{i,1} = \frac{C_i}{\rho^k_{i,1}} = \frac{(\rho u)_{i,1}}{\rho^k_{i,1}} = \frac{(\rho V)_{i,2}\cos\theta_i}{\rho^k_{i,1}} \tag{12-218}$$

$$v_{i,1} = \frac{D_i}{\rho^k_{i,1}} = \frac{(\rho v)_{i,1}}{\rho^k_{i,1}} = \frac{(\rho V)_{i,2}\sin\theta_i}{\rho^k_{i,1}} \tag{12-219}$$

where θ_i is the local surface inclination at grid point i.

(3) Determine the pressure at the surface either by simple extrapolation such as (12-215) or by solving Equation (H-11). A numerical scheme to solve Equation (H-11) is provided in Appendix H.

(4) Use Equation (12-217) to compute the density.

(5) Check the convergence according to

$$\text{CONV} = \sum_{i=1}^{i=IM} \left| \frac{\rho^{k+1}_{i,1} - \rho^k_{i,1}}{\rho^{k+1}_{i,1}} \right| \le \epsilon$$

where ϵ is a specified small number (typically 0.01).

(6) If the convergence criterion is not met, then

$$C_i = C_i \cdot R$$

$$D_i = D_i \cdot R$$

where

$$R = \frac{1}{\rho^k_{i,1}} \left[W_1 \rho^{k+1}_{i,1} + W_2 \rho_{i,2} \right] \tag{12-220}$$

W_1 and W_2 are weighting functions, such that $W_1 + W_2 = 1.0$. Now, $\rho^k_{i,1}$ is set equal to $\rho^{k+1}_{i,1}$, and steps 2 through 6 are repeated. The iteration will continue until a converged solution is reached.

The procedure can also be performed using the metrics as follow. A unit vector normal to the surface can be determined according to

$$\hat{n} = \frac{\nabla\eta}{|\nabla\eta|}$$

where

$$\nabla\eta = \eta_x \hat{\imath} + \eta_y \hat{\jmath} = J(-y_\xi \hat{\imath} + x_\xi \hat{\jmath})$$

The velocity vector for a 2-D flow in Cartesian coordinates is expressed as

$$\vec{V} = u\hat{\imath} + v\hat{\jmath}$$

For mutually perpendicular vectors,

$$\vec{V} \cdot \hat{n} = 0$$

Thus,

$$u\eta_x + v\eta_y = 0$$

which results in

$$\frac{u}{v} = -\frac{\eta_y}{\eta_x}$$

Now the values of C_i and D_i corresponding to (12-213) and (12-214) are determined according to

$$C_i = \frac{A_i \eta_y}{B_i}$$

$$D_i = -\frac{A_i \eta_x}{B_i}$$

where

$$B_i = \sqrt{(\eta_x^2 + \eta_y^2)_{i,1}}$$

The same iterative procedure is used for the solution. Another method by which the flow properties at the surface can be determined is simply by the application of the equations described above without incorporating an iterative method. In this case, the pressure at the surface is set as $p_{i,1} = p_{i,2}$, and C_i and D_i are determined from (12-213) and (12-214). Now, relation (12-217) can be used to compute $\rho_{i,1}$, and, subsequently, (12-213) and (12-214) are used to determine $u_{i,1}$ and $v_{i,1}$.

12.9.3.2 Symmetry: The symmetry boundary can be used for configurations and the associated flowfields which are symmetric or axisymmetric. The grid line $i = 1$ is set below the axis of symmetry, and $i = 2$ is set above the axis of symmetry, as shown in Figure 12-36. Due to the symmetry of the grid lines at $i = 1$ and at $i = 2$, the Jacobian of Transformation is equal, that is, $J_{1,j} = J_{2,j}$ for all j. Now, from a physical point of view, the following constraints hold for the flow and thermodynamic properties.

$$\rho_{1,j} = \rho_{2,j}$$

$$u_{1,j} = u_{2,j}$$

$$v_{1,j} = -v_{2,j}$$

$$(e_t)_{1,j} = (e_t)_{2,j}$$

Therefore,

$$\Delta\bar{Q}_{1,j} = \frac{1}{J_{1,j}} \begin{bmatrix} (\Delta\rho)_{1,j} \\ (\Delta\rho u)_{1,j} \\ (\Delta\rho v)_{1,j} \\ (\Delta\rho e_t)_{1,j} \end{bmatrix} = \frac{1}{J_{2,j}} \begin{bmatrix} (\Delta\rho)_{2,j} \\ (\Delta\rho u)_{2,j} \\ -(\Delta\rho v)_{2,j} \\ (\Delta\rho e_t)_{2,j} \end{bmatrix} \tag{12-221}$$

12.9.3.3 Inflow: The inflow boundary is defined as a location for which $\vec{V} \cdot \vec{n}$ is negative, where vector $\vec{n}$ is the unit vector normal to the boundary in an outward direction. The specification of inflow boundary conditions follows the discussion of boundary conditions presented in Section 12.4. For a problem for which all the eigenvalues are positive, all the boundary conditions at the inflow are specified. For example, for a two-dimensional supersonic flow with all positive eigenvalues, the values of u, v, ρ, and p can be specified. If the flow at the inflow is subsonic and one of the eigenvalues is negative, then three boundary conditions are specified and one is extrapolated from the interior domain. Typically, the inflow boundary is set at the freestream, let's say at $j = JM$. If the freestream is independent of time and is supersonic, then no changes in the flow properties should occur. Furthermore, if the grid system is independent of time such that the Jacobian of transformation is constant, then the boundary condition is specified by

$$\Delta\bar{Q}_{i,JM} = 0$$

When the inflow is subsonic, one component of $\Delta\bar{Q}_{i,JM}$ would be nonzero and is determined from the interior points.

12.9.3.4 Outflow: The outflow boundary is defined as a location for which $\vec{V} \cdot \vec{n}$ is positive. The specification of boundary conditions at the outflow follows similar procedure as described in Section 12.4. That is, if all the eigenvalues are positive, then the information at the boundary is received from the interior of the domain, and, therefore, no boundary condition is specified. If one of the eigenvalues is negative, then one boundary condition is specified at the boundary and the remaining flow properties are determined from the interior of the domain. Recall that the calculation of flow properties at the boundary from the interior domain is by extrapolation, as described in Section 12.4. For example, when all the eigenvalues are positive and a zero-order extrapolation is used, then

$$\Delta\bar{Q}_{IM,j} = \Delta\bar{Q}_{IMM1,j}$$

where $i = IM$ defines the location of outflow boundary.

12.9.3.5 Boundary Conditions Based on Characteristics: The boundary conditions at the inflow and/or outflow can be specified and/or determined based on the characteristic variables defined in Appendix G. In this case, local one-dimensionality is assumed, and Riemann invariants are used. The procedure is best illustrated by the following example. Consider specification and/or determination of boundary conditions at $\eta = \eta_{JM}$, as shown in Figure 12-35. The components of the velocity normal and tangent to lines of constant η are determined in Appendix G and are

$$V_{n\eta} = \frac{\eta_x u + \eta_y v}{(\eta_x^2 + \eta_y^2)^{\frac{1}{2}}} \tag{12-222}$$

$$V_{t\eta} = \frac{\eta_y u - \eta_x v}{(\eta_x^2 + \eta_y^2)^{\frac{1}{2}}} \tag{12-223}$$

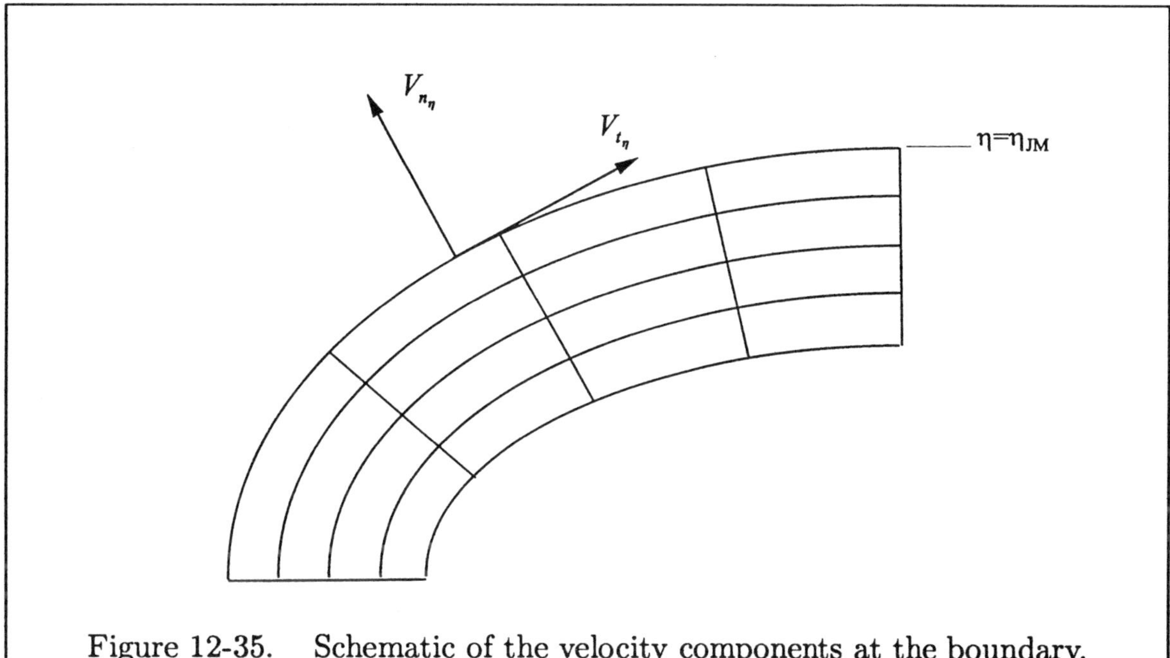

Figure 12-35. Schematic of the velocity components at the boundary.

Furthermore, the Riemann invariants defined by (F-62) and (F-63) can be written as

$$R^+ = V_{n\eta} + \frac{2a}{\gamma - 1} \tag{12-224}$$

and

$$R^- = V_{n\eta} - \frac{2a}{\gamma - 1} \tag{12-225}$$

and the two characteristic velocities are

$$\lambda_1 = V_{n\eta} + a \tag{12-226}$$

and

$$\lambda_2 = V_{n\eta} - a \tag{12-227}$$

Now the procedure to determine the variables at the boundary is as follows.

12.9.3.5.1 Inflow Boundary: A boundary is treated as an inflow boundary if $V_{n\eta} < 0$. As a consequence, $\lambda_2 = V_{n\eta} - a$ is also negative. Now $\lambda_1 = V_{n\eta} + a$ could be either positive or negative. If λ_1 is negative, then all the variables are set equal to the freestream values, that is,

$$R^- = R^-_\infty \ , \quad s = s_\infty \ , \quad V_{t\eta} = V_{t\eta_\infty} \ , \quad \text{and } R^+ = R^+_\infty$$

If λ_1 is positive, then R^+ is extrapolated from the interior of the domain, and R^-, s, and $V_{t\eta}$ are set to their freestream values as

$$R^- = R^-_\infty \ , \quad s = s_\infty \ , \quad \text{and } V_{t\eta} = V_{t\eta_\infty}$$

12.9.3.5.2 Outflow Boundary: A boundary is considered as outflow if $V_{n\eta} > 0$, in which case $\lambda_1 = V_{n\eta} + a$ is also positive. Now if $\lambda_2 = V_{n\eta} - a > 0$, then all the variables, that is, R^+, R^-, s, and $V_{t\eta}$ are extrapolated from the interior domain. When λ_2 is negative, then R^- is set to its freestream value, that is, $R^- = R^-_\infty$, and the remaining variables R^+, s, and $V_{t\eta}$ are extrapolated from the interior domain.

12.9.3.5.3 Determination of Flow Variables: Once the variables R^+, R^-, s, and $V_{t\eta}$ at the boundary have been determined, whether it is an inflow boundary or an outflow boundary, the flow variables u, v, p, and ρ at the boundary can be computed as follow:

$$V_{n\eta} = \frac{1}{2}(R^+ + R^-)$$

Now $V_{n\eta}$ and $V_{t\eta}$ given by (12-222) and (12-223) are used to compute u and v. Furthermore, the speed of sound can be determined from relations (12-224) and (12-225) according to

$$a = \frac{1}{4}(\gamma - 1)\,(R^+ - R^-) \tag{12-228}$$

In addition, the speed of sound is given by

$$a^2 = \gamma\frac{p}{\rho} \tag{12-229}$$

A combination of relations (12-228) and (12-229) provides

$$\frac{p}{\rho} = \frac{1}{\gamma}\left[\frac{1}{4}(\gamma - 1)\,(R^+ - R^-)\right]^2 \tag{12-230}$$

Furthermore,

$$ln\left(\frac{p}{\rho^{\gamma}}\right) = s \tag{12-231}$$

Now Equations (12-230) and (12-231) are used to determine the pressure and the density at the boundary.

12.9.4 Implicit Formulations

A general implicit formulation for the two-dimensional Euler equation can be written as

$$\frac{\Delta\bar{Q}}{\Delta\tau} + \left(\frac{\partial\bar{E}}{\partial\xi}\right)^{n+1} + \left(\frac{\partial\bar{F}}{\partial\eta}\right)^{n+1} + \alpha\left(\bar{H}\right)^{n+1} = 0 \tag{12-232}$$

In order to linearize Equation (12-232), relations (12-132) through (12-134) are substituted and rearranged as

$$\Delta\bar{Q} + \Delta\tau\left\{\frac{\partial}{\partial\xi}[\bar{E}^n + A^n\Delta\bar{Q}] + \frac{\partial}{\partial\eta}[\bar{F}^n + B^n\Delta\bar{Q}] + \alpha[\bar{H}^n + C^n\Delta\bar{Q}]\right\} = 0 \tag{12-233}$$

or

$$\left\{I + \Delta\tau\left[\frac{\partial}{\partial\xi}(A^n) + \frac{\partial}{\partial\eta}(B^n) + \alpha C^n\right]\right\}\Delta\bar{Q} = -\Delta\tau\left\{\frac{\partial\bar{E}^n}{\partial\xi} + \frac{\partial\bar{F}^n}{\partial\eta} + \alpha\bar{H}^n\right\} \tag{12-234}$$

Recall that when a scalar model equation for an unsteady two-dimensional equation was investigated [e.g., Equation (3-18)], it was concluded that the implicit formulation results in a system of equations with a pentadiagonal coefficient matrix. The solution of such a system is very time-consuming and, therefore, expensive. To overcome this situation, an approximate factorization procedure was introduced. This technique replaces the original FDE with a series of equations where the coefficient matrices are tridiagonal.

The approximate factorization approach is used in Equation (12-234) in order to reduce the pentadiagonal coefficient matrix into two tridiagonal systems.

The approximate factorization of Equation (12-234) is

$$\left[I + \Delta\tau\frac{\partial A^n}{\partial\xi}\right]\left[I + \Delta\tau\frac{\partial B^n}{\partial\eta} + \alpha\Delta\tau C^n\right]\Delta\bar{Q} = -\Delta\tau\left\{\frac{\partial\bar{E}^n}{\partial\xi} + \frac{\partial\bar{F}^n}{\partial\eta} + \alpha\bar{H}^n\right\} \tag{12-235}$$

Note that the terms on the right-hand side of this equation are evaluated at the known time level n. The unknown is the vector $\Delta\bar{Q}$.

Following the discussion of Section 12.3.3.1, the Steger and Warming flux vector splitting scheme is used to illustrate the development of an implicit scheme.

12.9.4.1 Steger and Warming Flux Vector Splitting: Consider Equation (12-235) where flux vector splitting is used. Hence,

$$\left\{I + \Delta\tau\left[\frac{\partial}{\partial\xi}(A^+ + A^-)\right]\right\}\left\{I + \Delta\tau\left[\frac{\partial}{\partial\eta}(B^+ + B^-)\right] + \alpha\Delta\tau C\right\}\Delta\bar{Q} =$$

$$-\Delta\tau\left\{\frac{\partial}{\partial\xi}(\bar{E}^+ + \bar{E}^-) + \frac{\partial}{\partial\eta}(\bar{F}^+ + \bar{F}^-) + \alpha\bar{H}\right\}$$

As discussed previously, a backward difference approximation is used for the positive terms, and a forward difference approximation is used for the negative terms. In the following, first-order approximations are used, and, therefore,

$$\left[I + \frac{\Delta\tau}{\Delta\xi}(A^+_{i,j} - A^+_{i-1,j} + A^-_{i+1,j} - A^-_{i,j})\right]\left[I + \frac{\Delta\tau}{\Delta\eta}(B^+_{i,j} - B^+_{i,j-1} + B^-_{i,j+1} - B^-_{i,j})\right.$$

$$\left. + \alpha\Delta\tau C_{i,j}\right]\Delta\bar{Q} = -\Delta\tau\left[\frac{1}{\Delta\xi}(\bar{E}^+_{i,j} - \bar{E}^+_{i-1,j} + \bar{E}^-_{i+1,j} - \bar{E}^-_{i,j})\right.$$

$$\left. + \frac{1}{\Delta\eta}(\bar{F}^+_{i,j} - \bar{F}^+_{i,j-1} + \bar{F}^-_{i,j+1} - \bar{F}^-_{i,j}) + \alpha\bar{H}_{i,j}\right] \qquad \text{(12-236)}$$

Equation (12-236) is solved in two stages as

$$\left[I + \frac{\Delta\tau}{\Delta\xi}(A^+_{i,j} - A^+_{i-1,j} + A^-_{i+1,j} - A^-_{i,j})\right]\Delta\bar{Q}^* =$$

$$-\Delta\tau\left[\frac{1}{\Delta\xi}(\bar{E}^+_{i,j} - \bar{E}^+_{i-1,j} + \bar{E}^-_{i+1,j} - \bar{E}^-_{i,j}) + \frac{1}{\Delta\eta}(\bar{F}^+_{i,j} - \bar{F}^+_{i,j-1} + \bar{F}^-_{i,j+1} - \bar{F}^-_{i,j}) + \alpha\bar{H}_{i,j}\right] \qquad \text{(12-237)}$$

and

$$\left[I + \frac{\Delta\tau}{\Delta\eta}(B^+_{i,j} - B^+_{i,j-1} + B^-_{i,j+1} - B^-_{i,j}) + \alpha\Delta\tau C_{i,j}\right]\Delta\bar{Q} = \Delta\bar{Q}^* \qquad \text{(12-238)}$$

where each equation will form a block tridiagonal coefficient matrix. These equations are rearranged as

$$\left(-\frac{\Delta\tau}{\Delta\xi}A^+_{i-1,j}\right)\Delta\bar{Q}^*_{i-1,j} + \left[I + \frac{\Delta\tau}{\Delta\xi}(A^+_{i,j} - A^-_{i,j})\right]\Delta\bar{Q}^*_{i,j} + \left(\frac{\Delta\tau}{\Delta\xi}A^-_{i+1,j}\right)\Delta\bar{Q}^*_{i+1,j} = (RHS)_{i,j} \qquad \text{(12-239)}$$

and

$$\left(-\frac{\Delta\tau}{\Delta\eta}B^+_{i,j-1}\right)\Delta\bar{Q}_{i,j-1} + \left[I + \frac{\Delta\tau}{\Delta\eta}(B^+_{i,j} - B^-_{i,j}) + \alpha\Delta\tau C_{i,j}\right]\Delta\bar{Q}_{i,j}$$

$$+ \left(\frac{\Delta\tau}{\Delta\eta}B^-_{i,j+1}\right)\Delta\bar{Q}_{i,j+1} = \Delta\bar{Q}^*_{i,j} \qquad \text{(12-240)}$$

The terms in Equations (12-239) and (12-240) are redefined as:

$$CAM = -\frac{\Delta\tau}{\Delta\xi} A^{+}_{i-1,j}$$

$$CA = \left[I + \frac{\Delta\tau}{\Delta\xi}(A^{+}_{i,j} - A^{-}_{i,j}) \right]$$

$$CAP = \frac{\Delta\tau}{\Delta\xi} A^{-}_{i+1,j}$$

$$CBM = -\frac{\Delta\tau}{\Delta\eta} B^{+}_{i,j-1}$$

$$CB = \left[I + \frac{\Delta\tau}{\Delta\eta}(B^{+}_{i,j} - B^{-}_{i,j}) + \alpha\Delta\tau C_{i,j} \right]$$

$$CBP = \frac{\Delta\tau}{\Delta\eta} B^{-}_{i,j+1}$$

Therefore, Equations (12-239) and (12-240) become

$$CAM_{i,j}\Delta\bar{Q}^{*}_{i-1,j} + CA_{i,j}\Delta\bar{Q}^{*}_{i,j} + CAP_{i,j}\Delta\bar{Q}^{*}_{i+1,j} = (RHS)_{i,j} \tag{12-241}$$

and

$$CBM_{i,j}\Delta\bar{Q}_{i,j-1} + CB_{i,j}\Delta\bar{Q}_{i,j} + CBP_{i,j}\Delta\bar{Q}_{i,j+1} = \Delta\bar{Q}^{*}_{i,j} \tag{12-242}$$

The grid system for Equations (12-241) and (12-242) may be generated by any one of the techniques introduced in Chapter 9. If a grid line is aligned along the stagnation streamline, which for the axisymmetric configuration at zero degree angle of attack is coincident with the body axis, some difficulty in convergence is observed. Therefore, the first constant ξ grid line ($i = 1$) is set below the body axis, and the second constant ξ line ($i = 2$) is set by symmetry as illustrated in Figure 12-36. The steps in the ξ and η sweeps in the computational domain are shown in Figure 12-37.

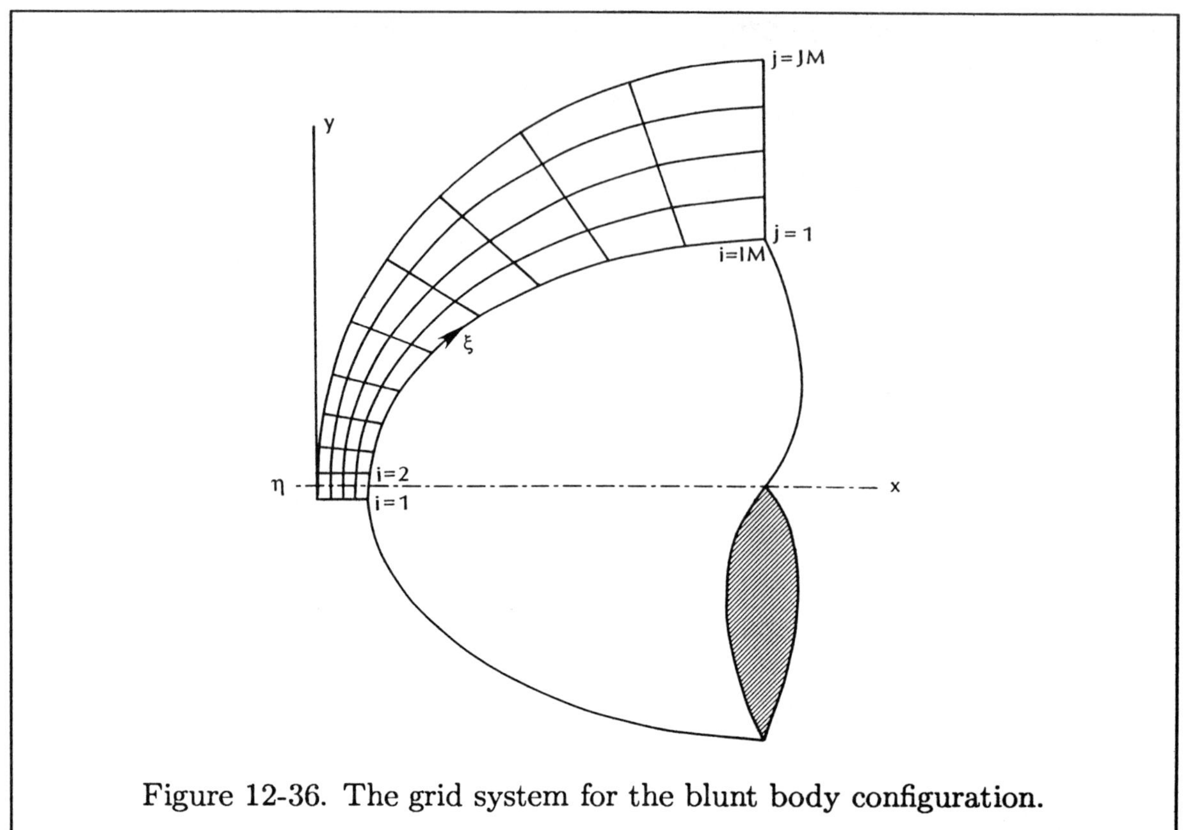

Figure 12-36. The grid system for the blunt body configuration.

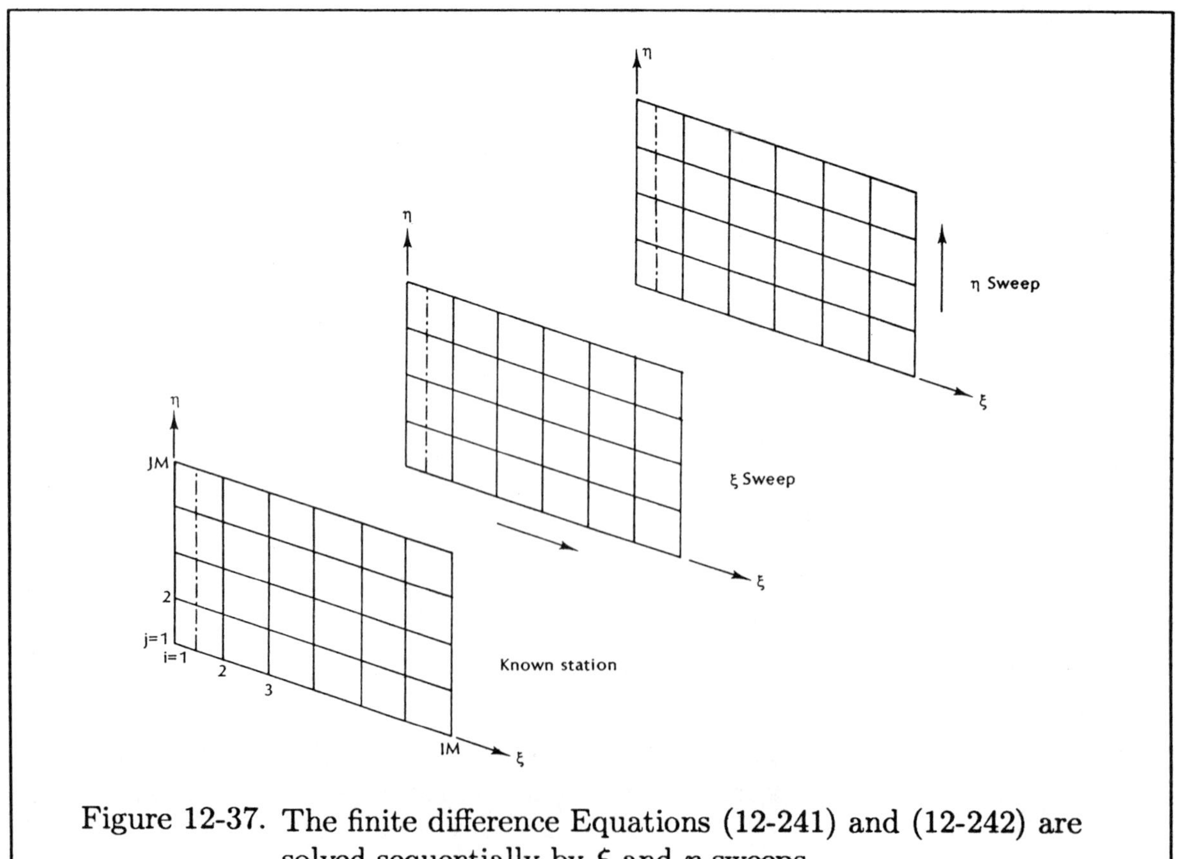

Figure 12-37. The finite difference Equations (12-241) and (12-242) are solved sequentially by ξ and η sweeps.

Before exploring the implementation of the boundary conditions, two issues with regard to evaluation of Jacobian matrices need to be addressed. The first is the matrix manipulations to provide for example A^+ or A^- at various grid points within the domain. The second issue must address what values of the flow properties (such as u, v, e, p) should be used to evaluate the elements of Jacobian matrices. These issues are considered in the following two sections.

12.9.4.1.1 Computation of the Jacobian Matrices: The Jacobian matrices appearing in Equation (12-241) are evaluated at the previous time level of n. Obviously, this is the result of linearization by lagging the coefficients. For example, to evaluate the coefficient

$$CAM_{i,j} = -\frac{\Delta\tau}{\Delta\xi} A^+_{i-1,j}$$

either relation (12-150) or A^+, as described in Section 12.9.2.1.1, is used. The values of u, v, and a which appear within the elements of A^+ are taken at the known values at time level n. And of course the metrics are provided by the grid generation routine.

Now consider Equation (12-242) and the evaluation of the coefficients involved, for example CBM, which is given by

$$CBM = -\frac{\Delta\tau}{\Delta\eta} B^+_{i,j-1}$$

Recognizing that Equation (12-241) has been solved, the values of $\Delta\bar{Q}^*$ and, subsequently, $\bar{Q}^*$ are known. Thus, it can be decomposed to provide the values of u, v, and a at the intermediate level designated by "$*$." It would seem logical to use these updated values to evaluate CBM. Another option would be to evaluate CBM at the time level of n. If one chooses to use this approach, decomposition of $\bar{Q}^*$ is not required. Based on limited numerical experimentation, either one can be used with no clear advantage of one over the other. However, there may be an advantage to using updated values if the time step is large. The solutions shown in this chapter were obtained by evaluating all the Jacobian matrices at the n time level.

12.9.4.1.2 Boundary Conditions: When Equation (12-241) is written at each η for all i from $i = 2$ to $i = IMM1$, a block tridiagonal system is obtained. This system is expressed as

$$\begin{bmatrix} [CA]_{2,j} & [CAP]_{2,j} & & & \\ [CAM]_{3,j} & [CA]_{3,j} & [CAP]_{3,j} & & \\ & & & & \\ & [CAM]_{IMM2,j} & [CA]_{IMM2,j} & [CAP]_{IMM2,j} & \\ & & [CAM]_{IMM1,j} & [CA]_{IMM1,j} & \end{bmatrix} \begin{bmatrix} \Delta\bar{Q}^*_{2,j} \\ \Delta\bar{Q}^*_{3,j} \\ | \\ \Delta\bar{Q}^*_{IMM2,j} \\ \Delta\bar{Q}^*_{IMM1,j} \end{bmatrix} =$$

$$= \begin{bmatrix} (RHS)_{2,j} - [CAM]_{2,j}\Delta\bar{Q}^*_{1,j} \\ (RHS)_{3,j} \\ | \\ (RHS)_{IMM2,j} \\ (RHS)_{IMM1,j} - [CAP]_{IMM1,j}\Delta\bar{Q}^*_{IM,j} \end{bmatrix} \tag{12-243}$$

If the boundary conditions are explicitly implemented, the values of $\Delta\bar{Q}^*_{1,j}$ and $\Delta\bar{Q}^*_{IM,j}$ are taken from the previous time level. For an implicit treatment of the boundary conditions, some modifications to the first and last equations must be introduced. The procedure is as follows.

At $i = 1$, symmetry is imposed, and the boundary conditions as described in Section 12.9.3.2 yield

$$\Delta\bar{Q}^*_{1,j} = \frac{1}{J_{1,j}} \begin{bmatrix} (\Delta\rho)^*_{1,j} \\ (\Delta\rho u)^*_{1,j} \\ (\Delta\rho v)^*_{1,j} \\ (\Delta\rho e_t)^*_{1,j} \end{bmatrix} = \frac{1}{J_{2,j}} \begin{bmatrix} (\Delta\rho)^*_{2,j} \\ (\Delta\rho u)^*_{2,j} \\ -(\Delta\rho v)^*_{2,j} \\ (\Delta\rho e_t)^*_{2,j} \end{bmatrix}$$

Hence, the first equation represented as the first row of the block tridiagonal system (12-243) is modified according to

$$\left[\overline{CA}\right]_{2,j} \Delta\bar{Q}^*_{2,j} + [CAP]_{2,j}\,\Delta\bar{Q}^*_{3,j} = (RHS)_{2,j}$$

where

$$\left[\overline{CA}\right]_{2,j} = [CA]_{2,j} + [CAM]_{2,j} \cdot \begin{bmatrix} 1 & 0 & 0 & 0 \\ 0 & 1 & 0 & 0 \\ 0 & 0 & -1 & 0 \\ 0 & 0 & 0 & 1 \end{bmatrix}$$

Assuming $i = IM$ is far downstream, such that the flow is supersonic, extrapolation is used. For simplicity, the following zero-order extrapolation is employed:

$$\Delta\bar{Q}^*_{IM,j} = \Delta\bar{Q}^*_{IMM1,j}$$

The reason for using this simple extrapolation scheme is that the higher order schemes may overestimate the properties causing instabilities and eventual failure of the solution. That is especially true for the first few iterations. With this simple extrapolation, the last equation of the block tridiagonal system (12-243) is modified according to

$$[CAM]_{IMM1,j}\,\Delta\bar{Q}^*_{IMM2,j} + \left[\overline{CA}\right]_{IMM1,j}\Delta\bar{Q}^*_{IMM1,j} = (RHS)_{IMM1,j}$$

where

$$\left[\overline{CA}\right]_{IMM1,j} = [CA]_{IMM1,j} + [CAP]_{IMM1,j}$$

Finally, the block tridiagonal system is expressed as

$$\begin{bmatrix} [\overline{CA}]_{2,j} & [CAP]_{2,j} & & & \\ [CAM]_{3,j} & [CA]_{3,j} & [CAP]_{3,j} & & \\ & & & & \\ & [CAM]_{IMM2,j} & [CA]_{IMM2,j} & [CAP]_{IMM2,j} & \\ & & [CAM]_{IMM1,j} & \left[\overline{CA}\right]_{IMM1,j} & \end{bmatrix} \begin{bmatrix} \Delta\bar{Q}^*_{2,j} \\ \Delta\bar{Q}^*_{3,j} \\ | \\ \Delta\bar{Q}^*_{IMM2,j} \\ \Delta\bar{Q}^*_{IMM1,j} \end{bmatrix} = \begin{bmatrix} (RHS)_{2,j} \\ (RHS)_{3,j} \\ | \\ (RHS)_{IMM2,j} \\ (RHS)_{IMM1,j} \end{bmatrix} \tag{12-244}$$

Once $\Delta\bar{Q}^*_{i,j}$ is computed, the RHS of Equation (12-242) is known. When this equation is applied at each ξ for all j from $j = 2$ to $j = JMM1$, one obtains the following block tridiagonal system:

$$\begin{bmatrix} [CB]_{i,2} & [CBP]_{i,2} & & & \\ [CBM]_{i,3} & [CB]_{i,3} & [CBP]_{i,3} & & \\ & & & & \\ & [CBM]_{i,JMM2} & [CB]_{i,JMM2} & [CBP]_{i,JMM2} & \\ & & [CBM]_{i,JMM1} & [CB]_{i,JMM1} & \end{bmatrix} \begin{bmatrix} \Delta\bar{Q}_{i,2} \\ \Delta\bar{Q}_{i,3} \\ | \\ \Delta\bar{Q}_{i,JMM2} \\ \Delta\bar{Q}_{i,JMM1} \end{bmatrix} =$$

$$\begin{bmatrix} \Delta\bar{Q}^*_{i,2} - [CBM]_{i,2}\Delta\bar{Q}_{i,1} \\ \Delta\bar{Q}^*_{i,3} \\ | \\ \Delta\bar{Q}^*_{i,JMM2} \\ \Delta\bar{Q}^*_{i,JMM1} - [CBP]_{i,JMM1}\Delta\bar{Q}_{i,JM} \end{bmatrix} \tag{12-245}$$

If the outer boundary of the domain (i.e., $j = JM$) is set at the freestream, $\Delta\bar{Q}_{i,JM}$ is zero. Note that for this statement to be valid, the spatial grid is assumed to be independent of time, i.e., grid points do not move; as a result, $J^n_{i,j} = J^{n+1}_{i,j}$. But what about $\Delta\bar{Q}_{i,1}$? For the time being, its value can be set to zero. Subsequently, this value can be computed and updated according to the procedure described in Section 12.9.3.1. By setting $\Delta\bar{Q}_{i,1}$ equal to zero, the system (12-245) may be solved for all the $\Delta\bar{Q}_{i,j}$ from $j = 2$ to $JMM1$ for each ξ from $i = 2$ to $i = IMM1$.

12.10 Application: Axisymmetric/Two-Dimensional Problems

In this section, the application of the axisymmetric/two-dimensional Euler equation is illustrated. In the first example, a channel which includes both compression and expansion corners is used to illustrate the formation of oblique shock and expansion waves and their reflection and interaction. In the second example, an axisymmetric blunt body is used to illustrate the development of the flowfield.

In order to develop a solution procedure, the following issues must be addressed. (1) The physical domain of the problem must be defined. This definition may be accomplished by specifying the boundaries of the domain. (2) A grid generation technique must be used to distribute the grid points on the boundaries and within the domain. Various grid generation schemes were investigated in Chapter 9. (3) A transformation from physical space to computational space must be performed. This procedure simplifies the application of the boundary conditions, and, in addition, the computational domain is rectangularly shaped and the grid points are equally spaced. The transformation of equations of fluid motion were addressed in Chapter 11. (4) Finally, a numerical algorithm must be selected to solve the equations of fluid motion.

12.10.1 Supersonic Channel Flow

A channel with a compression corner and an expansion corner located at the lower surface and a straight upper surface is considered. A supersonic flow enters the channel at the left entrance. The objective of this application is to use the Euler equation to determine the inviscid flow pattern within the channel. The domain of solution and the nomenclature are shown in Figure 12-38.

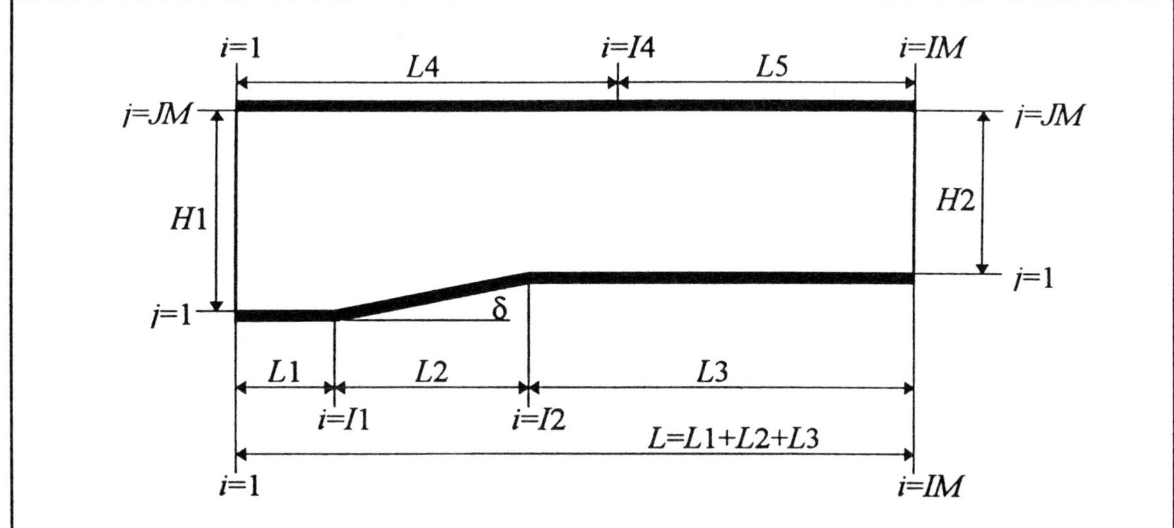

Figure 12-38. Illustration of the nomenclature for the two-dimensional channel flow.

12.10.1.1 Grid Generation: A simple algebraic grid generation routine is developed for this problem. In order to better resolve the oblique shock formed at the compression corner and the expansion wave at the expansion corner, as well as the shock reflection from the upper surface, grid point clustering is used to concentrate grid points at these locations. Furthermore, grid point clustering near the lower and upper surfaces is employed to better approximate the implementation of surface boundary conditions. The grid point clustering function given by Equation (9-57) is utilized in this application. The specific grid point clustering function for each segment of the domain is as follows.

1. $1 \leq i \leq I1$

$$x = L1 - L1\left\{1 - \beta1\left[\left(\frac{\beta1+1}{\beta1-1}\right)^{1-\frac{\xi_i}{\xi1}} - 1\right] \Big/ \left[\left(\frac{\beta1+1}{\beta1-1}\right)^{1-\frac{\xi_i}{\xi1}} + 1\right]\right\} \tag{12-246}$$

2. $I1 \leq i \leq I1 + I2/2$

$$x = L1 + L2\left\{1 - \beta2\left[\left(\frac{\beta2+1}{\beta2-1}\right)^{1-2\frac{(\xi_i-\xi1)}{\xi2}} - 1\right] \Big/ \left[\left(\frac{\beta2+1}{\beta2-1}\right)^{1-2\frac{(\xi_i-\xi1)}{\xi2}} + 1\right]\right\} \tag{12-247}$$

3. $I1 + I2/2 \le i \le I1 + I2$

$$x = L1 + L2 - L2\left\{1 - \beta 2\left[\left(\frac{\beta 2 + 1}{\beta 2 - 1}\right)^{1 - 2\frac{(\xi_i - \xi 1 - \xi 2/2)}{\xi 2}} - 1\right] \Big/ \left[\left(\frac{\beta 2 + 1}{\beta 2 - 1}\right)^{1 - 2\frac{(\xi_i - \xi 1 - \xi 2/2)}{\xi 2}} + 1\right]\right\} \qquad (12\text{-}248)$$

4. $I1 + I2 \le i \le IM$

$$x = L1 + L2 + L3\left\{1 - \beta 3\left[\left(\frac{\beta 3 + 1}{\beta 3 - 1}\right)^{1 - \frac{(\xi_i - \xi 1 - \xi 2)}{\xi 3}} - 1\right] \Big/ \left[\left(\frac{\beta 3 + 1}{\beta 3 - 1}\right)^{1 - \frac{(\xi_i - \xi 1 - \xi 2)}{\xi 3}} + 1\right]\right\} \qquad (12\text{-}249)$$

5. $1 \le i \le I4$

$$x = L4 - L4\left\{1 - \beta 4\left[\left(\frac{\beta 4 + 1}{\beta 4 - 1}\right)^{1 - \frac{\xi_i}{\xi 4}} - 1\right] \Big/ \left[\left(\frac{\beta 4 + 1}{\beta 4 - 1}\right)^{1 - \frac{\xi_i}{\xi 4}} + 1\right]\right\} \qquad (12\text{-}250)$$

6. $I4 \le i \le IM$

$$x = L4 + L5\left\{1 - \beta 5\left[\left(\frac{\beta 5 + 1}{\beta 5 - 1}\right)^{1 - \frac{(\xi_i - \xi 4)}{\xi 5}} - 1\right] \Big/ \left[\left(\frac{\beta 5 + 1}{\beta 5 - 1}\right)^{1 - \frac{(\xi_i - \xi 4)}{\xi 5}} + 1\right]\right\} \qquad (12\text{-}251)$$

7. $1 \le j \le JM$

$$y = \frac{H}{2}\left\{\left[(1 + \beta 6)\left(\frac{\beta 6 + 1}{\beta 6 - 1}\right)^{2\left(\frac{\eta_j}{\eta_l} - 1\right)} + 1 - \beta 6\right] \Big/ \left[\left(\frac{\beta 6 + 1}{\beta 6 - 1}\right)^{2\left(\frac{\eta_j}{\eta_l} - 1\right)} + 1\right]\right\} \qquad (12\text{-}252)$$

where $\beta 1$, $\beta 2$, $\beta 3$, $\beta 4$, $\beta 5$, and $\beta 6$ are the clustering parameters, and $\xi 1$, $\xi 2$, $\xi 3$, $\xi 4$, $\xi 5$, and $\eta 1$ are the lengths in the computational domain corresponding to the physical lengths of $L1$, $L2$, $L3$, $L4$, $L5$, and H, respectively. Note that H is replaced by $H1$ at $i = IMAX$ location. Expression (12-252) can also be used to determine grid points within the domain. In that case, the length of straight lines connecting the same i locations at the lower surface and at the upper surface are determined and denoted by δ_i. Now, expression (12-252) can be used to distribute grid points along each line where H is replaced by δ. Subsequently, the x and y coordinates of grid points are determined.

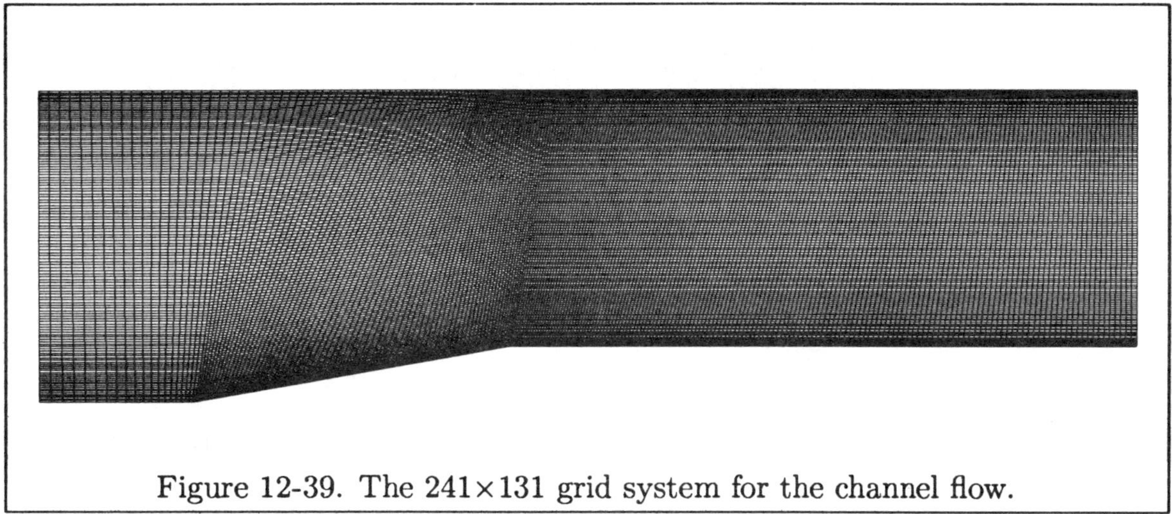

Figure 12-39. The 241×131 grid system for the channel flow.

12.10.1.2 Numerical Scheme: The numerical scheme selected for this application is the modified Runge-Kutta given by Equations (12-167) through (12-171) augmented with the Davis-Yee symmetric TVD model with limiters (12-205) and (12-208), that is

$$\bar{Q}_{i,j}^{(1)} = \bar{Q}_{i,j}^{n} \tag{12-253}$$

$$\bar{Q}_{i,j}^{(2)} = \bar{Q}_{i,j}^{n} - \frac{\Delta\tau}{4}\left[\left(\frac{\partial \bar{E}}{\partial \xi}\right)_{i,j}^{1} + \left(\frac{\partial \bar{F}}{\partial \eta}\right)_{i,j}^{(1)}\right] \tag{12-254}$$

$$\bar{Q}_{i,j}^{(3)} = \bar{Q}_{i,j}^{n} - \frac{\Delta\tau}{3}\left[\left(\frac{\partial \bar{E}}{\partial \xi}\right)_{i,j}^{2} + \left(\frac{\partial \bar{F}}{\partial \eta}\right)_{i,j}^{(2)}\right] \tag{12-255}$$

$$\bar{Q}_{i,j}^{(4)} = \bar{Q}_{i,j}^{n} - \frac{\Delta\tau}{2}\left[\left(\frac{\partial \bar{E}}{\partial \xi}\right)_{i,j}^{(3)} + \left(\frac{\partial \bar{F}}{\partial \eta}\right)_{i,j}^{(3)}\right] \tag{12-256}$$

$$\bar{Q}_{i,j}^{n+1} = \bar{Q}_{i,j}^{n} - \Delta\tau\left[\left(\frac{\partial \bar{E}}{\partial \xi}\right)_{i,j}^{(4)} + \left(\frac{\partial \bar{F}}{\partial \eta}\right)_{i,j}^{(4)}\right] \tag{12-257}$$

where convective terms in Equations (12-254) through (12-257) are approximated by second-order central difference expressions as follow

$$\left(\frac{\partial \bar{E}}{\partial \xi}\right)_{i,j} + \left(\frac{\partial \bar{F}}{\partial \eta}\right)_{i,j} = \frac{\bar{E}_{i+1,j} - \bar{E}_{i-1,j}}{2\Delta\xi} + \frac{\bar{F}_{i,j+1} - \bar{F}_{i,j-1}}{2\Delta\eta}$$

After the computation of Equation (12-257), the value of $\bar{Q}$ is updated according to

$$\begin{aligned}\bar{Q}^{n+1}_{i,j} &= \bar{Q}^{n+1}_{i,j} - \frac{1}{2}\frac{\Delta\tau}{\Delta\xi}\left[(X_A)^n_{i+\frac{1}{2},j}(\Phi_\xi)^n_{i+\frac{1}{2},j} - (X_A)^n_{i-\frac{1}{2},j}(\Phi_\xi)^n_{i-\frac{1}{2},j}\right] \\ &- \frac{1}{2}\frac{\Delta\tau}{\Delta\eta}\left[(X_B)^n_{i,j+\frac{1}{2}}(\Phi_\eta)^n_{i,j+\frac{1}{2}} - (X_B)^n_{i,j-\frac{1}{2}}(\Phi_\eta)^n_{i,j-\frac{1}{2}}\right] \qquad (12\text{-}258)\end{aligned}$$

The specific limiters are

$$\begin{aligned}(G_\xi)_{i+\frac{1}{2},j} &= \\ &\text{minmod}\left\{2(\delta_\xi)_{i-\frac{1}{2},j}\,,\ 2(\delta_\xi)_{i+\frac{1}{2},j}\,,\ 2(\delta_\xi)_{i+\frac{3}{2},j}\,,\ \frac{1}{2}\left[(\delta_\xi)_{i-\frac{1}{2},j} + (\delta_\xi)_{i+\frac{3}{2},j}\right]\right\} \qquad (12\text{-}259)\end{aligned}$$

$$\begin{aligned}(G_\xi)_{i,j+\frac{1}{2}} &= \\ &\text{minmod}\left\{2(\delta_\eta)_{i,j-\frac{1}{2}}\,,\ 2(\delta_\eta)_{i,j+\frac{1}{2}}\,,\ 2(\delta_\eta)_{i,j+\frac{3}{2}}\,,\ \frac{1}{2}\left[(\delta_\eta)_{i,j-\frac{1}{2}} + (\delta_\xi)_{i,j+\frac{3}{2}}\right]\right\} \qquad (12\text{-}260)\end{aligned}$$

12.10.1.3 Analytical Solution: The analytical solution is used to check the accuracy of the numerical solution. An analytical expression between the turning (compression) angle δ, the shock angle θ, and the freestream Mach number M_1 can be established as follows.

$$\tan\delta = \frac{2\cot\theta(M_1^2\sin^2\theta - 1)}{2 + (\gamma + \cos 2\theta)M_1^2} \qquad (12\text{-}261)$$

For a given value of δ and M_1, the analytical expression (12-261) must be solved iteratively for the shock angle θ. Solution can also be obtained by using charts where Equation (12-261) is presented graphically, typically for γ of 1.4. Once the shock angle is determined, a normal component of the Mach number M_{1n} is calculated from which the flow properties are computed from the normal shock relations (or tables). The corresponding equations are as follow.

$$\begin{aligned}M_{1n} &= M_1\sin\theta \\ \frac{p_2}{p_1} &= \frac{2\gamma}{\gamma+1}M_{1n}^2 - \frac{\gamma-1}{\gamma+1} \qquad (12\text{-}262)\end{aligned}$$

and

$$\frac{\rho_2}{\rho_1} = \frac{(\gamma+1)M_{1n}^2}{(\gamma-1)M_{1n}^2+2} \tag{12-263}$$

To calculate the flow properties downstream of an expansion wave, the Prandtl-Meyer function ν given by the following expression is used.

$$\nu(M) = \sqrt{\frac{\gamma+1}{\gamma-1}}\tan^{-1}\sqrt{\frac{\gamma+1}{\gamma-1}(M^2-1)} - \tan^{-1}\sqrt{M^2-1} \tag{12-264}$$

Again, given the value of ν, this expression needs to be solved iteratively for the corresponding Mach number. As in the case of oblique shock relations, tables are also available for γ of 1.4. Now, with the turning (expansion) angle θ and the upstream Mach number M_1, the downstream Mach number is determined from the Prandtl-Meyer function from

$$\nu(M_2) = \theta + \nu(M_1) \tag{12-265}$$

Once the downstream Mach number is determined, the remaining flow properties across the expansion wave can be calculated. Recall that the expansion wave is an isentropic process, and, therefore, the isentropic relations provide

$$\frac{T_2}{T_1} = \frac{1+\frac{\gamma-1}{2}M_1^2}{1+\frac{\gamma-1}{2}M_2^2} \tag{12-266}$$

$$\frac{p_2}{p_1} = \left(\frac{1+\frac{\gamma-1}{2}M_1^2}{1+\frac{\gamma-1}{2}M_2^2}\right)^{\frac{\gamma}{\gamma-1}} \tag{12-267}$$

and the density can be computed from the equation of state

$$\rho_2 = \frac{p_2}{RT_2} \tag{12-268}$$

12.10.1.4 The Physical Domain and Flow Conditions: The physical dimensions of the domain in this problem is specified as follows: $L_1 = 10$cm, $L_2 = 20$cm, $L_3 = 40$cm, $L_5 = 42$cm, $H_1 = 20$cm, and the compression/expansion angle is 10°. The number of gridpoints for the domain is specified as $IM = 241$ and $JM = 131$, where the specific values of i are set according to $I1 = 21$, $I2 = 91$, and $I4 = 61$.

The freestream pressure and temperature are 100 kPa and 300 K, respectively. The ratio of specific heats γ is 1.4, and the gas constant is given as 287.05 $(N \cdot m)$/kgK, values corresponding to air. Solutions are required for two different freestream Mach numbers of 2.0 and 3.0.

12.10.1.5 Initial Conditions and Time Step: Computations are initialized by specification of the inflow conditions over the entire domain. The local time stepping procedure of Section 12.7 is used to march in time toward steady state solution. The CFL number of 0.2 is specified for both cases. A convergence criterion is set according to

$$\text{CONV} = \sum_{\substack{i=1\\ j=1}}^{\substack{j=JMM1\\ i=IMM1}} \left[ABS(p_{i,j}^{n+1} - p_{i,j}^{n})\right]$$

The solution is assumed converged (steady state) when CONV$\leq$0.1.

12.10.1.6 Results: The pressure and density contours illustrating the formation of oblique shock, expansion wave, and their reflection and interaction are shown in Figures 12.40 through 12.43 for $M = 2.0$ and $M = 3.0$. The computed pressure and density distributions along the lower surface are compared to the analytical solutions in Figures 12.44 through 12.47. The computational values compare very well to those of the analytical solution.

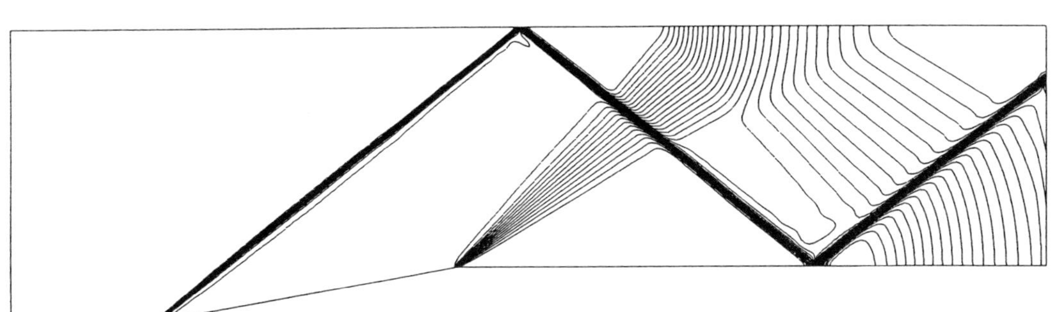

Figure 12.40. Pressure contours for $M_\infty = 2.0$.

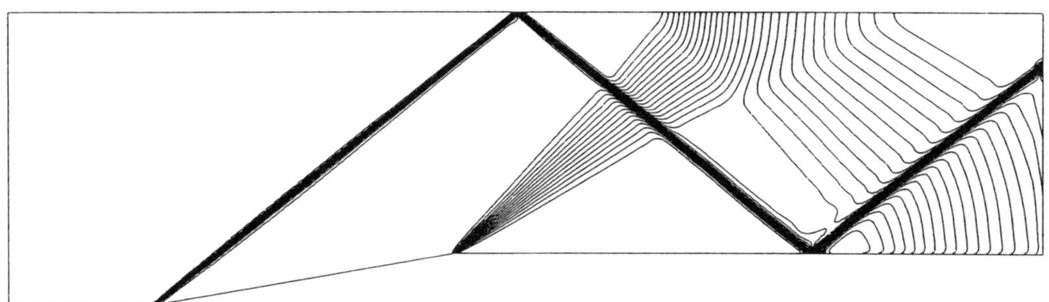

Figure 12.41. Density contours for $M_\infty = 2.0$.

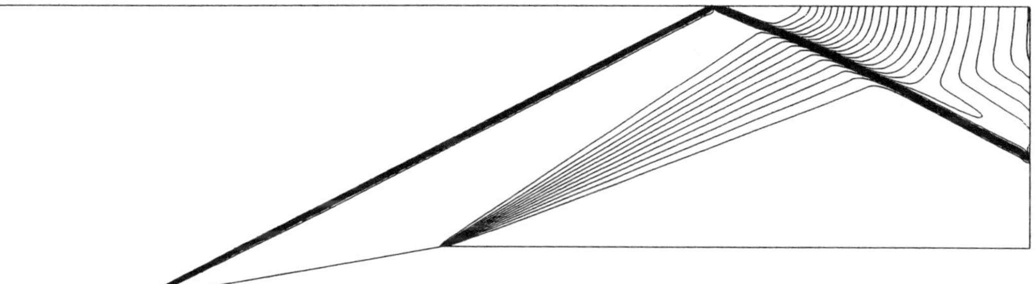

Figure 12.42. Pressure contours for $M_\infty = 3.0$.

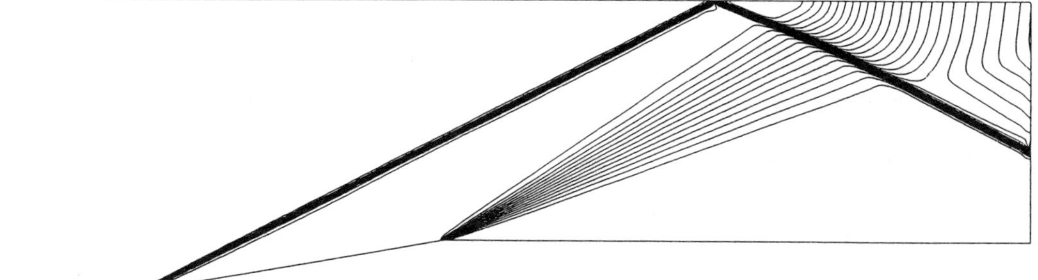

Figure 12.43. Density contours for $M_\infty = 3.0$.

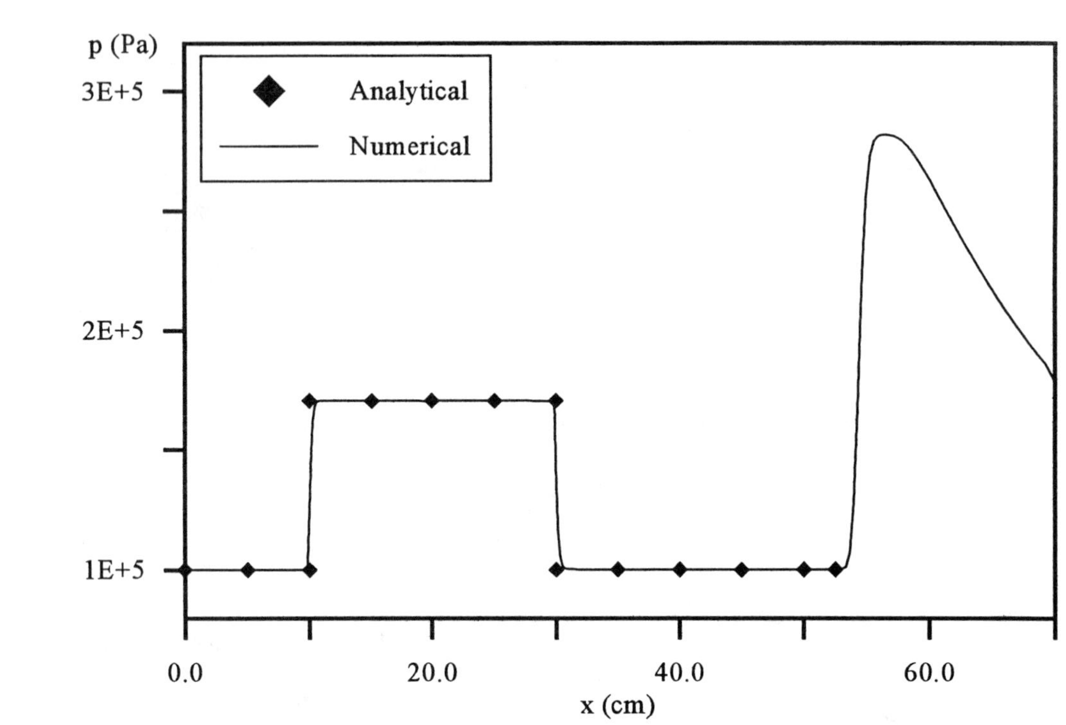

Figure 12-44. Comparison of the pressure distributions on the lower surface for $M_\infty = 2.0$.

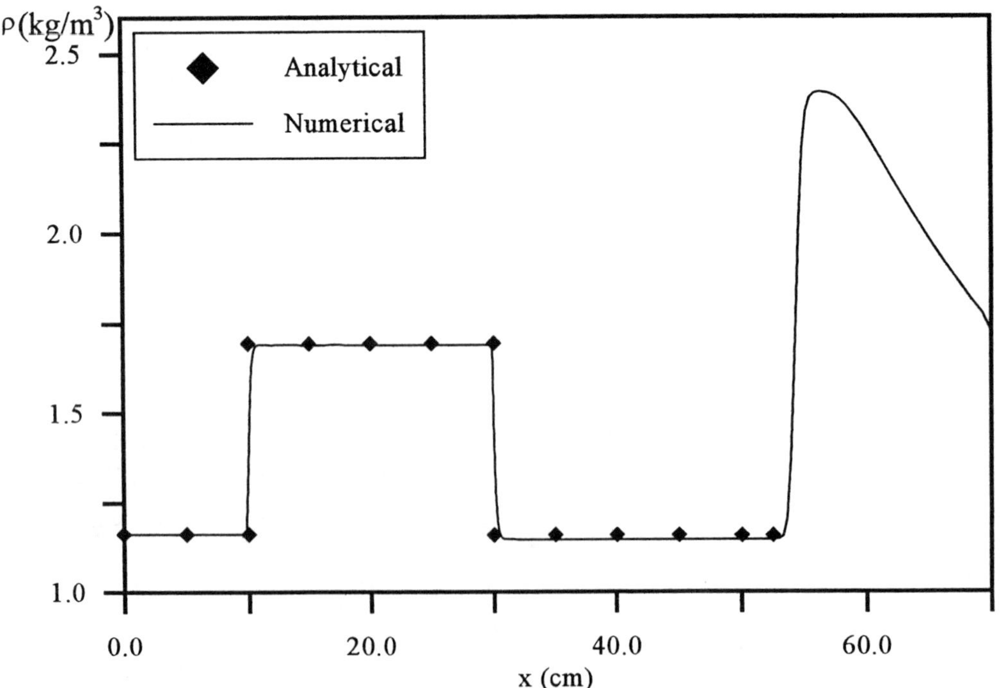

Figure 12-45. Comparison of the density distributions on the lower surface for $M_\infty = 2.0$.

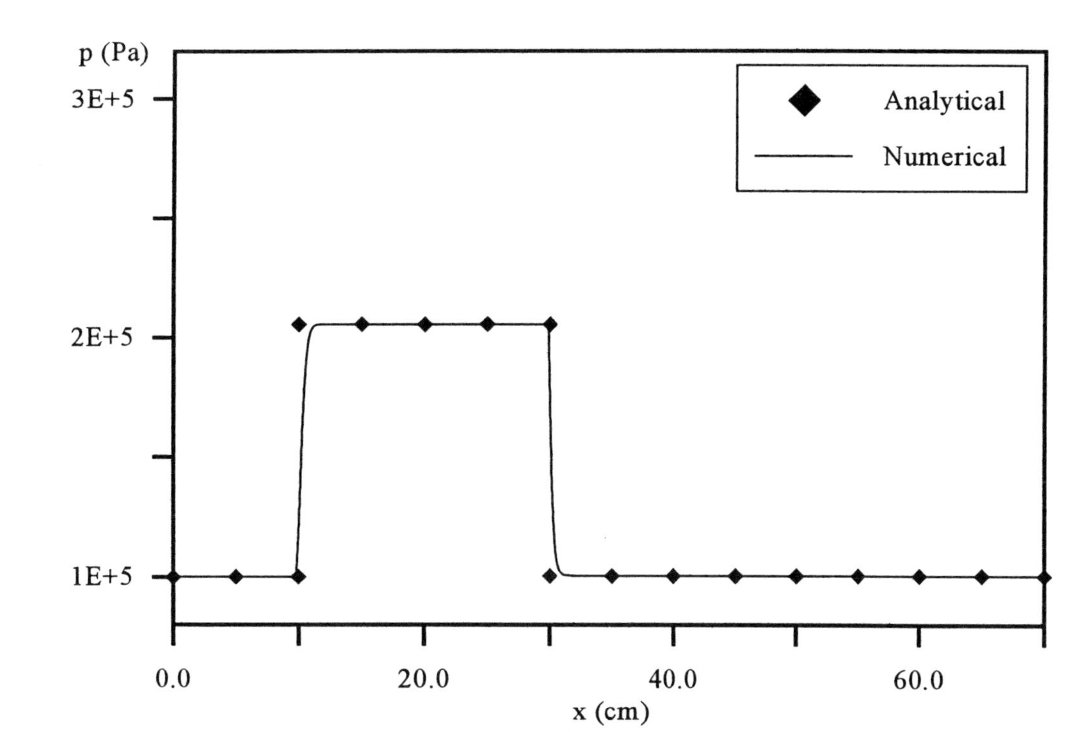

Figure 12-46. Comparison of the pressure distributions on the lower surface for $M_\infty = 3.0$.

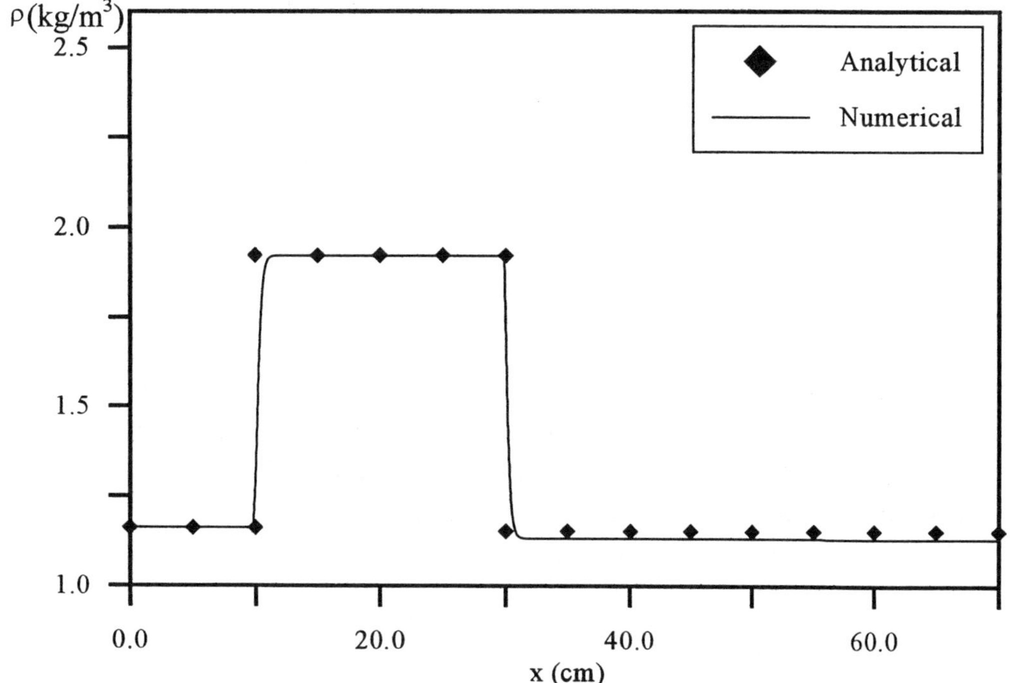

Figure 12-47. Comparison of the density distributions on the lower surface for $M_\infty = 3.0$.

12.10.2 Axisymmetric Blunt Body

Consider an axisymmetric blunt body defined geometrically by an ellipse as shown in Figure 12-48 where the semi-major axis is denoted by a_1, and the semi-minor axis is denoted by b_1. Assume a hypersonic flowfield where the freestream Mach number is 18.

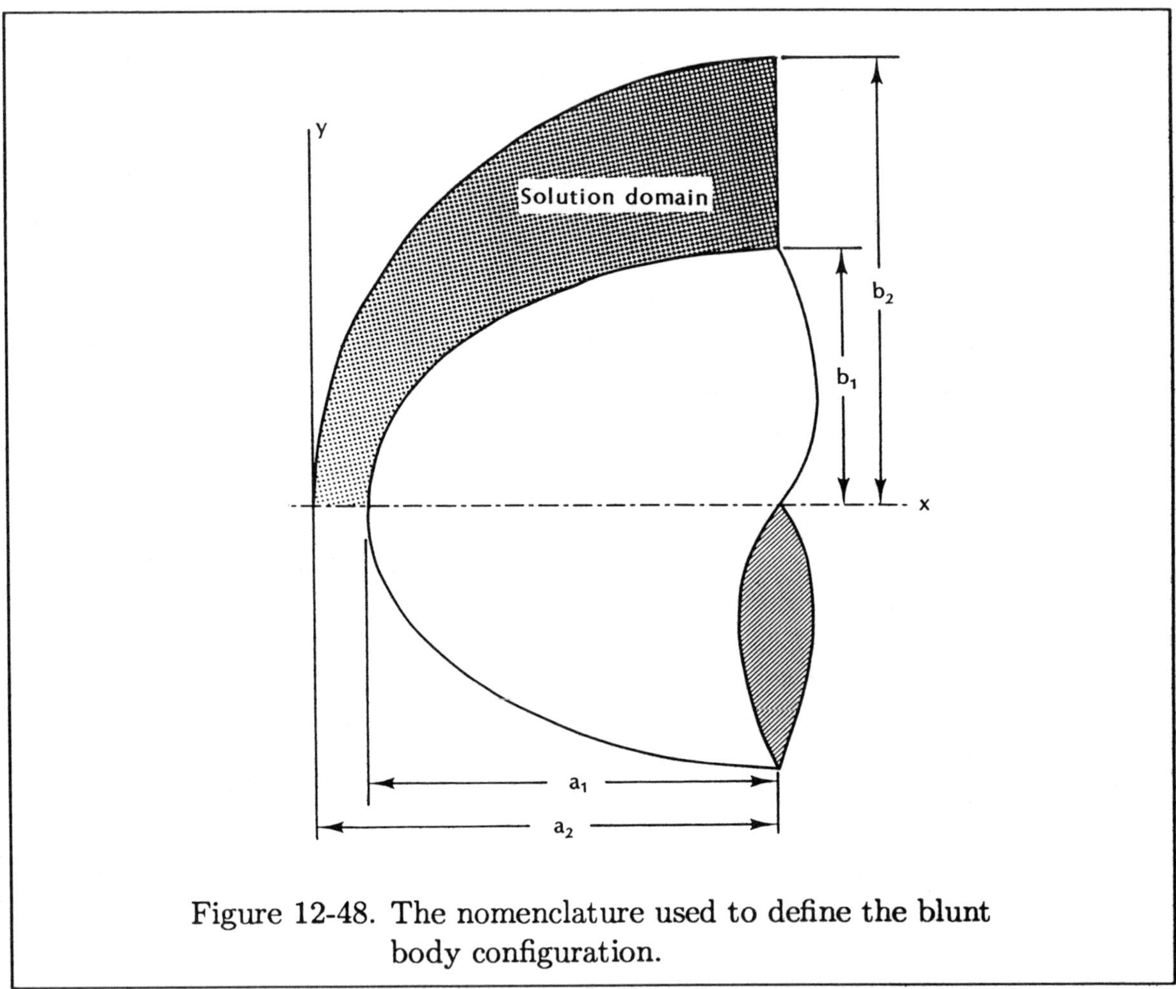

Figure 12-48. The nomenclature used to define the blunt body configuration.

Furthermore, the configuration is at zero degree angle of attack. It is required to compute the steady-state, inviscid flowfield. The implicit Steger and Warming flux vector splitting scheme given by Equations (12-241) and (12-242) is selected as the numerical scheme to solve the proposed problem.

The physical domain is specified by the following data

$$a_1 = 5.0' \;, \quad b_1 = 3.0' \;, \quad a_2 = 5.6' \;, \quad b_2 = 5.2'$$

A 71×53 grid system is created by the elliptic scheme of Section 9.7.1 and is shown in Figure 12-49. Observe that grid points are clustered near the stagnation region where the flow is expected to be subsonic. The CFL number for this problem is defined as

$$C_\xi = \lambda_\xi \frac{\Delta\tau}{\Delta\xi} \quad \text{and} \quad C_\eta = \lambda_\eta \frac{\Delta\tau}{\Delta\eta}$$

where maximum value of λ is used. Recall that the λs are given by relations (12-138) through (12-141) and (12-144) through (12-147). Thus, if a maximum CFL number is specified, then the maximum time step is easily determined. With the grid system specified, the tridiagonal systems (12-241) and (12-242) are sequentially solved.

Contours of constant pressure and Mach number are shown in Figures 12-50 and 12-51, respectively. A convergent solution was obtained after 550 iterations.

Note that convergent solution and steady-state solution are used to imply the same meaning. Similarly, the number of iterations and the number of time steps are used interchangeably. However, it should be pointed out that the Euler equation may be solved to provide a time dependent solution. In that instance, the procedure will be referred to as the time accurate solution.

As a second example of an axisymmetric flow, a dented configuration is used to investigate the shock/shock interaction phenomena. The configuration of interest and the grid system are shown in Figure 12-52. The grid system was created by an elliptic grid generator with a clustering option. The pressure contours for a Mach 18 flow are shown in Figure 12-53. It clearly illustrates the complex expansion/compression process within the domain of interest.

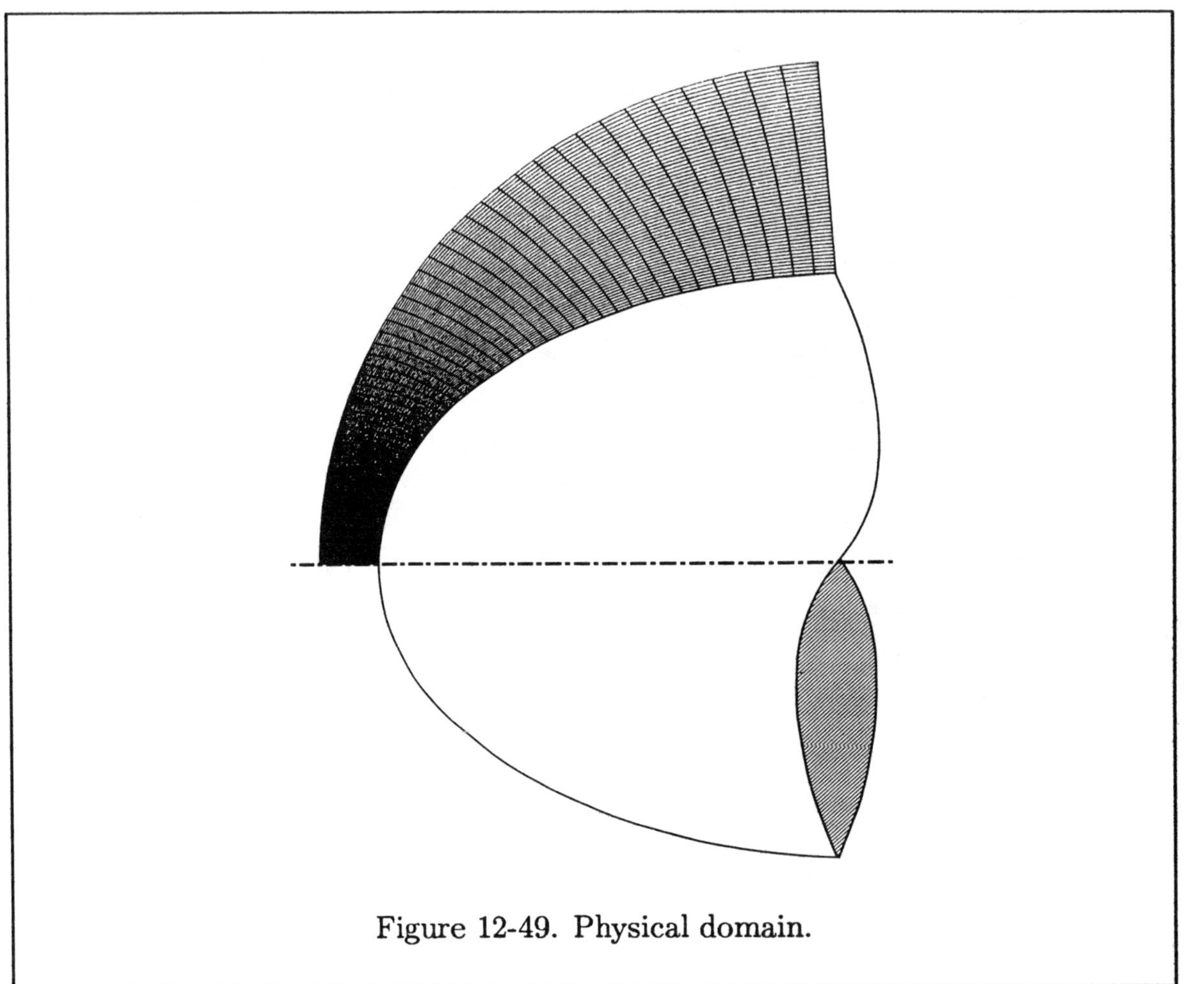

Figure 12-49. Physical domain.

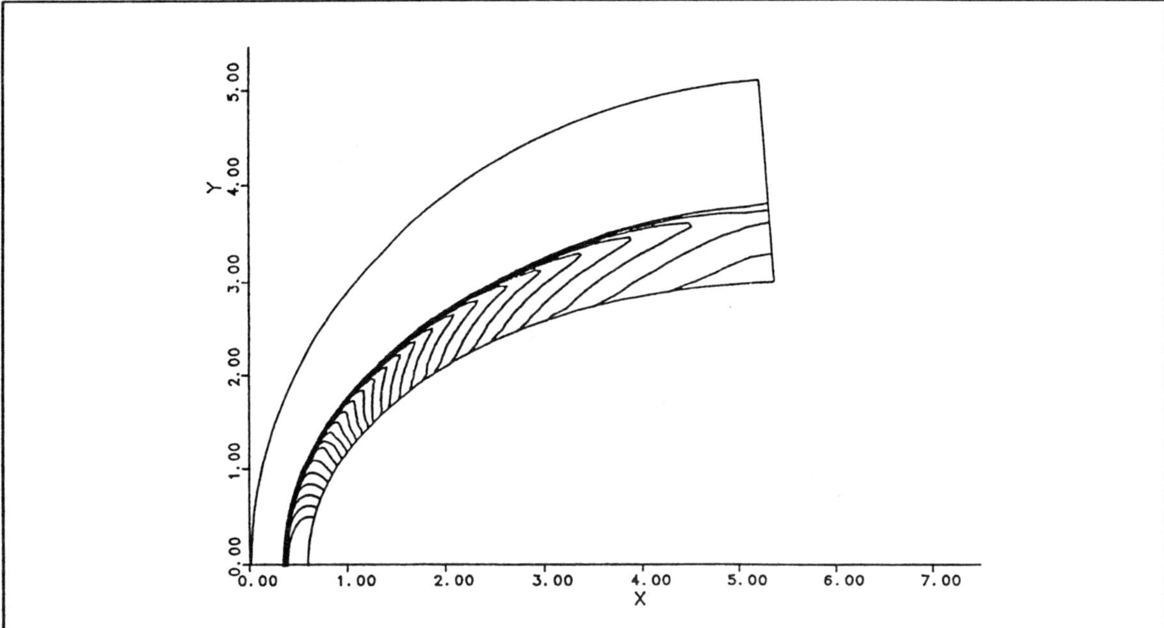

Figure 12-50. Pressure contours for the blunt body configuration at a freestream Mach number of 18.

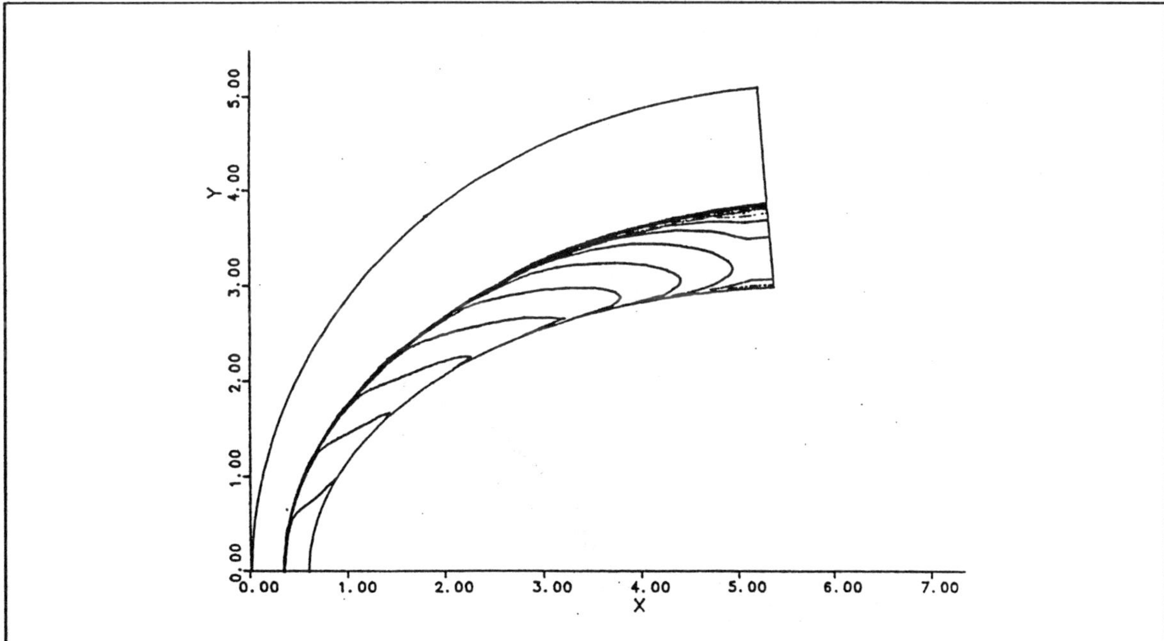

Figure 12-51. Mach contours for the blunt body configuration at a freestream Mach number of 18.

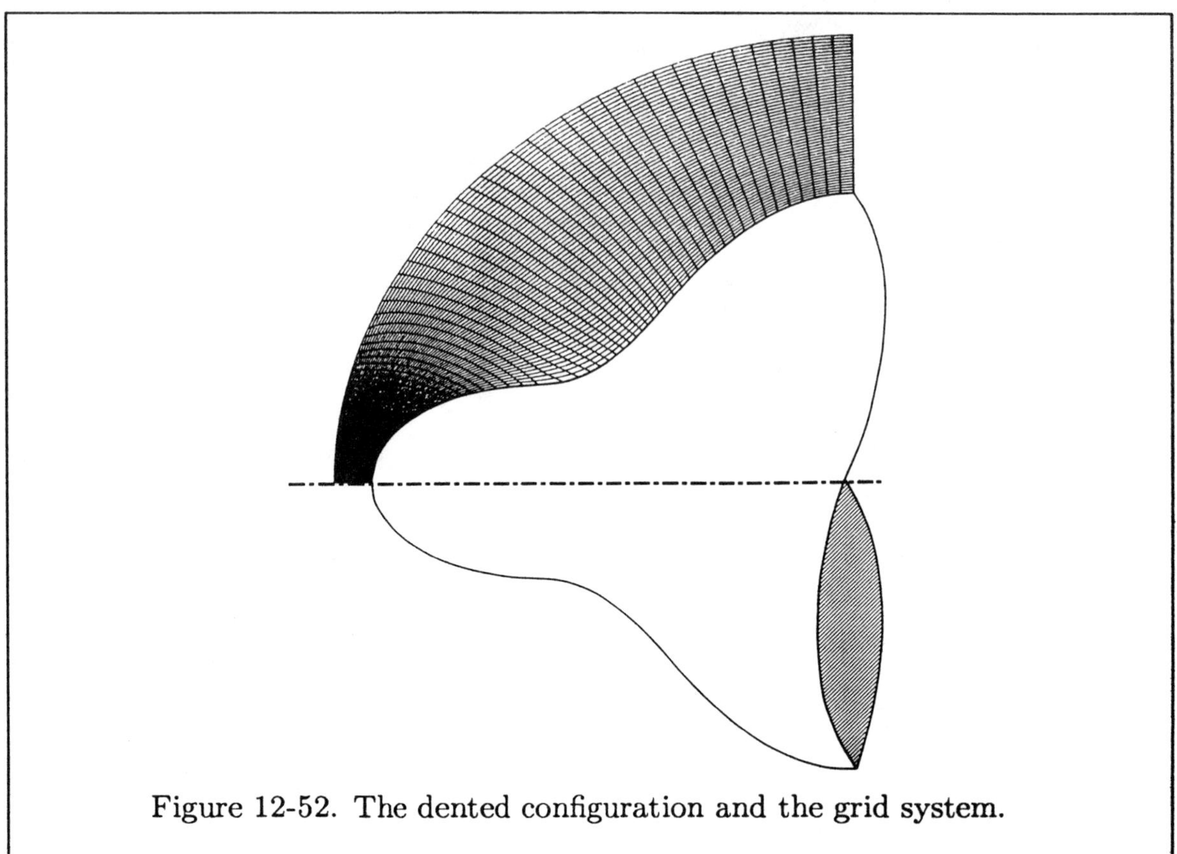

Figure 12-52. The dented configuration and the grid system.

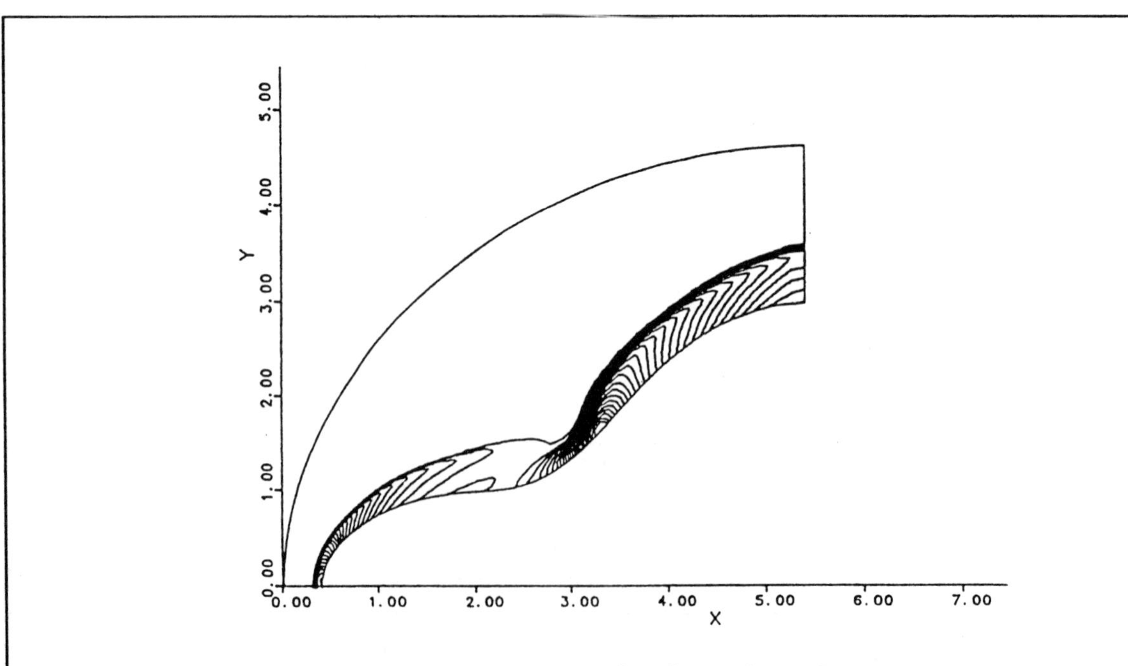

Figure 12-53. Pressure contours for the dented configuration at a freestream Mach number of 18.

12.11 Concluding Remarks

In this chapter an attempt is made to illustrate solution procedures for the Euler equation. Several numerical schemes were introduced, and the procedures were extensively discussed. It is hoped that, with a good understanding of the numerical procedure in this chapter, other procedures may be investigated with ease. Furthermore, the materials presented in the previous chapters are brought together in this chapter to illustrate the steps required for the solution of an inviscid flow field. Finally, it is important to emphasize again that the numerical procedures presented in this chapter will be implemented in the solution of the Navier-Stokes equations, particularly in Chapter 14.

12.12 Problems

12.1 Derive the Jacobian matrix A given by Equation (12-18).

12.2 Consider the nozzle described in Section 12.5 with the following supersonic inflow conditions at $x = 0.0$: $M_1 = 3.0$, $p_1 = 1000$ lb_f/ft^2, and $\rho_1 = 0.00237$ slugs/ft^3. For a supersonic flow at the exit ($x = 10.0$ ft), determine the flowfield within the domain, using the explicit first-order Steger and Warming flux vector splitting scheme. (a) Use $\Delta x = 0.1$ ft, and $\Delta t = 1 \times 10^{-5}$ sec, with a convergence criterion of 0.1 based on pressure, as given by Relation (12-93). Print the steady state solution and plot the comparison of the computed pressure distribution with the analytical solution. (b) Investigate the effect of the spatial step size on the solution by using Δx of 0.2 ft and 0.4 ft. Plot the error distributions of the three solutions obtained by Δx of 0.1, 0.2, and 0.4. Define the error as the difference between the numerical and analytical values of the pressure.

12.3 Repeat Problem 12.2 using the explicit second-order Steger and Warming flux-vector splitting scheme.

12.4 Repeat Problem 12.2 using the fourth-order Runge-Kutta scheme.

12.5 Repeat Problem 12.2 using the Harten-Yee TVD scheme.

12.6 Consider the nozzle described in Section 12.5 with the supersonic inflow specified by $M_1 = 3.0$, $p_1 = 1000$ lb_f/ft^2, and $\rho_1 = 0.00237$ slugs/ft^3. The flow at the exit plane is subsonic, and the flow properties correspond to conditions of a normal shock located just downstream of $x = 5.0$ ft. The analytical solution would provide the following conditions at the exit plane: $M_{IM} = 0.358$, $u_{IM} = 454.44$ fps, $p_{IM} = 8511.67$ lb_f/ft^2, $\rho_{IM} = 0.007389$ slugs/ft^3. However, recall that only one boundary condition at the exit needs to be specified. (a) Use $\Delta x = 0.1$ ft, and $\Delta t = 1 \times 10^{-5}$ sec, and CONV $= 0.1$ to obtain a steady state solution by the explicit first-order Steger and Warming flux vector splitting scheme. Print the converged solution and plot the comparison of the computed pressure distribution with the analytical solution. (b) Investigate the effect of the spatial step size on the solution by using Δx of 0.2 and 0.4. Plot the error distributions of the three solutions obtained by Δx of 0.1, 0.2, and 0.4. Define the error as the difference between the numerical and analytical values of the pressure.

12.7 Repeat Problem 12.5 using the Harten-Yee TVD scheme.

12.8 Consider Problem 12.5 with $\Delta x = 0.1$ ft. (a) Use the global time steps of 1×10^{-4} sec, 1×10^{-5} sec, and 1×10^{-6} sec to obtain steady state solutions. Compare the convergence histories of the solutions. (b) Use the concept of local time step to obtain steady state solutions with the following specified CFL numbers: 0.1, 0.2, 0.3. Compare the convergence histories.

Chapter 13

Parabolized Navier-Stokes Equations

13.1 Introductory Remarks

The design of a vehicle in a supersonic/hypersonic stream requires detailed analysis of the flowfield. Accurate calculation of the shock structure, pressure, skin friction, and heat transfer distributions are important parameters for the designers. Calculation procedures that range from simple methods to complex numerical techniques have been developed over the years. A traditional approach to the problem is to decouple the flowfield into an inviscid region governed by the inviscid-flow relations and a viscous region adjacent to the surface governed by the boundary layer equations. Once the inviscid flowfield is known, either from experimental measurements or from a theoretical solution, procedures can be developed to generate solutions for the thin boundary layer near the vehicle surface. These procedures vary in degree of sophistication from simple correlation to numerical programs which calculate the non-similar boundary layers for laminar, transitional, and/or turbulent flows.

Two problems in the boundary layer formulation are: (1) the uncertainties in the flow properties associated with having to determine the boundary layer edge required for the solution of the boundary layer equations, and (2) the iterative process for locating the boundary layer edge.

In addition, for applications where a strong interaction between the viscous and inviscid regions occurs, the decoupling of the flowfield is no longer an acceptable approach. This fact is particularly important for hypersonic flowfields where the shock layer is relatively thin and the viscous effects may influence a large portion of the shock layer. Flowfield computations of vehicles in a supersonic/hypersonic stream at large angles of attack, where cross flow separation dominates, also fall in this category. Examples of such flowfields are illustrated in Figure (13-1).

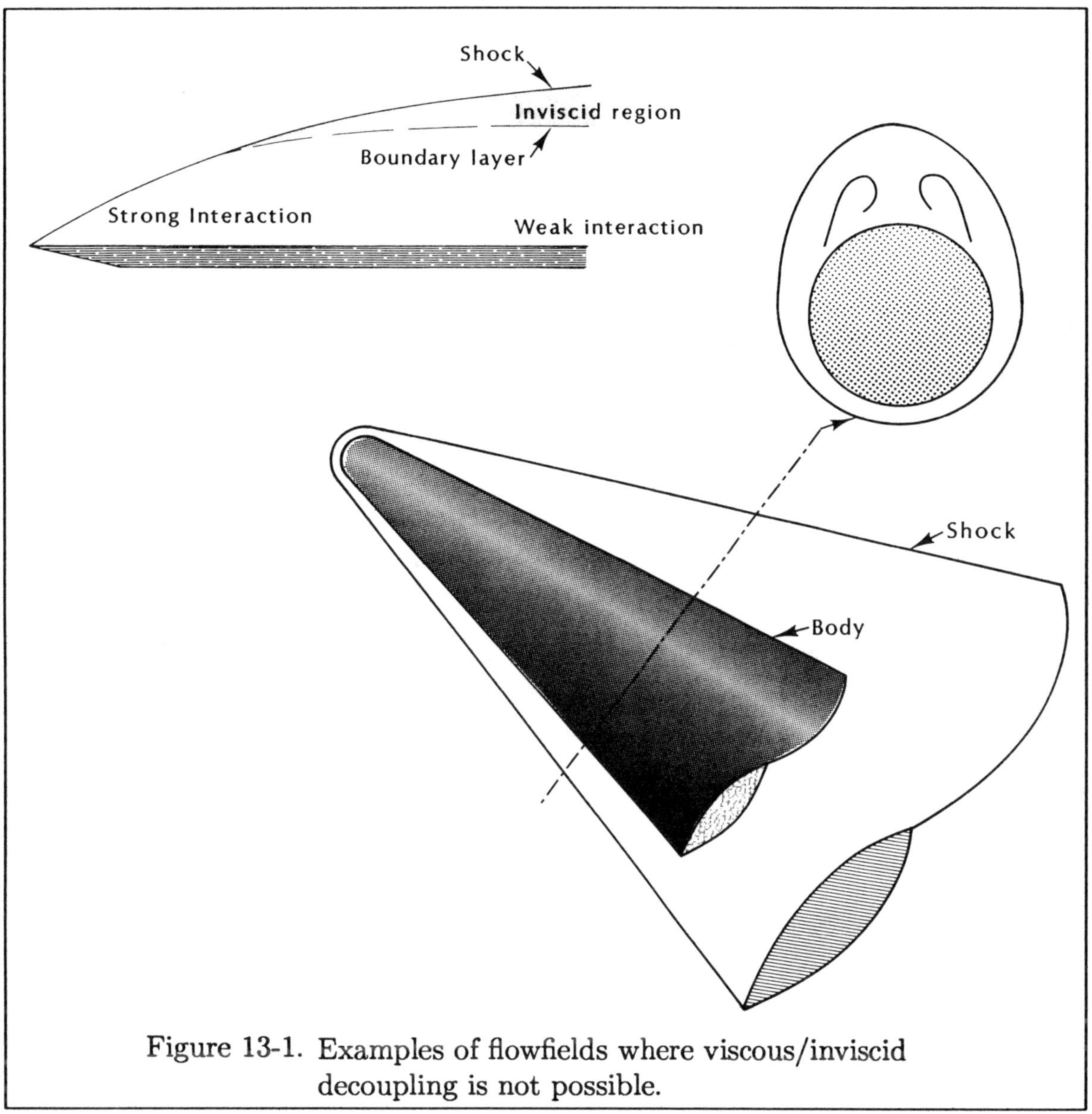

Figure 13-1. Examples of flowfields where viscous/inviscid decoupling is not possible.

To address these categories of flowfields, procedures are developed to solve the equations of motion for the entire flowfield. Obviously the Navier-Stokes equations may be employed to solve such flowfields. However, the numerical solution of the Navier-Stokes equations requires a substantial amount of computer time and storage. This difficulty is due to solving the unsteady Navier-Stokes equations in time until a converged steady-state solution is reached. Thus, it is desirable to reduce the Navier-Stokes equations to a form which can be solved efficiently while the physics of the problem is preserved. Therefore, among the issues to be considered are: (a) the reduction of the equations of fluid motion valid for the categories specified above, and (b) development of an efficient numerical scheme to solve the system of equations.

A popular method which has proven successful for the computations of these categories of flowfields is the parabolized Navier-Stokes (PNS) equations. The PNS equations are obtained from the full Navier-Stokes equations by the following assumptions: (a) steady state, (b) neglecting the streamwise viscous gradients, and (c) approximating the streamwise pressure gradient within the subsonic portion of the viscous flow near the surface. Thus, the Navier-Stokes equations are reduced to a set of parabolic equations which are marched in the direction of a space coordinate aligned along the streamwise flow direction from an initial plane of data.

A comparison of numerical marching for the NS equations and the PNS equations is illustrated in Figure 13-2. It is important to note that the NS equations are marched in time with a step Δt, whereas the PNS equations are marched in space with a step $\Delta\xi$. For either case a set of initial conditions is required. For the NS equations, freestream conditions are usually imposed everywhere within the domain and the steady-state solution is sought. For the PNS equations, a set of accurate initial data must be provided. This set of data can be obtained by solving the NS equations for the nose region of the configuration.

A vast number of researchers have investigated the PNS equations and various numerical algorithms to solve these equations within the last two decades. Extensive research in this area is still being conducted at the present. As a result, certain guidelines and conclusions have been established which are summarized below.

1. Transformation from physical space to computational space with clustering near the surface is necessary.
2. Implicit numerical schemes are preferred over explicit schemes for efficiency purposes.
3. The equations must be expressed in their conservative form for shock capturing purposes.
4. The outer bow shock is usually used as one of the boundaries and is calculated using the shock-fitting procedure.
5. Numerous approximations for the pressure gradient within the subsonic portion have been introduced.
6. Formulations which employ central difference approximation of the convective terms require the addition of damping terms. Thus, the addition of second-order and/or fourth-order dissipation terms to prevent oscillations in the flowfield has been used by many investigators.
7. Flux vector splitting schemes of the previous chapter may be employed to formulate the convective terms and thus eliminate the need to include damping terms.

In this chapter, the PNS equations, assumptions, approximations of the pressure gradient, initial data plane, boundary conditions, and a numerical algorithm are presented. Two-dimensional/axisymmetric equations are discussed first with the extension to 3-D to follow.

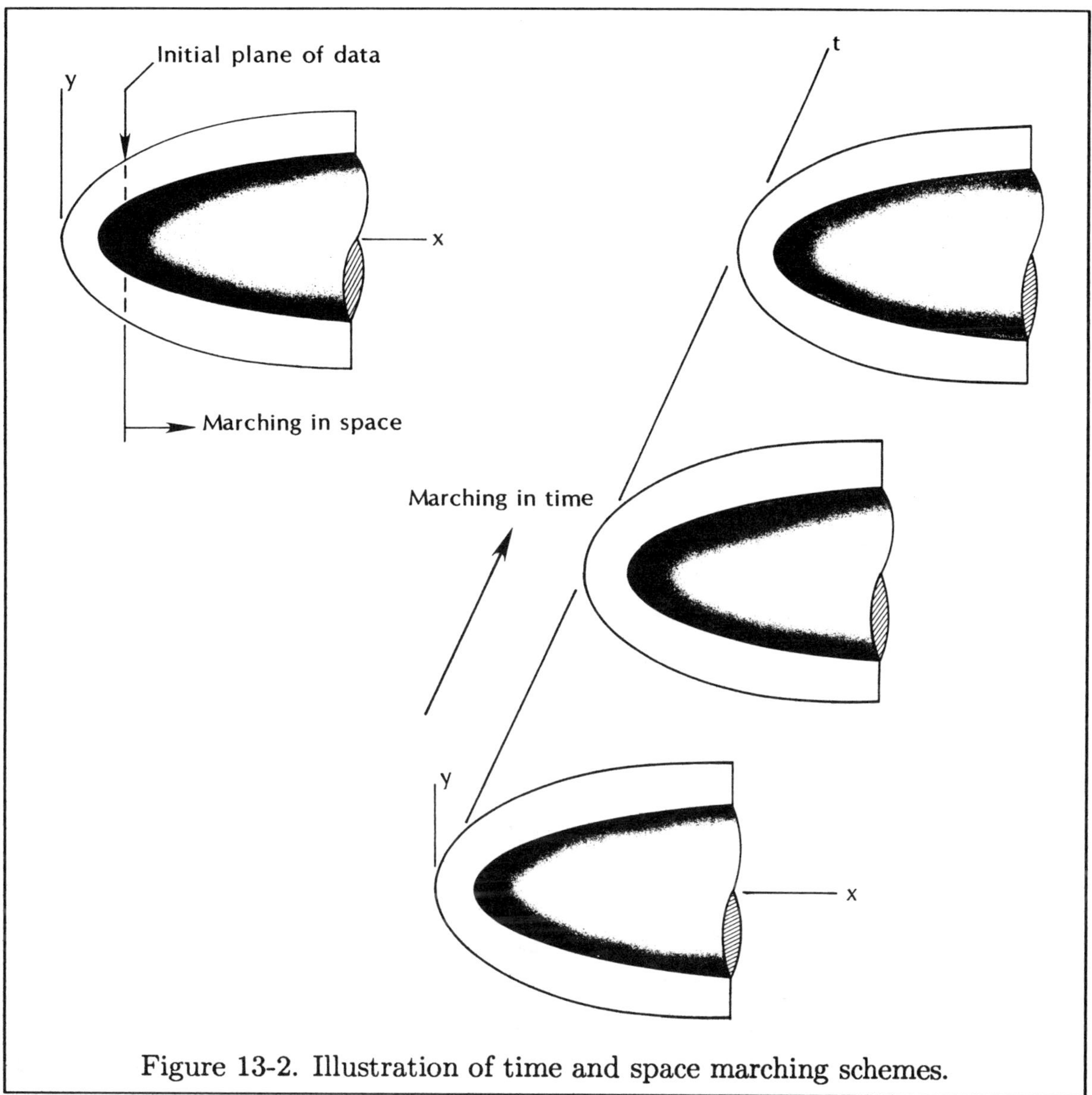

Figure 13-2. Illustration of time and space marching schemes.

13.2 Governing Equations of Motion

The governing equations of motion and various reduced forms were considered in Chapter 11. These equations were nondimensionalized and expressed in a conservative form. In addition, they were transformed from the physical space to the computational space. PNS equations were among those presented in that chapter. Therefore, only a brief review is presented in this section. In the following sections, two-dimensional problems are discussed in detail; and, subsequently, the approach is extended to three-dimensional problems.

The two-dimensional, axisymmetric Navier-Stokes equations in the computational space are expressed in a vector formulation as

$$\frac{\partial \bar{Q}}{\partial \tau} + \frac{\partial \bar{E}}{\partial \xi} + \frac{\partial \bar{F}}{\partial \eta} + \alpha \bar{H} = \frac{\partial \bar{E}_v}{\partial \xi} + \frac{\partial \bar{F}_v}{\partial \eta} + \alpha \bar{H}_v \tag{13-1}$$

The flux vectors are given in Chapter 11 by Equations (11-201) through (11-207). The first assumption in the reduction of equations is the steady-state condition, which mathematically states that

$$\frac{\partial \bar{Q}}{\partial \tau} = 0$$

The second assumption requires that we neglect the streamwise gradient of the viscous terms, i.e.,

$$\frac{\partial \bar{F}_v}{\partial \eta} \gg \frac{\partial \bar{E}_v}{\partial \xi}$$

Note that this reduction is imposed after the equations have been transformed into the computational domain. In addition, we have selected the coordinate ξ aligned in the streamwise direction where numerical marching will progress. With these assumptions, the Navier-Stokes equations given by (13-1) are reduced to

$$\frac{\partial \bar{E}}{\partial \xi} + \frac{\partial \bar{F}}{\partial \eta} + \alpha \bar{H} = \frac{\partial \bar{F}_{vP}}{\partial \eta} + \alpha \bar{H}_v \tag{13-2}$$

Since mixed partial derivatives which appear in $\frac{\partial \bar{F}_v}{\partial \eta}$ are dropped, as discussed in Chapter 11, the modified viscous flux vector $\bar{F}_v$ is redefined as $\bar{F}_{vP}$. The flux vector $\bar{E}$ includes pressure and, therefore, the streamwise pressure gradient appears in $\frac{\partial \bar{E}}{\partial \xi}$. With no modification of the streamwise pressure gradient within the subsonic portion of the viscous region, the equations are elliptic which would allow the propagation of a disturbance upstream. Therefore, a space marching procedure to solve the system cannot be incorporated. If a marching procedure is used, exponential growth or decay in the solution near the surface will occur, which will cause failure of the numerical scheme. This failed solution is known as the departure solution. To overcome this difficulty, the streamwise pressure gradient is modified such that the equations become parabolic. One approach for this approximation is to split the inviscid flux vector $\bar{E}$ as

$$\bar{E} = \bar{E}_P + \bar{E}_{PP} \tag{13-3}$$

where

$$\bar{E}_P = \frac{1}{J}[\xi_x E_P + \xi_y F_P] \tag{13-4}$$

and

$$\bar{E}_{PP} = \frac{1}{J}[\xi_x E_{PP} + \xi_y F_{PP}] \tag{13-5}$$

The flux vectors in (13-4) and (13-5) are:

$$E_P = \begin{bmatrix} \rho u \\ \rho u^2 + \omega p \\ \rho uv \\ (\rho e_t + p)u \end{bmatrix} \qquad F_P = \begin{bmatrix} \rho v \\ \rho vu \\ \rho v^2 + \omega p \\ (\rho e_t + p)v \end{bmatrix}$$

$$E_{PP} = \begin{bmatrix} 0 \\ (1-\omega)p \\ 0 \\ 0 \end{bmatrix} \qquad F_{PP} = \begin{bmatrix} 0 \\ 0 \\ (1-\omega)p \\ 0 \end{bmatrix}$$

Thus, Equation (13-2) is expressed in terms of these flux vectors as

$$\frac{\partial \bar{E}_P}{\partial \xi} + \frac{\partial \bar{E}_{PP}}{\partial \xi} + \frac{\partial \bar{F}}{\partial \eta} + \alpha \bar{H} = \frac{\partial \bar{F}_{vP}}{\partial \eta} + \alpha \bar{H}_v \tag{13-6}$$

In the supersonic portion of the flowfield, $\omega = 1$ and no modification of the pressure gradient is required. Various approximations for the streamwise pressure gradient within the subsonic portion of the flow will be discussed shortly.

Equation (13-6) is a mixed hyperbolic/parabolic system of equations. Stability analysis [13-1] indicates that for a stable solution the following restrictions must be satisfied.

1. The component of the local velocity in the streamwise direction must be positive.

2. The Mach number determined by the component of the velocity in the streamwise direction must be supersonic everywhere. However, once an approximation to the streamwise pressure gradient has been incorporated, this condition will allow subsonic flow in the viscous regions.

Note that the first condition eliminates streamwise flow separation; however, cross flow separation (in three-dimensional problems) is permitted. Next various approximation techniques will be considered for the streamwise pressure gradient; and, subsequently, a numerical algorithm will be investigated for the solution of the PNS equations.

13.3 Streamwise Pressure Gradient

In order to suppress the elliptic nature of Equation (13-6), the pressure gradient within the subsonic portion must be approximated. Numerous schemes have been introduced for this purpose, some of which are reviewed in this section.

1. The obvious and simplest approximation is to drop the pressure gradient within the subsonic portion of the flow. This crude approximation will introduce inaccuracies in the solution of the flowfields where large pressure gradients are present and therefore has limited use.

2. The pressure gradient is evaluated explicitly by a backward difference approximation, i.e.,

$$\left.\frac{\partial p}{\partial \xi}\right|_{i+1} = \frac{p_i - p_{i-1}}{\Delta \xi}$$

   When this approximation is used, the stability requirement imposes a lower limit on the selection of the marching step size, i.e., $\Delta\xi$ must be larger than some $(\Delta\xi)_{\text{min}}$ which is provided by the stability analysis.

3. In the development of the boundary layer equations it is assumed that the normal gradient of pressure within the boundary layer is negligible. Consistent with this assumption, the streamwise pressure gradient in the PNS equations is computed at the first supersonic point and imposed upon the subsonic portion. This procedure is known as the sublayer approximation. It has been observed that this approximation introduces instability for some cases.

4. Earlier the flux vector $\bar{E}$, which includes the pressure, was decomposed by introducing a parameter "ω". Based on an eigenvalue stability analysis, a fraction ω of the streamwise pressure gradient may be retained and evaluated implicitly while the remaining $(1-\omega)$ fraction is evaluated explicitly. Therefore,

$$\frac{\partial p}{\partial \xi} = \omega \left.\frac{\partial p}{\partial \xi}\right|_{\text{implicit}} + (1-\omega)\left.\frac{\partial p}{\partial \xi}\right|_{\text{explicit}} \tag{13-7}$$

   The parameter ω is determined by the stability analysis and is given by

$$\omega = \sigma\left[\frac{\gamma M_\xi^2}{1 + (\gamma - 1)M_\xi^2}\right] \tag{13-8}$$

   where σ is a safety factor, usually assigned a value of 0.8, and M_ξ is the local Mach number in the ξ-direction. Note that in regions where the flow is supersonic, $M_\xi > 1$, $\omega = 1$, and no approximation is incorporated.

The approximations reviewed above are only a few among the many reported in the literature. In the remaining discussion and in the results to be presented shortly, the fourth approximation for the streamwise pressure gradient has been incorporated.

13.4 Numerical Algorithm

Following the same procedure used for the Euler equations in Chapter 12, the delta formulation of the PNS equations are derived as

$$\frac{1}{\Delta\xi}(\Delta\bar{E}_P)+\frac{\partial}{\partial\eta}(\Delta\bar{F})-\frac{\partial}{\partial\eta}(\Delta\bar{F}_{vP})=-\left.\frac{\partial\bar{F}}{\partial\eta}\right|_i+\left.\frac{\partial\bar{F}_{vP}}{\partial\eta}\right|_i$$

$$-\left.\frac{\partial\bar{E}_{PP}}{\partial\xi}\right|_i-\bar{H}_i+(\bar{H}_v)_i \tag{13-9}$$

where a first-order backward difference approximation for $\partial\bar{E}_P/\partial\xi$ is employed. The right-hand side is evaluated at the known streamwise location "i". Equation (13-9) is nonlinear and, therefore, a linearization procedure similar to that introduced in Chapter 11 is considered. To illustrate the linearization procedure, consider the flux vector $\bar{E}_P$. Recall that

$$\bar{E}_P=f(\bar{Q},\xi_x,\xi_y) \tag{13-10}$$

A Taylor series expansion in the ξ direction yields

$$\bar{E}_{P_{i+1}}=\bar{E}_{P_i}+\frac{\partial\bar{E}_P}{\partial\xi}\Delta\xi+O(\Delta\xi)^2 \tag{13-11}$$

The chain rule of differentiation applied to (13-10) results in

$$\frac{\partial\bar{E}_P}{\partial\xi}=\frac{\partial\bar{E}_P}{\partial\bar{Q}}\frac{\partial\bar{Q}}{\partial\xi}+\frac{\partial\bar{E}_P}{\partial\xi_x}\frac{\partial\xi_x}{\partial\xi}+\frac{\partial\bar{E}_P}{\partial\xi_y}\frac{\partial\xi_y}{\partial\xi} \tag{13-12}$$

This equation is substituted into (13-11) to give

$$\bar{E}_{P_{i+1}}=\bar{E}_{P_i}+\left[\frac{\partial\bar{E}_P}{\partial\bar{Q}}\frac{\partial\bar{Q}}{\partial\xi}+\frac{\partial\bar{E}_P}{\partial\xi_x}\frac{\partial\xi_x}{\partial\xi}+\frac{\partial\bar{E}_P}{\partial\xi_y}\frac{\partial\xi_y}{\partial\xi}\right]\Delta\xi+O(\Delta\xi)^2 \tag{13-13}$$

The ξ derivatives in (13-13) are approximated by a first-order backward difference approximation, for example,

$$\frac{\partial\bar{Q}}{\partial\xi}=\frac{\bar{Q}_{i+1}-\bar{Q}_i}{\Delta\xi}=\frac{\Delta\bar{Q}}{\Delta\xi}$$

$$\frac{\partial\xi_x}{\partial\xi}=\frac{\xi_{x_{i+1}}-\xi_{x_i}}{\Delta\xi}=\frac{\Delta\xi_x}{\Delta\xi}$$

After substitution into (13-13), the following is obtained:

$$\Delta\bar{E}_P=\left[\frac{\partial\bar{E}_P}{\partial\bar{Q}}\frac{\Delta\bar{Q}}{\Delta\xi}+\frac{\partial\bar{E}_P}{\partial\xi_x}\frac{\Delta\xi_x}{\Delta\xi}+\frac{\partial\bar{E}_P}{\partial\xi_y}\frac{\Delta\xi_y}{\Delta\xi}\right]\Delta\xi+O(\Delta\xi)^2$$

or

$$\Delta \bar{E}_P = \frac{\partial \bar{E}_P}{\partial \bar{Q}} \Delta \bar{Q} + \frac{\partial \bar{E}_P}{\partial \xi_x} \Delta \xi_x + \frac{\partial \bar{E}_P}{\partial \xi_y} \Delta \xi_y + O(\Delta \xi)^2 \tag{13-14}$$

Similarly,

$$\Delta \bar{F} = \frac{\partial \bar{F}}{\partial \bar{Q}} \Delta \bar{Q} + \frac{\partial \bar{F}}{\partial \eta_x} \Delta \eta_x + \frac{\partial \bar{F}}{\partial \eta_y} \Delta \eta_y + O(\Delta \xi)^2 \tag{13-15}$$

and

$$\Delta \bar{F}_{vP} = \frac{\partial \bar{F}_{vP}}{\partial \bar{Q}} \Delta \bar{Q} + \frac{\partial \bar{F}_{vP}}{\partial \eta_x} \Delta \eta_x + \frac{\partial \bar{F}_{vP}}{\partial \eta_y} \Delta \eta_y + O(\Delta \xi)^2 \tag{13-16}$$

Note that the $\partial \bar{E}_P/\partial \bar{Q}$, $\partial \bar{F}/\partial \bar{Q}$, and $\partial \bar{F}_{vP}/\partial \bar{Q}$ appearing in Equations (13-14) through (13-16) are the Jacobian matrices derived in Chapter 11 and are given by Equations (11-228) through (11-230). The remaining terms may be evaluated as follows.

Recall that

$$\bar{E}_P = \frac{1}{J}[\xi_x E_P + \xi_y F_P] \tag{13-17}$$

Therefore, the partial derivatives $\frac{\partial \bar{E}_P}{\partial \xi_x}$ and $\frac{\partial \bar{E}_P}{\partial \xi_y}$ in Equation (13-14) can be determined from (13-17) as

$$\frac{\partial \bar{E}_P}{\partial \xi_x} = \frac{E_P}{J} \tag{13-18a}$$

$$\frac{\partial \bar{E}_P}{\partial \xi_y} = \frac{F_P}{J} \tag{13-18b}$$

Similarly, from the expressions for $\bar{F}$ and $\bar{F}_{vP}$,

$$\frac{\partial \bar{F}}{\partial \eta_x} = \frac{E}{J} \tag{13-18c}$$

$$\frac{\partial \bar{F}}{\partial \eta_y} = \frac{F}{J} \tag{13-18d}$$

$$\frac{\partial \bar{F}_{vP}}{\partial \eta_x} = \frac{E_v}{J} \tag{13-18e}$$

$$\frac{\partial \bar{F}_{vP}}{\partial \eta_y} = \frac{F_v}{J} \tag{13-18f}$$

Returning to the delta formulation:

$$\Delta \bar{E}_P + \Delta \xi \frac{\partial}{\partial \eta}(\Delta \bar{F} - \Delta \bar{F}_{vP}) = -\Delta \xi \frac{\partial}{\partial \eta}(\bar{F}_i - \bar{F}_{vP_i}) - \Delta \xi \left. \frac{\partial \bar{E}_{PP}}{\partial \xi} \right|_i - \Delta \xi (\bar{H}_i - \bar{H}_{v_i})$$

Substitution of the linearized approximations given by (13-14) through (13-16) into this equation yields:

$$\frac{\partial \bar{E}_P}{\partial \bar{Q}}\Delta\bar{Q} + \frac{\partial \bar{E}_P}{\partial \xi_x}\Delta\xi_x + \frac{\partial \bar{E}_P}{\partial \xi_y}\Delta\xi_y + \Delta\xi\frac{\partial}{\partial\eta}\left\{\left[\frac{\partial \bar{F}}{\partial \bar{Q}}\Delta\bar{Q} + \left(\frac{\partial \bar{F}}{\partial \eta_x}\right)\Delta\eta_x + \left(\frac{\partial \bar{F}}{\partial \eta_y}\right)\Delta\eta_y\right] - \left[\frac{\partial \bar{F}_{vP}}{\partial \bar{Q}}\Delta\bar{Q} + \frac{\partial \bar{F}_{vP}}{\partial \eta_x}\Delta\eta_x + \frac{\partial \bar{F}_{vP}}{\partial \eta_y}\Delta\eta_y\right]\right\} =$$

$$-\Delta\xi\frac{\partial}{\partial\eta}(\bar{F}_i - \bar{F}_{v_i}) - \Delta\xi\frac{\partial \bar{E}_{PP}}{\partial\xi} - \Delta\xi(\bar{H}_i - \bar{H}_{v_i})$$

From which

$$\frac{\partial \bar{E}_P}{\partial \bar{Q}}\Delta\bar{Q} + \Delta\xi\frac{\partial}{\partial\eta}\left\{\frac{\partial \bar{F}}{\partial \bar{Q}}\Delta\bar{Q} - \frac{\partial \bar{F}_{vP}}{\partial \bar{Q}}\Delta\bar{Q}\right\} =$$

$$-\Delta\xi\frac{\partial}{\partial\eta}(\bar{F}_i - \bar{F}_{vP_i}) - \Delta\xi\frac{\partial \bar{E}_{PP}}{\partial\xi} - \Delta\xi(\bar{H}_i - \bar{H}_{v_i})$$

$$-\left(\frac{\partial \bar{E}_P}{\partial \xi_x}\Delta\xi_x + \frac{\partial \bar{E}_P}{\partial \xi_y}\Delta\xi_y\right)$$

$$-\Delta\xi\frac{\partial}{\partial\eta}\left\{\left[\frac{\partial \bar{F}}{\partial \eta_x} - \frac{\partial \bar{F}_{vP}}{\partial \eta_x}\right]\Delta\eta_x + \left[\frac{\partial \bar{F}}{\partial \eta_y} - \frac{\partial \bar{F}_{vP}}{\partial \eta_y}\right]\Delta\eta_y\right\}$$

The left-hand side (LHS) is rearranged as

$$\left\{\frac{\partial \bar{E}_P}{\partial \bar{Q}} + \Delta\xi\frac{\partial}{\partial\eta}\left(\frac{\partial \bar{F}}{\partial \bar{Q}} - \frac{\partial \bar{F}_{vP}}{\partial \bar{Q}}\right)\right\}\Delta\bar{Q} = \text{ Right-hand side (RHS)}$$

Note that a similar operator was used in Chapter 12 for the Euler equations. To emphasize how this operator is defined, note that

$$\left\{\frac{\partial}{\partial\eta}\left(\frac{\partial \bar{F}}{\partial \bar{Q}} - \frac{\partial \bar{F}_{vP}}{\partial \bar{Q}}\right)\right\}\Delta\bar{Q}$$

implies

$$\frac{\partial}{\partial\eta}\left[\left(\frac{\partial \bar{F}}{\partial \bar{Q}}\right)\Delta\bar{Q}\right] - \frac{\partial}{\partial\eta}\left[\left(\frac{\partial \bar{F}_{vP}}{\partial \bar{Q}}\right)\Delta\bar{Q}\right]$$

Now, the RHS is modified as follows. Expressions (13-18a) through (13-18f) are substituted into the RHS to provide

$$\text{RHS} = -\Delta\xi\frac{\partial}{\partial\eta}(\bar{F}_i - \bar{F}_{vP_i}) - \Delta\xi\frac{\partial \bar{E}_{PP}}{\partial\xi} - \Delta\xi(\bar{H}_i - \bar{H}_{v_i}) - \left[\frac{E_P}{J}\Delta\xi_x + \frac{F_P}{J}\Delta\xi_y\right]$$

$$-\Delta\xi\frac{\partial}{\partial\eta}\left\{\left[\frac{E}{J} - \frac{E_v}{J}\right]\Delta\eta_x + \left[\frac{F}{J} - \frac{F_v}{J}\right]\Delta\eta_y\right\}$$

In order to reduce this expression further, consider a Taylor series expansion of a function such as f

$$f_{i+1} = f_i + \frac{\partial f}{\partial x}\Delta x + O(\Delta x)^2$$

or

$$f_{i+1} - f_i = \frac{\partial f}{\partial x}\Delta x + O(\Delta x)^2$$

from which

$$\Delta f = \frac{\partial f}{\partial x}\Delta x + O(\Delta x)^2 \tag{13-19}$$

Using this approximation, the following relations may be written:

$$\Delta\xi_x = \frac{\partial \xi_x}{\partial \xi}\Delta\xi + O(\Delta\xi)^2 \tag{13-20a}$$

$$\Delta\xi_y = \frac{\partial \xi_y}{\partial \xi}\Delta\xi + O(\Delta\xi)^2 \tag{13-20b}$$

$$\Delta\eta_x = \frac{\partial \eta_x}{\partial \xi}\Delta\xi + O(\Delta\xi)^2 \tag{13-20c}$$

$$\Delta\eta_y = \frac{\partial \eta_y}{\partial \xi}\Delta\xi + O(\Delta\xi)^2 \tag{13-20d}$$

Incorporating these approximations into the RHS results in the following:

$$RHS = -\Delta\xi\frac{\partial}{\partial\eta}(\bar{F}_i - \bar{F}_{vP_i}) - \Delta\xi\frac{\partial \bar{E}_{PP}}{\partial\xi} - \Delta\xi(\bar{H}_i - \bar{H}_{v_i})$$

$$-\left(\frac{E_P}{J}\frac{\partial\xi_x}{\partial\xi}\Delta\xi + \frac{F_P}{J}\frac{\partial\xi_y}{\partial\xi}\delta\xi\right) - \Delta\xi\frac{\partial}{\partial\eta}\left\{\left(\frac{E}{J} - \frac{E_v}{J}\right)\frac{\partial\eta_x}{\partial\xi}\Delta\xi\right.$$

$$\left. + \left(\frac{F}{J} - \frac{F_v}{J}\right)\frac{\partial\eta_y}{\partial\xi}\Delta\xi\right\} = -\Delta\xi\frac{\partial}{\partial\eta}(\bar{F}_i - \bar{F}_{vP_i})$$

$$-\Delta\xi\frac{\partial \bar{E}_{PP}}{\partial\xi} - \Delta\xi(\bar{H}_i - \bar{H}_{v_i}) - \left(\frac{E_P}{J}\frac{\partial\xi_x}{\partial\xi} + \frac{F_P}{J}\frac{\partial\xi_y}{\partial\xi}\right)\Delta\xi$$

$$-\frac{\partial}{\partial\eta}\left\{\left(\frac{E}{J}\frac{\partial\eta_x}{\partial\xi} + \frac{F}{J}\frac{\partial\eta_y}{\partial\xi}\right) - \left(\frac{E_v}{J}\frac{\partial\eta_x}{\partial\xi} + \frac{F_v}{J}\frac{\partial\eta_y}{\partial\xi}\right)\right\}(\Delta\xi)^2$$

Note that the last term in this relation is second order. Since the numerical method is a first-order scheme, this second-order term is dropped. Hence, the formulation becomes

$$\left\{\frac{\partial \bar{E}_P}{\partial \bar{Q}} + \Delta\xi\frac{\partial}{\partial\eta}\left[\frac{\partial \bar{F}}{\partial \bar{Q}} - \frac{\partial \bar{F}_{vP}}{\partial \bar{Q}}\right]\right\}\Delta\bar{Q} = -\Delta\xi\frac{\partial}{\partial\eta}(\bar{F}_i - \bar{F}_{vP_i})$$

$$-\left(\frac{\partial \bar{E}_P}{\partial\xi}\right)_{\bar{Q}}\Delta\xi - \Delta\xi\left(\frac{\partial \bar{E}_{PP}}{\partial\xi}\right) - \Delta\xi(\bar{H}_i - \bar{H}_{v_i}) \tag{13-21}$$

where

$$\left(\frac{\partial \bar{E}_P}{\partial \xi}\right)_{\bar{Q}} = \frac{E_P}{J}\frac{\partial \xi_x}{\partial \xi} + \frac{F_P}{J}\frac{\partial \xi_y}{\partial \xi}$$

A second-order central difference approximation is used for the η derivatives on the left-hand side of Equation (13-21). The viscous and inviscid terms will be investigated separately. The inviscid term is approximated by

$$\frac{\partial}{\partial \eta}\left[\left(\frac{\partial \bar{F}}{\partial \bar{Q}}\right)\Delta \bar{Q}\right] = \frac{\left[\left(\frac{\partial \bar{F}}{\partial \bar{Q}}\right)\Delta \bar{Q}\right]_{i,j+1} - \left[\left(\frac{\partial \bar{F}}{\partial \bar{Q}}\right)\Delta \bar{Q}\right]_{i,j-1}}{2\Delta \eta}$$

The approximation of the viscous term is complicated due to the presence of embedded gradients. These gradients include flow properties as well as the metrics and Jacobian of transformation. To identify the problem clearly, consider element 2,1 of the viscous Jacobian matrix, i.e.,

$$\left(\frac{\partial \bar{F}_{vP}}{\partial \bar{Q}}\right)_{2,1} = -\frac{\mu}{Re_\infty J}\left[b_1\left(J\frac{u}{\rho}\right)_\eta + b_3\left(J\frac{v}{\rho}\right)_\eta\right]$$

In order to write a general formulation to present the terms of the elements, consider the first term of the relation above, i.e.,

$$\left(\frac{\mu}{Re_\infty J}b_1\right)\left(J\frac{u}{\rho}\right)_\eta$$

and express it as

$$\left(\frac{\mu}{Re_\infty J}b_1\right)\left(J\frac{u}{\rho}\right)_\eta = (L)(M)_\eta$$

where

$$L = \frac{\mu}{Re_\infty J}b_1 \quad \text{and} \quad M = J\frac{u}{\rho}$$

Note that each element in the Jacobian matrix $\partial \bar{F}_{vP}/\partial \bar{Q}$ has the same general form and can be expressed as $L(M)_\eta$. Now proceed with the approximation of the gradient of the Jacobian matrix, where the general form just defined is used. Thus, consider an approximation of

$$\frac{\partial}{\partial \eta}\left[\left(\frac{\partial \bar{F}_{vP}}{\partial \bar{Q}}\right)\Delta \bar{Q}\right] = \frac{\partial}{\partial \eta}\left[\left(L\frac{\partial M}{\partial \eta}\right)\Delta \bar{Q}\right]$$

A second-order central difference approximation at i is used providing

$$\frac{\partial}{\partial \eta}\left[\left(L\frac{\partial M}{\partial \eta}\right)\Delta \bar{Q}\right] = \frac{\left[\left(L\frac{\partial M}{\partial \eta}\right)\Delta \bar{Q}\right]_{i,j+\frac{1}{2}} - \left[\left(L\frac{\partial M}{\partial \eta}\right)\Delta \bar{Q}\right]_{i,j-\frac{1}{2}}}{2(\Delta \eta/2)}$$

Grid points $(i, j+1/2)$ and $(i, j-1/2)$ are considered as dummy grid points which are eliminated in the next step. The coefficient L is evaluated by averaging as

$$L_{i,j+\frac{1}{2}} = \tfrac{1}{2}(L_{i,j} + L_{i,j+1})$$

and

$$L_{i,j-\frac{1}{2}} = \tfrac{1}{2}(L_{i,j} + L_{i,j-1})$$

The gradient $\partial M/\partial \eta$ is evaluated by a second-order central difference approximation as well. Hence,

$$\frac{\partial}{\partial \eta}\left[\left(L\frac{\partial M}{\partial \eta}\right)\Delta \bar{Q}\right] =$$

$$\frac{\tfrac{1}{2}(L_{i,j} + L_{i,j+1})\left[\frac{(M\Delta\bar{Q})_{i,j+1} - (M\Delta\bar{Q})_{i,j}}{2(\Delta\eta/2)}\right] - \tfrac{1}{2}(L_{i,j} + L_{i,j-1})\left[\frac{(M\Delta\bar{Q})_{i,j} - (M\Delta\bar{Q})_{i,j-1}}{2(\Delta\eta/2)}\right]}{\Delta\eta}$$

or

$$\frac{\partial}{\partial \eta}\left[\left(L\frac{\partial M}{\partial \eta}\right)\Delta \bar{Q}\right] =$$

$$\frac{(L_{i,j} + L_{i,j+1})\left[(M\Delta\bar{Q})_{i,j+1} - (M\Delta\bar{Q})_{i,j}\right] - (L_{i,j} + L_{i,j-1})\left[(M\Delta\bar{Q})_{i,j} - (M\Delta\bar{Q})_{i,j-1}\right]}{2(\Delta\eta)^2}$$

$$= \frac{1}{2(\Delta\eta)^2}\left[(L_{i,j+1} + L_{i,j})(M_{i,j+1})\Delta\bar{Q}_{j+1} - (L_{i,j+1} + 2L_{i,j} + L_{i,j-1})(M_{i,j})\Delta\bar{Q}_j\right.$$

$$\left.+(L_{i,j} + L_{i,j-1})(M_{i,j-1})\Delta\bar{Q}_{j-1}\right] \tag{13-22}$$

In order to write this equation in a compact form, define

$$\begin{aligned}
\bar{L}_{i,j+1} &= L_{i,j+1} + L_{i,j} \\
\bar{L}_{i,j} &= L_{i,j+1} + 2L_{i,j} + L_{i,j-1} \\
\bar{L}_{i,j-1} &= L_{i,j} + L_{i,j-1}
\end{aligned}$$

Hence,

$$\frac{\partial}{\partial \eta}\left[\left(L\frac{\partial M}{\partial \eta}\right)\Delta \bar{Q}\right] = \frac{1}{2(\Delta\eta)^2}\left[(\bar{L}_{i,j+1}M_{i,j+1})\Delta\bar{Q}_{j+1}\right.$$

$$\left.-(\bar{L}_{i,j}M_{i,j})\Delta\bar{Q}_j + (\bar{L}_{i,j-1}M_{i,j-1})\Delta\bar{Q}_{j-1}\right]$$

With the approximations for the gradients of the viscous and inviscid Jacobian matrices identified, the finite difference representation of the LHS for Equation

(13-21) is

$$\left.\frac{\partial \bar{E}_P}{\partial \bar{Q}}\right|_{i,j} \Delta\bar{Q}_j + \frac{\Delta\xi}{2\Delta\eta}\left[\left.\frac{\partial \bar{F}}{\partial \bar{Q}}\right|_{i,j+1}\Delta\bar{Q}_{j+1} - \left.\frac{\partial \bar{F}}{\partial \bar{Q}}\right|_{i,j-1}\Delta\bar{Q}_{j-1}\right]$$
$$-\frac{\Delta\xi}{2(\Delta\eta)^2}\left[(M_{i,j+1}\bar{L}_{i,j+1})\Delta\bar{Q}_{j+1} - (M_{i,j}\bar{L}_{i,j})\Delta\bar{Q}_j + (M_{i,j-1}\bar{L}_{i,j-1})\Delta\bar{Q}_{j-1}\right]$$
$$= \text{RHS}$$

which is rearranged as

$$\left[-\frac{\Delta\xi}{2\Delta\eta}\left.\frac{\partial \bar{F}}{\partial \bar{Q}}\right|_{i,j-1} - \frac{\Delta\xi}{2(\Delta\eta)^2}(M_{i,j-1}\bar{L}_{i,j-1})\right]\Delta\bar{Q}_{j-1} +$$
$$\left[\left.\frac{\partial \bar{E}_P}{\partial \bar{Q}}\right|_{i,j} + \frac{\Delta\xi}{2(\Delta\eta)^2}(M_{i,j}\bar{L}_{i,j})\right]\Delta\bar{Q}_j +$$
$$\left[\frac{\Delta\xi}{2\Delta\eta}\left.\frac{\partial \bar{F}}{\partial \bar{Q}}\right|_{i,j+1} - \frac{\Delta\xi}{2(\Delta\eta)^2}(M_{i,j+1}\bar{L}_{i,j+1})\right]\Delta\bar{Q}_{j+1} = \text{RHS} \tag{13-23}$$

The brackets in this equation are defined as

$$AA_{i,j} = -\frac{\Delta\xi}{2\Delta\eta}\left.\frac{\partial \bar{F}}{\partial \bar{Q}}\right|_{i,j-1} - \frac{\Delta\xi}{2(\Delta\eta)^2}M_{i,j-1}\bar{L}_{i,j-1} \tag{13-24}$$

$$BB_{i,j} = \left.\frac{\partial \bar{E}_P}{\partial \bar{Q}}\right|_{i,j} + \frac{\Delta\xi}{2(\Delta\eta)^2}M_{i,j}\bar{L}_{i,j} \tag{13-25}$$

$$CC_{i,j} = \frac{\Delta\xi}{2\Delta\eta}\left.\frac{\partial \bar{F}}{\partial \bar{Q}}\right|_{i,j+1} - \frac{\Delta\xi}{2(\Delta\eta)^2}M_{i,j+1}\bar{L}_{i,j+1} \tag{13-26}$$

Hence, Equation (13-23) is expressed as

$$AA_{i,j}\Delta\bar{Q}_{j-1} + BB_{i,j}\Delta\bar{Q}_j + CC_{i,j}\Delta\bar{Q}_{j+1} = \text{RHS}_{i,j} \tag{13-27}$$

At this point the RHS, which is evaluated at the known streamwise station "i", will be investigated. Recall that

$$\text{RHS} = -\Delta\xi\frac{\partial}{\partial\eta}(\bar{F}_i - \bar{F}_{vP_i}) - \left(\frac{\partial \bar{E}_P}{\partial\xi}\right)_{\bar{Q}}\Delta\xi - \Delta\xi\left.\frac{\partial \bar{E}_{PP}}{\partial\xi}\right|_i - \Delta\xi(\bar{H}_i - \bar{H}_{v_i})$$

The gradient of the inviscid flux vector is approximated by a second-order finite difference expression, i.e.,

$$\frac{\partial \bar{F}_i}{\partial\eta} = \frac{\bar{F}_{i,j+1} - \bar{F}_{i,j-1}}{2\Delta\eta}$$

The gradient of the viscous flux vector is approximated in a similar manner as the implicit viscous terms on the LHS. To derive a general formulation for these terms, note that the viscous terms may be expressed as

$$\frac{\partial}{\partial \eta}\left[L\frac{\partial M}{\partial \eta}\right]$$

Following a similar procedure previously employed which produced approximation (13-22), the following may be written:

$$\frac{\partial}{\partial \eta}\left[L\frac{\partial M}{\partial \eta}\right] = \frac{(L_{i,j}+L_{i,j+1})\left(\frac{(M_{i,j+1}-M_{i,j}}{\Delta\eta}\right) - (L_{i,j-1}+L_{i,j})\left(\frac{(M_{i,j}-M_{i,j-1}}{\Delta\eta}\right)}{2\Delta\eta}$$

$$= \frac{(L_{i,j}+L_{i,j+1})(M_{i,j+1}-M_{i,j}) - (L_{i,j-1}+L_{i,j})(M_{i,j}-M_{i,j-1})}{2(\Delta\eta)^2}$$

The third term, $\left(\frac{\partial \bar{E}_P}{\partial \xi}\right)_{\bar{Q}}$, is

$$\left(\frac{\partial \bar{E}_P}{\partial \xi}\right)_{\bar{Q}} = \frac{E_P}{J}\frac{\partial \xi_x}{\partial \xi} + \frac{F_P}{J}\frac{\partial \xi_y}{\partial \xi}$$

The gradients of the metrics ξ_x and ξ_y are evaluated in the grid generation subroutine (or a program, if it is performed externally) with E_P and F_P evaluated at the grid point (i, j). The pressure gradient term is determined explicitly as

$$\left.\frac{\partial p}{\partial \xi}\right|_{i,j} = \frac{p_{i,j} - p_{i-1,j}}{\Delta\xi}$$

At this point, return to the implicit formulation given by Equation (13-27). This equation is applied at a new streamwise station "$i+1$" for all j grid points from $j = 1$ to $j = JM$. The domain of solution may be specified in two different fashions. Obviously, one boundary of the domain is the body surface. The outer boundary may be selected far into the freestream, such that the bow shock is included in the domain; or the bow shock itself is taken as the boundary, in which case a shock-fitting procedure is used. In the formulations to follow, the inner boundary is selected at the surface where $j = 1$, and the outer boundary is at the freestream where $j = JM$. The physical and computational domains are illustrated in Figure (13-3). The shock fitting procedure is discussed in Section 13.8.

Now, Equation (13-27) is applied to each j grid point from $j = 1$ to $j = JM$, providing the following finite difference equations:

$$BB_{i,1}\Delta\bar{Q}_1 + CC_{i,1}\Delta\bar{Q}_2 = R_{i,1}$$

$$AA_{i,2}\Delta\bar{Q}_1 + BB_{i,2}\Delta\bar{Q}_2 + CC_{i,2}\Delta\bar{Q}_3 = R_{i,2}$$

$$AA_{i,3}\Delta\bar{Q}_2 + BB_{i,3}\Delta\bar{Q}_3 + CC_{i,3}\Delta\bar{Q}_4 = R_{i,3}$$

$$AA_{i,4}\Delta\bar{Q}_3 + BB_{i,4}\Delta\bar{Q}_4 + CC_{i,4}\Delta\bar{Q}_5 = R_{i,4}$$

$$\vdots$$

$$AA_{i,JMM2}\Delta\bar{Q}_{JMM3} + BB_{i,JMM2}\Delta\bar{Q}_{JMM2} + CC_{i,JMM2}\Delta\bar{Q}_{JMM1} = R_{i,JMM2}$$

$$AA_{i,JMM1}\Delta\bar{Q}_{JMM2} + BB_{i,JMM1}\Delta\bar{Q}_{JMM1} + CC_{i,JMM1}\Delta\bar{Q}_{JM} = R_{i,JMM1}$$

$$AA_{i,JM}\Delta\bar{Q}_{JMM1} + BB_{i,JM}\Delta\bar{Q}_{JM} = R_{i,JM}$$

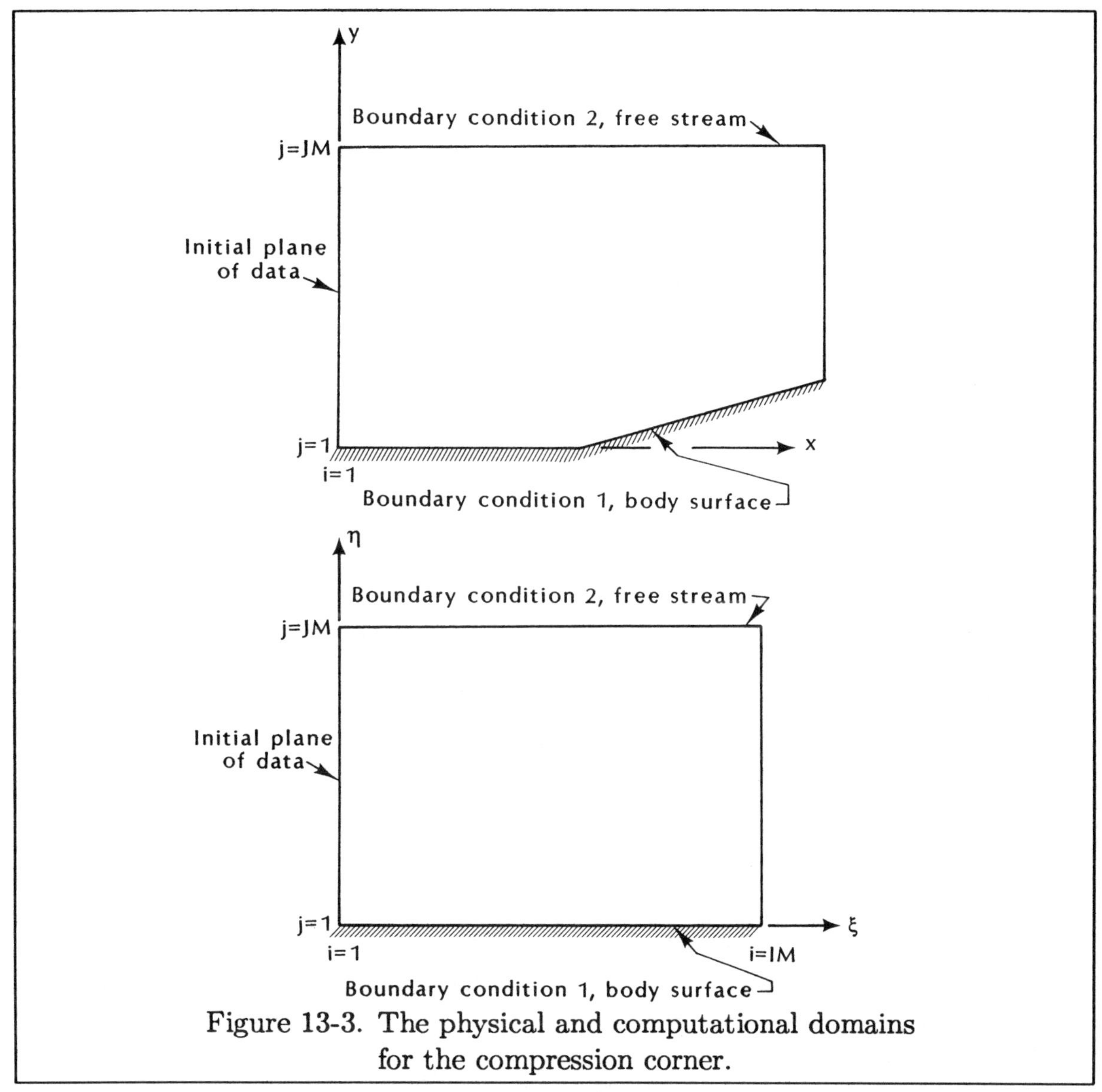

Figure 13-3. The physical and computational domains for the compression corner.

This system of equations is written in a matrix form as:

$$\begin{bmatrix} \overline{BB}_{i,1} & \overline{CC}_{i,1} & & & & & & \\ AA_{i,2} & BB_{i,2} & CC_{i,2} & & & & & \\ & AA_{i,3} & BB_{i,3} & CC_{i,3} & & & & \\ & & AA_{i,4} & BB_{i,4} & CC_{i,4} & & & \\ & & \ddots & \ddots & \ddots & & & \\ & & & AA_{i,JMM2} & BB_{i,JMM2} & CC_{i,JMM2} & & \\ & & & & AA_{i,JMM1} & BB_{i,JMM1} & CC_{i,JMM1} & \\ & & & & & \overline{AA}_{i,JM} & \overline{BB}_{i,JM} & \end{bmatrix}$$

$$\begin{bmatrix} \Delta\bar{Q}_1 \\ \Delta\bar{Q}_2 \\ \Delta\bar{Q}_3 \\ \Delta\bar{Q}_4 \\ \vdots \\ \Delta\bar{Q}_{JMM3} \\ \Delta\bar{Q}_{JMM2} \\ \Delta\bar{Q}_{JMM1} \\ \Delta\bar{Q}_{JM} \end{bmatrix} = \begin{bmatrix} R_{i,1} \\ R_{i,2} \\ R_{i,3} \\ R_{i,4} \\ \vdots \\ R_{i,JMM3} \\ R_{i,JMM2} \\ R_{i,JMM1} \\ R_{i,JM} \end{bmatrix} \tag{13-28}$$

Notice that the elements on the first and the last rows of the coefficient matrix are distinguished by an overbar from the remaining elements. The reason for this distinction is due to the imposition of the boundary conditions which will modify these elements. The boundary conditions are discussed in the next section. Before we proceed to implement the boundary conditions, consider the system of equations given by (13-28) and briefly review its features.

1. The system is expressed as a block tridiagonal formulation. Efficient numerical schemes to solve this system have been developed.

2. The elements of the coefficient matrix are 4 x 4 matrices for a two-dimensional problem and 5 x 5 matrices for three-dimensional problems.

3. The unknowns in this system of equations are the $\Delta\bar{Q}$ vectors, which are 4 x 1 for two-dimensional problems and 5 x 1 for three-dimensional problems.

4. The right-hand side is computed at the known station. Note that these elements are 4 x 1 vectors for a two-dimensional problem and 5 x 1 vectors for a three-dimensional problem.

5. The coefficient matrix is evaluated at the known station "i", whereas the unknowns are at the "$i+1$" location. Once the unknown vector $\Delta\bar{Q}$ has been

obtained, the vector $\bar{Q}$ at "$i+1$" is evaluated by

$$\bar{Q}_{i+1} = \bar{Q}_i + \Delta\bar{Q}$$

6. For the first level of computation, the known station is provided by the specified initial data set. Subsequently, the newly computed values are stored and used for the next level of computation.

13.5 Boundary Conditions

Boundary conditions are required at the surface and at an outer boundary set at the freestream or at the bow shock. Graphical representation of a domain with boundaries at the surface and the freestream along with an initial plane of data is shown in Figure (13-3).

The initial plane of data is specified at "$i = 1$". Therefore, Equation (13-28) is first solved for the unknowns at "$i = 2$" for all the grid points from $j = 1$ to $j = JM$. Some of the dependent variables at the boundaries may be specified from physical laws pertinent to the problem. The remaining variables at the boundaries must be computed as a part of the overall solution. For this purpose additional (assumed) boundary conditions must be specified. For example, the no-slip condition is applied at the surface for viscous flows. Therefore, the velocity at the surface is set equal to zero. On the other hand, the value of the density at the surface is unknown and must be obtained from the solution of Equation (13-28). Along the surface either the value for the temperature or its gradient may be specified. For an adiabatic flow the temperature gradient is set to zero. The assumed boundary condition is usually specified by setting the pressure gradient normal to the surface equal to zero.

At the outer boundary, designated as boundary 2 in Figure 13-3, the freestream conditions are imposed if the domain is extended to the freestream. A second option of specifying the outer boundary is to use the bow-shock as boundary 2. Such a domain is shown in Figure 13-4. In this problem the location of the shock is unknown and is computed as a part of the overall solution. For this purpose, Rankine-Hugoniot relations are used. The mathematical implementation of various types of boundary conditions and the manner in which they effect the matrices, $\overline{BB}_{i,1}$, $\overline{CC}_{i,1}$, $\overline{AA}_{i,JM}$, and $\overline{BB}_{i,JM}$, are discussed next.

Consider first the wall boundary conditions. Assuming no flow injection or suction, the no-slip condition is imposed at the surface. Therefore,

$$(\Delta\bar{Q}_2)_{i,1} = (\Delta\bar{\rho}\bar{u})_{i,1} = 0$$

and

$$(\Delta\bar{Q}_3)_{i,1} = (\Delta\bar{\rho}\bar{v})_{i,1} = 0$$

The inner subscripts in $\bar{Q}$ denote the component of the vector $\bar{Q}$, i.e.,

$$\bar{Q} = \begin{bmatrix} \bar{Q}_1 \\ \bar{Q}_2 \\ \bar{Q}_3 \\ \bar{Q}_4 \end{bmatrix} = \frac{1}{J} \begin{bmatrix} \rho \\ \rho u \\ \rho v \\ \rho e_t \end{bmatrix} = \begin{bmatrix} \bar{\rho} \\ \bar{\rho} u \\ \bar{\rho} v \\ \bar{\rho} e_t \end{bmatrix}$$

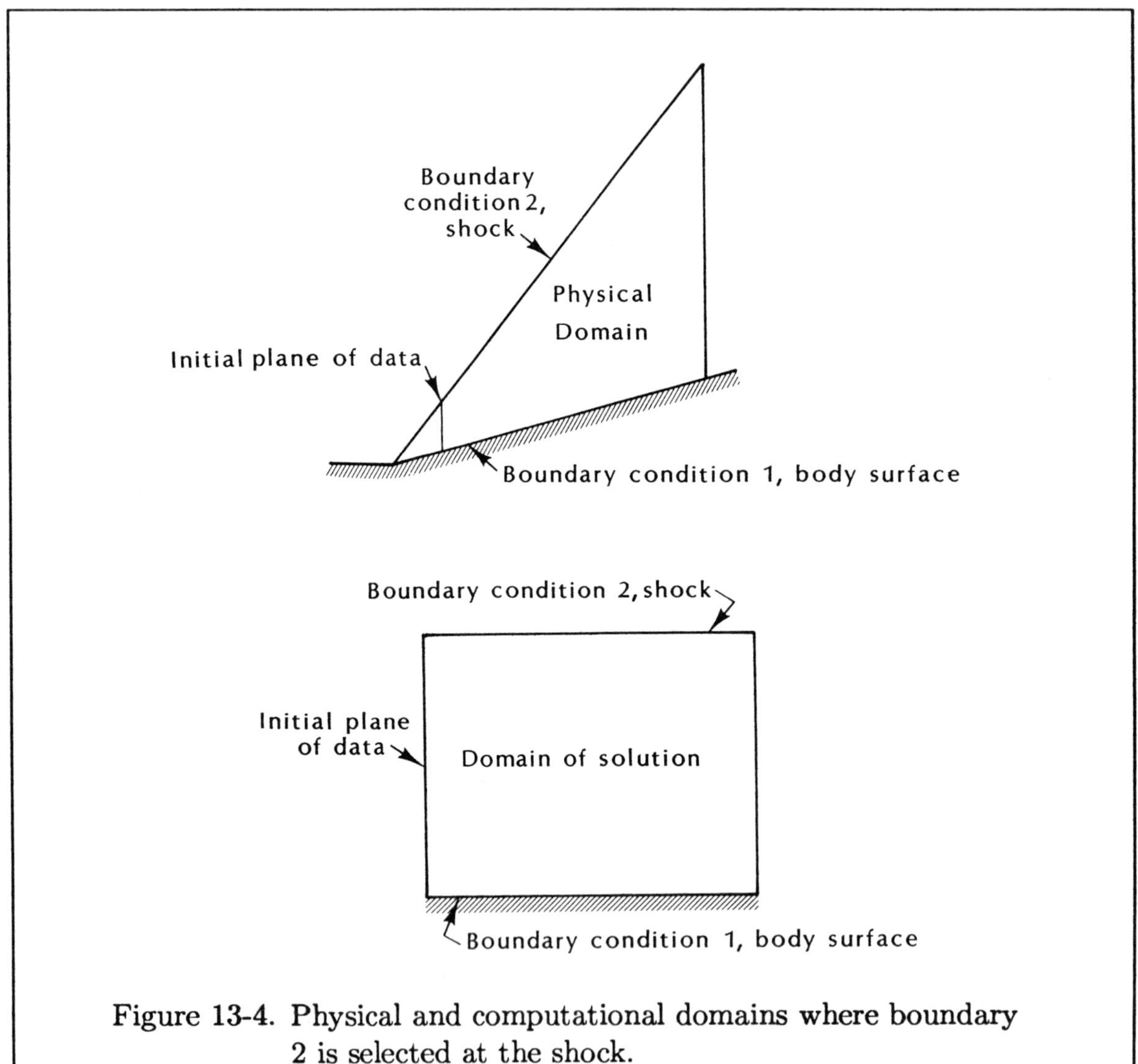

Figure 13-4. Physical and computational domains where boundary 2 is selected at the shock.

To describe the density at the surface, note that $\bar{Q}_1 = \bar{\rho} = \rho/J$. By definition

$$(\Delta\bar{\rho})_{i,1} = \bar{\rho}_{i+1,1} - \bar{\rho}_{i,1} = (\frac{\rho}{J})_{i+1,1} - (\frac{\rho}{J})_{i,1}$$

The term $\psi_{i+1}(\Delta\bar{\rho})_{i,2}$ is subtracted from this equation, where ψ_{i+1} is the ratio of the

Jacobian of transformation defined as

$$\psi_{i+1} = \frac{J_{i+1,2}}{J_{i+1,1}} = \left(\frac{J_2}{J_1}\right)_{i+1}$$

Note that the term ψ is provided by the grid generation routine. Hence,

$$(\Delta\bar{\rho})_{i,1} - \psi_{i+1}(\Delta\bar{\rho})_{i,2} = \left[(\frac{\rho}{J})_{i+1,1} - (\frac{\rho}{J})_{i,1}\right] - \psi_{i+1}\left[(\frac{\rho}{J})_{i+1,2} - (\frac{\rho}{J})_{i,2}\right] \tag{13-29}$$

If an adiabatic wall is assumed, then

$$T_{i+1,1} = T_{i+1,2}$$

In addition, it is assumed that $p_{i+1,1} = p_{i+1,2}$. This assumption implies that $\partial p/\partial\eta = 0$ at the surface, which falls within the overall assumptions used in PNS equations. Therefore, for an adiabatic wall boundary condition,

$$\rho_{i+1,1} = \rho_{i+1,2}$$

Now, Equation (13-29) is rearranged as

$$(\Delta\bar{\rho})_{i,1} - \psi_{i+1}(\Delta\bar{\rho})_{i,2} = \psi_{i+1}(\frac{\rho}{J})_{i,2} - (\frac{\rho}{J})_{i,1} + (\frac{\rho}{J})_{i+1,1} - \psi_{i+1}(\frac{\rho}{J})_{i+1,2} \tag{13-30}$$

The last two terms add up to zero! This statement is proven below. The terms are rearranged as

$$(\frac{\rho}{J})_{i+1,1} - \psi_{i+1}(\frac{\rho}{J})_{i+1,2} = (\frac{\rho}{J})_{i+1,1} - \frac{J_{i+1,2}}{J_{i+1,1}}(\frac{\rho}{J})_{i+1,2} = (\frac{\rho}{J})_{i+1,1} - \frac{\rho_{i+1,2}}{J_{i+1,1}}$$

But for an adiabatic wall,

$$\rho_{i+1,1} = \rho_{i+1,2}$$

Hence,

$$\frac{\rho_{i+1,1}}{J_{i+1,1}} - \frac{\rho_{i+1,2}}{J_{i+1,1}} = 0$$

As a result, Equation (13-30) is reduced to

$$(\Delta\bar{\rho})_{i,1} - \psi_{i+1}(\Delta\bar{\rho})_{i,2} = \psi_{i+1}(\frac{\rho}{J})_{i,2} - (\frac{\rho}{J})_{i,1} \tag{13-31}$$

The last term of this equation is modified as follows:

$$\left(\frac{\rho}{J}\right)_{i,1} = \frac{\rho_{i,1}}{J_{i,1}}\frac{J_{i,2}}{J_{i,2}} = \frac{J_{i,2}}{J_{i,1}}\frac{\rho_{i,1}}{J_{i,2}} = \psi_i\frac{\rho_{i,2}}{J_{i,2}}$$

Note that $\rho_{i,1} = \rho_{i,2}$ is used , which is valid for an adiabatic wall condition as described earlier. Finally, Equation (13-31) is expressed as

$$(\Delta\bar{\rho})_{i,1} - \psi_i(\Delta\bar{\rho})_{i,2} = (\psi_{i+1} - \psi_i)\frac{\rho_{i,2}}{J_{i,2}} \tag{13-32}$$

In this equation $(\Delta\bar{\rho})_{i,1}$ and $(\Delta\bar{\rho})_{i,2}$ are the unknowns, whereas $\rho_{i,2}$ is known and the Jacobian of transformation at any grid point is provided by the grid generation routine.

Similar mathematical manipulation is used to obtain a relation for the energy. To derive this relation, start by writing the following expressions:

$$(\Delta\bar{\rho}e_t)_{i,1} - \psi_{i+1}(\Delta\bar{\rho}e_t)_{i,2} = \left[\frac{(\rho e_t)_{i+1,1}}{J_{i+1,1}} - \frac{(\rho e_t)_{i,1}}{J_{i,1}}\right]$$

$$-\psi_{i+1}\left[\frac{(\rho e_t)_{i+1,2}}{J_{i+1,2}} - \frac{(\rho e_t)_{i,2}}{J_{i,2}}\right] \tag{13-33a}$$

$$(\psi_{i+1}u_{i,2})(\Delta\bar{\rho}u)_{i,2} = (\psi_{i+1}u_{i,2})\left[\frac{(\rho u)_{i+1,2}}{J_{i+1,2}} - \frac{(\rho u)_{i,2}}{J_{i,2}}\right] \tag{13-33b}$$

$$(\psi_{i+1}v_{i,2})(\Delta\bar{\rho}v)_{i,2} = (\psi_{i+1}v_{i,2})\left[\frac{(\rho v)_{i+1,2}}{J_{i+1,2}} - \frac{(\rho v)_{i,2}}{J_{i,2}}\right] \tag{13-33c}$$

and

$$-\frac{1}{2}\psi_{i+1}(u_{i,2}^2 + v_{i,2}^2)(\Delta\bar{\rho})_{i,2} = -\frac{1}{2}\psi_{i+1}(u_{i,2}^2 + v_{i,2}^2)\left[\frac{\rho_{i+1,2}}{J_{i+1,2}} - \frac{\rho_{i,2}}{J_{i,2}}\right] \tag{13-33d}$$

Equations (13-33a) through (13-33d) are summed, and the result expressed as LHS and RHS is

$$\begin{aligned} \text{LHS} &= \Delta(\bar{\rho}e_t)_{i,1} - \psi_{i+1}(\Delta\bar{\rho}e_t)_{i,2} + (\psi_{i+1}u_{i,2})(\Delta\bar{\rho}u)_{i,2} \\ &\quad +(\psi_{i+1}v_{i,2})(\Delta\bar{\rho}v)_{i,2} - \frac{1}{2}\psi_{i+1}(u_{i,2}^2 + v_{i,2}^2)(\Delta\bar{\rho})_{i,2} \end{aligned}$$

and

$$\text{RHS} = \frac{(\rho e_t)_{i+1,1}}{J_{i+1,1}} - \frac{(\rho e_t)_{i,1}}{J_{i,1}} - \frac{1}{J_{i+1,1}}[\,(\rho e_t)_{i+1,2} - u_{i,2}(\rho u)_{i+1,2} +$$

$$-v_{i,2}(\rho v)_{i+1,2} + \tfrac{1}{2}(u_{i,2}^2 + v_{i,2}^2)(\rho_{i+1,2})\ ]$$

$$+\psi_{i+1}\frac{1}{J_{i,2}}\ [\ (\rho e_t)_{i,2} - u_{i,2}(\rho u)_{i,2} - v_{i,2}(\rho v)_{i,2}$$

$$+\tfrac{1}{2}(u_{i,2}^2 + v_{i,2}^2)(\rho_{i,2})\ ]$$

The quantities in the RHS expression need to be expressed in terms of the known values of i. For this purpose the following relation is employed:

$$\rho e_t = \frac{p}{\gamma - 1} + \tfrac{1}{2}\rho(u^2 + v^2)$$

Therefore,

$$\begin{aligned}
\text{RHS} \quad = \quad & \frac{1}{J_{i+1,1}}\left[\frac{p_{i+1,1}}{\gamma-1} + \tfrac{1}{2}\rho_{i+1,1}(u_{i+1,1}^2 + v_{i+1,1}^2)\right] \\
& -\frac{1}{J_{i,1}}\left[\frac{p_{i,1}}{\gamma-1} + \tfrac{1}{2}\rho_{i,1}(u_{i,1}^2 + v_{i,1}^2)\right] - \frac{1}{J_{i+1,1}}\left[\frac{p_{i+1,2}}{\gamma-1} + \tfrac{1}{2}\rho_{i+1,2}(u_{i+1,2}^2\right. \\
& \left.+v_{i,2}^2) - u_{i,2}(\rho u)_{i+1,2} - v_{i,2}(\rho v)_{i+1,2} + \tfrac{1}{2}\rho_{i+1,2}(u_{i,2}^2 + v_{i,2}^2)\right] \\
& +\psi_{i+1}\frac{1}{J_{i,2}}\left[\frac{p_{i,2}}{\gamma-1} + \tfrac{1}{2}\rho_{i,2}(u_{i,2}^2 + v_{i,2}^2) - u_{i,2}(\rho u)_{i,2}\right. \\
& \left.-v_{i,2}(\rho v)_{i,2} + \tfrac{1}{2}\rho_{i,2}(u_{i,2}^2 + v_{i,2}^2)\right]
\end{aligned}$$

To simplify this expression, consider the following terms:

$$\begin{aligned}
& \tfrac{1}{2}\rho_{i+1,2}u_{i+1,2}^2 + \tfrac{1}{2}\rho_{i+1,2}v_{i+1,2}^2 - \rho_{i+1,2}u_{i,2}u_{i+1,2} - \rho_{i+1,2}v_{i,2}v_{i+1,2} \\
& +\tfrac{1}{2}\rho_{i+1,2}u_{i,2}^2 + \tfrac{1}{2}\rho_{i+1,2}v_{i+1,2}^2 = \tfrac{1}{2}\rho_{i+1,2}(u_{i+1,2}^2 - 2u_{i,2}u_{i+1,2} + u_{i,2}^2) \\
& \qquad +\tfrac{1}{2}\rho_{i+1,2}(v_{i+1,2}^2 - 2v_{i,2}v_{i+1,2} + v_{i,2}^2)
\end{aligned}$$

The terms in the parenthesis are assumed small and, therefore, they are omitted. Hence,

$$\begin{aligned}
\text{RHS} \quad = \quad & \frac{1}{J_{i+1,1}}\,\frac{1}{\gamma-1}p_{i+1,1} - \frac{1}{J_{i,1}}\,\frac{1}{\gamma-1}p_{i,1} - \frac{1}{J_{i+1,1}}\,\frac{1}{\gamma-1}p_{i+1,2} \\
& +\psi_{i+1}\frac{1}{\gamma-1}\,\frac{1}{J_{i,2}}p_{i,2} = \frac{1}{\gamma-1}\,\frac{1}{J_{i+1,1}}(p_{i+1,1} - p_{i+1,2}) \\
& +\frac{1}{\gamma-1}\psi_{i+1}\frac{1}{J_{i,2}}p_{i,2} - \frac{1}{\gamma-1}\,\frac{1}{J_{i,1}}p_{i,1}
\end{aligned}$$

But, it was previously stated that $p_{i,1} = p_{i,2}$ for all i, therefore,

$$\begin{aligned}
\text{RHS} &= \frac{1}{\gamma-1}\left[\frac{1}{J_{i,2}}(\psi_{i+1}p_{i,2}) - \frac{1}{J_{i,1}}p_{i,1}\right] \\
&= \frac{1}{\gamma-1}\left[\frac{1}{J_{i,2}}(\psi_{i+1}p_{i,2}) - \frac{p_{i,1}}{J_{i,1}}\frac{J_{i,2}}{J_{i,2}}\right] \\
&= \frac{1}{\gamma-1}\left[\psi_{i+1} - \psi_i\frac{1}{J_{i,2}}p_{i,2}\right]
\end{aligned}$$

Now, the final form of the expression for the total energy e_t may be written as

$$\begin{aligned}
&(\Delta\bar{\rho}e_t)_{i,1} - \psi_{i+1}(\Delta\bar{\rho}e_t)_{i,2} + \psi_{i+1}u_{i,2}(\Delta\bar{\rho}u)_{i,2} + \psi_{i+1}v_{i,2}(\Delta\bar{\rho}v)_{i,2} \\
&\quad -\frac{1}{2}\psi_{i+1}(u_{i,2}^2 + v_{i,2}^2)(\Delta\bar{\rho})_{i,2} = \frac{1}{\gamma-1}(\psi_{i+1} - \psi_i)\frac{p_{i,2}}{J_{i,2}}
\end{aligned}$$

The four expressions to be used at the wall boundary and the imposed assumptions are summarized below.

1. Assumptions
   (a) adiabatic wall
   (b) normal gradient of the pressure at the surface is zero
   (c) no slip at the surface
2. Equations
   (a) $(\Delta\bar{\rho})_{i,1} - \psi_{i+1}(\Delta\bar{\rho})_{i,2} = (\psi_{i+1} - \psi_i)\left(\frac{\rho_{i,2}}{J_{i,2}}\right)$
   (b) $(\Delta\bar{\rho}u)_{i,1} = 0$
   (c) $(\Delta\bar{\rho}v)_{i,1} = 0$
   (d) $(\Delta\bar{\rho}e_t)_{i,1} - \psi_{i+1}[\frac{1}{2}(u_{i,2}^2+v_{i,2}^2)(\Delta\bar{\rho})_{i,2} - u_{i,2}(\Delta\bar{\rho}u)_{i,2} - v_{i,2}(\Delta\bar{\rho}v)_{i,2} + \Delta(\bar{\rho}e_t)_{i,2}] = \frac{1}{\gamma-1}(\psi_{i+1} - \psi_i)\frac{p_{i,2}}{J_{i,2}}$

These equations are expressed in a matrix formulation as

$$\begin{bmatrix} 1 & 0 & 0 & 0 \\ 0 & 1 & 0 & 0 \\ 0 & 0 & 1 & 0 \\ 0 & 0 & 0 & 1 \end{bmatrix}\begin{bmatrix} (\Delta\bar{\rho})_{i,1} \\ (\Delta\bar{\rho}u)_{i,1} \\ (\Delta\bar{\rho}v)_{i,1} \\ (\Delta\bar{\rho}e_t)_{i,1} \end{bmatrix} + \psi_{i+1}\begin{bmatrix} -1 & 0 & 0 & 0 \\ 0 & 0 & 0 & 0 \\ 0 & 0 & 0 & 0 \\ -\frac{1}{2}(u_{i,2}^2 + v_{i,2}^2) & u_{i,2} & v_{i,2} & -1 \end{bmatrix}\begin{bmatrix} (\Delta\bar{\rho})_{i,2} \\ (\Delta\bar{\rho}u)_{i,2} \\ (\Delta\bar{\rho}v)_{i,2} \\ (\Delta\bar{\rho}e_t)_{i,2} \end{bmatrix}$$

$$= \begin{bmatrix} (\psi_{i+1} - \psi_i)\bar{p}_{i,2} \\ 0 \\ 0 \\ \frac{1}{\gamma-1}(\psi_{i+1} - \psi_i)\bar{p}_{i,2} \end{bmatrix}$$

or

$$\overline{BB}_{i,1}\Delta\bar{Q}_1 + \overline{CC}_{i,1}\Delta\bar{Q}_2 = R_{i,1}$$

The derivation just described completes the required modifications at the surface for an adiabatic wall condition. Next the constant wall temperature condition will be investigated.

By definition

$$(\Delta\bar{\rho}e_t)_{i,1} = (\bar{\rho}e_t)_{i+1,1} - (\bar{\rho}e_t)_{i,1}$$

At the surface

$$\bar{\rho}e_t = \bar{\rho}e$$

because the velocity at the wall is zero. In addition,

$$e = \frac{T_w}{\gamma(\gamma-1)M_\infty^2}$$

Therefore,

$$(\Delta\bar{\rho}e_t)_{i,1} = \bar{\rho}_{i+1,1}\frac{T_{w_{i+1}}}{\gamma(\gamma-1)M_\infty^2} - \bar{\rho}_{i,1}\frac{T_{w_i}}{\gamma(\gamma-1)M_\infty^2}$$

For a constant, uniform surface temperature

$$T_{w_{i+1}} = T_{w_i} = T_w$$

Therefore,

$$(\Delta\bar{\rho}e_t)_{i,1} = \phi'(\bar{\rho}_{i+1,1} - \bar{\rho}_{i,1}) \tag{13-34}$$

where ϕ' is defined as

$$\phi' = \frac{T_w}{\gamma(\gamma-1)M_\infty^2}$$

Equation (13-34) may be rearranged as

$$\frac{1}{\phi'}(\Delta\bar{\rho}e_t)_{i,1} = \bar{\rho}_{i+1,1} - \bar{\rho}_{i,1}$$

or

$$\phi(\Delta\bar{\rho}e_t)_{i,1} = \bar{\rho}_{i+1,1} - \bar{\rho}_{i,1} \tag{13-35}$$

where

$$\phi = \frac{1}{\phi'} = \frac{\gamma(\gamma-1)M_\infty^2}{T_w} \tag{13-36}$$

By definition, the following may also be written:

$$(\Delta\bar{\rho})_{i,1} = \bar{\rho}_{i+1,1} - \bar{\rho}_{i,1} \tag{13-37}$$

Now, Equation (13-35) is subtracted from Equation (13-37) to yield:

$$(\Delta\bar{\rho})_{i,1} - \phi(\Delta\bar{\rho}e_t)_{i,1} = \bar{\rho}_{i+1,1} - \bar{\rho}_{i,1} - (\bar{\rho}_{i+1,1} - \bar{\rho}_{i,1}) = 0$$

The required equations applied at the surface for the constant temperature boundary condition may be summarized as:

$$\begin{aligned}(\Delta\bar{\rho})_{i,1} - \phi(\Delta\bar{\rho}e_t)_{i,1} &= 0\\ (\Delta\bar{\rho u})_{i,1} &= 0\\ (\Delta\bar{\rho v})_{i,1} &= 0\end{aligned}$$

$$(\Delta\bar{\rho}e_t)_{i,1} - \psi_{i+1}\left[\tfrac{1}{2}(u_{i,2}^2 + v_{i,2}^2)(\Delta\bar{\rho})_{i,2} - u_{i,2}(\Delta\bar{\rho u})_{i,2} - v_{i,2}(\Delta\bar{\rho v})_{i,2} + (\Delta\bar{\rho}e_t)_{i,2}\right] = \frac{1}{\gamma-1}(\psi_{i+1} - \psi_i)\frac{p_{i,2}}{J_{i,2}}$$

The imposed assumptions which lead to these equations are:

1. Constant wall temperature.
2. Normal gradient of the pressure at the surface is zero.
3. No slip at the surface.

When these equations are expressed in a matrix form, the following is obtained:

$$\begin{bmatrix} 1 & 0 & 0 & -\phi \\ 0 & 1 & 0 & 0 \\ 0 & 0 & 1 & 0 \\ 0 & 0 & 0 & 1 \end{bmatrix}\begin{bmatrix} (\Delta\bar{\rho})_{i,1} \\ (\Delta\bar{\rho u})_{i,1} \\ (\Delta\bar{\rho v})_{i,1} \\ (\Delta\bar{\rho}e_t)_{i,1} \end{bmatrix} + \psi_{i+1}\begin{bmatrix} 0 & 0 & 0 & 0 \\ 0 & 0 & 0 & 0 \\ 0 & 0 & 0 & 0 \\ -\frac{1}{2}(u_{i,2}^2 + v_{i,2}^2) & u_{i,2} & v_{i,2} & -1 \end{bmatrix}\begin{bmatrix} (\Delta\bar{\rho})_{i,2} \\ (\Delta\bar{\rho u})_{i,2} \\ (\Delta\bar{\rho v})_{i,2} \\ (\Delta\bar{\rho}e_t)_{i,2} \end{bmatrix}$$

$$= \begin{bmatrix} 0 \\ 0 \\ 0 \\ \frac{1}{\gamma-1}(\psi_{i+1} - \psi_i)\bar{p}_{i,2} \end{bmatrix}$$

or

$$\overline{BB}_{i,1}\Delta\bar{Q}_1 + \overline{CC}_{i,1}\Delta\bar{Q}_2 = R_{i,1}$$

Now, consider the outer boundary set at the freestream. At this location, flow properties do not change and, therefore, Q's at "$i+1$" and "i" are equal, i.e., $Q_{i+1,JM} = Q_{i,JM}$ which results in $\Delta Q_{i,JM} = 0$. However, in general $\Delta\bar{Q}_{i,JM}$ is not necessarily zero, since $J_{i+1,JM}$ does not have to be equal to $J_{i,JM}$. To proceed with mathematical development, one may write

$$(\Delta\bar{\rho})_{i,JM} = \bar{\rho}_{i+1,JM} - \bar{\rho}_{i,JM} = \frac{\rho_{i+1,JM}}{J_{i+1,JM}} - \frac{\rho_{i,JM}}{J_{i,JM}}$$

At the freestream

$$\rho_{i+1,JM} = \rho_{i,JM}$$

Therefore,

$$(\Delta\bar{\rho})_{i,JM} = \rho_{i,JM}\left(\frac{1}{J_{i+1,JM}} - \frac{1}{J_{i,JM}}\right) = \Gamma_{JM}(\rho_{i,JM}) \tag{13-38}$$

where it has been defined that

$$\frac{1}{J_{i+1,JM}} - \frac{1}{J_{i,JM}} = \Gamma_{JM}$$

Note that $(\Delta\bar{\rho})_{i,JM}$ would be equal to zero only if $J_{i+1,JM} = J_{i,JM}$. This condition may be achieved under certain transformations, but it is not required for these derivations. Following similar procedures which lead to Equation (13-38), it may be concluded that:

$$(\Delta\bar{\rho u})_{i,JM} = \Gamma_{JM}(\rho u)_{i,JM} \tag{13-39}$$

$$(\Delta\bar{\rho v})_{i,JM} = \Gamma_{JM}(\rho v)_{i,JM} \tag{13-40}$$

$$(\Delta\bar{\rho e_t})_{i,JM} = \Gamma_{JM}(\rho e_t)_{i,JM} \tag{13-41}$$

Equations (13-38) through (13-41) are expressed in a matrix form as

$$\begin{bmatrix} 0 & 0 & 0 & 0 \\ 0 & 0 & 0 & 0 \\ 0 & 0 & 0 & 0 \\ 0 & 0 & 0 & 0 \end{bmatrix} \begin{bmatrix} (\Delta\bar{\rho})_{i,JMM1} \\ (\Delta\bar{\rho u})_{i,JMM1} \\ (\Delta\bar{\rho v})_{i,JMM1} \\ (\Delta\bar{\rho e_t})_{i,JMM1} \end{bmatrix} + \begin{bmatrix} 1 & 0 & 0 & 0 \\ 0 & 1 & 0 & 0 \\ 0 & 0 & 1 & 0 \\ 0 & 0 & 0 & 1 \end{bmatrix} \begin{bmatrix} (\Delta\bar{\rho})_{i,JM} \\ (\Delta\bar{\rho u})_{i,JM} \\ (\Delta\bar{\rho v})_{i,JM} \\ (\Delta\bar{\rho e_t})_{i,JM} \end{bmatrix} =$$

$$\Gamma_{JM}\begin{bmatrix} \rho_{i,JM} \\ (\rho u)_{i,JM} \\ (\rho v)_{i,JM} \\ (\rho e_t)_{i,JM} \end{bmatrix}$$

or

$$\overline{AA}_{i,JM}\Delta\bar{Q}_{JMM1} + \overline{BB}_{i,JM}\Delta\bar{Q}_{JM} = R_{i,JM}$$

This completes the specification of boundary conditions at the surface and the freestream.

13.6 Extension to Three-Dimensions

The PNS equations expressed in a generalized coordinate system were given in Chapter 11. These equations are written in a flux vector formulation as

$$\frac{\partial \bar{E}_p}{\partial \xi} + \frac{\partial \bar{E}_{PP}}{\partial \xi} + \frac{\partial \bar{F}}{\partial \eta} + \frac{\partial \bar{G}}{\partial \zeta} = \frac{\partial \bar{F}_{vP}}{\partial \eta} + \frac{\partial \bar{G}_{vP}}{\partial \zeta} \tag{13-42}$$

The flux vectors are given by Equations (11-158) through (11-163).

The assumptions and modifications imposed on the Navier-Stokes equations which resulted in Equation (13-42) are briefly reviewed here. They are:

1. Steady state.
2. Streamwise gradient of the viscous terms are omitted.
3. Flow must be supersonic in the streamwise direction; however, modification of the streamwise pressure gradient allows subsonic flow within the viscous layer.
4. All mixed partial derivatives are assumed negligible and are omitted from the formulation.

Various modifications of the streamwise pressure gradient discussed previously are valid for three-dimensional problems as well and, therefore, they are not repeated here. The same approximation of the streamwise pressure gradient incorporated for the two-dimensional problem will be employed for the three-dimensional studies. Specification of initial and boundary conditions is also similar to the previous discussion. However, the numerical algorithm to solve the equations is different and, therefore, it is presented in detail.

13.6.1 Numerical Algorithm

Using a first-order backward difference approximation in the streamwise direction, Equation (13-42) is expressed in delta formulation as

$$\frac{1}{\Delta\xi}(\Delta\bar{E}_P) + \frac{\partial}{\partial\eta}(\Delta\bar{F}) + \frac{\partial}{\partial\zeta}(\Delta\bar{G}) - \frac{\partial}{\partial\eta}(\Delta\bar{F}_{vP}) - \frac{\partial}{\partial\zeta}(\Delta\bar{G}_{vP}) =$$
$$-\left.\frac{\partial \bar{E}_{PP}}{\partial \xi}\right|_i - \left.\frac{\partial \bar{F}}{\partial \eta}\right|_i - \left.\frac{\partial \bar{G}}{\partial \zeta}\right|_i + \left.\frac{\partial \bar{F}_{vP}}{\partial \eta}\right|_i + \left.\frac{\partial \bar{G}_{vP}}{\partial \zeta}\right|_i$$

or

$$\Delta\bar{E}_P + \Delta\xi\frac{\partial}{\partial\eta}\left[\Delta\bar{F} - \Delta\bar{F}_{vP}\right] + \Delta\xi\frac{\partial}{\partial\zeta}\left[\Delta\bar{G} - \Delta\bar{G}_{vP}\right] =$$
$$\Delta\xi\left[-\left.\frac{\partial \bar{E}_{PP}}{\partial \xi}\right|_i - \left.\frac{\partial \bar{F}}{\partial \eta}\right|_i - \left.\frac{\partial \bar{G}}{\partial \zeta}\right|_i + \left.\frac{\partial \bar{F}_{vP}}{\partial \eta}\right|_i + \left.\frac{\partial \bar{G}_{vP}}{\partial \zeta}\right|_i\right] \tag{13-43}$$

where the pressure term on the right-hand side is computed explicitly. Equation (13-43) is nonlinear and, therefore, a linearization procedure similar to the previously discussed approximations is utilized. The approach is briefly outlined here. Each flux vector in Equation (13-43) is expanded in a Taylor series in the ξ (streamwise) direction. For example,

$$\bar{E}_{P_{i+1}} = \bar{E}_{P_i} + \frac{\partial \bar{E}_P}{\partial \xi}\Delta\xi + O(\Delta\xi)^2 \tag{13-44}$$

Recall that

$$\bar{E}_P = f(\bar{Q},\ \xi_x,\ \xi_y,\ \xi_z)$$

and, therefore, the chain-rule of differentiation provides

$$\frac{\partial \bar{E}_P}{\partial \xi} = \frac{\partial \bar{E}_P}{\partial \bar{Q}}\frac{\partial \bar{Q}}{\partial \xi} + \frac{\partial \bar{E}_P}{\partial \xi_x}\frac{\partial \xi_x}{\partial \xi} + \frac{\partial \bar{E}_P}{\partial \xi_y}\frac{\partial \xi_y}{\partial \xi} + \frac{\partial \bar{E}_P}{\partial \xi_z}\frac{\partial \xi_z}{\partial \xi} \tag{13-45}$$

After substituting this equation into Equation (13-44), one obtains

$$\Delta\bar{E}_P = \left(\frac{\partial \bar{E}_P}{\partial \bar{Q}}\frac{\partial \bar{Q}}{\partial \xi} + \frac{\partial \bar{E}_P}{\partial \xi_x}\frac{\partial \xi_x}{\partial \xi} + \frac{\partial \bar{E}_P}{\partial \xi_y}\frac{\partial \xi_y}{\partial \xi} + \frac{\partial \bar{E}_P}{\partial \xi_z}\frac{\partial \xi_z}{\partial \xi}\right)\Delta\xi + O(\Delta\xi)^2$$

or

$$\Delta\bar{E}_P = \frac{\partial \bar{E}_P}{\partial \bar{Q}}\Delta\bar{Q} + \frac{\partial \bar{E}_P}{\partial \xi_x}\Delta\xi_x + \frac{\partial \bar{E}_P}{\partial \xi_y}\Delta\xi_y + \frac{\partial \bar{E}_P}{\partial \xi_z}\Delta\xi_z + O(\Delta\xi)^2 \tag{13-46a}$$

Similarly,

$$\Delta\bar{F} = \frac{\partial \bar{F}}{\partial \bar{Q}}\Delta\bar{Q} + \frac{\partial \bar{F}}{\partial \eta_x}\Delta\eta_x + \frac{\partial \bar{F}}{\partial \eta_y}\Delta\eta_y + \frac{\partial \bar{F}}{\partial \eta_z}\Delta\eta_z + O(\Delta\xi)^2 \tag{13-46b}$$

$$\Delta\bar{G} = \frac{\partial \bar{G}}{\partial \bar{Q}}\Delta\bar{Q} + \frac{\partial \bar{G}}{\partial \zeta_x}\Delta\zeta_x + \frac{\partial \bar{G}}{\partial \zeta_y}\Delta\zeta_y + \frac{\partial \bar{G}}{\partial \zeta_z}\Delta\zeta_z + O(\Delta\xi)^2 \tag{13-46c}$$

$$\Delta\bar{F}_{vP} = \frac{\partial \bar{F}_{vP}}{\partial \bar{Q}}\Delta\bar{Q} + \frac{\partial \bar{F}_{vP}}{\partial \eta_x}\Delta\eta_x + \frac{\partial \bar{F}_{vP}}{\partial \eta_y}\Delta\eta_y + \frac{\partial \bar{F}_{vP}}{\partial \eta_z}\Delta\eta_z + O(\Delta\xi)^2 \tag{13-46d}$$

$$\Delta\bar{G}_{vP} = \frac{\partial \bar{G}_{vP}}{\partial \bar{Q}}\Delta\bar{Q} + \frac{\partial \bar{G}_{vP}}{\partial \zeta_x}\Delta\zeta_x + \frac{\partial \bar{G}_{vP}}{\partial \zeta_y}\Delta\zeta_y + \frac{\partial \bar{G}_{vP}}{\partial \zeta_z}\Delta\zeta_z + O(\Delta\xi)^2 \tag{13-46e}$$

The inviscid flux Jacobian matrices, $\partial\bar{E}_P/\partial\bar{Q}$, $\partial\bar{F}/\partial\bar{Q}$, and $\partial\bar{G}/\partial\bar{Q}$, and the viscous flux Jacobian matrices, $\partial\bar{F}_{vP}/\partial\bar{Q}$ and $\partial\bar{G}_{vP}/\partial\bar{Q}$, which appear in these equations are given by Equations (11-187) through (11-191). The remaining terms in Equations (13-46a) through (13-46e) are evaluated in the same manner as the two-dimensional case, i.e.,

$$\frac{\partial \bar{E}_P}{\partial \xi_x} = \frac{E_P}{J} \tag{13-47a}$$

$$\frac{\partial \bar{E}_P}{\partial \xi_y} = \frac{F_P}{J} \tag{13-47b}$$

$$\frac{\partial \bar{E}_P}{\partial \xi_z} = \frac{G_P}{J} \tag{13-47c}$$

$$\frac{\partial \bar{F}}{\partial \eta_x} = \frac{\partial \bar{G}}{\partial \zeta_x} = \frac{E}{J} \tag{13-47d}$$

$$\frac{\partial \bar{F}}{\partial \eta_y} = \frac{\partial \bar{G}}{\partial \zeta_y} = \frac{F}{J} \tag{13-47e}$$

$$\frac{\partial \bar{F}}{\partial \eta_z} = \frac{\partial \bar{G}}{\partial \zeta_z} = \frac{G}{J} \tag{13-47f}$$

$$\frac{\partial \bar{F}_{vP}}{\partial \eta_x} = \frac{\partial \bar{G}_{vP}}{\partial \zeta_x} = \frac{E_v}{J} \tag{13-47g}$$

$$\frac{\partial \bar{F}_{vP}}{\partial \eta_y} = \frac{\partial \bar{G}_{vP}}{\partial \zeta_y} = \frac{F_v}{J} \tag{13-47h}$$

$$\frac{\partial \bar{F}_{vP}}{\partial \eta_z} = \frac{\partial \bar{G}_{vP}}{\partial \zeta_z} = \frac{G_v}{J} \tag{13-47i}$$

Now, Equations (13-46a) through (13-46e) are substituted into Equation (13-43) to yield:

$$\begin{aligned}
&\frac{\partial \bar{E}_P}{\partial \bar{Q}}\Delta\bar{Q} + \frac{\partial \bar{E}_P}{\partial \xi_x}\Delta\xi_x + \frac{\partial \bar{E}_P}{\partial \xi_y}\Delta\xi_y + \frac{\partial \bar{E}_P}{\partial \xi_z}\Delta\xi_z + \\
&\Delta\xi\frac{\partial}{\partial \eta}\left[\left(\frac{\partial \bar{F}}{\partial \bar{Q}}\Delta\bar{Q} + \frac{\partial \bar{F}}{\partial \eta_x}\Delta\eta_x + \frac{\partial \bar{F}}{\partial \eta_y}\Delta\eta_y + \frac{\partial \bar{F}}{\partial \eta_z}\Delta\eta_z\right) - \left(\frac{\partial \bar{F}_{vP}}{\partial \bar{Q}}\Delta\bar{Q}\right.\right. \\
&\left.\left. + \frac{\partial \bar{F}_{vP}}{\partial \eta_x}\Delta\eta_x + \frac{\partial \bar{F}_{vP}}{\partial \eta_y}\Delta\eta_y + \frac{\partial \bar{F}_{vP}}{\partial \eta_z}\Delta\eta_z\right)\right] + \Delta\xi\frac{\partial}{\partial \zeta}\left[\left(\frac{\partial \bar{G}}{\partial \bar{Q}}\Delta\bar{Q} +\right.\right. \\
&\left. + \frac{\partial \bar{G}}{\partial \zeta_x}\Delta\zeta_x + \frac{\partial \bar{G}}{\partial \zeta_y}\Delta\zeta_y + \frac{\partial \bar{G}}{\partial \zeta_z}\Delta\zeta_z\right) - \left(\frac{\partial \bar{G}_{vP}}{\partial \bar{Q}}\Delta\bar{Q} + \frac{\partial \bar{G}_{vP}}{\partial \zeta_x}\Delta\zeta_x\right. \\
&\left.\left. + \frac{\partial \bar{G}_{vP}}{\partial \zeta_y}\Delta\zeta_y + \frac{\partial \bar{G}_{vP}}{\partial \zeta_z}\Delta\zeta_z\right)\right] = \Delta\xi\left[-\frac{\partial \bar{E}_{PP}}{\partial \xi} - \frac{\partial \bar{F}}{\partial \eta} - \frac{\partial \bar{G}}{\partial \zeta}\right. \\
&\left. + \frac{\partial \bar{F}_{vP}}{\partial \eta} + \frac{\partial \bar{G}_{vP}}{\partial \zeta}\right]
\end{aligned} \tag{13-48}$$

This equation is rearranged following the same procedure as for the two-dimensional problem. Therefore,

$$\left\{\frac{\partial \bar{E}_P}{\partial \bar{Q}} + \Delta\xi \frac{\partial}{\partial \eta}\left[\frac{\partial \bar{F}}{\partial \bar{Q}} - \frac{\partial \bar{F}_{vP}}{\partial \bar{Q}}\right] + \Delta\xi \frac{\partial}{\partial \zeta}\left[\frac{\partial \bar{G}}{\partial \bar{Q}} - \frac{\partial \bar{G}_{vP}}{\partial \bar{Q}}\right]\right\}\Delta \bar{Q} = \text{RHS}$$

The remaining terms of Equation (13-48) are included in the RHS, which is modified using approximation (13-19). Hence,

$$\begin{aligned}
\text{RHS} \;=\; & -\Delta\xi \frac{\partial \bar{E}_{PP}}{\partial \xi} - \Delta\xi \frac{\partial}{\partial \eta}(\bar{F} - \bar{F}_{vP}) - \Delta\xi \frac{\partial}{\partial \zeta}(\bar{G} - \bar{G}_{vP}) \\
& -(\Delta\xi)\left\{\frac{E_P}{J}\frac{\partial \xi_x}{\partial \xi} + \frac{F_P}{J}\frac{\partial \xi_y}{\partial \xi} + \frac{G_P}{J}\frac{\partial \xi_z}{\partial \xi}\right\} \\
& -(\Delta\xi)^2 \frac{\partial}{\partial \eta}\left\{\left(\frac{E}{J} - \frac{E_v}{J}\right)\frac{\partial \eta_x}{\partial \xi} + \left(\frac{F}{J} - \frac{F_v}{J}\right)\frac{\partial \eta_y}{\partial \xi}\right. \\
& \left. + \left(\frac{G}{J} - \frac{G_v}{J}\right)\frac{\partial \eta_z}{\partial \xi}\right\} - (\Delta\xi)^2 \frac{\partial}{\partial \zeta}\left\{\left(\frac{E}{J} - \frac{E_v}{J}\right)\frac{\partial \zeta_x}{\partial \xi}\right. \\
& \left. + \left(\frac{F}{J} - \frac{F_v}{J}\right)\frac{\partial \zeta_y}{\partial \xi} + \left(\frac{G}{J} - \frac{G_v}{J}\right)\frac{\partial \zeta_z}{\partial \xi}\right\}
\end{aligned}$$

Note that the last two brackets are second-order in the streamwise direction ξ. Since a first-order method has been employed, these terms are dropped.

Finally, the delta formulation of the PNS equation is written as

$$\begin{aligned}
& \left\{\frac{\partial \bar{E}_P}{\partial \bar{Q}} + \Delta\xi \frac{\partial}{\partial \eta}\left[\frac{\partial \bar{F}}{\partial \bar{Q}} - \frac{\partial \bar{F}_{vP}}{\partial \bar{Q}}\right] + \Delta\xi \frac{\partial}{\partial \zeta}\left[\frac{\partial \bar{G}}{\partial \bar{Q}} - \frac{\partial \bar{G}_{vP}}{\partial \bar{Q}}\right]\right\}\Delta \bar{Q} = \\
& -\Delta\xi \frac{\partial \bar{E}_{PP}}{\partial \xi} - \Delta\xi \frac{\partial}{\partial \eta}[\bar{F} - \bar{F}_{vP}] - \Delta\xi \frac{\partial}{\partial \zeta}[\bar{G} - \bar{G}_{vP}] - \Delta\xi \left(\frac{\partial \bar{E}_P}{\partial \xi}\right)_{\bar{Q}} \qquad (13\text{-}49)
\end{aligned}$$

where

$$\left(\frac{\partial \bar{E}_P}{\partial \xi}\right)_{\bar{Q}} = \frac{E_P}{J}\frac{\partial \xi_x}{\partial \xi} + \frac{F_P}{J}\frac{\partial \xi_y}{\partial \xi} + \frac{G_P}{J}\frac{\partial \xi_z}{\partial \xi}$$

Direct approximation of Equation (13-49) by a second-order central difference expression for the η and ζ derivatives will result in a block pentadiagonal system. As mentioned previously, numerical solution of such a system is very time consuming. To overcome this deficiency, approximate factorization is used.

In order to write Equation (13-49) in a factored form, a second-order term, i.e., $O(\Delta\xi)^2$, is added to this equation. The addition of this term does not affect the order of accuracy of the method, because the scheme is first order. Subsequently, Equation (13-49) is factored as

$$\left\{\left[\left(\frac{\partial \bar{E}_P}{\partial \bar{Q}}\right) + \Delta\xi \frac{\partial}{\partial \zeta}\left(\frac{\partial \bar{G}}{\partial \bar{Q}} - \frac{\partial \bar{G}_{vP}}{\partial \bar{Q}}\right)\right]\left(\frac{\partial \bar{E}_P}{\partial \bar{Q}}\right)^{-1}\right.$$

$$\left[\left(\frac{\partial \bar{E}_P}{\partial \bar{Q}}\right) + \Delta\xi \frac{\partial}{\partial \eta}\left(\frac{\partial \bar{F}}{\partial \bar{Q}} - \frac{\partial \bar{F}_{vP}}{\partial \bar{Q}}\right\}\right]\Bigg\}\Delta \bar{Q} \quad = \quad \text{RHS} \tag{13-50}$$

Now this equation is split as

$$\left[\frac{\partial \bar{E}_P}{\partial \bar{Q}} + \Delta\xi \frac{\partial}{\partial \zeta}\left(\frac{\partial \bar{G}}{\partial \bar{Q}} - \frac{\partial \bar{G}_{vP}}{\partial \bar{Q}}\right)\right] \Delta \bar{Q}^* = \text{RHS} \tag{13-51a}$$

and

$$\left[\frac{\partial \bar{E}_P}{\partial \bar{Q}} + \Delta\xi \frac{\partial}{\partial \eta}\left(\frac{\partial \bar{F}}{\partial \bar{Q}} - \frac{\partial \bar{F}_{vP}}{\partial \bar{Q}}\right)\right] \Delta \bar{Q} = \Delta \bar{Q}^{**} \tag{13-51b}$$

where

$$\Delta \bar{Q}^{**} = \left(\frac{\partial \bar{E}_P}{\partial \bar{Q}}\right) \Delta \bar{Q}^* \tag{13-51c}$$

The solution proceeds as follows:

1. The RHS in Equation (13-51a) is computed. Note that all of the terms on the RHS are at the "i" location, where all the dependent variables are known.

2. Equation (13-51a) is solved for the unknown vectors $\Delta \bar{Q}^*$. Note that this system of equations is block tridiagonal. This system is solved along each constant "η" line from $j = 1$ to JM, progressing from $k = 1$ to KM. This step is referred to as the ζ sweep.

3. Equation (13-51c) is used to evaluate $\Delta \bar{Q}^{**}$.

4. Equation (13-51b) is used to solve for the unknown vectors $\Delta \bar{Q}$. This system of equations is also tridiagonal, and is solved along the constant "ζ" line from $k = 1$ to KM, progressing from $j = 1$ to JM. This step is referred to as the η sweep.

5. The unknown flux vector $\bar{Q}_{i+1}$ is evaluated from its known value at "i" and the incremental change $\Delta \bar{Q}$, which was computed in step 4. Thus,

$$\bar{Q}_{i+1} = \bar{Q}_i + \Delta \bar{Q}$$

6. The five steps just outlined are repeated from one streamwise surface to another until the entire domain has been evaluated.

13.7 Numerical Damping Terms

Previously in Chapter 6, it was illustrated that the second-order central difference scheme inherently introduces dispersion errors into the solution. These high frequency oscillations may produce unacceptable results and are highly visible in regions where natural viscosity is absent. For example, these oscillations appear in the vicinity of shock waves in the inviscid regions. Note that in the viscous regions, the oscillations are usually damped out by natural viscosity. To eliminate or reduce the oscillations in the inviscid regime, numerical viscosity, herein defined as *damping terms*, are added to the governing equations. The addition of such terms should be such that the order of accuracy of the scheme is not altered. Usually a fourth-order damping term will satisfy this condition. However, when a damping term is included implicitly, second-order terms are usually employed. The following operators are commonly introduced to represent the difference approximation within the damping terms. For a second-order term in η

$$(\nabla_\eta \Delta_\eta) f = f_{j+1} - 2f_j + f_{j-1}$$

and for a fourth-order term

$$(\nabla_\eta \Delta_\eta)^2 f = f_{j+2} - 4f_{j+1} + 6f_j - 4f_{j-1} + f_{j-2}$$

The damping terms may be added implicitly or explicitly. When the inclusion is implicit, i.e., the damping term is added to the left-hand side, a second-order term is used. That is done to preserve the tridiagonal nature of the system. For explicit addition, fourth-order damping is used.

For the PNS formulation just prescribed, the modification to include damping terms is as follows:

$$\left[\frac{\partial \bar{E}_P}{\partial \bar{Q}} + \Delta\xi \frac{\partial}{\partial \zeta}\left(\frac{\partial \bar{G}}{\partial \bar{Q}} - \frac{\partial \bar{G}_{vP}}{\partial \bar{Q}}\right) + (D_\zeta)_{\text{imp}}\right] \Delta \bar{Q}^* = \text{RHS} + D_{\text{exp}}$$

and

$$\left[\frac{\partial \bar{E}_P}{\partial \bar{Q}} + \Delta\xi \frac{\partial}{\partial \eta}\left(\frac{\partial \bar{F}}{\partial \bar{Q}} - \frac{\partial \bar{F}_{vP}}{\partial \bar{Q}}\right) + (D_\eta)_{\text{imp}}\right] \Delta \bar{Q} = \Delta \bar{Q}^{**}$$

where

$$\begin{aligned} (D_\zeta)_{\text{imp}} &= -(\epsilon_\zeta)_{\text{imp}} \Delta\xi J^{-1} (\nabla_\zeta \Delta_\zeta) J \\ (D_\eta)_{\text{imp}} &= -(\epsilon_\eta)_{\text{imp}} \Delta\xi J^{-1} (\nabla_\eta \Delta_\eta) J \end{aligned}$$

and

$$D_{\text{exp}} = -\epsilon_e \Delta\xi J^{-1} \left[(\nabla_\zeta \Delta_\zeta)^2 + (\nabla_\eta \Delta_\eta)^2\right] J \bar{Q}^n$$

In the damping terms above, $(\epsilon_\zeta)_{\text{imp}}$ and $(\epsilon_\eta)_{\text{imp}}$ are the implicit damping coefficients and ϵ_e is the explicit damping coefficient. The values of these coefficients are input.

Obviously they are problem-dependent and must equal less than one. They should be selected as small as possible, since the addition of too much damping clearly will overwhelm the solution, i.e., flows in the high gradient regions, such as shock, are destroyed. Generally speaking, the implicit damping coefficients are specified as twice the explicit damping coefficient.

13.8 Shock Fitting Procedure

The bow shock generated by an object in a supersonic/hypersonic flowfield may be selected as the outer boundary of the domain and determined as a part of the overall solution. Obviously this procedure has some advantages compared to an outer boundary set at the freestream where the bow shock must be captured. First, the number of grid points in the domain may be decreased. That is due to the fact that additional grid points must be located in the freestream if the bow shock is to be captured. Second, shock smearing (of the bow shock), which is a consequence of shock capturing, will not appear in the solution when shock fitting is used. Of course, these advantages are accompanied by some disadvantages. Most shock fitting procedures are explicit and, therefore, some additional stability requirement is imposed. Furthermore, additional sets of equations must be used to determine the shock location. Usually Rankine-Hugoniot relations are used for this purpose. In this section, the shock fitting procedure and the derivations of the pertinent equations are explored.

The initial conditions at "$i = 1$" provide all the required data including the shock slope. To generate the grid at the next station, i.e., "$i = 2$", the shock is linearly extrapolated. Therefore, the grid system at "$i = 2$" is generated and the PNS equations are solved for $j = 1$ (at the surface) to $j = JM$ (at the shock, i.e., JM is the grid point just behind the shock). Note that the finite difference equations of the PNS equations must be modified and a one-sided difference approximation employed at the shock. Once the PNS equations at "$i = 2$" are solved, all the flow properties including the pressure are known. However, recall that the shock location at "$i = 2$" was extrapolated from the previous station. Therefore, an updating procedure must be used to modify the shock properties and compute a new shock slope at "$i = 2$". The procedure is described in two steps. First, an equation for the shock slope is derived, and subsequently Rankine-Hugoniot equations are introduced. In the following, the subscripts n and t denote the normal and tangential directions to the shock wave. The detailed derivations are given for 2-D problems with extension to 3-D to follow.

The normal component of the freestream velocity is

$$\vec{V}_{n_\infty} = (\vec{V}_\infty \cdot \hat{n}_s)\hat{n}_s$$

where $\hat{n}_s$ is a unit vector normal to the shock and is oriented outward. The nomenclature is shown in Figure 13-5.

The unit normal is expressed as

$$\vec{n}_s = \frac{\nabla\eta}{|\nabla\eta|} = \frac{\eta_x\hat{\imath} + \eta_y\hat{\jmath}}{(\eta_x^2 + \eta_y^2)^{1/2}}$$

The freestream velocity vector is

$$\vec{V}_\infty = u_\infty\hat{\imath} + v_\infty\hat{\jmath}$$

Thus, the normal component of the freestream velocity is expressed as

$$\vec{V}_{n_\infty} = \frac{(u_\infty\eta_x + v_\infty\eta_y)(\eta_x\hat{\imath} + \eta_y\hat{\jmath})}{|\nabla\eta|^2}$$

whereas its magnitude is

$$V_{n_\infty} = \frac{u_\infty\eta_x + v_\infty\eta_y}{|\nabla\eta|} \tag{13-52}$$

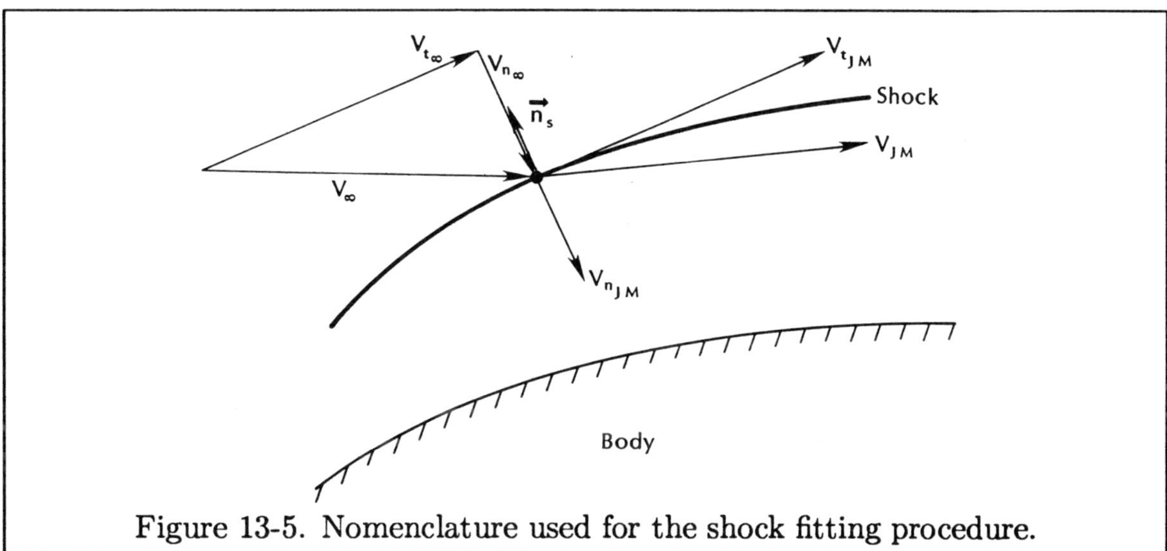

Figure 13-5. Nomenclature used for the shock fitting procedure.

The tangential component may now be expressed as

$$\vec{V}_{t_\infty} = \vec{V}_\infty - \vec{V}_{n_\infty} = (u_\infty\hat{\imath} + v_\infty\hat{\jmath}) - \frac{(u_\infty\eta_x + v_\infty\eta_y)(\eta_x\hat{\imath} + \eta_y\hat{\jmath})}{|\nabla\eta|^2}$$

The normal component of the velocity behind the oblique shock is

$$\vec{V}_{n_{JM}} = -V_{n_{JM}}\hat{n}_s = -V_{n_{JM}}\frac{\eta_x\hat{\imath} + \eta_y\hat{\jmath}}{|\nabla\eta|}$$

The tangential component of the momentum equation requires that

$$\vec{V}_{t_{JM}} = \vec{V}_{t_\infty}$$

Thus,

$$\vec{V}_{JM} = \vec{V}_{n_{JM}} + \vec{V}_{t_{JM}} = -V_{n_{JM}} \frac{\eta_x \hat{\imath} + \eta_y \hat{\jmath}}{|\nabla \eta|} + u_\infty \hat{\imath} + v_\infty \hat{\jmath} - \frac{(u_\infty \eta_x + v_\infty \eta_y)}{|\nabla \eta|^2} (\eta_x \hat{\imath} + \eta_y \hat{\jmath})$$

From which

$$\begin{aligned} u_{JM} &= u_\infty - \frac{V_{n_{JM}} \eta_x}{|\nabla \eta|} - \frac{u_\infty \eta_x + v_\infty \eta_y}{|\nabla \eta|^2} \eta_x \\ v_{JM} &= v_\infty - \frac{V_{n_{JM}} \eta_y}{|\nabla \eta|} - \frac{u_\infty \eta_x + v_\infty \eta_y}{|\nabla \eta|^2} \eta_y \end{aligned}$$

These equations are modified by using Equation (13-52) to provide:

$$u_{JM} = u_\infty - \frac{V_{n_{JM}} \eta_x}{|\nabla \eta|} - \frac{V_{n_\infty} \eta_x}{|\nabla \eta|} = u_\infty + \frac{V_{n_\infty} \eta_x}{|\nabla \eta|} \left(1 - \frac{V_{n_{JM}}}{V_{n_\infty}}\right) \tag{13-53}$$

$$v_{JM} = v_\infty - \frac{V_{n_{JM}} \eta_y}{|\nabla \eta|} - \frac{V_{n_\infty} \eta_y}{|\nabla \eta|} = v_\infty + \frac{V_{n_\infty} \eta_y}{|\nabla \eta|} \left(1 - \frac{V_{n_{JM}}}{V_{n_\infty}}\right) \tag{13-54}$$

Recall that the conservation of mass across an oblique shock is written as

$$\rho_\infty V_{n_\infty} = \rho_{JM} V_{n_{JM}}$$

or

$$\frac{V_{n_{JM}}}{V_{n_\infty}} = \frac{\rho_\infty}{\rho_{JM}} \tag{13-55}$$

Before proceeding further, the required equations are nondimensionalized. Equation (13-55) is nondimensionalized as

$$\frac{\dfrac{V_{n_{JM}}}{V_\infty}}{\dfrac{V_{n_\infty}}{V_\infty}} = \frac{\dfrac{\rho_\infty}{\rho_\infty}}{\dfrac{\rho_{JM}}{\rho_\infty}}$$

or

$$\frac{V^*_{n_{JM}}}{V^*_{n_\infty}} = \frac{1}{\rho^*_{JM}} \tag{13-56}$$

and Equation (13-53) becomes

$$\frac{u_{JM}}{V_\infty} = \frac{u_\infty}{V_\infty} + \frac{V_{n_\infty}}{V_\infty} \frac{\eta_x}{|\nabla \eta|} \left(1 - \frac{V_{n_{JM}}}{V_\infty} \frac{V_\infty}{V_{n_\infty}}\right)$$

or

$$u^*_{JM} = u^*_\infty + V^*_{n_\infty} \frac{\eta_x}{|\nabla\eta|}\left(1 - \frac{V^*_{n_{JM}}}{V^*_{n_\infty}}\right)$$

Substitution of Equation (13-56) into the equation above yields:

$$u^*_{JM} = u^*_\infty + V^*_{n_\infty} \frac{\eta_x}{|\nabla\eta|}\left(1 - \frac{1}{\rho^*_{JM}}\right) \tag{13-57}$$

Similarly, Equation (13-54) is expressed as

$$v^*_{JM} = v^*_\infty + V^*_{n_\infty} \frac{\eta_y}{|\nabla\eta|}\left(1 - \frac{1}{\rho^*_{JM}}\right) \tag{13-58}$$

Next, consider the computation of η_x and η_y. Recall that

$$\eta_x = -Jy_\xi$$

and

$$\eta_y = Jx_\xi$$

The coordinates of the grid points at constant ξ locations may be expressed as

$$x = x_b + \delta S n_1 \tag{13-59}$$

$$y = y_b + \delta S n_2 \tag{13-60}$$

where

$$n_1 = \hat{e}_\eta \cdot \hat{\imath}$$

$$n_2 = \hat{e}_\eta \cdot \hat{\jmath}$$

and $\hat{e}_\eta$ is defined as a unit vector along η, δ is the shock standoff distance, and S is a grid clustering function.

Now Equations (13-59) and (13-60) are differentiated with respect to ξ to provide:

$$x_\xi = x_{b_\xi} + S(n_{1_\xi}\delta + n_1\delta_\xi) = (x_{b_\xi} + Sn_{1_\xi}\delta) + Sn_1\delta_\xi$$

$$y_\xi = y_{b_\xi} + S(n_{2_\xi}\delta + n_2\delta_\xi) = (y_{b_\xi} + Sn_{2_\xi}\delta) + Sn_2\delta_\xi$$

In order to write these equations in a compact form, the following definitions are used:

$$\alpha_1 = x_{b_\xi} + Sn_{1_\xi}\delta \tag{13-61}$$

$$\alpha_2 = y_{b_\xi} + Sn_{2_\xi}\delta \tag{13-62}$$

$$\beta_1 = Sn_1 \tag{13-63}$$

$$\beta_2 = Sn_2 \tag{13-64}$$

Thus,

$$x_\xi = \alpha_1 + \beta_1 \delta_\xi$$
$$y_\xi = \alpha_2 + \beta_2 \delta_\xi$$

These equations are substituted into the metrics η_x and η_y to provide:

$$\eta_x = -J(\alpha_2 + \beta_2 \delta_\xi) \tag{13-65}$$
$$\eta_y = J(\alpha_1 + \beta_1 \delta_\xi) \tag{13-66}$$

Now, return back to the normal component of the freestream velocity given by Equation (13-52)

$$V_{n_\infty} = \frac{u_\infty \eta_x + v_\infty \eta_y}{(\eta_x^2 + \eta_y^2)^{1/2}}$$

This equation is rearranged by squaring the terms; hence,

$$V_{n_\infty}^2 (\eta_x^2 + \eta_y^2) = \eta_x^2 u_\infty^2 + \eta_y^2 v_\infty^2 + 2 u_\infty v_\infty \eta_x \eta_y \tag{13-67}$$

Now, substitute Equations (13-65) and (13-66) into the equation above to get

$$V_{n_\infty}^2 [(\beta_1^2 + \beta_2^2)\delta_\xi^2 + 2(\beta_2 + \beta_2)\delta_\xi + (\alpha_1^2 + \alpha_2^2)] =$$
$$(\beta_2^2 u_\infty^2 + \beta_1^2 v_\infty^2 + 2\beta_1\beta_2 u_\infty v_\infty)\delta_\xi^2$$
$$+[2\beta_2 u_\infty^2 + 2\beta_1 v_\infty^2 + 2(\alpha_2\beta_1 + \alpha_1\beta_2)u_\infty v_\infty]\delta_\xi$$
$$+(\alpha_2^2 u_\infty^2 + \alpha_1^2 v_\infty^2 + 2\alpha_1\alpha_2 u_\infty v_\infty)$$

This equation is written in a compact form as

$$(A1L)\delta_\xi^2 + (A2L)\delta_\xi + (A3L) = (A1R)\delta_\xi^2 + (A2R)\delta_\xi + (A3R) \tag{13-68}$$

where

$$A1L = (\beta_1^2 + \beta_2^2)V_{n_\infty}^2$$
$$A2L = 2(\beta_1 + \beta_2)V_{n_\infty}^2$$
$$A3L = (\alpha_1^2 + \alpha_2^2)V_{n_\infty}^2$$
$$A1R = (\beta_2 u_\infty + \beta_1 v_\infty)^2$$
$$A2R = 2[\beta_2 u_\infty^2 + \beta_1 v_\infty^2 + (\alpha_2\beta_1 + \alpha_1\beta_2)u_\infty v_\infty]$$
$$A3R = (\alpha_2 u_\infty + \alpha_1 v_\infty)^2$$

Now, Equation (13-68) is expressed as

$$A_1\delta_\xi^2 + A_2\delta_\xi + A_3 = 0 \tag{13-69}$$

where

$$\begin{aligned} A_1 &= A1L - A1R \\ A_2 &= A2L - A2R \\ A_3 &= A3L - A3R \end{aligned}$$

Equation (13-69) is solved for δ_ξ, where the positive root is selected; hence,

$$\delta_\xi = \frac{-A_2 + \sqrt{A_2^2 - 4A_1A_3}}{2A_1} \tag{13-70}$$

In order to determine the parameters A_1 through A_3, Equations (13-61) through (13-64) are employed. These equations require data about the body configuration, grid system, and shock standoff distance. Of course, the body configuration is known, the grid system with a specified clustering has been selected, and δ has been extrapolated from the known station "i". Thus, Equation (13-70) provides a new shock slope at the new location "$i+1$".

The Rankine-Hugoniot relations are now used to update the flow properties behind the shock. These newly computed values will replace the values of the flow properties behind the shock at "JM". Note that these values were initially computed by the PNS equations. A brief review of the Rankine-Hugoniot equations to be employed here is provided next.

Conservation of mass, normal component of momentum, and energy across an oblique shock are:

$$\rho_1 V_{n_1} = \rho_2 V_{n_2} \tag{13-71}$$

$$\rho_1 V_{n_1}^2 + p_1 = \rho_2 V_{n_2}^2 + p_2 \tag{13-72}$$

$$h_1 + \frac{1}{2}V_{n_1}^2 = h_2 + \frac{1}{2}V_{n_2}^2 \tag{13-73}$$

where the subscripts 1 and 2 denote the flow properties ahead and behind the shock. When Equations (13-71) and (13-72) are combined to eliminate V_{n_2}, the result is

$$V_{n_1}^2 = \frac{p_2 - p_1}{\rho_1}\left(\frac{1}{1 - \dfrac{\rho_1}{\rho_2}}\right) \tag{13-74}$$

Continuity is combined with the energy equation to provide

$$h_1 + \frac{1}{2}V_{n_1}^2 = h_2 + \frac{1}{2}\left(\frac{\rho_1}{\rho_2}\right)^2 V_{n_1}^2$$

This equation is rearranged as

$$h_2 = h_1 + \frac{1}{2}\left[1 - \left(\frac{\rho_1}{\rho_2}\right)^2\right] V_{n_1}^2$$

Now, Equation (13-74) is substituted into the equation above to provide

$$h_2 = h_1 + \frac{1}{2}\left[1 - \left(\frac{\rho_1}{\rho_2}\right)^2\right]\left[\frac{p_2 - p_1}{\rho_1}\left(\frac{1}{1 - \frac{\rho_1}{\rho_2}}\right)\right]$$

which is rearranged as

$$h_2 = h_1 + \frac{p_2 - p_1}{2\rho_1}\left(1 + \frac{\rho_1}{\rho_2}\right) \tag{13-75}$$

Recall that by definition

$$h = e + \frac{p}{\rho}$$

and if the perfect gas assumption is invoked, then

$$p = \rho e(\gamma - 1)$$

Thus,

$$h = \frac{\gamma}{\gamma - 1}\frac{p}{\rho}$$

Enthalpy in Equation (13-75) is replaced by this equation. Therefore,

$$\frac{\gamma}{\gamma - 1}\frac{p_2}{\rho_2} = \frac{\gamma p_1}{(\gamma - 1)\rho_1} + \frac{p_2 - p_1}{2\rho_1}\left(1 + \frac{\rho_1}{\rho_2}\right)$$

which may be rearranged as

$$\frac{\gamma p_2}{(\gamma - 1)\rho_2} = \frac{\gamma(p_1 + p_2) + p_1 - p_2}{2(\gamma - 1)\rho_1} + \frac{p_2 - p_1}{2\rho_2} \tag{13-76}$$

The nondimensional form of this equation is

$$\frac{\gamma p_2^*}{(\gamma - 1)\rho_2^*} = \frac{\gamma(p_1^* + p_2^*) + p_1^* - p_2^*}{2(\gamma - 1)} + \frac{p_2^* - p_1^*}{2\rho_2^*} \tag{13-77}$$

When this equation is solved for ρ_2^*, the following is obtained:

$$\rho_2^* = \frac{(\gamma + 1)p_2^* + (\gamma - 1)p_1^*}{(\gamma + 1)p_1^* + (\gamma - 1)p_2^*} \tag{13-78}$$

This equation is the one which is used to update the density behind the shock at "JM".

The normal component of the velocity V_{n_1} given by Equation (13-74) in nondimensional form may be expressed as:

$$V_{n_1}^{*^2} = \frac{p_2^* - p_1^*}{1 - \dfrac{1}{\rho_2^*}} \tag{13-79}$$

Now, the continuity may be used to compute $V_{n_2}^*$. Since $V_{t_1}^* = V_{t_2}^*$, the total velocity at "JM", and as a result its components u^* and v^*, may be computed. These values replace the previously computed values at JM. The energy term, e_t, is then updated according to

$$e_{t_2}^* = \frac{p_2^*}{(\gamma - 1)\rho_2^*} + \frac{1}{2}(u_2^{*^2} + v_2^{*^2})$$

Once this updating procedure has been completed, the procedure is repeated at the next streamwise station and proceeds forward throughout the domain.

13.8.1 Extension to Three Dimensions

The shock fitting procedure just described for a two-dimensional problem will now be extended to three-dimensional problems. Since detailed mathematical manipulation has been explored, only a summary of results is given in the section.

The unit normal (to the shock) is given by:

$$\hat{n}_s = \frac{\nabla\eta}{|\nabla\eta|} = \frac{\eta_x\hat{\imath} + \eta_y\hat{\jmath} + \eta_z\hat{k}}{(\eta_x^2 + \eta_y^2 + \eta_z^2)^{1/2}}$$

The normal and tangential components of the freestream velocity are:

$$\vec{V}_{n_\infty} = \frac{(u_\infty\eta_x + v_\infty\eta_y + w_\infty\eta_z)(\eta_x\hat{\imath} + \eta_y\hat{\jmath} + \eta_z\hat{k})}{|\nabla\eta|^2}$$

$$\vec{V}_{t_\infty} = (u_\infty\hat{\imath} + v_\infty\hat{\jmath} + w_\infty\hat{k}) - \frac{(u_\infty\eta_x + v_\infty\eta_y + w_\infty\eta_z)(\eta_x\hat{\imath} + \eta_y\hat{\jmath} + \eta_z\hat{k})}{|\nabla\eta|^2}$$

The velocity component behind the shock at $j = JM$ is

$$\vec{V}_{JM} = -V_{n_{JM}}\frac{\eta_x\hat{\imath} + \eta_y\hat{\jmath} + \eta_z\hat{k}}{|\nabla\eta|} + u_\infty\hat{\imath} + v_\infty\hat{\jmath} + w_\infty\hat{k} - \frac{(u_\infty\eta_x + v_\infty\eta_y + w_\infty\eta_z)(\eta_x\hat{\imath} + \eta_y\hat{\jmath} + \eta_z\hat{k})}{|\nabla\eta|^2}$$

from which:

$$
\begin{aligned}
u_{JM} &= u_\infty - \frac{V_{n_{JM}}\eta_x}{|\nabla\eta|} - \frac{u_\infty\eta_x + v_\infty\eta_y + w_\infty\eta_z}{|\nabla\eta|^2}\eta_x \\
v_{JM} &= v_\infty - \frac{V_{n_{JM}}\eta_y}{|\nabla\eta|} - \frac{u_\infty\eta_x + v_\infty\eta_y + w_\infty\eta_z}{|\nabla\eta|^2}\eta_y \\
w_{JM} &= w_\infty - \frac{V_{n_{JM}}\eta_z}{|\nabla\eta|} - \frac{u_\infty\eta_x + v_\infty\eta_y + w_\infty\eta_z}{|\nabla\eta|^2}\eta_z
\end{aligned}
$$

These equations are now modified by using continuity and are subsequently nondimensionalized to provide:

$$u_2^* = u_\infty^* + V_{n_\infty}^* \frac{\eta_x}{|\nabla\eta|}\left(1 - \frac{1}{\rho_{JM}^*}\right) \tag{13-80}$$

$$v_2^* = v_\infty^* + V_{n_\infty}^* \frac{\eta_y}{|\nabla\eta|}\left(1 - \frac{1}{\rho_{JM}^*}\right) \tag{13-81}$$

$$w_2^* = w_\infty^* + V_{n_\infty}^* \frac{\eta_z}{|\nabla\eta|}\left(1 - \frac{1}{\rho_{JM}^*}\right) \tag{13-82}$$

As described previously, the coordinates of the grid points may be written as

$$x = x_b + \delta S n_1 \tag{13-83}$$

$$y = y_b + \delta S n_2 \tag{13-84}$$

$$z = z_b + \delta S n_3 \tag{13-85}$$

where

$$
\begin{aligned}
n_1 &= \hat{e}_\eta \cdot \hat{\imath} \\
n_2 &= \hat{e}_\eta \cdot \hat{\jmath} \\
n_3 &= \hat{e}_\eta \cdot \hat{k}
\end{aligned}
$$

When Equations (13-83) through (13-85) are differentiated with respect to ξ, the following are obtained:

$$
\begin{aligned}
x_\xi &= (x_{b_\xi} + Sn_{1_\xi}\delta) + Sn_1\delta_\xi \\
y_\xi &= (y_{b_\xi} + Sn_{2_\xi}\delta) + Sn_2\delta_\xi \\
z_\xi &= (z_{b_\xi} + Sn_{3_\xi}\delta) + Sn_3\delta_\xi
\end{aligned}
$$

These equations are written in a compact form by using the following definitions:

$$\begin{aligned}
\alpha_1 &= x_{b_\xi} + Sn_{1_\xi}\delta \\
\alpha_2 &= y_{b_\xi} + Sn_{2_\xi}\delta \\
\alpha_3 &= z_{b_\xi} + Sn_{3_\xi}\delta \\
\beta_1 &= Sn_1 \\
\beta_2 &= Sn_2 \\
\beta_3 &= Sn_3
\end{aligned}$$

Therefore,

$$x_\xi = \alpha_1 + \beta_1\delta_\xi \tag{13-86}$$

$$y_\xi = \alpha_2 + \beta_2\delta_\xi \tag{13-87}$$

$$z_\xi = \alpha_3 + \beta_3\delta_\xi \tag{13-88}$$

Recall that the metrics η_x, η_y, and η_z are given by:

$$\begin{aligned}
\eta_x &= -J(y_\xi z_\zeta - y_\zeta z_\xi) \\
\eta_y &= J(x_\xi z_\zeta - x_\zeta z_\xi) \\
\eta_z &= -J(x_\xi y_\zeta - x_\zeta y_\xi)
\end{aligned}$$

Substitution of Equations (13-86) through (13-88) into the metrics given above yields:

$$\eta_x = J[(\alpha_3 + \beta_3\delta_\xi)y_\zeta - (\alpha_2 + \beta_2\delta_\xi)z_\zeta] \tag{13-89}$$

$$\eta_y = J[(\alpha_1 + \beta_1\delta_\xi)z_\zeta - (\alpha_3 + \beta_3\delta_\xi)x_\zeta] \tag{13-90}$$

$$\eta_z = J[(\alpha_2 + \beta_2\delta_\xi)x_\zeta - (\alpha_1 + \beta_1\delta_\xi)y_\zeta] \tag{13-91}$$

The normal component of the freestream velocity may be squared to provide

$$\begin{aligned}
V_{n_\infty}^2(\eta_x^2 + \eta_y^2 + \eta_z^2) &= \eta_x^2 u_\infty^2 + \eta_y^2 v_\infty^2 + \eta_z^2 w_\infty^2 + 2\eta_x\eta_y u_\infty v_\infty \\
&\quad + 2\eta_x\eta_z u_\infty w_\infty + 2\eta_y\eta_z v_\infty w_\infty
\end{aligned} \tag{13-92}$$

To determine η_x^2, Equation (13-89) is squared and rearranged as

$$\begin{aligned}
\eta_x^2 &= (\beta_2 z_\zeta - \beta_3 y_\zeta)^2\delta_\xi^2 + 2(\alpha_2 z_\zeta - \alpha_3 y_\zeta)(\beta_2 z_\zeta - \beta_3 y_\zeta)\delta_\xi \\
&\quad + (\alpha_2 z_\zeta - \alpha_3 y_\zeta)^2
\end{aligned}$$

This equation may be expressed as

$$\eta_x^2 = (\epsilon_{11} + \epsilon_{12}\delta_\xi)^2 \tag{13-93}$$

Similarly,

$$\eta_y^2 = (\epsilon_{21} + \epsilon_{22}\delta_\xi)^2 \tag{13-94}$$

$$\eta_z^2 = (\epsilon_{31} + \epsilon_{32}\delta_\xi)^2 \tag{13-95}$$

$$\eta_x\eta_y = \epsilon_{12}\epsilon_{22}\delta_\xi^2 + (\epsilon_{11}\epsilon_{22} + \epsilon_{12}\epsilon_{21})\delta_\xi + \epsilon_{11}\epsilon_{21} \tag{13-96}$$

$$\eta_x\eta_z = \epsilon_{12}\epsilon_{32}\delta_\xi^2 + (\epsilon_{11}\epsilon_{32} + \epsilon_{12}\epsilon_{31})\delta_\xi + \epsilon_{11}\epsilon_{31} \tag{13-97}$$

$$\eta_y\eta_z = \epsilon_{22}\epsilon_{32}\delta_\xi^2 + (\epsilon_{21}\epsilon_{32} + \epsilon_{31}\epsilon_{22})\delta_\xi + \epsilon_{21}\epsilon_{31} \tag{13-98}$$

where the following definitions are incorporated:

$$\epsilon_{11} = (\alpha_2 z_\zeta - \alpha_3 y_\zeta) \tag{13-99}$$

$$\epsilon_{12} = (\beta_2 z_\zeta - \beta_3 y_\zeta) \tag{13-100}$$

$$\epsilon_{21} = (\alpha_1 z_\zeta - \alpha_3 x_\zeta) \tag{13-101}$$

$$\epsilon_{22} = (\beta_1 z_\zeta - \beta_3 x_\zeta) \tag{13-102}$$

$$\epsilon_{31} = (\alpha_1 y_\zeta - \alpha_2 x_\zeta) \tag{13-103}$$

$$\epsilon_{32} = (\beta_1 y_\zeta - \beta_2 x_\zeta) \tag{13-104}$$

Substitution of Equations (13-93) through (13-98) into Equation (13-92) yields:

$$A_1\delta_\xi^2 + A_2\delta_\xi + A_3 = 0 \tag{13-105}$$

where

$$A_1 = A1L - A1R$$

$$A_2 = A2L - A2R$$

$$A_3 = A3L - A3R$$

and

$$A1L = (\epsilon_{12}^2 + \epsilon_{22}^2 + \epsilon_{32}^2)V_{n_\infty}^2$$

$$A2L = 2(\epsilon_{11}\epsilon_{12} + \epsilon_{21}\epsilon_{22} + \epsilon_{31}\epsilon_{32})V_{n_\infty}^2$$

$$A3L = (\epsilon_{11}^2 + \epsilon_{21}^2 + \epsilon_{31}^2) V_{n_\infty}^2$$

$$A1R = (\epsilon_{12} u_\infty + \epsilon_{22} v_\infty)^2 + \epsilon_{32} w_\infty^2 + 2\epsilon_{12}\epsilon_{32} u_\infty w_\infty - 2\epsilon_{22}\epsilon_{32} v_\infty w_\infty$$

$$A2R = 2[\epsilon_{11}\epsilon_{12} u_\infty^2 + \epsilon_{22}\epsilon_{21} v_\infty^2 + \epsilon_{31}\epsilon_{32} w_\infty^2 + (\epsilon_{11}\epsilon_{22} + \epsilon_{21}\epsilon_{12}) u_\infty v_\infty$$
$$+(\epsilon_{11}\epsilon_{32} + \epsilon_{31}\epsilon_{12}) u_\infty w_\infty + (\epsilon_{21}\epsilon_{32} + \epsilon_{31}\epsilon_{22}) v_\infty w_\infty]$$

$$A3R = (\epsilon_{11} u_\infty + \epsilon_{21} v_\infty)^2 + \epsilon_{31} w_\infty^2 + 2\epsilon_{11}\epsilon_{31} u_\infty w_\infty - 2\epsilon_{21}\epsilon_{31} v_\infty w_\infty$$

Now Equation (13-105) is solved for the shock slope to provide

$$\delta_\xi = \frac{-A_2 + \sqrt{A_2^2 - 4A_1 A_3}}{2A_1}$$

The Rankine-Hugoniot relations given previously are now used to update the flow properties behind the shock.

13.9 Application

To illustrate the solution obtained by the PNS equations, the following simple configuration is proposed. The configuration is a sharp cone of 10° half angle at an angle of attack of 24°. The freestream conditions are $M_\infty = 7.95$, $P_\infty = 3.984$ psi, $T_\infty = 99.7$°R with a wall temperature of 557.6°R. The shock fitting procedure described previously is used to obtain the outer shock. In addition, due to symmetry of the flowfield, only half of the domain is used in the computation. However, the grid system and the solution to be presented shortly are shown for the entire domain. The domain at a cross-section is shown in Figure 13-6.

To start the PNS computations, a plane of data (for a first-order scheme) or two sets of data (for a second-order scheme) is required. Since the configuration in this example is simple, approximate methods may be used to provide the initial condition. Otherwise the Navier-Stokes equations must be solved to provide the required data. The approximate initial data set may be obtained as follows. Assume an initial shock shape, for example an elliptical shock. The windward shock angle is determined by the charts in standard text [12-7] or NACA 1135 [12-8]. For the leeside, the limiting Mach angle is used. Now, the Rankine-Hugoniot relations are employed to determine the flow properties behind the shock. A sine function is used to represent the velocity within the viscous region of the flowfield. The pressure and density are assumed constant within the interior domain as provided by their values obtained by the Rankine-Hugoniot relations.

The initial data is used to march downstream using the PNS code, usually 20–30 streamwise stations. Then, the properties within the domain are scaled back using conical flow assumption to provide a new starting solution. This iterative procedure is continued until a convergent solution is obtained. This set(s) of initial data is used to march the PNS over the entire domain.

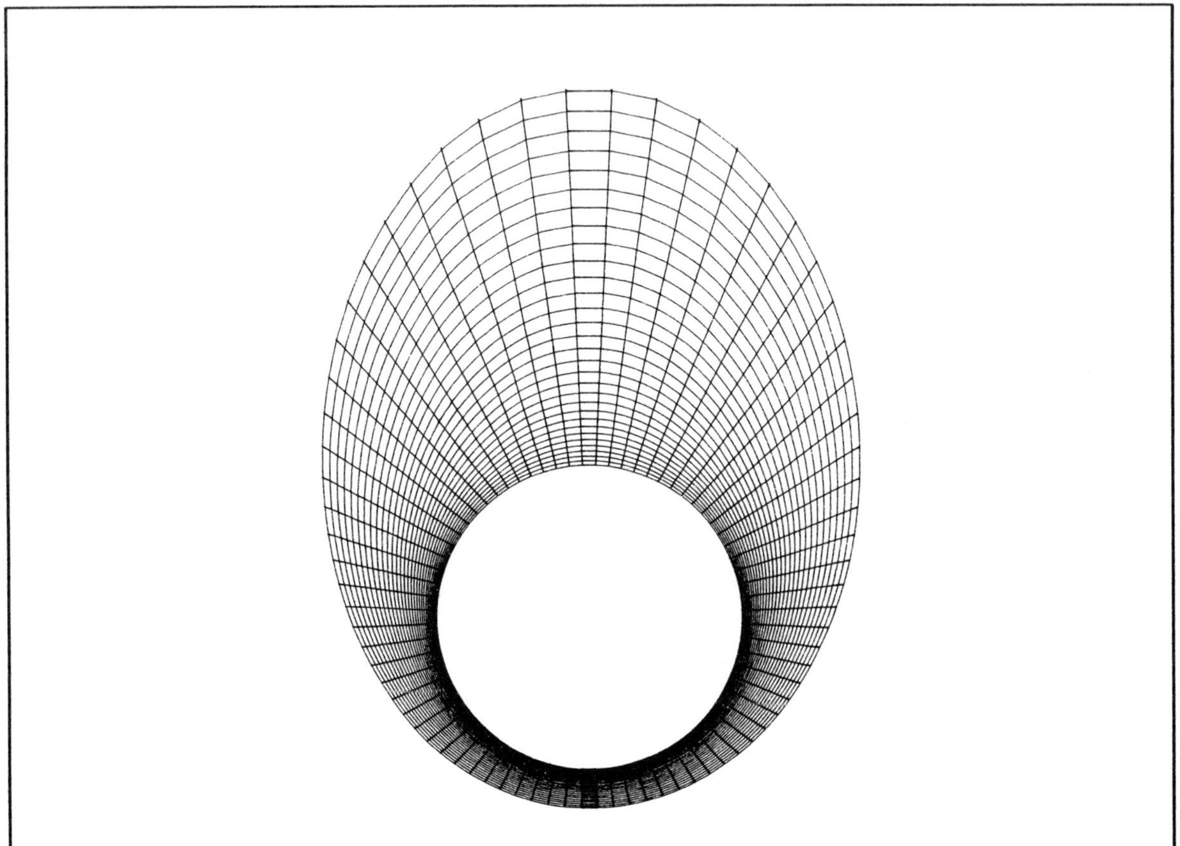

Figure 13-6. Physical domain at a cross-section where symmetry is applied at $i = 1$ and 2 and $i = IM$ and $IMM1$.

A typical solution at a cross-section is illustrated in Figures 13-7 through 13-9.

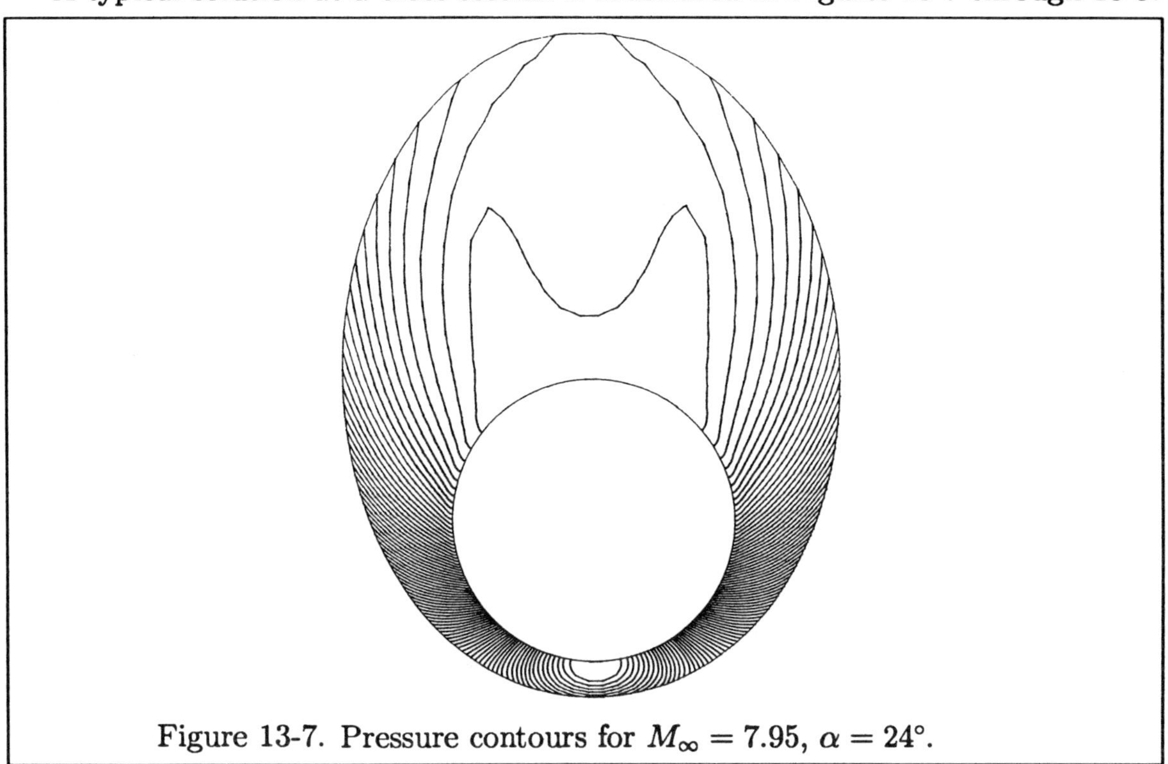

Figure 13-7. Pressure contours for $M_\infty = 7.95$, $\alpha = 24°$.

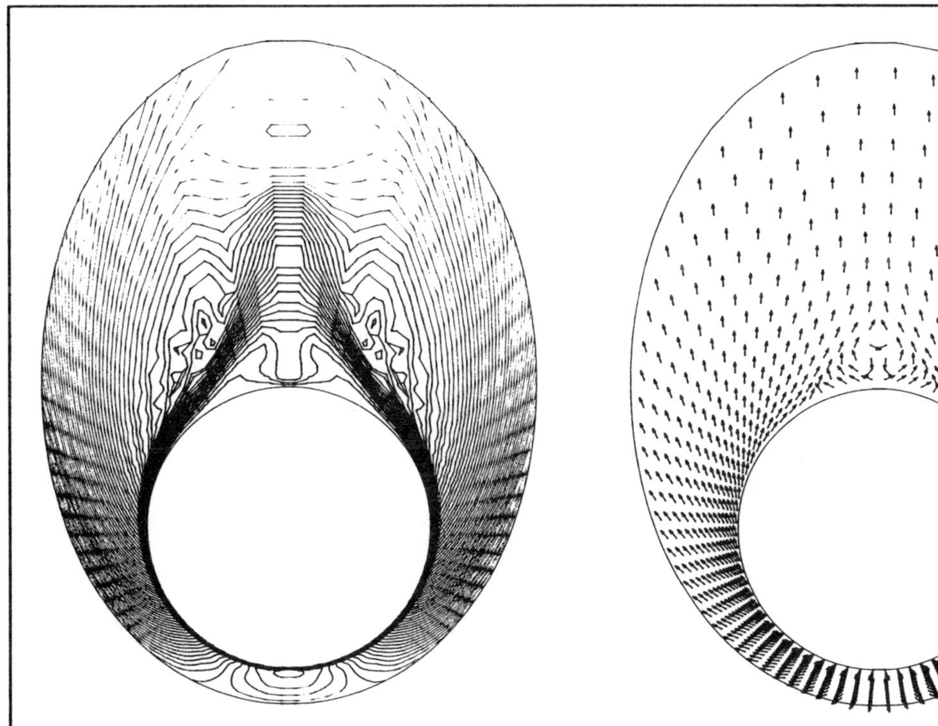

Figure 13-8. Contours of total enthalpies for $M_\infty = 7.95$, $\alpha = 24°$.

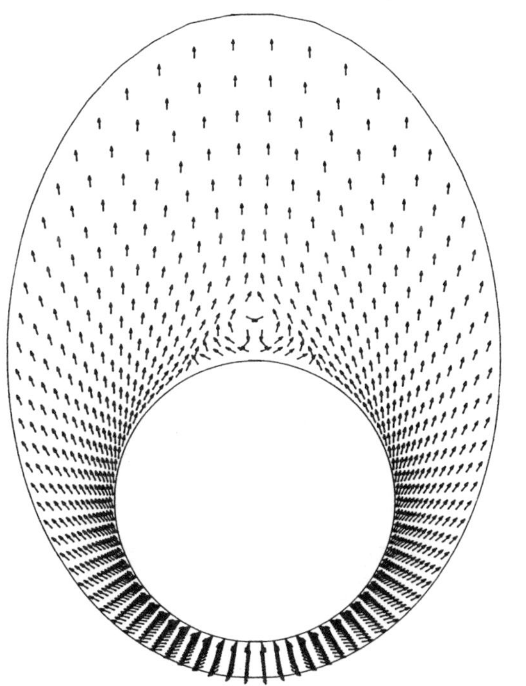

Figure 13-9. Velocity vector plot for $M_\infty = 7.95$, $\alpha = 24°$.

Pressure contours are shown in Figure 13-7, where the expansion which occurs as the flow accelerates in the circumferential direction is clearly evident. Figure 13-8 represents contours of total enthalpies. In this figure the growth of the viscous boundary layer in the circumferential direction and eventual separation on the lee-side is illustrated. Finally, in Figure 13-9, the flow separation on the leeside and formation of vortices are illustrated by the velocity vector plot.

In this chapter, the fundamental aspects of the Parabolized Navier-Stokes equations and a typical numerical scheme for its solution have been presented. Additional consideration which requires some modification of the equation is the incorporation of turbulence and equilibrium or non-equilibrium chemistry effects. These individual topics will be addressed briefly in Chapters 17 and 18.

13.10 Summary Objectives

After completing this chapter, you should be able to discuss:

1. Mathematical assumptions introduced in the PNS equations
2. Shock-fitting and shock-capturing procedures
3. Various assumptions used for the streamwise pressure gradient term
4. Numerical scheme used to solve the PNS equations
5. Implementation of the boundary conditions

13.11 Problems

13.1 You are required to solve the flowfield around a reentry vehicle at moderate angle of attack ($\sim 20°$–$30°$). You have developed a PNS code for this purpose. A colleague has an NS code which he claims will solve the entire flowfield, i.e., the NS equations are solved by marching in time until the required steady-state solution has been achieved. Present a convincing argument as to why your PNS code is more economical compared to the NS code for this application. Your discussion should also include the starter solution for the PNS code and how your colleague's code may be helpful in this regard.

13.2 List the mathematical assumptions involved in the PNS equations and their physical implications.

Chapter 14
Navier-Stokes Equations

14.1 Introductory Remarks

For complex flowfields with separations or strong viscous/inviscid interaction, the reduced form of the equations of fluid motion do not provide adequate solution. Therefore, to accurately compute the flowfield, the Navier-Stokes equations must be considered. On many occasions where the flow separation is not severe, Thin-Layer Navier-Stokes equations are used.

The Navier-Stokes equations may be solved to either provide a steady-state solution or a time-accurate solution. For problems involving complex interactions, separation, and mixed flowfields composed of subsonic and supersonic regimes, the Navier-Stokes equations are initiated with an assumed data set within the domain and marched in time to steady state. As remarked previously, the intermediate solutions have no physical meaning. Hence, when a steady-state (or converged) solution is sought, the maximum allowable time step dictated by the stability requirement is used to reach the solution with the minimum number of time steps and, therefore, computation time.

For time-accurate solutions, a physically correct set of data is used to initiate the solution. The time step must not only satisfy the stability requirements of the numerical scheme, but must be consistent with the physical requirement of the problem.

In this chapter, selected numerical schemes to solve the Navier-Stokes equations are introduced.

14.2 Navier-Stokes Equations

The Navier-Stokes equations in a flux-vector formulation in the computational space is given by Eq. (11-60) as

$$\frac{\partial \bar{Q}}{\partial \tau} + \frac{\partial \bar{E}}{\partial \xi} + \frac{\partial \bar{F}}{\partial \eta} + \frac{\partial \bar{G}}{\partial \zeta} = \frac{\partial \bar{E}_v}{\partial \xi} + \frac{\partial \bar{F}_v}{\partial \eta} + \frac{\partial \bar{G}_v}{\partial \zeta} \tag{14-1}$$

This three-dimensional equation may be reduced to two-dimensional planar or axisymmetric applications as

$$\frac{\partial \bar{Q}}{\partial \tau} + \frac{\partial \bar{E}}{\partial \xi} + \frac{\partial \bar{F}}{\partial \eta} + \alpha \bar{H} = \frac{\partial \bar{E}_v}{\partial \xi} + \frac{\partial \bar{F}_v}{\partial \eta} + \alpha \bar{H}_v \tag{14-2}$$

Before proceeding with the numerical algorithms, a few guidelines established in the previous chapters which may be applied to the Navier-Stokes equations are summarized.

1. Coordinate transformation from the physical space to computational space is necessary. This transformation drastically simplifies the applications of the boundary conditions and may include various options on grid point clustering and orthogonality, both being extremely important for the solution of the Navier-Stokes equations. Obviously grid point clustering near the surface for viscous flows is required in order to resolve the flow gradient. The orthogonality at the surface is desirable to facilitate the computation of the normal gradients. Recall that the normal gradients are used to compute the heat transfer and skin friction.

2. For efficiency purposes, multi-dimensional problems expressed in implicit formulations are split such that each equation has a block tridiagonal coefficient.

3. When the convective terms are approximated by the second-order central difference expression, the addition of damping terms or TVD terms may be required to prevent oscillations in the inviscid region of the flowfield where strong pressure gradients may exist.

4. The use of first-order forward or backward difference approximation for the convective terms, while preserving the physics of the problem, overcomes the difficulty associated with the central differencing. Mathematically, this leads to flux vector splitting. However, the first-order approximation may lead to excessive dissipation error, thereby destroying the solutions. For this reason, higher order approximations are considered.

5. Implicit formulations in general possess less restrictive stability requirements than explicit formulations.

6. Since the Navier-Stokes equations are nonlinear, a linearization procedure based on Taylor series expansion is introduced.

7. The viscous terms in the Navier-Stokes equations are almost always approximated by a second-order central difference expression. On the other hand, there are several algorithms for the approximation of the convective terms, as seen previously in Chapter 12.

These established guidelines will be considered in the investigation of numerical schemes for the Navier-Stokes equations.

14.3 Thin-Layer Navier-Stokes Equations

For problems where the flow separation is moderate, the normal gradient of the stress terms are much larger than the streamwise and circumferential gradients. Therefore, the Navier-Stokes equations are reduced to

$$\frac{\partial \bar{Q}}{\partial \tau} + \frac{\partial \bar{E}}{\partial \xi} + \frac{\partial \bar{F}}{\partial \eta} + \frac{\partial \bar{G}}{\partial \zeta} = \frac{\partial \bar{F}_{v_T}}{\partial \eta} \tag{14-3}$$

Usually the mixed partial derivatives on the right-hand side of (14-3) are also dropped; hence the flux vector $\bar{F}_v$ is redefined as $\bar{F}_{v_T}$, which is given by (11-156).

A factor which may dictate the selection of the Thin-Layer Navier-Stokes equation over the full Navier-Stokes equation is the computer capability. When Navier-Stokes equations are used, additional grid points in the streamwise and circumferential directions must be included. This large number of points is important to resolve the viscous gradients in those directions. As a result, the solution of the Navier-Stokes equations requires more grid points, and therefore more memory and storage must be available. All computers have limits; hence, computer hardware imposes an upper limit on the number of grid points within the domain of interest. In addition, computation time will increase due to an increase in the number of nodes. Therefore, based on the physical considerations as well as computer limitations, Thin-Layer Navier-Stokes equations are used extensively.

The numerical algorithms to be developed next are based on two-dimensional planar or axisymmetric Navier-Stokes equations. Reduction of the equations to Thin-Layer Navier-Stokes equations is obvious, and the extension to three dimensions is addressed in Sec. 14.5.

14.4 Numerical Algorithms

Several numerical algorithms were presented in Chapter 6 to solve the hyperbolic equations, and the schemes were extended to a hyperbolic system of equations in Chapter 12. These equations typically include a time-dependent derivative and convective terms. Subsequently, in Chapter 7, a diffusion term was added to the equation, and several schemes were investigated. The procedure was extended to a system of equation in Chapter 8 where the incompressible Navier-Stokes equation was explored. In this chapter, the numerical schemes investigated previously are extended to the Navier-Stokes equation. Thus, even though the formulations appear to be more complex, they should be familiar to us at point. The emphasis in this chapter will be placed on the approximation of the right-hand side of the Navier-Stokes equations, since the left-hand side forms the Euler equation which was discussed in Chapter 12. A summary of possible approximations of terms in the Navier-Stokes equations is provided in Table 14.1.

Time derivative ⇒	a.	Forward difference in time ⇒ Explicit formulation
	b.	Backward difference in time ⇒ Implicit formulation
	c.	Runge-Kutta (with damping terms or TVD terms)
Convective terms ⇒	a.	Flux vector splitting
	b.	Total variation diminishing
	c.	Central difference approximation with damping terms
Diffusion terms ⇒	a.	Central difference approximation

Table 14.1: Approximation of terms in the Navier-Stokes equations.

14.4.1 Explicit Formulations

A first order in time, explicit formulation of the Navier-Stokes equations can be written as

$$\bar{Q}^{n+1} = \bar{Q}^n - \Delta t \left(\frac{\partial \bar{E}^n}{\partial \xi} + \frac{\partial \bar{F}^n}{\partial \eta} \right) + \Delta t \left(\frac{\partial \bar{E}_v^n}{\partial \xi} + \frac{\partial \bar{F}_v^n}{\partial \eta} \right) - \alpha(\bar{H}^n + \bar{H}_v^n) \qquad (14\text{-}4)$$

At this point, several schemes are available for the convection terms which were explored in Chapter 12. The diffusion terms, which include the viscous and heat conduction terms, are typically approximated by a second-order central difference expression. This approximation may be at the grid points or may also include midpoints as well. Both methods are investigated in this section.

The explicit schemes for the Navier-Stokes equations which are reviewed in this section include the MacCormack explicit, flux vector splitting, modified Runge-Kutta, and TVD formulations.

14.4.1.1 MacCormack Explicit Formulation: In Chapters 6 and 7, the MacCormack explicit scheme was introduced to solve the inviscid and viscous Burgers equations. The procedure is a multi-level (predictor/corrector) scheme which uses forward difference approximation in one level (predictor step) followed by backward difference approximation in the next level (corrector step). This order may be altered, i.e., backward followed by forward differencing. The overall accuracy of the scheme is second-order in time and space. The scheme is subject to stability requirements and, for the wave equation, leads to the requirement of $c \leq 1$. Stability analysis of the viscous Burgers equation does not lead to a simple expression. Therefore, based on numerical experimentation requirements such as

$$\left(\left|u\right|\frac{\Delta t}{\Delta x} + 2\mu\frac{\Delta t}{(\Delta x)^2}\right) \leq 1 \tag{14-5}$$

or

$$\Delta t \leq \min\left[\frac{\Delta x}{|u|}\ ,\ \frac{(\Delta x)^2}{2\mu}\right] \tag{14-6}$$

have been reported in the literature.

Now the scheme is applied to the two-dimensional Navier-Stokes equation given by Equation (14-2). For the first level, forward differencing is used to provide

$$\frac{\bar{Q}^*_{i,j} - \bar{Q}^n_{i,j}}{\Delta\tau} + \frac{\bar{E}^n_{i+1,j} - \bar{E}^n_{i,j}}{\Delta\xi} + \frac{\bar{F}^n_{i,j+1} - \bar{F}^n_{i,j}}{\Delta\eta} + \alpha\bar{H}^n_{i,j} =$$

$$\frac{(\bar{E}^n_v)_{i+1,j} - (\bar{E}^n_v)_{i,j}}{\Delta\xi} + \frac{(\bar{F}^n_v)_{i,j+1} - (\bar{F}^n_v)_{i,j}}{\Delta\eta} + \alpha(\bar{H}^n_v)_{i,j}$$

from which

$$\begin{aligned}\bar{Q}^*_{i,j} &= \bar{Q}^n_{i,j} - \frac{\Delta\tau}{\Delta\xi}\left[\bar{E}^n_{i+1,j} - \bar{E}^n_{i,j}\right] - \frac{\Delta\tau}{\Delta\eta}\left[\bar{F}^n_{i,j+1} - \bar{F}^n_{i,j}\right] \\ &+ \frac{\Delta\tau}{\Delta\xi}\left[(\bar{E}^n_v)_{i+1,j} - (\bar{E}^n_v)_{i,j}\right] + \frac{\Delta\tau}{\Delta\eta}\left[(\bar{F}^n_v)_{i,j+1} - (\bar{F}^n_v)_{i,j}\right] \\ &- \Delta\tau\alpha\left[\bar{H}^n_{i,j} - (\bar{H}^n_v)_{i,j}\right]\end{aligned} \tag{14-7}$$

For the second level (corrector step), a backward differencing is used to result in

$$\begin{aligned}\bar{Q}^{n+1}_{i,j} &= \frac{1}{2}\Big\{\bar{Q}^n_{i,j} + \bar{Q}^*_{i,j} - \frac{\Delta\tau}{\Delta\xi}\left[\bar{E}^*_{i,j} - \bar{E}^*_{i-1,j}\right] - \frac{\Delta\tau}{\Delta\eta}\left[\bar{F}^*_{i,j} - \bar{F}^*_{i,j-1}\right] \\ &+ \frac{\Delta\tau}{\Delta\xi}\left[(\bar{E}^*_v)_{i,j} - (\bar{E}^*_v)_{i-1,j}\right] + \frac{\Delta\tau}{\Delta\eta}\left[(\bar{F}^*_v)_{i,j} - (\bar{F}^*_v)_{i,j-1}\right] \\ &- \Delta\tau\alpha\left[\bar{H}^*_{i,j} - (\bar{H}^*_v)_{i,j}\right]\Big\}\end{aligned} \tag{14-8}$$

A point to consider at this stage is the manner by which the gradients embedded in the viscous terms are approximated. Consider, for example, the second component of $\bar{E}_v$ given by Equation (11-208) as

$$(\bar{E}_v)_2 = \frac{\mu}{JRe_\infty}\left[a_1 u_\xi + a_3 v_\xi + c_1 u_\eta + c_3 v_\eta\right]$$

When considering the gradient $\partial \bar{E}_v/\partial\xi$, the derivatives in $\bar{E}_v$ in the ξ direction are approximated in the opposite direction of that used in $\partial \bar{E}_v/\partial\xi$, i.e., since forward approximation was used for $\partial \bar{E}_v/\partial\xi$; thus a backward approximation is employed for the ξ gradients within $\bar{E}_v$. For example,

$$\frac{\partial u}{\partial \xi} = \frac{u_{i-1,j} - u_{i,j}}{\Delta\xi}$$

The η derivatives in $\bar{E}_v$ will result in mixed partial derivatives in $\partial \bar{E}_v/\partial\xi$. These derivatives are evaluated by central difference approximation; for example,

$$\frac{\partial u}{\partial \eta} = \frac{u_{i,j+1} - u_{i,j-1}}{2\Delta\eta}$$

The MacCormack explicit scheme is subject to stability requirements. A suggested (empirical) requirement is

$$\Delta\tau \leq \frac{\sigma(\Delta\tau)_{inv}}{1 + 2/Re_c} \tag{14-9}$$

where σ is a safety factor (usually about 0.8), and $(\Delta\tau)_{inv}$ is the time step associated with the inviscid Courant number such that

$$(\Delta\tau)_{inv} \leq \left\{ \frac{1}{\frac{|U|}{\Delta\xi} + \frac{|V|}{\Delta\eta} + \frac{|W|}{\Delta\zeta} + a\left[\frac{1}{(\Delta\xi)^2} + \frac{1}{(\Delta\eta)^2} + \frac{1}{(\Delta\zeta)^2}\right]^{.5}} \right\} \tag{14-10}$$

Re_c is the minimum of cell Reynolds numbers associated with ξ, η, or ζ directions, i.e.,

$$Re_c = \min(Re_\xi\ ,\quad Re_\eta\ ,\quad Re_\zeta)$$

where

$$Re_\xi = \frac{\rho|U|\Delta\xi}{\mu}$$

$$Re_\eta = \frac{\rho|V|\Delta\eta}{\mu}$$

$$Re_\zeta = \frac{\rho|W|\Delta\zeta}{\mu}$$

14.4.1.2 Flux Vector Splitting: The convective terms can be approximated by either a first-order or a second-order scheme, as given by Equations (12-155) or (12-156), respectively. The diffusion term $(\partial \bar{E}_v/\partial \xi + \partial \bar{F}_v/\partial \eta)$ is approximated by a second-order central difference scheme as follows.

The viscous flux vector $\bar{E}_v$ is given by (11-208) for which the second component is

$$\bar{E}_{v2} = \frac{\mu}{Re_\infty J}(a_1 u_\xi + a_3 v_\xi + c_1 u_\eta + c_3 v_\eta)$$

Consider the first term given by

$$(\bar{E}_{v2})_{1St} = \left(\frac{\mu}{Re_\infty J} a_1\right)(u_\xi)$$

In fact, all the terms in $\bar{E}_v$ have a similar form. Thus, a general expression can be defined as $(L)\,(M_\xi)$, where, for the first term

$$L = \frac{\mu}{Re_\infty J}\, a_1 \qquad \text{and} \qquad M = u$$

Now, recall that the objective at this point is to develop an approximation for $\partial \bar{E}_v/\partial \xi$. For the time being, consider only the ξ derivatives in $\bar{E}_v$, defined by $\bar{E}_{v\xi}$. Thus, using the general formulation, one can write

$$\begin{aligned}
\frac{\partial \bar{E}_v}{\partial \xi} &= \frac{\partial}{\partial \xi}\left(L\frac{\partial M}{\partial \xi}\right) = \frac{\left(L\frac{\partial M}{\partial \xi}\right)_{i+\frac{1}{2},j} - \left(L\frac{\partial M}{\partial \xi}\right)_{i-\frac{1}{2},j}}{2\left(\frac{\Delta \xi}{2}\right)} \\
&= \frac{\frac{1}{2}(L_{i+1,j} + L_{i,j})\left(\frac{M_{i+1,j} - M_{i,j}}{\Delta \xi}\right) - \frac{1}{2}(L_{i,j} + L_{i-1,j})\frac{M_{i,j} - M_{i-1,j}}{\Delta \xi}}{\Delta \xi} \\
&= \frac{(L_{i+1,j} + L_{i,j})\,(M_{i+1,j} - M_{i,j}) - (L_{i,j} + L_{i-1,j})\,(M_{i,j} - M_{i-1,j})}{2(\Delta \xi)^2} \\
&= \frac{1}{2(\Delta \xi)^2}\Big[(L_{i+1,j} + L_{i,j})\,M_{i+1,j} - (L_{i+1,j} + 2L_{i,j} + L_{i-1,j})\,M_{i,j} \\
&\qquad + (L_{i,j} + L_{i-1,j})\,M_{i-1,j}\Big]
\end{aligned} \tag{14-11}$$

Define

$$\hat{L}_{i+1,j} = L_{i+1,j} + L_{i,j} \tag{14-12}$$

$$\hat{L}_{i,j} = L_{i+1,j} + 2L_{i,j} + L_{i-1,j} \tag{14-13}$$

$$\hat{L}_{i-1,j} = L_{i,j} + L_{i-1,j} \tag{14-14}$$

Then

$$\frac{\partial}{\partial \xi}\left(L\frac{\partial M}{\partial \xi}\right) = \frac{1}{2(\Delta\xi)^2}\left[\hat{L}_{i+1,j}M_{i+1,j} - \hat{L}_{i,j}M_{i,j} + \hat{L}_{i-1,j}M_{i-1,j}\right] \tag{14-15}$$

In the formulations to follow, all the mixed partial derivatives in $\partial\bar{E}_v/\partial\xi$ and $\partial\bar{F}_v/\partial\eta$ are omitted. Therefore,

$$\frac{\partial \bar{E}_v}{\partial \xi} + \frac{\partial \bar{F}_v}{\partial \eta} = \frac{1}{2(\Delta\xi)^2}\left[\hat{e}_{i+1,j}\hat{E}_{i+1,j} - \hat{e}_{i,j}\hat{E}_{i,j} + \hat{e}_{i-1,j}\hat{E}_{i-1,j}\right]$$

$$+ \frac{1}{2(\Delta\eta)^2}\left[\hat{f}_{i,j+1}\hat{F}_{i,j+1} - \hat{f}_{i,j}\hat{F}_{i,j} + \hat{f}_{i,j-1}\hat{F}_{i,j-1}\right] \tag{14-16}$$

Note that $\hat{e}$ and $\hat{f}$ are used to represent $\hat{L}$, as defined in the general formulations given by (14-12) through (14-14), $\hat{E}$ and $\hat{F}$ are used for M, and L_e and L_f will be used for L. Now consider $\bar{E}_{v\xi}$ and $\bar{F}_{v\eta}$, that is

$$\bar{E}_{v\xi} = \frac{\mu}{Re_\infty J}\begin{bmatrix} 0 \\ a_1 u_\xi + a_3 v_\xi \\ a_3 u_\xi + a_2 v_\xi \\ \frac{1}{2}a_1(u^2)_\xi + \frac{1}{2}a_2(v^2)_\xi + a_3(uv)_\xi + \frac{\gamma}{Pr}a_4 e_\xi \end{bmatrix} \tag{14-17}$$

and

$$\bar{F}_{v\eta} = \frac{\mu}{Re_\infty J}\begin{bmatrix} 0 \\ b_1 u_\eta + b_3 v_\eta \\ b_3 u_\eta + b_2 v_\eta \\ \frac{1}{2}b_1(u^2)_\eta + \frac{1}{2}b_2(v^2)_\eta + b_3(uv)_\eta + \frac{\gamma}{Pr}b_4 e_\eta \end{bmatrix} \tag{14-18}$$

Now the corresponding terms can be easily identified as follow

Component of $\bar{E}_{v\xi}$ →	$\bar{E}_{v\xi 1}$	$\bar{E}_{v\xi 2}$		$\bar{E}_{v\xi 3}$		$\bar{E}_{v\xi 4}$			
Terms in each component →	0	1	2	1	2	1	2	3	4
$\left(\frac{Re_\infty J}{\mu}\right) L_e$	0	a_1	a_3	a_3	a_2	$\frac{1}{2}a_1$	$\frac{1}{2}a_2$	a_3	$\frac{\gamma}{Pr}a_4$
$\hat{E}$	0	u	v	u	v	u^2	v^2	uv	e

Component of $\bar{F}_{v\eta} \rightarrow$	$\bar{F}_{v\eta 1}$	$\bar{F}_{v\eta 2}$		$\bar{F}_{v\eta 3}$		$\bar{F}_{v\eta 4}$			
Terms in each component $\rightarrow$	0	1	2	1	2	1	2	3	4
$\left(\frac{Re_\infty J}{\mu}\right) L_f$	0	b_1	b_3	b_3	b_2	$\frac{1}{2}b_1$	$\frac{1}{2}b_2$	b_3	$\frac{\gamma}{Pr}b_4$
$\hat{F}$	0	u	v	u	v	u^2	v^2	uv	e

If mixed partial derivatives are included, the following general formulations can be used

$$\frac{\partial}{\partial \eta}\left(L\frac{\partial M}{\partial \xi}\right) = \frac{1}{8\Delta\eta\Delta\xi}\Big[\left(L_{i,j}+L_{i,j+1}\right)\left(M_{i+1,j}-M_{i-1,j}+M_{i+1,j+1}-M_{i-1,j+1}\right)$$
$$-\left(L_{i,j}+L_{i,j-1}\right)\left(M_{i+1,j}-M_{i-1,j}+M_{i+1,j-1}-M_{i-1,j-1}\right)\Big] \tag{14-19}$$

and

$$\frac{\partial}{\partial \xi}\left(L\frac{\partial M}{\partial \eta}\right) = \frac{1}{8\Delta\xi\Delta\eta}\Big[\left(L_{i,j}+L_{i+1,j}\right)\left(M_{i,j+1}-M_{i,j-1}+M_{i+1,j+1}-M_{i+1,j-1}\right)$$
$$-\left(L_{i,j}+L_{i-1,j}\right)\left(M_{i,j+1}-M_{i,j-1}+M_{i-1,j+1}-M_{i-1,j-1}\right)\Big] \tag{14-20}$$

Now the first-order scheme is written as

$$\bar{Q}^{n+1} = \bar{Q}^n - \Delta\tau\Big[\frac{1}{\Delta\xi}\left(\bar{E}^+_{i,j}-\bar{E}^+_{i-1,j}+\bar{E}^-_{i+1,j}-\bar{E}^-_{i,j}\right)$$
$$+\frac{1}{\Delta\eta}\left(\bar{F}^+_{i,j}-\bar{F}^+_{i,j-1}+\bar{F}^-_{i,j+1}-\bar{F}^-_{i,j}\right)+\alpha\bar{H}_{i,j}\Big]$$
$$+\frac{\Delta\tau}{2(\Delta\xi)^2}\left(\hat{e}_{i+1,j}\hat{E}_{i+1,j}-\hat{e}_{i,j}\hat{E}_{i,j}+\hat{e}_{i-1,j}\hat{E}_{i-1,j}\right)$$
$$+\frac{\Delta\tau}{2(\Delta\eta)^2}\left(\hat{f}_{i,j+1}\hat{F}_{i,j+1}-\hat{f}_{i,j}\hat{F}_{i,j}+\hat{f}_{i,j-1}\hat{F}_{i,j-1}\right)+\alpha\Delta\tau\left(\bar{H}_v\right)_{i,j} \tag{14-21}$$

and the second-order scheme is

$$\bar{Q}^{n+1} = \bar{Q}^n - \Delta\tau\Big[\frac{1}{2\Delta\xi}\left(\bar{E}^+_{i-2,j}-4\bar{E}^+_{i-1,j}+3\bar{E}^+_{i,j}-3\bar{E}^-_{i,j}-3\bar{E}^-_{i,j}+4\bar{E}^-_{i+1,j}-\bar{E}^-_{i+2,j}\right)$$
$$+\frac{1}{2\Delta\eta}\left(\bar{F}^+_{i,j-2}-4\bar{F}^+_{i,j-1}+3\bar{F}^+_{i,j}-4\bar{F}^-_{i,j+1}-\bar{F}^-_{i,j+2}\right)+\alpha\bar{H}_{i,j}\Big]$$

$$+\frac{\Delta\tau}{2(\Delta\xi)^2}\left(\hat{e}_{i+1,j}\hat{E}_{i+1,j}-\hat{e}_{i,j}\hat{E}_{i,j}+\hat{e}_{i-1,j}\hat{E}_{i-1,j}\right)$$

$$+\frac{\Delta\tau}{2(\Delta\eta)^2}\left(\hat{f}_{i,j+1}\hat{F}_{i,j+1}-\hat{f}_{i,j}\hat{F}_{i,j}+\hat{f}_{i,j-1}\hat{F}_{i,j-1}\right)+\alpha\Delta\tau\left(\bar{H}_v\right)_{i,j} \tag{14-22}$$

The inviscid fluxes E^+, E^-, F^+, and F^- are specified for the Steger and Warming flux vector splitting in Section 12.9.2.1 and for the van Leer flux vector splitting in Section 12.9.2.2.

These formulations are relatively simple to program; however, the schemes are subject to stability requirements. Determination of an expression for the stability requirement of Equations (14-21) or (14-22) is not easily accomplished. However, as a general rule, a CFL number of around one will be required. It is cautioned that the exact value (of the CFL number) will depend on the grid system as well as the particular flow conditions. For example, in regions where grid spacing is very small, a smaller CFL number may be required.

One way to avoid the use of small CFL numbers is to utilize a mixed explicit/implicit formulation. For example, if grid clustering near the surface is being used, i.e., in the η direction, an implicit formulation may be utilized in that direction to eliminate the severe stability requirements of the explicit formulation. If the grid spacing in the streamwise direction ξ is not severe, then the explicit formulation may be employed along that direction.

14.4.1.3 Modified Runge-Kutta Scheme: The modified Runge-Kutta scheme introduced in Section 12.8.2.3 for the Euler equation can be used to develop a fourth-order Runge-Kutta scheme for the Navier-Stokes equation. The formulations following (12-167) through (12-171) are

$$\bar{Q}^{(1)}_{i,j}=\bar{Q}^{n}_{i,j} \tag{14-23}$$

$$\bar{Q}^{(2)}_{i,j}=\bar{Q}^{n}_{i,j}-\frac{\Delta\tau}{4}\left[\left(\frac{\partial\bar{E}}{\partial\xi}\right)^{(1)}_{i,j}+\left(\frac{\partial\bar{F}}{\partial\eta}\right)^{(1)}_{i,j}-\left(\frac{\partial\bar{E}_v}{\partial\xi}\right)^{(1)}_{i,j}-\left(\frac{\partial\bar{F}_v}{\partial\eta}\right)^{(1)}_{i,j}+\alpha(\bar{H}^{(1)}_{i,j}-\bar{H}v^{(1)}_{i,j})\right] \tag{14-24}$$

$$\bar{Q}^{(3)}_{i,j}=\bar{Q}^{n}_{i,j}-\frac{\Delta\tau}{3}\left[\left(\frac{\partial\bar{E}}{\partial\xi}\right)^{(2)}_{i,j}+\left(\frac{\partial\bar{F}}{\partial\eta}\right)^{(2)}_{i,j}-\left(\frac{\partial\bar{E}_v}{\partial\xi}\right)^{(2)}_{i,j}-\left(\frac{\partial\bar{F}_v}{\partial\eta}\right)^{(2)}_{i,j}+\alpha(\bar{H}^{(2)}_{i,j}-\bar{H}v^{(2)}_{i,j})\right] \tag{14-25}$$

$$\bar{Q}^{(4)}_{i,j}=\bar{Q}^{n}_{i,j}-\frac{\Delta\tau}{2}\left[\left(\frac{\partial\bar{E}}{\partial\xi}\right)^{(3)}_{i,j}+\left(\frac{\partial\bar{F}}{\partial\eta}\right)^{(3)}_{i,j}-\left(\frac{\partial\bar{E}_v}{\partial\xi}\right)^{(3)}_{i,j}-\left(\frac{\partial\bar{F}_v}{\partial\eta}\right)^{(3)}_{i,j}+\alpha(\bar{H}^{(3)}_{i,j}-\bar{H}v^{(3)}_{i,j})\right] \tag{14-26}$$

$$\bar{Q}_{i,j}^{n+1} = \bar{Q}_{i,j}^{n} - \Delta\tau \left[\left(\frac{\partial \bar{E}}{\partial \xi}\right)_{i,j}^{(4)} + \left(\frac{\partial \bar{F}}{\partial \eta}\right)_{i,j}^{(4)} - \left(\frac{\partial \bar{E}_v}{\partial \xi}\right)_{i,j}^{(4)} - \left(\frac{\partial \bar{F}_v}{\partial \eta}\right)_{i,j}^{(4)} + \alpha(\bar{H}_{i,j}^{(4)} - \bar{H}v_{i,j}^{(4)}) \right] \tag{14-27}$$

The convective terms are approximated by a second-order central difference expression as

$$\left(\frac{\partial \bar{E}}{\partial \xi}\right)_{i,j} + \left(\frac{\partial \bar{F}}{\partial \eta}\right)_{i,j} = \frac{\bar{E}_{i+1,j} - \bar{E}_{i-1,j}}{2\Delta\xi} + \frac{\bar{F}_{i,j+1} - \bar{F}_{i,j-1}}{2\Delta\eta} \tag{14-28}$$

and the diffusion terms are approximated according to the formulation of (14-16). To reduce computational efforts, the diffusion term may be evaluated only at the first stage, that is,

$$\left(\frac{\partial \bar{E}_v}{\partial \xi}\right)_{i,j}^{(1)} + \left(\frac{\partial F_v}{\partial \eta}\right)_{i,j}^{(1)}$$

and used subsequently at the next three successive stages. As discussed in Section 12.9.2.3, the addition of damping terms or TVD may be required to reduce any oscillations which may develop in the vicinity of sharp flow gradients due to dispersion errors.

14.4.2 Boundary Conditions

The boundary conditions described in Section 12.9.3 for the Euler equation are extended to the Navier-Stokes equation, with the modification of velocity at the surface. Recall that the slip condition was imposed for the velocity at the surface for inviscid flow. The no-slip condition is specified at the surface (nonporous) for the Navier-Stokes equations, namely, the velocity, and therefore velocity components are zero at the surface. Typically, zero-order extrapolation is used to obtain pressure at the surface. Thus, the boundary conditions at the surface are specified as

$$\begin{aligned} u_{i,1} &= 0 \\ v_{i,1} &= 0 \\ p_{i,1} &= p_{i,2} \end{aligned}$$

Furthermore, note that at the surface the total energy e_t is reduced to the internal energy, that is,

$$e_t = e + \frac{1}{2}(u^2 + v^2) = e$$

If the wall temperature distribution is specified, then

$$e = \frac{T_w}{\gamma(\gamma-1)M_\infty^2} \tag{14-29}$$

where T_w is the nondimensional wall temperature. If the wall is specified as adiabatic, then

$$\frac{\partial T}{\partial n} = 0$$

and therefore

$$T_{i,1} = T_{i,2}$$

and furthermore

$$\rho_{i,1} = \rho_{i,2}$$

Now the total energy e_t can be determined as follows

$$e_t = e + \frac{1}{2}(u^2 + v^2)$$

or

$$\rho e_t = \rho e + \frac{1}{2}\rho(u^2 + v^2)$$

Since for a perfect gas $p = \rho e(\gamma - 1)$,

$$\rho e_t = \frac{p}{\gamma - 1} + \frac{1}{2}\rho(u^2 + v^2)$$

and at the wall ($u = v = 0$); therefore, $\rho e_t = p/(\gamma - 1)$, or

$$(\rho e_t)_{i,1} = \frac{p_{i,1}}{\gamma - 1} = \frac{p_{i,2}}{\gamma - 1} \tag{14-30}$$

When a boundary is located at the freestream where the flow can be described as inviscid, the inviscid boundary conditions described in Section 12.9.3 are used.

Implementation of the boundary conditions and modification of equations at the boundaries for implict formulations will be further deliberated in Section 14.4.3.

14.4.3 Implicit Formulations

A first-order backward approximation in time provides the general implicit formulation for the Navier-Stokes equations as follow

$$\frac{\bar{Q}^{n+1} - \bar{Q}^n}{\Delta\tau} + \left(\frac{\partial \bar{E}}{\partial \xi}\right)^{n+1} + \left(\frac{\partial \bar{F}}{\partial \eta}\right)^{n+1} - \left(\frac{\partial \bar{E}_v}{\partial \xi}\right)^{n+1} - \left(\frac{\partial \bar{F}_v}{\partial \eta}\right)^{n+1} + \alpha\left(\bar{H} - \bar{H}_v\right)^{n+1} = 0 \tag{14-31}$$

Equation (14-31) is nonlinear and, therefore, a linearization procedure must be employed. Previously a linearization scheme based on a Taylor series expansion was introduced and used in the linearization of the incompressible Navier-Stokes, Euler, and PNS equations. Similar approximations are used here for the Navier-Stokes equations providing the following relations:

$$\bar{E}^{n+1} = \bar{E}^n + \frac{\partial \bar{E}}{\partial \bar{Q}} \Delta \bar{Q} = \bar{E}^n + A \Delta \bar{Q} \tag{14-32}$$

$$\bar{F}^{n+1} = \bar{F}^n + \frac{\partial \bar{F}}{\partial \bar{Q}} \Delta \bar{Q} = \bar{F}^n + B \Delta \bar{Q} \tag{14-33}$$

$$\bar{H}^{n+1} = \bar{H}^n + \frac{\partial \bar{H}}{\partial \bar{Q}} \Delta \bar{Q} = \bar{H}^n + C \Delta \bar{Q} \tag{14-34}$$

$$\bar{E}_v^{n+1} = \bar{E}_v^n + \frac{\partial \bar{E}_v}{\partial \bar{Q}} \Delta \bar{Q} = \bar{E}_v^n + A_v \Delta \bar{Q} \tag{14-35}$$

$$\bar{F}_v^{n+1} = \bar{F}_v^n + \frac{\partial \bar{F}_v}{\partial \bar{Q}} \Delta \bar{Q} = \bar{F}_v^n + B_v \Delta \bar{Q} \tag{14-36}$$

$$\bar{H}_v^{n+1} = \bar{H}_v^n + \frac{\partial \bar{H}_v}{\partial \bar{Q}} \Delta \bar{Q} = \bar{H}_v^n + C_v \Delta \bar{Q} \tag{14-37}$$

The Jacobian matrices A, B, C, A_v, B_v, and C_v are given by Equations (11-213) through (11-218).

The linearized expressions (14-32) through (14-37) are substituted into Equation (14-31) to yield

$$\frac{\Delta \bar{Q}}{\Delta \tau} + \frac{\partial}{\partial \xi}\left(\bar{E}^n + A\Delta\bar{Q}\right) + \frac{\partial}{\partial \eta}\left(\bar{F}^n + B\Delta\bar{Q}\right) - \frac{\partial}{\partial \xi}\left(\bar{E}_v^n + A_v\Delta\bar{Q}\right)$$
$$- \frac{\partial}{\partial \eta}\left(\bar{F}_v^n + B_v\Delta\bar{Q}\right) + \alpha\left[\bar{H}^n + C\Delta\bar{Q} - \left(\bar{H}_v^n + C_v\Delta\bar{Q}\right)\right] = 0 \tag{14-38}$$

This equation is rearranged as

$$\left\{I + \Delta\tau\left[\frac{\partial}{\partial \xi}(A) + \frac{\partial}{\partial \eta}(B) - \frac{\partial}{\partial \xi}(A_v) - \frac{\partial}{\partial \eta}(B_v) + \alpha(C - C_v)\right]\right\}\Delta\bar{Q} =$$
$$- \Delta\tau\left[\frac{\partial \bar{E}^n}{\partial \xi} + \frac{\partial \bar{F}^n}{\partial \eta} - \frac{\partial \bar{E}_v^n}{\partial \xi} - \frac{\partial \bar{F}_v^n}{\partial \eta} + \alpha\left(\bar{H}^n - \bar{H}_v^n\right)\right] \tag{14-39}$$

Based on the experience gained through the solution of the multidimensional problems, it is obvious that the coefficient matrix of this system will be a block pentadiagonal system. To increase the efficiency, the system is approximated by two sets of equations solved sequentially; i.e., approximate factorization is used.

For two-dimensional problems, an approximate factorization scheme provides a stable solution. However, for three-dimensional problems approximate factorization is unstable. To overcome the stability problem, artificial viscosity in the form of damping terms or TVD must be added. A scheme known as the LU decomposition is stable for two-dimensional problems as well as for three-dimensional problems. Therefore, it will be discussed in this chapter. For now, the procedure using approximate factorization is investigated. Equation (14-39) is factored as

$$\left\{I + \Delta\tau\left[\frac{\partial}{\partial\xi}(A) - \frac{\partial}{\partial\xi}(A_v)\right]\right\}\left\{I + \Delta\tau\left[\frac{\partial}{\partial\eta}(B) - \frac{\partial}{\partial\eta}(B_v)\right.\right.$$

$$\left.\left. + \Delta\tau\alpha(C - C_v)\right]\right\}\Delta\bar{Q} = -\Delta\tau\left[\frac{\partial\bar{E}^n}{\partial\xi} + \frac{\partial\bar{F}^n}{\partial\eta}\right.$$

$$\left. - \frac{\partial\bar{E}^n_v}{\partial\xi} - \frac{\partial\bar{F}^n_v}{\partial\eta} + \alpha\left(\bar{H}^n - \bar{H}^n_v\right)\right] \tag{14-40}$$

Again, there are several methods by which the convection (inviscid) terms can be approximated, whereas the diffusion terms are approximated by the second-order central difference expression.

14.4.3.1 Flux Vector Splitting:

To retain correct differencing associated with signal propagation, the inviscid fluxes are split as discussed in Chapter 12. Thus, the inviscid Jacobian matrices A and B and flux vectors $\bar{E}$ and $\bar{F}$ are split according to the sign of the eigenvalues, resulting in the following equation:

$$\left[I + \Delta\tau\frac{\partial}{\partial\xi}(A^+ + A^-) - \Delta\tau\frac{\partial A_v}{\partial\xi}\right]\left[I + \Delta\tau\frac{\partial}{\partial\eta}(B^+ + B^-)\right.$$

$$\left. - \Delta\tau\frac{\partial B_v}{\partial\eta} + \Delta\tau\alpha(C - C_v)\right]\Delta\bar{Q} =$$

$$- \Delta\tau\left[\frac{\partial}{\partial\xi}(\bar{E}^+ + \bar{E}^-) + \frac{\partial}{\partial\eta}(\bar{F}^+ + \bar{F}^-) - \frac{\partial\bar{E}_v}{\partial\xi} - \frac{\partial\bar{F}_v}{\partial\eta} + \alpha(\bar{H} - \bar{H}_v)\right] \tag{14-41}$$

Equation (14-41) is solved in sequential stages according to

$$\left[I + \Delta\tau\frac{\partial}{\partial\xi}(A^+ + A^-) - \Delta\tau\frac{\partial A_v}{\partial\xi}\right]\Delta\bar{Q}^* = -\Delta\tau\left[\frac{\partial}{\partial\xi}(\bar{E}^+ + \bar{E}^-)\right.$$

$$\left. + \frac{\partial}{\partial\eta}(\bar{F}^+ + \bar{F}^-) - \frac{\partial\bar{E}_v}{\partial\xi} - \frac{\partial\bar{F}_v}{\partial\eta} + \alpha(\bar{H} - \bar{H}_v)\right] \tag{14-42}$$

and

$$\left[I + \Delta\tau\frac{\partial}{\partial\eta}(B^+ + B^-) - \Delta\tau\frac{\partial B_v}{\partial\eta} + \Delta\tau\alpha(C - C_v)\right]\Delta\bar{Q} = \Delta\bar{Q}^* \tag{14-43}$$

At this point the finite difference approximations of the derivatives in Equations (14-42) and (14-43) are considered. The viscous terms are approximated by a second-order central difference relation. Note that due to the existence of natural viscosity, use of central differencing for the viscous terms does not introduce oscillations into the solution. That is, any dispersion error which may cause oscillation is damped out by viscosity.

The approximations of $\partial A_v/\partial\xi$ and $\partial B_v/\partial\eta$ are similar to that of $\partial \bar{E}_v/\partial\xi$ and $\partial F_v/\partial\eta$ described in Section 14.4.1.2. Again, all the terms in matrices A_v and B_v are similar; therefore, a general expression is derived to represent the elements of these matrices. Recall that matrix A_v is given by (11-216) where element (2,1) is

$$A_{v_{2,1}} = \left(\frac{\partial \bar{E}_v}{\partial \bar{Q}}\right)_{2,1} = \frac{\mu}{Re_\infty J}\left\{-\left[a_1\left(J\frac{u}{\rho}\right)_\xi + a_3\left(J\frac{v}{\rho}\right)_\xi\right]\right.$$

$$\left. - \left[c_1\left(J\frac{u}{\rho}\right)_\eta + c_3\left(J\frac{v}{\rho}\right)_\eta\right]\right\}$$

Now consider the first term, that is,

$$\left(\frac{\mu}{Re_\infty J}\right)(-a_1)\left(J\frac{u}{\rho}\right)_\xi = -\left(a_1\frac{\mu}{Re_\infty J}\right)\left(J\frac{u}{\rho}\right)_\xi = (K)(N)_\xi = K\frac{\partial N}{\partial \xi} \qquad (14\text{-}44)$$

First consider the viscous terms, realizing that the viscous term that appears in the formulation (14-42) is of the form

$$\frac{\partial}{\partial\xi}(A_v)\Delta\bar{Q} = \frac{\partial}{\partial\xi}\left[\left(K\frac{\partial N}{\partial\xi}\right)\Delta\bar{Q}\right] \qquad (14\text{-}45)$$

A second-order central difference approximation at point i, j provides

$$\frac{\partial}{\partial\xi}\left[\left(K\frac{\partial N}{\partial\xi}\right)\Delta\bar{Q}\right] = \frac{\left[K\frac{\partial N}{\partial\xi}\Delta\bar{Q}\right]_{i+\frac{1}{2},j} - \left[K\frac{\partial N}{\partial\xi}\Delta\bar{Q}\right]_{i-\frac{1}{2},j}}{2\left(\frac{\Delta\xi}{2}\right)}$$

$$= \frac{1}{\Delta\xi}\left\{\frac{1}{2}(K_{i,j} + K_{i+1,j})\left[\frac{(N\Delta\bar{Q})_{i+1,j} - (N\Delta\bar{Q})_{i,j}}{2(\Delta\xi/2)}\right]\right.$$

$$\left. - \frac{1}{2}(K_{i,j} + K_{i-1,j})\left[\frac{(N\Delta\bar{Q})_{i,j} - (N\Delta\bar{Q})_{i-1,j}}{2(\Delta\xi/2)}\right]\right\}$$

$$= \frac{1}{2(\Delta\xi)^2}\left\{\frac{1}{2}(K_{i,j} + K_{i+1,j})\left[(N\Delta\bar{Q})_{i+1,j} - (N\Delta\bar{Q})_{i,j}\right]\right.$$

$$-\frac{1}{2}(K_{i,j} + K_{i-1,j}) \left[(N\Delta\bar{Q}_{i,j} - (N\Delta\bar{Q})_{i-1,j}\right] \Big\}$$

$$= \frac{1}{2(\Delta\xi)^2} \Big[(K_{i,j} + K_{i+1,j})(N\Delta\bar{Q})_{i+1,j} - (K_{i+1,j} + 2K_{i,j} + K_{i-1,j})\ (N\Delta\bar{Q})_{i,j}$$

$$+ (K_{i,j} + K_{i-1,j})(N\Delta\bar{Q})_{i-1,j}\Big] \quad (14\text{-}46)$$

Define

$$\hat{K}_{i+1,j} = K_{i,j} + K_{i+1,j} \quad (14\text{-}47)$$

$$\hat{K}_{i,j} = K_{i+1,j} + 2K_{i,j} + K_{i-1,j} \quad (14\text{-}48)$$

$$\hat{K}_{i-1,j} = K_{i,j} + K_{i-1,j} \quad (14\text{-}49)$$

Then

$$\frac{\partial}{\partial\xi}\left[\left(K\frac{\partial M}{\partial\xi}\right)\Delta\bar{Q}\right] = \frac{1}{2(\Delta\xi)^2}\Big[\left(\hat{K}_{i+1,j}N_{i+1,j}\right)\Delta\bar{Q}_{i+1,j} - \left(\hat{K}_{i,j}N_{i,j}\right)\Delta\bar{Q}_{i,j}$$

$$+ \left(\hat{K}_{i-1,j}N_{i-1,j}\right)\Delta\bar{Q}_{i-1,j}\Big] \quad (14\text{-}50)$$

The convective terms in Equations (14-42) and (14-43) are approximated by forward or backward difference expressions. The procedure is the same as the one introduced in Chapter 12. It has been noted that when first-order approximate expressions are used for the convective terms, some accuracy is lost. That is due to excessive dissipative error of the first-order approximation. To overcome this problem, second-order or third-order approximations for the convective terms can be used. However, a difficulty is introduced, namely, the tridiagonal system is distorted because new points such as $i-2$ will also appear in the formulation.

First-order approximation of the convective terms is considered first and subsequently extended to higher order. Before writing the finite difference equations, one more point must be considered. Recall that the elements of the viscous Jacobian involve both gradients, i.e., derivatives with respect to ξ and η. Therefore, when the gradient $\partial A_v/\partial\xi$ is considered, some mixed partial derivatives will appear, i.e., $\partial^2/\partial\xi\partial\eta$. Inclusion of this and similar terms implicitly will distort the tridiagonal nature of the coefficient matrix. To overcome this problem, the mixed partial derivatives are moved to the right-hand side and evaluated explicitly. As discussed in the previous section, some investigators suggest to drop the mixed partial derivatives completely and, indeed, inclusion of these terms does not affect the accuracy for the majority of applications. Therefore, in the formulation to follow, the mixed partial derivatives are dropped.

Now approximation (14-50) is utilized for the viscous terms. The following notation is used for this purpose,

$$\frac{\partial A_v}{\partial \xi} = \frac{\partial}{\partial \xi}\left(\hat{r}\frac{\partial R}{\partial \xi}\right) = \frac{1}{2(\Delta\xi)^2}\left[\hat{r}_{i+1,j}R_{i+1,j} - \hat{r}_{i,j}R_{i,j} + \hat{r}_{i-1,j}R_{i-1,j}\right] \tag{14-51}$$

Similarly,

$$\frac{\partial B_v}{\partial \eta} = \frac{\partial}{\partial \eta}\left(\hat{s}\frac{\partial S}{\partial \eta}\right) = \frac{1}{2(\Delta\eta)^2}\left[\hat{s}_{i,j+1}S_{i,j+1} - \hat{s}_{i,j}S_{i,j} + \hat{s}_{i,j-1}S_{i,j-1}\right] \tag{14-52}$$

where $\hat{r}$ and R represent $\hat{K}$ and N associated with A_v, and $\hat{s}$ and S represent $\hat{K}$ and N associated with B_v.

Observe that the mixed partial derivatives have been dropped in the approximations above. Now, the finite difference approximation of Equation (14-42) is written as

$$\left[I + \frac{\Delta\tau}{\Delta\xi}\left(A^+_{i,j} - A^+_{i-1,j} + A^-_{i+1,j} - A^-_{i,j}\right) - \frac{\Delta\tau}{2(\Delta\xi)^2}\left(\hat{r}_{i+1,j}R_{i+1,j} - \hat{r}_{i,j}R_{i,j}\right.\right.$$
$$\left.\left. + \hat{r}_{i-1,j}R_{i-1,j}\right)\right]\Delta\bar{Q}^* = \text{RHS}$$

which is rearranged as

$$\left\{-\frac{\Delta\tau}{\Delta\xi}A^+_{i-1,j} - \frac{\Delta\tau}{2(\Delta\xi)^2}\hat{r}_{i-1,j}R_{i-1,j}\right\}\Delta\bar{Q}^*_{i-1,j}$$
$$+\left\{I + \frac{\Delta\tau}{\Delta\xi}\left(A^+_{i,j} - A^-_{i,j}\right) + \frac{\Delta\tau}{2(\Delta\xi)^2}\hat{r}_{i,j}R_{i,j}\right\}\Delta\bar{Q}^*_{i,j}$$
$$+\left\{\frac{\Delta\tau}{\Delta\xi}A^-_{i+1,j} - \frac{\Delta\tau}{2(\Delta\xi)^2}\hat{r}_{i+1,j}R_{i+1,j}\right\}\Delta\bar{Q}^*_{i+1,j} = \text{RHS} \tag{14-53}$$

Similarly, the finite difference equation for (14-43) is expressed as

$$\left\{-\frac{\Delta\tau}{\Delta\eta}B^+_{i,j-1} - \frac{\Delta\tau}{2(\Delta\eta)^2}\hat{s}_{i,j-1}S_{i,j-1}\right\}\Delta\bar{Q}_{i,j-1}$$
$$+\left\{I + \frac{\Delta\tau}{\Delta\eta}\left(B^+_{i,j} - B^-_{i,j}\right) + \frac{\Delta\tau}{2(\Delta\eta)^2}\hat{s}_{i,j}S_{i,j} + \alpha\Delta\tau\left[C_{i,j} - (C_v)_{i,j}\right]\right\}\Delta\bar{Q}_{i,j}$$
$$+\left\{\frac{\Delta\tau}{\Delta\eta}B^-_{i,j+1} - \frac{\Delta\tau}{2(\Delta\eta)^2}\hat{s}_{i,j+1}S_{i,j+1}\right\}\Delta\bar{Q}_{i,j+1} = \Delta\bar{Q}^*_{i,j} \tag{14-54}$$

The evaluation of the RHS of Equation (14-53) is simple and is briefly described as follows. The RHS is

$$\text{RHS} = -\Delta\tau\left[\frac{\partial}{\partial\xi}(E^{+} + \bar{E}^{-}) + \frac{\partial}{\partial\eta}(\bar{F}^{+} + \bar{F}^{-}) - \frac{\partial \bar{E}_v}{\partial\xi} - \frac{\partial \bar{F}_v}{\partial\eta} + \alpha(\bar{H} - \bar{H}_v)\right]$$

The gradients of the viscous terms include embedded derivatives with respect to ξ and η. These gradients may be expressed in a general form as

$$\frac{\partial}{\partial\eta}\left(L\frac{\partial M}{\partial\eta}\right) \quad \text{or} \quad \frac{\partial}{\partial\eta}\left(L\frac{\partial M}{\partial\xi}\right)$$

where, as before, L represents some combination of the metrics, Jacobian of transformation J, Re_∞, and μ; whereas M represents flow variables. The mixed partial derivatives do not create any difficulty since they are evaluated explicitly, or, for most applications, they are omitted. The approximation given by (14-16) is used to evaluate the viscous terms as described in Section 14.4.1.2.

The gradients of the inviscid terms are evaluated by forward or backward difference approximations in the same fashion as the Euler equations. In order to write Equations (14-53) and (14-54) in a compact form, the coefficients are defined as follows:

$$CAM = -\left[\frac{\Delta\tau}{\Delta\xi}A^{+}_{i-1,j} + \frac{\Delta\tau}{2(\Delta\xi)^2}\hat{r}_{i-1,j}R_{i-1,j}\right] \tag{14-55}$$

$$CA = \left[I + \frac{\Delta\tau}{\Delta\xi}\left(A^{+}_{i,j} - A^{-}_{i,j}\right) + \frac{\Delta\tau}{2(\Delta\xi)^2}\hat{r}_{i,j}R_{i,j}\right] \tag{14-56}$$

$$CAP = \left[\frac{\Delta\tau}{\Delta\xi}A^{-}_{i+1,j} - \frac{\Delta\tau}{2(\Delta\xi)^2}\hat{r}_{i+1,j}R_{i+1,j}\right] \tag{14-57}$$

and

$$CBM = -\left[\frac{\Delta\tau}{\Delta\eta}B^{+}_{i,j-1} + \frac{\Delta\tau}{2(\Delta\eta)^2}\hat{s}_{i,j-1}S_{i,j-1}\right] \tag{14-58}$$

$$CB = \left[I + \frac{\Delta\tau}{\Delta\eta}\left(B^{+}_{i,j} - B^{-}_{i,j}\right) + \frac{\Delta\tau}{2(\Delta\eta)^2}\hat{s}_{i,j}S_{i,j} + \alpha\Delta\tau[C_{i,j} - (C_v)_{i,j}]\right] \tag{14-59}$$

$$CBP = \left[\frac{\Delta\tau}{\Delta\eta}B^{-}_{i,j+1} - \frac{\Delta\tau}{2(\Delta\eta)^2}\hat{s}_{i,j+1}S_{i,j+1}\right] \tag{14-60}$$

where

$$\text{RHS} = -\Delta\tau\left\{\frac{1}{\Delta\xi}\left(\bar{E}^{+}_{i,j} - E^{+}_{i-1,j} + \bar{E}^{-}_{i+1,j} - \bar{E}^{-}_{i,j}\right) + \frac{1}{\Delta\eta}\left(\bar{F}^{+}_{i,j}\right.\right.$$

$$- \bar{F}^{+}_{i,j-1} + \bar{F}^{-}_{i,j+1} - \bar{F}^{-}_{i,j}\Big) - \frac{1}{2(\Delta\xi)^2}\left(\hat{e}_{i+1,j}\hat{E}_{i+1,j} - \hat{e}_{i,j}\hat{E}_{i,j} + \hat{e}_{i-1,j}\hat{E}_{i-1,j}\right)$$

$$- \frac{1}{2(\Delta\eta)^2}\left(\hat{f}_{i,j+1}\hat{F}_{i,j+1} - \hat{f}_{i,j}\hat{F}_{i,j} + \hat{f}_{i,j-1}\hat{F}_{i,j-1}\right) + \alpha\left(\bar{H}_{i,j} - (\bar{H}_v)_{i,j}\right)\Big\}$$

Note that in the formulation of the RHS given above the mixed partial derivatives are excluded.

Thus, Equations (14-53) and (14-54) become

$$CAM_{i,j}\Delta\bar{Q}^*_{i-1,j} + CA_{i,j}\Delta\bar{Q}^*_{i,j} + CAP_{i,j}\Delta\bar{Q}^*_{i+1,j} = \text{RHS}_{i,j} \tag{14-61}$$

and

$$CBM_{i,j}\Delta\bar{Q}_{i,j-1} + CB_{i,j}\Delta\bar{Q}_{i,j} + CBP_{i,j}\Delta\bar{Q}_{i,j+1} = \Delta\bar{Q}^*_{i,j} \tag{14-62}$$

Equations (14-61) and (14-62) are solved sequentially as the ξ and η sweeps. First, Equation (14-61) is applied at each η line (j constant) for all i from $i = 2$ to $i = IM1$, resulting in the following set of equations:

$$i = 2: \quad CAM_{2,j}\Delta\bar{Q}^*_{1,j} + CA_{2,j}\Delta\bar{Q}^*_{2,j} + CAP_{2,j}\Delta\bar{Q}^*_{3,j} = (\text{RHS})_{2,j} \tag{14-63}$$

$$i = 3: \quad CAM_{3,j}\Delta\bar{Q}^*_{2,j} + CA_{3,j}\Delta\bar{Q}^*_{3,j} + CAP_{3,j}\Delta\bar{Q}^*_{4,j} = (\text{RHS})_{3,j} \tag{14-64}$$

$$\vdots$$

$$i = IM2: \quad CAM_{IM2,j}\Delta\bar{Q}^*_{IM3,j} + CA_{IM2,j}\Delta\bar{Q}^*_{IM2,j}\Delta\bar{Q}^*_{IM2,j}$$
$$+ CAP_{IM2,j}\Delta\bar{Q}^*_{IM1,j} = (\text{RHS})_{IM2,j} \tag{14-65}$$

$$i = IM1: \quad CAM_{IM1,j}\Delta\bar{Q}^*_{IM2,j} + CA_{IM1,j}\Delta\bar{Q}^*_{IM1,j}$$
$$+ CAP_{IM1,j}\Delta\bar{Q}^*_{IM,j} = (\text{RHS})_{IM1,j} \tag{14-66}$$

If the boundary points at $i = 1$ and $i = IM$ are treated explicitly, then the system of Equations (14-63) through (14-66) is expressed in a matrix form as

$$\begin{bmatrix} CA_{2,j} & CAP_{2,j} & & & \\ CAM_{3,j} & CA_{3,j} & CAP_{3,j} & & \\ & & & & \\ & CAM_{IM2,j} & CA_{IM2,j} & CAP_{IM2,j} & \\ & & CAM_{IM1,j} & CA_{IM1,j} & \end{bmatrix} \begin{bmatrix} \Delta\bar{Q}^*_{2,j} \\ \Delta\bar{Q}^*_{3,j} \\ | \\ \Delta\bar{Q}^*_{IM2,j} \\ \Delta\bar{Q}^*_{IM1,j} \end{bmatrix} =$$

$$\begin{bmatrix} \text{RHS}_{2,j} - CAM_{2,j}\Delta\bar{Q}^*_{1,j} \\ \text{RHS}_{3,j} \\ | \\ \text{RHS}_{IM2,j} \\ \text{RHS}_{IM1,j} - CAP_{IM1,j}\Delta\bar{Q}^*_{IM,j} \end{bmatrix} \tag{14-67}$$

Once a new solution is obtained, the flow variables at the boundary points are updated. If the boundary conditions are to be treated implicitly, the following modifications are introduced. Boundary point $i = 1$, which appears in Equation (14-63), is repeated here as:

$$CAM_{2,j}\Delta\bar{Q}^*_{1,j} + CA_{2,j}\Delta\bar{Q}^*_{2,j} + CAP_{2,j}\Delta\bar{Q}^*_{3,j} = \text{RHS}_{2,j}$$

For the two-dimensional problem, the grid system is selected such that grid lines $i = 1$ and $i = 2$ are symmetrical with respect to the body axis. Thus,

$$\rho^*_{1,j} = \rho^*_{2,j}$$

$$u^*_{1,j} = u^*_{2,j}$$

$$v^*_{1,j} = -v^*_{2,j}$$

$$(e^*_t)_{1,j} = (e^*_t)_{2,j}$$

Therefore,

$$\Delta\bar{Q}^*_{1,j} = \frac{1}{J_{1,j}} \begin{bmatrix} (\Delta\rho)^*_{1,j} \\ (\Delta\rho u)^*_{1,j} \\ (\Delta\rho v)^*_{1,j} \\ (\Delta\rho e_t)^*_{1,j} \end{bmatrix} = \frac{1}{J_{2,j}} \begin{bmatrix} (\Delta\rho)^*_{2,j} \\ (\Delta\rho u)^*_{2,j} \\ -(\Delta\rho v)^*_{2,j} \\ (\Delta\rho e_t)^*_{2,j} \end{bmatrix}$$

Now, Equation (14-63) is modified as

$$[CAM_{2,j} \cdot II + CA_{2,j}]\,\Delta\bar{Q}^*_{2,j} + CAP_{2,j}\Delta\bar{Q}^*_{3,j} = \text{RHS}_{2,j} \tag{14-68}$$

where

$$II = \begin{bmatrix} 1 & 0 & 0 & 0 \\ 0 & 1 & 0 & 0 \\ 0 & 0 & -1 & 0 \\ 0 & 0 & 0 & 1 \end{bmatrix}$$

Defining

$$CAM_{2,j} \cdot II + CA_{2,j} = \overline{CA}_{2,j}$$

Then Equation (14-68) becomes

$$\overline{CA}_{2,j}\Delta\bar{Q}^*_{2,j} + CAP_{2,j}\Delta\bar{Q}^*_{3,j} = \text{RHS}_{2,j} \tag{14-69}$$

At the outflow boundary set at $i = IM$, extrapolation is used, i.e.,

$$\Delta\bar{Q}^*_{IM,j} = \Delta\bar{Q}^*_{IM1,j} \tag{14-70}$$

Now, Equation (14-66) is modified by using Equation (14-70) to provide

$$CAM_{IM1,j}\Delta\bar{Q}^*_{IM2,j} + (CA_{IM1,j} + CAP_{IM1,j})\Delta\bar{Q}^*_{IM1,j} = \text{RHS}_{IM1,j} \tag{14-71}$$

With the following definition,

$$\overline{CA}_{IM1,j} = CA_{IM1,j} + CAP_{IM1,j}$$

Equation (14-71) becomes

$$CAM_{IM1,j}\Delta\bar{Q}^*_{IM2,j} + \overline{CA}_{IM1,j}\Delta\bar{Q}^*_{IM1,j} = \text{RHS}_{IM1,j}$$

Thus, the matrix formulation is modified as

$$\begin{bmatrix} \overline{CA}_{2,j} & CAP_{2,j} & & & \\ CAM_{3,j} & CA_{3,j} & CAP_{3,j} & & \\ & & & & \\ & CAM_{IM2,j} & CA_{IM2,j} & CAP_{IM2,j} \\ & & CAM_{IM1,j} & \overline{CA}_{IM1,j} \end{bmatrix} \begin{bmatrix} \Delta\bar{Q}^*_{2,j} \\ \Delta\bar{Q}^*_{3,j} \\ | \\ \Delta\bar{Q}^*_{IM2,j} \\ \Delta\bar{Q}^*_{IM1,j} \end{bmatrix} = \begin{bmatrix} \text{RHS}_{2,j} \\ \text{RHS}_{3,j} \\ | \\ \text{RHS}_{IM2,j} \\ \text{RHS}_{IM1,j} \end{bmatrix} \tag{14-72}$$

Now consider Equation (14-62). This equation is applied along each constant ξ line for all j from $j = 2$ to $j = JM1$. Thus,

$$j = 2: \quad CBM_{i,2}\Delta\bar{Q}_{i,1} + CB_{i,2}\Delta\bar{Q}_{i,2} + CBP_{i,2}\Delta\bar{Q}_{i,3} = \Delta\bar{Q}^*_{i,2} \tag{14-73}$$

$$j = 3: \quad CBM_{i,3}\Delta\bar{Q}_{i,2} + CB_{i,3}\Delta\bar{Q}_{i,3} + CBP_{i,3}\Delta\bar{Q}_{i,4} = \Delta\bar{Q}^*_{i,3} \tag{14-74}$$

$$\vdots$$

$$j = JM2: \quad CBM_{i,JM2}\Delta\bar{Q}_{i,JM3} + CB_{i,JM2}\Delta\bar{Q}_{i,JM2} + CBP_{i,JM2}\Delta\bar{Q}_{i,JM1} = \Delta\bar{Q}^*_{i,JM2} \tag{14-75}$$

$$j = JM1: \quad CBM_{i,JM1}\Delta\bar{Q}_{i,JM2} + CB_{i,JM1}\Delta\bar{Q}_{i,JM1} + CBP_{i,JM1}\Delta\bar{Q}_{i,JM} = \Delta\bar{Q}^*_{i,JM1} \tag{14-76}$$

Before these equations are written in a matrix form, consider the equations at $j = 2$ and $j = JM1$. Boundary conditions at $j = 1$ (wall) and $j = JM$ (freestream) appear in these equations. At the wall, the no-slip condition requires

$$u = 0 \quad , \quad v = 0$$

Thus the total energy is reduced to the internal energy, i.e.,

$$e_t = e + \frac{1}{2}(u^2 + v^2) = e$$

Therefore, the unknown vector $\Delta\bar{Q}$ at the wall is

$$\Delta\bar{Q}_{i,1} = \begin{bmatrix} \Delta(\bar{\rho}) \\ 0 \\ 0 \\ \Delta(\overline{\rho e}) \end{bmatrix} \tag{14-77}$$

Usually, the pressure gradient normal to the surface is set to zero; thus,

$$\frac{\partial p}{\partial n} = 0$$

which results in

$$p_{i,1} = p_{i,2}$$

For a constant temperature wall, the internal energy is expressed as

$$e = \frac{T_w}{\gamma(\gamma - 1)M_\infty^2} \tag{14-78}$$

where T_w is the nondimensionalized wall temperature.

The simplest procedure is to explicitly implement the wall boundary condition. Thus, Equation (14-73) at $j = 2$ is written as

$$CB_{i,2}\Delta\bar{Q}_{i,2} + CBP_{i,2}\Delta\bar{Q}_{i,3} = \Delta\bar{Q}^*_{i,2} - CBM_{i,2}\Delta\bar{Q}_{i,1}$$

Once the solution is obtained, ρ and e at the wall are updated according to the following

$$e_{i,1} = \frac{(T_w)_i}{\gamma(\gamma - 1)M_\infty^2} \quad \text{and,} \quad \rho_{i,1} = \frac{p_{i,1}}{(\gamma - 1)e_{i,1}}$$

which may be modified as

$$\rho_{i,1} = \frac{p_{i,2}}{\gamma - 1}\frac{\gamma(\gamma - 1)M_\infty^2}{(T_w)_i} = \frac{\gamma M_\infty^2}{(T_w)_i}p_{i,2} \tag{14-79}$$

For an adiabatic wall condition

$$T_{i,1} = T_{i,2}$$

In addition, imposing the zero pressure gradient at the wall yields

$$p_{i,1} = p_{i,2}$$

As a result,

$$\rho_{i,1} = \rho_{i,2} \tag{14-80}$$

and

$$(\rho e_t)_{i,1} = \frac{p_{1,2}}{\gamma - 1} \tag{14-81}$$

After the solution is obtained, Equations (14-80) and (14-81) are used to update the values at the wall. At $j = JM$, which is set at the freestream,

$$\Delta \bar{Q}_{i,JM} = 0$$

Thus, the equations are written in a matrix formulation as

$$\begin{bmatrix} CB_{i,2} & CBP_{i,2} & & & \\ CBM_{i,3} & CB_{i,3} & CBP_{i,3} & & \\ & & & & \\ & CBM_{i,JM2} & CB_{i,JM2} & CBP_{i,JM2} \\ & & CBM_{i,JM1} & CB_{i,JM1} \end{bmatrix} \begin{bmatrix} \Delta \bar{Q}_{i,2} \\ \Delta \bar{Q}_{i,3} \\ | \\ \Delta \bar{Q}_{i,JM2} \\ \Delta \bar{Q}_{i,JM1} \end{bmatrix} =$$

$$\begin{bmatrix} \Delta \bar{Q}^*_{i,2} - CBM_{i,2} \Delta \bar{Q}_{i,1} \\ \Delta \bar{Q}^*_{i,3} \\ | \\ \Delta \bar{Q}^*_{i,JM2} \\ \Delta \bar{Q}^*_{i,JM1} \end{bmatrix} \tag{14-82}$$

14.4.3.2 Higher-Order Flux-Vector Splitting: In the previous method, first-order approximations for the convective terms were utilized. The dissipation error associated with the first-order relation may require the use of higher-order approximations. For this purpose, a second-order or a third-order approximation may be used. For example, the backward approximations which may be considered are as follows.

$$\frac{\partial f}{\partial x} = \frac{f_{i-2} - 4f_{i-1} + 3f_i}{2\Delta x} + O(\Delta x)^2 \tag{14-83}$$

or

$$\frac{\partial f}{\partial x} = \frac{f_{i+1} - f_{i-1}}{2\Delta x} - \frac{f_{i+1} - 3f_i + 3f_{i-1} - f_{i-2}}{6\Delta x} + O(\Delta x)^3 \tag{14-84}$$

The introduction of point $i-2$ causes some difficulty—namely, that it distorts the tridiagonal nature of the system. To overcome this difficulty, point $i-2$ and point $i+2$ which would appear in the backward and forward difference approximations may be evaluated explicitly, thus preserving the tridiagonal formulation. The modifications of Equations (14-53) and (14-54) to include the higher-order approximations given by (14-83) or (14-84) are straightforward. For example, the use of second-order approximation yields the following FDE for Equation (14-42),

$$\Bigg[I + \frac{\Delta\tau}{2\Delta\xi}\Big(A^+_{i-2,j} - 4A^+_{i-1,j} + 3A^+_{i,j} - 3A^-_{i,j} + 4A^-_{i+1,j} - A^-_{i+2,j}\Big)$$

$$- \frac{\Delta\tau}{2(\Delta\xi)^2}\Big(\hat{r}_{i+1,j}R_{i+1,j} - \hat{r}_{i,j}R_{i,j} + \hat{r}_{i-1,j}R_{i-1,j}\Big)\Bigg]\Delta\bar{Q}^* = \text{RHS} \qquad (14\text{-}85)$$

where

$$\text{RHS} = -\,\Delta\tau\Bigg\{\frac{1}{2\Delta\xi}\Big(\bar{E}^+_{i-2,j} - 4\bar{E}^+_{i-1,j} + 3\bar{E}^+_{i,j} - 3\bar{E}^-_{i,j} + 4\bar{E}^-_{i+1,j} - \bar{E}^-_{i+2,j}\Big)$$

$$+ \frac{1}{2\Delta\eta}\Big(\bar{F}^+_{i,j-2} - 4\bar{F}^+_{i,j-1} + 3\bar{F}^+_{i,j} - 3\bar{F}^-_{i,j} + 4\bar{F}^-_{i,j+1} - \bar{F}^-_{i,j+2}\Big)$$

$$- \frac{1}{2(\Delta\xi)^2}\Big(\hat{e}_{i+1,j}\hat{E}_{i+1,j} - \hat{e}_{i,j}\hat{E}_{i,j} + \hat{e}_{i-1,j}\hat{E}_{i-1,j}\Big)$$

$$- \frac{1}{2(\Delta\eta)^2}\Big(\hat{f}_{i,j+1}\hat{F}_{i,j+1} - \hat{f}_{i,j}\hat{F}_{i,j} + \hat{f}_{i,j-1}\hat{F}_{i,j-1}\Big) + \alpha\Big(\bar{H}_{i,j} - (\bar{H}_v)_{i,j}\Big)\Bigg\} \quad (14\text{-}86)$$

To retain the tridiagonal nature of the formulation, variables at $(i-2,j)$ and $(i+2,j)$ are evaluated explicitly; thus,

$$\Bigg\{-2\frac{\Delta\tau}{\Delta\xi}A^+_{i-1,j} - \frac{\Delta\tau}{2(\Delta\xi)^2}\hat{r}_{i-1,j}R_{i-1,j}\Bigg\}\Delta\bar{Q}^*_{i-1,j}$$

$$+\Bigg\{I + \frac{3\Delta\tau}{2\Delta\xi}\Big(A^+_{i,j} - A^-_{i,j}\Big) + \frac{\Delta\tau}{2(\Delta\xi)^2}\hat{r}_{i,j}R_{i,j}\Bigg\}\Delta\bar{Q}^*_{i,j}$$

$$+\Bigg\{2\frac{\Delta\tau}{\Delta\xi}A^-_{i+1,j} - \frac{\Delta\tau}{2(\Delta\xi)^2}\hat{r}_{i+1,j}R_{i+1,j}\Bigg\}\Delta\bar{Q}^*_{i+1,j} =$$

$$\text{RHS} - \frac{\Delta\tau}{2\Delta\xi}\Big[A^+_{i-2,j}\Delta\bar{Q}^{n-1}_{i-2,j} - A^-_{i+2,j}\Delta\bar{Q}^{n-1}_{i+2,j}\Big] \qquad (14\text{-}87)$$

Similarly, the finite difference equation for (14-43) becomes

$$\Bigg\{-2\frac{\Delta\tau}{\Delta\eta}B^+_{i,j-1} - \frac{\Delta\tau}{2(\Delta\eta)^2}\hat{s}_{i,j-1}S_{i,j-1}\Bigg\}\Delta\bar{Q}_{i,j-1}$$

$$+\Bigg\{I + \frac{3\Delta\tau}{2\Delta\eta}\Big(B^+_{i,j} - B^-_{i,j}\Big) + \frac{\Delta\tau}{2(\Delta\eta)^2}\hat{s}_{i,j}S_{i,j} + \alpha\Delta\tau\Big(C_{i,j} - (C_v)_{i,j}\Big)\Bigg\}\Delta\bar{Q}_{i,j}$$

$$+\left\{2\frac{\Delta\tau}{\Delta\eta}B^{-}_{i,j+1} - \frac{\Delta\tau}{2(\Delta\eta)^2}\hat{s}_{i,j+1}S_{i,j+1}\right\}\Delta\bar{Q}_{i,j+1}$$

$$= \Delta\bar{Q}^{*}_{i,j} - \frac{\Delta\tau}{2\Delta\eta}\left[B^{+}_{i,j-2}\Delta\bar{Q}^{*}_{i,j-2} - B^{-}_{i,j+2}\Delta\bar{Q}^{*}_{i,j+2}\right] \tag{14-88}$$

The convective terms in the Navier-Stokes equations may also be approximated by a second-order central difference expression. Indeed, in the PNS formulation of the previous chapter, the central difference approximation was utilized. The formulation utilizing central difference approximation of the convective terms in the Navier-Stokes equations is addressed in Sec. 14.5. However, note that if the central difference approximation is used, the addition of damping terms or TVD terms most likely will be required.

14.4.3.3 Second-Order Accuracy in Time: The next issue to consider is the accuracy of the time derivative. Previously, a first-order accurate approximation for the time derivative was used. When one is interested in obtaining a steady-state solution, a first-order scheme will be sufficient. However, when a transient solution is sought, second-order approximation of the time derivative may be required. As a result, the following relation is incorporated into the Navier-Stokes equations

$$\frac{\partial\bar{Q}^{n+1}}{\partial\tau} = \frac{\bar{Q}^{n-1} - 4\bar{Q}^{n} + 3\bar{Q}^{n+1}}{2\Delta\tau} + O(\Delta\tau)^2$$

$$= \frac{3\Delta\bar{Q}^{n} - \Delta\bar{Q}^{n-1}}{2\Delta\tau} + O(\Delta\tau)^2 \tag{14-89}$$

where

$$\Delta\bar{Q}^{n} = \bar{Q}^{n+1} - \bar{Q}^{n}$$

$$\Delta\bar{Q}^{n-1} = \bar{Q}^{n} - \bar{Q}^{n-1}$$

It is obvious that two sets of data are required to march the solution in time. As expected, this formulation will increase the memory and storage requirements. The second-order backward finite difference approximation (14-89) is introduced into Equation (14-2) to yield

$$\frac{3\Delta\bar{Q}^{n} - \Delta\bar{Q}^{n-1}}{2\Delta\tau} + \left(\frac{\partial\bar{E}}{\partial\xi}\right)^{n+1} + \left(\frac{\partial\bar{F}}{\partial\eta}\right)^{n+1} - \left(\frac{\partial\bar{E}_v}{\partial\xi}\right)^{n+1} - \left(\frac{\partial\bar{F}_v}{\partial\eta}\right)^{n+1}$$

$$+\,\alpha\left(\bar{H}^{n+1} - H_v^{n+1}\right) = 0 \tag{14-90}$$

Linearized expressions (14-32) through (14-37) are substituted into (14-90) to provide

$$\Delta\bar{Q} + \frac{2}{3}\Delta\tau\left\{\frac{\partial}{\partial\xi}(\bar{E}^n + A\Delta\bar{Q}) + \frac{\partial}{\partial\eta}(\bar{F}^n + B\Delta\bar{Q}) - \frac{\partial}{\partial\xi}(E_v^n + A_v\Delta\bar{Q})\right.$$

$$\left. - \frac{\partial}{\partial\eta}(\bar{F}_v^n + B_v\Delta\bar{Q}) + \alpha\left[\bar{H}^n + C\Delta\bar{Q} - (\bar{H}_v^n + C_v\Delta\bar{Q})\right]\right\} - \frac{1}{3}\Delta\bar{Q}^{n-1} = 0 \quad (14\text{-}91)$$

Note that the superscript n has been dropped from $\Delta\bar{Q}^n$. Now Equation (14-91) is rearranged as

$$\left\{I + \frac{2}{3}\Delta\tau\left[\frac{\partial}{\partial\xi}(A) + \frac{\partial}{\partial\eta}(B) - \frac{\partial}{\partial\xi}(A_v) - \frac{\partial}{\partial\eta}(B_v) + \alpha(C - C_v)\right]\right\}\Delta\bar{Q} =$$

$$\frac{1}{3}\Delta\bar{Q}^{n-1} - \frac{2}{3}\Delta\tau\left[\frac{\partial\bar{E}^n}{\partial\xi} + \frac{\partial\bar{F}^n}{\partial\eta} - \frac{\partial\bar{E}_v^n}{\partial\xi} - \frac{\partial\bar{F}_v^n}{\partial\eta} + \alpha(\bar{H}^n - \bar{H}_v^n)\right] \quad (14\text{-}92)$$

Equations (14-39) and (14-92), which are obtained by replacing the time derivative by a first-order or a second-order backward difference approximation, may be expressed in a combined form as

$$\left\{I + \frac{\Delta\tau}{1+\phi}\left[\frac{\partial}{\partial\xi}(A) + \frac{\partial}{\partial\eta}(B) - \frac{\partial}{\partial\xi}(A_v) - \frac{\partial}{\partial\eta}(B_v) + \alpha(C - C_v)\right]\right\}\Delta\bar{Q} =$$

$$\frac{\phi}{1+\phi}\Delta\bar{Q}^{n-1} - \frac{\Delta\tau}{1+\phi}\left[\frac{\partial\bar{E}_v^n}{\partial\xi} + \frac{\partial\bar{F}_v^n}{\partial\eta} - \frac{\partial\bar{E}^n}{\partial\xi} - \frac{\partial\bar{F}^n}{\partial\eta} + \alpha(\bar{H}^n - \bar{H}_v^n)\right] \quad (14\text{-}93)$$

where $\phi = \left\{\begin{array}{lcl} 0 & \Rightarrow & \text{First-order} \\ \frac{1}{2} & \Rightarrow & \text{Second-order} \end{array}\right\}.$

The approximation of the spatial gradients in Equation (14-92) has already been investigated.

14.4.3.4 LU Decomposition:

The linearized Navier-Stokes equation in two-spatial dimensions is given by (14-39). The flux vector splitting technique applied to the convective terms of Equation (14-39) yields

$$\left\{I + \Delta\tau\left[\frac{\partial}{\partial\xi}(A^+ + A^-) + \frac{\partial}{\partial\eta}(B^+ + B^-) - \frac{\partial A_v}{\partial\xi} - \frac{\partial B_v}{\partial\eta}\right.\right.$$

$$\left.\left. + \alpha(C - C_v)\right]\right\}\Delta\bar{Q} = -\Delta\tau\left[\frac{\partial}{\partial\xi}(\bar{E}^+ + \bar{E}^-) + \frac{\partial}{\partial\eta}(\bar{F}^+ + \bar{F}^-)\right.$$

$$\left. - \frac{\partial\bar{E}_v}{\partial\xi} - \frac{\partial\bar{F}_v}{\partial\eta} + \alpha(\bar{H} - \bar{H}_v)\right] \quad (14\text{-}94)$$

At this point, define the following difference operators:

$$\delta_x^+ f = \frac{f_{i+1} - f_i}{\Delta x} \quad : \text{ First-order forward difference operator} \tag{14-95}$$

$$\delta_x^- f = \frac{f_i - f_{i-1}}{\Delta x} \quad : \text{ First-order backward difference operator} \tag{14-96}$$

$$\delta_x f = \frac{f_{i+1} - f_{i-1}}{2\Delta x} \quad : \text{ Second-order central difference operator}$$

$$= \frac{1}{2}(\delta_x^+ + \delta_x^-) f \tag{14-97}$$

The procedure to follow is illustrated by using first-order approximation of the convective terms and second-order central difference approximation of the viscous terms. However, as discussed previously, the first-order approximation of the convective terms may introduce an excessive amount of numerical viscosity into the equation. Therefore, higher-order approximations may be required. Extension to higher-order schemes was illustrated previously and is therefore not repeated here. Returning to Equation (14-94), the finite difference formulation in terms of the operators defined by (14-95) through (14-97) becomes

$$\Big\{ I + \Delta\tau \Big[\delta_\xi^- A^+ + \delta_\xi^+ A^- + \delta_\eta^- B^+ + \delta_\eta^+ B^- - \delta_\xi A_v - \delta_\eta B_v + \alpha (C - C_v) \Big] \Big\} \Delta \bar{Q} = -\Delta\tau \Big[\delta_\xi^- \bar{E}^+ + \delta_\xi^+ \bar{E}^- + \delta_\eta^- \bar{F}^+ + \delta_\eta^+ \bar{F}^- - \delta_\xi \bar{E}_v - \delta_\eta \bar{F}_v + \alpha (\bar{H} - \bar{H}_v) \Big] \tag{14-98}$$

Note that backward difference approximation is used for the positive values, whereas forward difference approximation is used for the negative values.

Substitution of (14-97) and rearranging terms yields

$$\Big\{ I + \Delta\tau \Big[\delta_\xi^- A^+ + \delta_\eta^- B^+ - \frac{1}{2}\delta_\xi^- A_v - \frac{1}{2}\delta_\eta^- B_v \Big] + \Delta\tau \Big[\delta_\xi^+ A^- + \delta_\eta^+ B^- - \frac{1}{2}\delta_\xi^+ A_v - \frac{1}{2}\delta_\eta^+ B_v \Big] + \alpha \Delta\tau (C - C_v) \Big\} \Delta \bar{Q} = \text{RHS} \tag{14-99}$$

Note that RHS is evaluated at the known time level. Now, Equation (14-99) is factored as

$$\Big\{ I + \Delta\tau \Big[\delta_\xi^- A^+ + \delta_\eta^- B^+ - \frac{1}{2}\delta_\xi^- A_v - \frac{1}{2}\delta_\eta^- B_v \Big] \Big\} \Big\{ I + \Delta\tau \Big[\delta_\xi^+ A^- + \delta_\eta^+ B^- - \frac{1}{2}\delta_\xi^+ A_v - \frac{1}{2}\delta_\eta^+ B_v \Big] + \alpha \Delta\tau (C - C_v) \Big\} \Delta \bar{Q} = \text{RHS} \tag{14-100}$$

and is split as

$$\left\{I + \Delta\tau\left[\delta_\xi^- A^+ + \delta_\eta^- B^+ - \frac{1}{2}\delta_\xi^- A_v - \frac{1}{2}\delta_\eta^- B_v\right]\right\}\Delta\bar{Q}^* = \text{RHS} \tag{14-101}$$

and

$$\left\{I + \Delta\tau\left[\delta_\xi^+ A^- + \delta_\eta^+ B^- - \frac{1}{2}\delta_\xi^+ A_v - \frac{1}{2}\delta_\eta^+ B_v\right] + \alpha\Delta\tau(C - C_v)\right\}\Delta\bar{Q} = \Delta\bar{Q}^* \tag{14-102}$$

Substitution of the finite difference approximation given by (14-96) yields

$$\left\{I + \Delta\tau\left[\frac{A_{i,j}^+ - A_{i-1,j}^+}{\Delta\xi} + \frac{B_{i,j}^+ - B_{i,j-1}^+}{\Delta\eta} - \frac{1}{2}\frac{(A_v)_{i,j} - (A_v)_{i-1,j}}{\Delta\xi} - \frac{1}{2}\frac{(B_v)_{i,j} - (B_v)_{i,j-1}}{\Delta\eta}\right]\right\}\Delta\bar{Q}^* = \text{RHS}_{i,j} \tag{14-103}$$

which may be rearranged as

$$\frac{\Delta\tau}{\Delta\xi}\left[-A_{i-1,j}^+ + \frac{1}{2}(A_v)_{i-1,j}\right]\Delta\bar{Q}_{i-1,j}^* + \left\{I + \frac{\Delta\tau}{\Delta\xi}\left[A_{i,j}^+ - \frac{1}{2}(A_v)_{i,j}\right] + \frac{\Delta\tau}{\Delta\eta}\left[B_{i,j}^+ - \frac{1}{2}(B_v)_{i,j}\right]\right\}\Delta\bar{Q}_{i,j}^* + \frac{\Delta\tau}{\Delta\eta}\left[-B_{i,j-1}^+ + \frac{1}{2}(B_v)_{i,j-1}\right]\Delta\bar{Q}_{i,j-1}^* = \text{RHS}_{i,j} \tag{14-104}$$

This equation may be written in a compact form as

$$APL_{i,j}\Delta\bar{Q}_{i-1,j}^* + AP_{i,j}\Delta\bar{Q}_{i,j}^* + APR_{i,j}\Delta\bar{Q}_{i,j-1}^* = \text{RHS}_{i,j} \tag{14-105}$$

where

$$APL = \frac{\Delta\tau}{\Delta\xi}\left[-A_{i-1,j}^+ + \frac{1}{2}(A_v)_{i-1,j}\right] \tag{14-106}$$

$$AP = I + \frac{\Delta\tau}{\Delta\xi}\left[A_{i,j}^+ - \frac{1}{2}(A_v)_{i,j}\right] + \frac{\Delta\tau}{\Delta\eta}\left[B_{i,j}^+ - \frac{1}{2}(B_v)_{i,j}\right] \tag{14-107}$$

$$APR = \frac{\Delta\tau}{\Delta\eta}\left[-B_{i,j-1}^+ + \frac{1}{2}(B_v)_{i,j-1}\right] \tag{14-108}$$

Similarly, the finite difference formulation of Equation (14-102) is expressed as

$$AML_{i,j}\Delta\bar{Q}_{i+1,j} + AM_{i,j}\Delta\bar{Q}_{i,j} + AMR_{i,j}\Delta\bar{Q}_{i,j+1} = \Delta\bar{Q}_{i,j}^* \tag{14-109}$$

where the coefficients are defined by

$$AML = \frac{\Delta\tau}{\Delta\xi}\Big[A^-_{i+1,j} - \frac{1}{2}(A_v)_{i+1,j}\Big] \tag{14-110}$$

$$AM = \Big\{I + \frac{\Delta\tau}{\Delta\xi}\Big[-A^-_{i,j} + \frac{1}{2}(A_v)_{i,j}\Big] + \frac{\Delta\tau}{\Delta\eta}\Big[-B^-_{i,j} + \frac{1}{2}(B_v)_{i,j}\Big]$$
$$+\,\alpha\Delta\tau\Big[C_{i,j} - (C_v)_{i,j}\Big]\Big\} \tag{14-111}$$

$$AMR = \frac{\Delta\tau}{\Delta\eta}\Big[B^-_{i,j+1} - \frac{1}{2}(B_v)_{i,j+1}\Big] \tag{14-112}$$

Once Equation (14-105) is written for each grid point, only one unknown per equation appears. Therefore, Equation (14-105) is solved point by point, sweeping along each constant ξ-line, i.e.,

$$\Delta\bar{Q}^*_{i,j} = (AP_{i,j})^{-1}\Big[-APL_{i,j}\Delta\bar{Q}^*_{i-1,j} - APR_{i,j}\Delta\bar{Q}^*_{i,j-1} + \text{RHS}_{i,j}\Big] \tag{14-113}$$

Subsequently, Equation (14-109) is solved point by point along lines of constant η, i.e.,

$$\Delta\bar{Q}_{i,j} = (AM_{i,j})^{-1}\Big[-AML_{i,j}\Delta\bar{Q}_{i+1,j} - AMR_{i,j}\Delta\bar{Q}_{i,j+1} + \Delta\bar{Q}^*_{i,j}\Big] \tag{14-114}$$

Specification and application of boundary conditions are similar to the ones discussed in the previous section.

14.5 Extension to Three Dimensions

In the previous sections, the two-dimensional/axisymmetric Navier-Stokes equations were investigated. A selected number of commonly used numerical schemes, used to solve the Navier-Stokes equation, along with the implementation of boundary conditions, were introduced. In this section, the procedures are extended to three dimensions. Recall that the governing equation in vector form and in the computational space is given by

$$\frac{\partial\bar{Q}}{\partial t} + \frac{\partial\bar{E}}{\partial\xi} + \frac{\partial\bar{F}}{\partial\eta} + \frac{\partial\bar{G}}{\partial\zeta} = \frac{\partial\bar{E}_v}{\partial\xi} + \frac{\partial\bar{F}_v}{\partial\eta} + \frac{\partial\bar{G}_v}{\partial\zeta} \tag{14-115}$$

where all the quantities are non-dimensional. The Navier-Stokes equation given by (14-115) can be reduced to the thin-layer Navier-Stokes equation as

$$\frac{\partial \bar{Q}}{\partial t} + \frac{\partial \bar{E}}{\partial \xi} + \frac{\partial \bar{F}}{\partial \eta} + \frac{\partial \bar{G}}{\partial \zeta} = \frac{\partial \bar{F}_{vT}}{\partial \eta} \tag{14-116}$$

where the modified viscous flux vector $\bar{F}_v$, used in the thin-layer formulation, is redefined by $\bar{F}_{vT}$ and given by (11-156). The numerical schemes explored in this section are primarily applied to the thin-layer equation. The extension to Navier-Stokes equations is straightforward, since the gradient of viscous terms in almost all schemes is approximated by second-order central difference expressions. Initially, an explicit finite difference formulation based on the Steger and Warming flux vector splitting technique is investigated. Subsequently, typical implicit formulations are explored. Consistent with previous chapters, the indices, i, j, and k are used to denote grid points associated with x, y, and z in the physical space and with ξ, η, and ζ in the computational space.

14.5.1 Explicit Flux Vector Splitting Scheme

Application of a first-order forward difference approximation to Equation (14-116) provides

$$\frac{\bar{Q}^{n+1}_{i,j,k} - \bar{Q}^{n}_{i,j,k}}{\Delta t} = -\left(\frac{\partial \bar{E}}{\partial \xi} + \frac{\partial \bar{F}}{\partial \eta} + \frac{\partial \bar{G}}{\partial \zeta}\right)^n_{i,j,k} + \left(\frac{\partial \bar{F}_{vT}}{\partial \eta}\right)^n_{i,j,k}$$

The formulation is written in a delta form as

$$\Delta \bar{Q}_{i,j,k} = \Delta t \left[-\left(\frac{\partial \bar{E}}{\partial \xi} + \frac{\partial \bar{F}}{\partial \eta} + \frac{\partial \bar{G}}{\partial \zeta}\right)^n_{i,j,k} + \left(\frac{\partial \bar{F}_{vT}}{\partial \eta}\right)^n_{i,j,k}\right] \tag{14-117}$$

The Steger and Warming flux vector splitting scheme introduced in Chapter 12 is utilized to approximate the convective terms. To generalize the formulation, both the first-order and the second-order expressions are included. The inviscid flux vectors are decomposed into positive and negative vectors associated with the corresponding eigenvalues. The mathematical procedure is similar to that of the two-dimensional case introduced previously. Therefore, only the details of finite difference approximation for the three dimensions is considered. The procedure is illustrated for the flux vector $\bar{E}$. Thus

$$\left(\frac{\partial \bar{E}}{\partial \xi}\right)_{i,j,k} = \left(\frac{\partial \bar{E}^+}{\partial \xi} + \frac{\partial \bar{E}^-}{\partial \xi}\right)_{i,j,k}$$

The corresponding finite difference formulations are given by

$$\frac{\partial \bar{E}^+}{\partial \xi} = \begin{cases} \dfrac{\bar{E}^+_{i,j,k} - \bar{E}^+_{i-1,j,k}}{\Delta\xi} & : \text{ First-order} \\ \dfrac{3\bar{E}^+_{i,j,k} - 4\bar{E}^+_{i-1,j,k} + \bar{E}^+_{i-2,j,k}}{2\Delta\xi} & : \text{ Second-order} \end{cases}$$

and

$$\frac{\partial \bar{E}^-}{\partial \xi} = \begin{cases} \dfrac{\bar{E}^-_{i+1,j,k} - \bar{E}^-_{i,j,k}}{\Delta\xi} & : \text{ First-order} \\ \dfrac{-3\bar{E}^-_{i,j,k} + 4\bar{E}^-_{i+1,j,k} - \bar{E}^-_{i+2,j,k}}{2\Delta\xi} & : \text{ Second-order} \end{cases}$$

The convective terms $\dfrac{\partial \bar{F}}{\partial \eta}$ and $\dfrac{\partial \bar{G}}{\partial \zeta}$ are represented in a similar fashion. To proceed with the flux vector splitting scheme, the eigenvalues and the associated eigenvectors of the Jacobian matrices A, B, and C must be determined. Following the procedure discussed in Chapter 11, the eigenvalues of A are determined to be

$$\lambda_{\xi 1} = \lambda_{\xi 2} = \lambda_{\xi 3} = \xi_t + U \tag{14-118}$$

$$\lambda_{\xi 4} = \xi_t + U + a\sqrt{a_4} \tag{14-119}$$

$$\lambda_{\xi 5} = \xi_t + U - a\sqrt{a_4} \tag{14-120}$$

where the contravariant velocity U is given by (11-90a) and $a_4 = \xi_x^2 + \xi_y^2 + \xi_z^2$ as defined by (11-101). The associated eigenvectors for the repeated eigenvalues may be expressed as

$$X_{A1} = X_{A2} = X_{A3} = \begin{bmatrix} \alpha_1 \\ \dfrac{1}{\xi_x}\left[(\alpha_1 w - \alpha_2)\xi_z + (\alpha_1 v - \alpha_3)\xi_y + \alpha_1 u \xi_x\right] \\ \alpha_3 \\ \alpha_2 \\ \dfrac{1}{\xi_x}\Big[2\alpha_1 uw - 2\alpha_2 u)\xi_z + (2\alpha_1 uv - 2\alpha_3 u)\xi_y \\ \quad + (-\alpha_1 w^2 + 2\alpha_2 w - \alpha_1 v^2 + 2\alpha_3 v + \alpha_1 u^2)\xi_x\Big] \end{bmatrix}$$

where α_1, α_2, and α_3 are arbitrary constants. For each set of values of these constants — (0,0,1), (1,0,0), and (0,1,0) — three eigenvectors associated with each of the repeated eigenvalues can be obtained. The eigenvector associated with the fourth eigenvalue $\lambda_{\xi 4}$ is determined to be

$$X_{A4} = \begin{bmatrix} 1 \\ \frac{\xi_x}{\sqrt{a_4}} a + u \\ \frac{\xi_y}{\sqrt{a_4}} a + v \\ \frac{\xi_z}{\sqrt{a_4}} a + w \\ \frac{Ua}{\sqrt{a_4}} + \frac{1}{2}(q^2 + 2\gamma e) \end{bmatrix}$$

Similarly, the eigenvector associated with the eigenvalue $\lambda_{\xi 5}$ is

$$X_{A5} = \begin{bmatrix} 1 \\ -\frac{\xi_x}{\sqrt{a_4}} a + u \\ -\frac{\xi_y}{\sqrt{a_4}} a + v \\ -\frac{\xi_z}{\sqrt{a_4}} a + w \\ -\frac{Ua}{\sqrt{a_4}} + \frac{1}{2}(q^2 + 2\gamma e) \end{bmatrix}$$

The eigenvector matrix associated with Jacobian matrix A is

$$X_A = \begin{bmatrix} 0 & 1 & 0 & 1 & 1 \\ -\frac{\xi_y}{\xi_x} & \frac{U}{\xi_x} & -\frac{\xi_z}{\xi_x} & u + \frac{\xi_x a}{\sqrt{a_4}} & u - \frac{\xi_x a}{\sqrt{a_4}} \\ 1 & 0 & 0 & v + \frac{\xi_y a}{\sqrt{a_4}} & v - \frac{\xi_y a}{\sqrt{a_4}} \\ 0 & 0 & 1 & w + \frac{\xi_z a}{\sqrt{a_4}} & w - \frac{\xi_z a}{\sqrt{a_4}} \\ X_{A(5,1)} & X_{A(5,2)} & X_{A(5,3)} & X_{A(5,4)} & X_{A(5,5)} \end{bmatrix} \tag{14-121}$$

where

$$X_{A(5,1)} = -\frac{-\xi_y u + \xi_x v}{\xi_x}$$

$$X_{A(5,2)} = \frac{u(\xi_y v + \xi_z w)}{\xi_x} - \frac{1}{2}(v^2 + w^2 - u^2)$$

$$X_{A(5,3)} = \frac{\xi_x w - \xi_z u}{\xi_x}$$

$$X_{A(5,4)} = \gamma e + \frac{1}{2}q^2 + \frac{aU}{\sqrt{a_4}}$$

$$X_{A(5,5)} = \gamma e + \frac{1}{2}q^2 - \frac{aU}{\sqrt{a_4}}$$

and the inverse X_A^{-1} is

$$X_A^{-1} = \begin{bmatrix} \frac{\xi_y U}{a_4} - \frac{v}{2\gamma e}q^2 & \frac{uv}{\gamma e} - \frac{\xi_x \xi_y}{a_4} & \frac{v^2}{\gamma e} + \frac{\xi_x^2 + \xi_y^2}{a_4} & \frac{vw}{\gamma e} - \frac{\xi_y \xi_z}{a_4} & -\frac{v}{2\gamma e} \\ 1 - \frac{q^2}{2\gamma e} & \frac{u}{\gamma e} & \frac{v}{\gamma e} & \frac{w}{\gamma e} & -\frac{1}{\gamma e} \\ \frac{\xi_z U}{a_4} - \frac{w}{2\gamma e}q^2 & \frac{uw}{\gamma e} - \frac{\xi_x \xi_z}{a_4} & \frac{vw}{\gamma e} - \frac{\xi_y \xi_z}{a_4} & \frac{w^2}{\gamma e} + \frac{\xi_x^2 + \xi_y^2}{a_4} & -\frac{w}{\gamma e} \\ -\frac{U}{2a\sqrt{a_4}} + \frac{1}{4\gamma e}q^2 & \frac{\xi_x}{2a\sqrt{a_4}} - \frac{u}{2\gamma e} & \frac{\xi_z}{2a\sqrt{a_4}} - \frac{v}{2\gamma e} & \frac{\xi_z}{2a\sqrt{a_4}} - \frac{w}{2\gamma e} & \frac{1}{\gamma e} \\ \frac{U}{2a\sqrt{a_4}} + \frac{1}{4\gamma e}q^2 & -\frac{\xi_x}{2a\sqrt{a_4}} - \frac{u}{2\gamma e} & -\frac{\xi_y}{2a\sqrt{a_4}} - \frac{v}{2\gamma e} & -\frac{\xi_z}{2a\sqrt{a_4}} - \frac{w}{2\gamma e} & \frac{1}{\gamma e} \end{bmatrix} \tag{14-122}$$

Now that thc eigenvalues, eigenvectors, and the associated eigenvector matrix have been identified, one may proceed with the flux splitting. For this purpose, the following combinations of the eigenvalues are investigated:

(a) If $\lambda_{\xi 1} > 0$, then obviously $\lambda_{\xi 2}$ and $\lambda_{\xi 3}$ are positive due to (14-118). Furthermore, $\lambda_{\xi 4}$ will be positive as well, while $\lambda_{\xi 5}$ could be either positive or negative. Thus the following two categories must be considered.

(i) If $\lambda_{\xi 5} > 0$, then all the eigenvalues of A are positive and

$$A^+ = A \quad , \quad \bar{E}^+ = \bar{E}$$

$$A^- = 0 \quad , \quad \bar{E}^- = 0$$

(ii) If $\lambda_{\xi 5} < 0$, then

$$A^- = [X_A] \begin{bmatrix} 0 & & & & \\ & 0 & & & \\ & & 0 & & \\ & & & 0 & \\ & & & & \lambda_{\xi 5} \end{bmatrix} [X_A^{-1}] = \lambda_{\xi 5} [X_A] \begin{bmatrix} 0 & & & & \\ & 0 & & & \\ & & 0 & & \\ & & & 0 & \\ & & & & 1 \end{bmatrix} [X_A^{-1}]$$

and

$$A^+ = A - A^-$$

Subsequently, the flux vector $\bar{E}^-$ is determined as follows

$$\bar{E}^- = [A^-]\bar{Q} = [A^-]\left[\frac{\rho}{J}\right]\begin{bmatrix} 1 \\ u \\ v \\ w \\ e + \frac{1}{2}q^2 \end{bmatrix} = \lambda_{\xi 5}\frac{\rho}{J}\begin{bmatrix} \frac{1}{2\gamma} \\ \frac{u}{2\gamma} - \frac{a\xi_x}{2\gamma\sqrt{a_4}} \\ \frac{v}{2\gamma} - \frac{a\xi_y}{2\gamma\sqrt{a_4}} \\ \frac{w}{2\gamma} - \frac{a\xi_z}{2\gamma\sqrt{a_4}} \\ \frac{2\gamma e + q^2}{4\gamma} - \frac{aU}{2\gamma\sqrt{a_4}} \end{bmatrix}$$

or

$$\bar{E}^- = \lambda_{\xi 5}\frac{\rho}{2\gamma J}\begin{bmatrix} 1 \\ u - \frac{a\xi_x}{\sqrt{a_4}} \\ v - \frac{a\xi_y}{\sqrt{a_4}} \\ w - \frac{a\xi_z}{\sqrt{a_4}} \\ \frac{q^2}{2} + \frac{a^2}{(\gamma - 1)} - \frac{aU}{\sqrt{a_4}} \end{bmatrix} \tag{14-123}$$

and

$$\bar{E}^+ = \bar{E} - \bar{E}^-$$

(b) If $\lambda_{\xi 1}$, $\lambda_{\xi 2}$, $\lambda_{\xi 3} \leq 0$, then, $\lambda_{\xi 5}$ is also negative. However, the fourth eigenvalue $\lambda_{\xi 4}$ could be either positive or negative. Thus, consider the following two cases.

(i) If $\lambda_{\xi 4} < 0$, then all the eigenvalues are negative and

$$A^+ = 0 \quad , \quad \bar{E}^+ = 0$$

$$A^- = A \quad , \quad \bar{E}^- = \bar{E}$$

(ii) If $\lambda_{\xi 4} > 0$, A^+ is evaluated as follows.

$$A^+ = [X_A]\begin{bmatrix} 0 & & & & \\ & 0 & & & \\ & & 0 & & \\ & & & \lambda_{\xi 4} & \\ & & & & 0 \end{bmatrix}[X_A^{-1}] = \lambda_{\xi 4}\,[X_A]\begin{bmatrix} 0 & & & & \\ & 0 & & & \\ & & 0 & & \\ & & & 1 & \\ & & & & 0 \end{bmatrix}[X_A^{-1}]$$

and

$$A^- = A - A^+$$

Subsequently, the flux vectors $\bar{E}^+$ and $\bar{E}^-$ are determined by

$$\bar{E}^+ = [A^+]\bar{Q} = \lambda_{\xi 4}\frac{\rho}{2\gamma J}\begin{bmatrix} 1 \\ u + \dfrac{a\xi_x}{\sqrt{a_4}} \\ v + \dfrac{a\xi_y}{\sqrt{a_4}} \\ w + \dfrac{a\xi_z}{\sqrt{a_4}} \\ \dfrac{q^2}{2} + \dfrac{a^2}{(\gamma - 1)} + \dfrac{aU}{\sqrt{a_4}} \end{bmatrix} \tag{14-124}$$

and

$$\bar{E}^- = \bar{E} - \bar{E}^+$$

The flux vectors $\bar{F}^+$, $\bar{F}^-$, $\bar{G}^+$, and $\bar{G}^-$ have similar forms to those defined for $\bar{E}^+$ and $\bar{E}^-$. Simply replace ξ by η, the contravariant velocity U by V, and a_4 by b_4, i.e., (11-108). Thus, the eigenvalues ($\lambda_{\eta 1}$, $\lambda_{\eta 2}$, $\lambda_{\eta 3}$, $\lambda_{\eta 4}$, and $\lambda_{\eta 5}$), the eigenvector matrix and its inverse (X_B and X_B^{-1}), the decomposed matrices (B^+ and B^-), and the flux vectors ($\bar{F}^+$ and $\bar{F}^-$) are known. Similarly, the eigenvalues ($\lambda_{\zeta 1}$, $\lambda_{\zeta 2}$, $\lambda_{\zeta 3}$, $\lambda_{\zeta 4}$, and $\lambda_{\zeta 5}$), the eigenvector matrix and its inverse (X_C and X_C^{-1}), the decomposed matrices (C^+ and C^-), and the flux vectors ($\bar{G}^+$ and $\bar{G}^-$) are obtained by replacing the following: ξ by ζ, the contravariant velocity U by W, and a_4 by c_4, as given by (11-115). At this point, the splitting of the convective terms is completed. Next, the appropriate forward or backward finite difference approximations are used, thus providing the left-hand side of the finite difference equation. The right-hand side of the Navier-Stokes equation includes the viscous terms which are approximated by the second-order central difference expression applied at mid point. This procedure was introduced previously. However, due to its importance, it is repeated here and applied to the viscous term $\bar{F}_{vT}$. As discussed earlier, the elements of flux vector $\bar{F}_{vT}$ can be written in a generalized form as

$$\frac{\partial}{\partial \eta}(L^m M_\eta^m) \tag{14-125}$$

For example, the first element of the second component of viscous flux vector $\bar{F}_{vT}$ is $\dfrac{\mu}{Re_\infty J} b_1 u_\eta$ and its gradient is

$$\frac{\partial}{\partial \eta}\left(\frac{\mu}{Re_\infty J} b_1 u_\eta\right) = \frac{\partial}{\partial \eta}\left[\left(\frac{\mu}{Re_\infty J} b_1\right)(u_\eta)\right]$$

According to the generalized relation (14-125), one has

$$L = \frac{\mu}{Re_\infty J} b_1 \quad \text{and} \quad M = u$$

To simplify the mathematics, rewrite the viscous flux vector $\bar{F}_{vT}$ given by (11-156) in a generalized form in terms of the parameters L and M. Thus,

$$\bar{F}_{vT} = \begin{bmatrix} 0 \\ \sum\limits_{m=1}^{3} (L^m M_\eta^m) \\ \sum\limits_{m=4}^{6} (L^m M_\eta^m) \\ \sum\limits_{m=7}^{9} (L^m M_\eta^m) \\ \sum\limits_{m=10}^{16} (L^m M_\eta^m) \end{bmatrix} \tag{14-126}$$

where

m	1	2	3	4	5	6	7	8	9
$L^m / \left(\dfrac{\mu}{Re_\infty J}\right)$	b_1	b_5	b_7	b_5	b_2	b_6	b_7	b_6	b_3
M^m	u	v	w	u	v	w	u	v	w

m	10	11	12	13	14	15	16
$L^m / \left(\dfrac{\mu}{Re_\infty J}\right)$	$\frac{1}{2} b_1$	$\frac{1}{2} b_2$	$\frac{1}{2} b_3$	b_5	b_6	b_7	$\frac{\gamma}{Pr} b_4$
M^m	u^2	v^2	w^2	uv	vw	uw	e

Now the viscous gradient $\dfrac{\partial \bar{F}_{vT}}{\partial \eta}$ appearing on the right-hand side of the thin-layer Navier-Stokes equation is approximated by a second-order central difference expression applied at the mid point. Thus, the general expression becomes

$$\frac{\partial}{\partial\eta}(L^m M^m_\eta)_{i,j,k} = \frac{(L^m M^m_\eta)_{i,j+\frac{1}{2},k} - (L^m M^m_\eta)_{i,j-\frac{1}{2},k}}{2\left(\frac{\Delta\eta}{2}\right)}$$

$$= \frac{\frac{1}{2}(L^m_{i,j,k} + L^m_{i,j+1,k})\frac{M^m_{i,j+1,k} - M^m_{i,j,k}}{2(\Delta\eta/2)} - \frac{1}{2}(L^m_{i,j,k} + L^m_{i,j-1,k})\frac{M^m_{i,j,k} - M^m_{i,j-1,k}}{2(\Delta\eta/2)}}{\Delta\eta}$$

$$= \frac{1}{2(\Delta\eta)^2}\Big[(L^m_{i,j,k} + L^m_{i,j+1,k})(M^m_{i,j+1,k}) - (L^m_{i,j-1,k} + 2L^m_{i,j,k} + L^m_{i,j+1,k})(M^m_{i,j,k})$$

$$+(L^m_{i,j,k} + L^m_{i,j-1,k})(M^m_{i,j-1,k})\Big]$$

Consistent with the previous notation, define

$$\hat{L}_{i,j+1,k} = L_{i,j,k} + L_{i,j+1,k} \tag{14-127}$$

$$\hat{L}_{i,j,k} = L_{i,j-1,k} + 2L_{i,j,k} + L_{i,j+1,k} \tag{14-128}$$

$$\hat{L}_{i,j-1,k} = L_{i,j,k} + L_{i,j-1,k} \tag{14-129}$$

Therefore,

$$\frac{\partial}{\partial\eta}(L^m M^m_\eta)_{i,j,k} = \frac{1}{2(\Delta\eta)^2}\left[\hat{L}^m_{i,j+1,k}M^m_{i,j+1,k} - \hat{L}^m_{i,j,k}M^m_{i,j,k} + \hat{L}^m_{i,j-1,k}M^m_{i,j-1,k}\right] \tag{14-130}$$

Thus, the viscous gradient is expressed by

$$\left(\frac{\partial \bar{F}_{vT}}{\partial\eta}\right)_{i,j,k} = \frac{1}{2(\Delta\eta)^2}\begin{bmatrix} 0 \\ \sum_{m=1}^{3}\left(\hat{L}^m_{i,j+1,k}M^m_{i,j+1,k} - \hat{L}^m_{i,j,k}M^m_{i,j,k} + \hat{L}^m_{i,j-1,k}M^m_{i,j-1,k}\right) \\ \sum_{m=4}^{6}\left(\hat{L}^m_{i,j+1,k}M^m_{i,j+1,k} - \hat{L}^m_{i,j,k}M^m_{i,j,k} + \hat{L}^m_{i,j-1,k}M^m_{i,j-1,k}\right) \\ \sum_{m=7}^{9}\left(\hat{L}^m_{i,j+1,k}M^m_{i,j+1,k} - \hat{L}^m_{i,j,k}M^m_{i,j,k} + \hat{L}^m_{i,j-1,k}M^m_{i,j-1,k}\right) \\ \sum_{m=10}^{16}\left(\hat{L}^m_{i,j+1,k}M^m_{i,j+1,k} - \hat{L}^m_{i,j,k}M^m_{i,j,k} + \hat{L}^m_{i,j-1,k}M^m_{i,j-1,k}\right) \end{bmatrix} \tag{14-131}$$

Now, either first-order or second-order finite difference approximations for the convective terms, along with the approximation of the viscous term given by (14-131), are substituted into Equation (14-117) and solved for the unknown $\Delta\bar{Q}$.

14.5.2 Implicit Formulation

Consider again the Thin-Layer Navier-Stokes equation given by (14-3), i.e.,

$$\frac{\partial \bar{Q}}{\partial t} + \frac{\partial \bar{E}}{\partial \xi} + \frac{\partial \bar{F}}{\partial \eta} + \frac{\partial \bar{G}}{\partial \zeta} = \frac{\partial \bar{F}_{vT}}{\partial \eta}$$

Following the procedure described in Chapter 3, a combined formulation of first-order in time (i.e., Euler) and second-order in time (i.e., Crank-Nicolson) is expressed by

$$\frac{\Delta \bar{Q}}{\Delta t} + \beta \left[\left(\frac{\partial \bar{E}}{\partial \xi} + \frac{\partial \bar{F}}{\partial \eta} + \frac{\partial \bar{G}}{\partial \zeta} - \frac{\partial \bar{F}_{vT}}{\partial \eta} \right)^{n+1} \right] + (1-\beta) \left[\left(\frac{\partial \bar{E}}{\partial \xi} + \frac{\partial \bar{F}}{\partial \eta} + \frac{\partial \bar{G}}{\partial \zeta} - \frac{\partial \bar{F}_{vT}}{\partial \eta} \right)^{n} \right] = 0 \tag{14-132}$$

where

$\beta = 0$; the formulation is FTCS explicit

$\beta = \frac{1}{2}$; the formulation is Crank-Nicolson implicit

and

$\beta = 1$; the formulation is Euler implicit

The linearization of the nonlinear terms as introduced previously provides the following equation

$$\frac{\Delta \bar{Q}}{\Delta t} + \beta \left[\frac{\partial}{\partial \xi}(A \Delta \bar{Q}) + \frac{\partial}{\partial \eta}(B \Delta \bar{Q}) + \frac{\partial}{\partial \zeta}(C \Delta \bar{Q}) - \frac{\partial}{\partial \eta}(B_v \Delta \bar{Q}) \right] = -\left(\frac{\partial \bar{E}}{\partial \xi} + \frac{\partial \bar{F}}{\partial \eta} + \frac{\partial \bar{G}}{\partial \zeta} - \frac{\partial \bar{F}_{vT}}{\partial \eta} \right)^{n} \tag{14-133}$$

where $B_v = \dfrac{\partial \bar{F}_{vT}}{\partial \bar{Q}}$. Note that B_v can easily be obtained from (11-217) where only the terms involving η derivatives are retained.

Approximate factorization applied to (14-133) yields

$$\left\{ I + \Delta t \beta \left[\frac{\partial}{\partial \eta}(A) \right] \right\} \left\{ I + \Delta t \beta \left[\frac{\partial}{\partial \eta}(B) - \frac{\partial}{\partial \eta}(B_v) \right] \right\} \left\{ I + \Delta t \beta \left[\frac{\partial}{\partial \zeta}(C) \right] \right\} \Delta \bar{Q} = -\Delta t \left(\frac{\partial \bar{E}}{\partial \xi} + \frac{\partial \bar{F}}{\partial \eta} + \frac{\partial \bar{G}}{\partial \zeta} - \frac{\partial \bar{F}_{vT}}{\partial \eta} \right)^{n} = \text{RHS} \tag{14-134}$$

The formulation is solved sequentially in three steps as follows:

$$\left\{I + \Delta t\beta\left[\frac{\partial}{\partial\xi}(A) + D_{\text{imp}(\xi)}\right]\right\}\Delta\bar{Q}^* = \text{RHS} + D_{\text{exp}} \tag{14-135}$$

$$\left\{I + \Delta t\beta\left[\frac{\partial}{\partial\eta}(B - B_v) + D_{\text{imp}(\eta)}\right]\right\}\Delta\bar{Q}^{**} = \Delta\bar{Q}^* \tag{14-136}$$

$$\left\{I + \Delta t\beta\left[\frac{\partial}{\partial\zeta}(C) + D_{\text{imp}(\zeta)}\right]\right\}\Delta\bar{Q} = \Delta\bar{Q}^{**} \tag{14-137}$$

where second-order implicit and fourth-order explicit damping terms have been added. The fourth-order damping is generally given by

$$D_e = -\epsilon_e \Delta t \frac{1}{J}\left[(\nabla_\xi\Delta_\xi)^2 + (\nabla_\eta\Delta_\eta)^2 + (\nabla_\zeta\Delta_\zeta)^2\right] J\bar{Q}^n$$

and the second-order damping is given by

$$D_{\text{imp}(\xi)} = -\epsilon_i \Delta t \frac{1}{J}(\nabla_\xi\Delta_\xi J)$$

$$D_{\text{imp}(\eta)} = -\epsilon_i \Delta t \frac{1}{J}(\nabla_\eta\Delta_\eta J)$$

$$D_{\text{imp}(\zeta)} = -\epsilon_i \Delta t \frac{1}{J}(\nabla_\zeta\Delta_\zeta J)$$

The difference opcrators used in the equations above are associated with the central difference approximations of second- and fourth-order derivatives. For example,

$$(\nabla_\xi\Delta_\xi)Q = Q_{i+1} - 2Q_i + Q_{i-1}$$

and

$$(\nabla_\xi\Delta_\xi)^2 Q = Q_{i+2} - 4Q_{i+1} + 6Q_i - 4Q_{i-1} + Q_{i-2}$$

To maintain stability, in general $\epsilon_i \geq 2\epsilon_e$. Typically ϵ_i would be set to about $3\epsilon_e$. The specified values of ϵ_e must be as small as possible such that a stable solution can be obtained.

Typical central difference approximation of second order is applied to Equation (14-135) through (14-137), providing block tridiagonal systems which are solved sequentially.

14.6 Concluding Remarks

An attempt is made in this chapter to introduce the readers to selected finite difference equations to solve the Navier-Stokes equations. Some of the discussions were restricted primarily to Thin-Layer Navier-Stokes equations; however, extension to Navier-Stokes equations is straightforward. It is hoped that the various aspects of the numerical schemes discussed in this chapter will facilitate the understanding of other algorithms presented elsewhere.

14.7 Problems

14.1 Select a recently published journal article of your interest where the Navier-Stokes equations (or the Thin-Layer Navier-Stokes equations) are solved by a finite difference scheme. After careful review, relate the following topics in the article to the discussions of this chapter. Your discussion should address the following topics: (a) coordinate transformation, (b) grid system used, (c) linearization procedures, (d) numerical scheme, (e) boundary conditions, (f) stability requirements, and (g) specific application.

Chapter 15

Grid Generation – Unstructured Grids

15.1 Introductory Remarks

Discretization of a domain can be accomplished either directly in the physical space or on the transformed computational space. The choice will primarily depend on the numerical scheme to be utilized as well as the domains of solution. As seen previously, the finite difference equations approximating the partial differential equations are solved within a rectangular, equally spaced grid system. For non-rectangular physical domain, a coordinate transformation to computational space is required. The grid points are defined at the intersection of equally distanced parallel lines within the rectangular (2-D) or cubical (3-D) computational domain. There are corresponding grid points within the physical space established by algebraic relations or differential equations. The grid points can be easily identified and are usually designated by the indices, i, j, and k in an orderly manner along the grid lines. This type of grid is known as *structured grid,* which was the subject of Chapter 9.

In addition to finite difference schemes, two other numerical schemes are available for the solution of the conservation laws. These schemes are finite volume schemes, to be introduced in Chapter 16, and finite element schemes. Both of these schemes are integral methods, that is, the original differential equations are integrated on the physical domain and, subsequently, are solved numerically. Therefore, the grid system for the finite volume or finite element schemes are usually generated directly within the physical space. There exist various choices in the selection of the volumes or elements. Thus, the domain of solution is usually divided into triangles or quadrilaterals (or any other kind of polygon) in 2-D, whereas pyramids or tetrahedrals are used in 3-D. It is obvious that the grid points, in general, cannot be associated with grid lines. Therefore, the identification of the grid points must be individually specified. Such a grid system is known as an *unstructured grid system*. The main advantage of the unstructured grid is that it can be used easily to fit irregular, singly-connected domains, as well as multiply-connected domains. The unstructured grid also can be coupled with grid refinement techniques for the adaptive methods. However, unstructured grid generation is more difficult to program, that is, the programmer needs a sound background in the data structure arrangement and experience in the data book-keeping skills.

The objective of this chapter is to introduce the unstructured grid systems and the schemes by which they are generated. The discussion will be limited to the two-dimensional triangulation techniques, i.e., the physical domain is to be discretized with triangles. This selection is due to the fact that triangular elements are generally the most flexible shape to fit any type of boundary. The particular methods utilized to generate such a grid are the "Advancing Front" and the "Delaunay" methods. Both of these schemes can be extended to three-dimensional domains and are the most popular methods used today.

15.2 Domain Nodalization

The first step in triangulation of a physical domain is to distribute grid points within the interior domain, as well as the boundaries of the domain which will be referred to as *nodalization*. Subsequently, the interior and boundary nodes are connected to each other, forming the required elements. The detailed description of element formulation will be presented in the next section. As for nodalization, it is emphasized that it is currently an art rather than science, i.e., there is no "best" scheme for generating the required nodes. Nevertheless, the simple and direct method introduced in Ref. [15-1] will be employed to describe the fundamental concept of nodalization. It is also obvious that any scheme employed to distribute grid points within a domain by the algebraic or differential methods described in Chapter 9 can be used for this purpose. The procedure described in this section is, however, more or less an automated scheme with the least amount of user interference.

Consider an irregular domain shown in Figure 15-1 which is defined by the outer boundary nodes 1, 2, 3, ... 14 [observe that they are numbered in counterclockwise (ccw) order] and the inner boundary nodes 15, 16, 17, ... 20 [observe that they are arranged in clockwise (cw) order]. Now, define a line segment formed by connecting the two successive nodes as "edge." Thus an edge can be defined by its two end nodes such as (i, k), where i and k represent the end points. One may also represent the edge by designating an edge number such as j, where j can be related to either one of the end points.

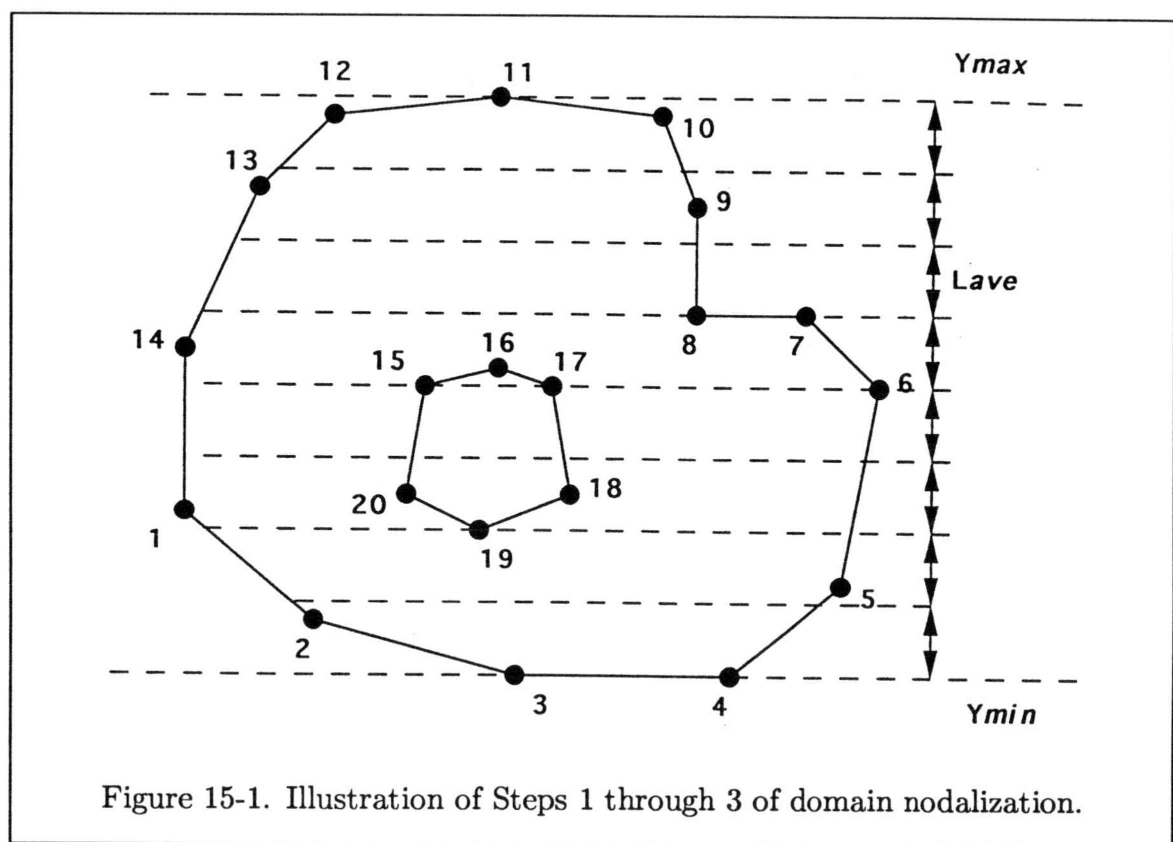

Figure 15-1. Illustration of Steps 1 through 3 of domain nodalization.

Now, using the domain shown in Figure 15-1, the details of procedure for nodalization are as follows.

1. The minimum and maximum y locations of the domain, i.e., $y_{\min}$ and $y_{\max}$, are determined as shown in Figure 15-1.

2. An average edge length is calculated, for example $L_{\text{avg}} = \sum_{j=1}^{N}(L_j/N)$, where L_j is the j-th edge length and N is the total number of edges.

3. A set of imaginary horizontal lines at different levels between $y_{\min}$ and $y_{\max}$ across the domain is created. The spacing between two successive horizontal lines can be set equal to L_{avg} (or any other suitable value). Thus, there are NL imaginary lines, where $NL = (y_{\min} - y_{\max})/L_{\text{avg}}$.

4. Determine the intersection points of a horizontal line (say line $y = D$) and the boundaries. For this purpose one must check to see whether intersection points exist. For an edge (i, k), there would be an intersection point if

(a) $(y_i - D)(y_k - D) < 0$

or

(b) $(y_i - D)(y_k - D) = 0$ and $D > y_i$ or $D > y_k$

The x coordinate of the intersection point is determined according to

$$x = x_i + (D - y_i)(x_k - x_i)/(y_k - y_i)$$

The y coordinate is obviously equal to D.

Lines which cut the boundaries with an even number of points will be referred to as *qualified useful lines.* A second category of horizontal lines could be encountered where an edge does not satisfy either one of the conditions (a) or (b), i.e., the edge is located along the horizontal line. In the same category one may include a horizontal line where the number of intersection points is odd. This category of lines which includes either one of the situations described above, will be referred to as *unqualified lines.* An example of each type is shown in Figure 15-2, where the edge $(7, 8)$ is coincident with the horizontal line $y = M$,

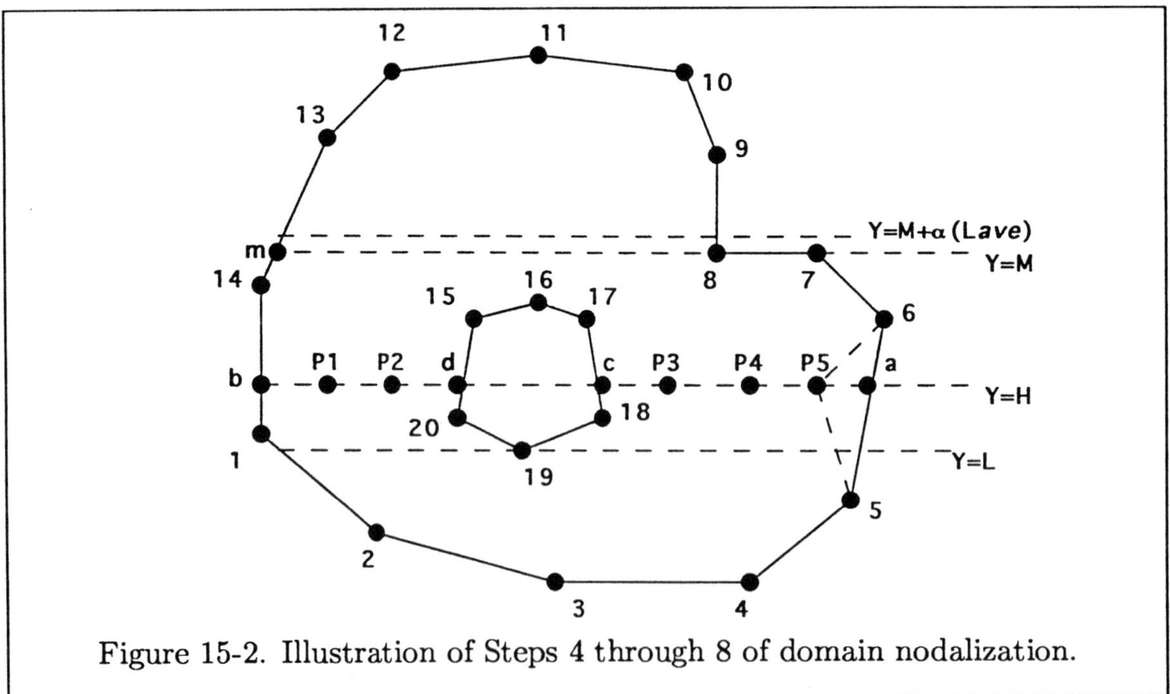

Figure 15-2. Illustration of Steps 4 through 8 of domain nodalization.

such that neither of the conditions described by (a) or (b) can be satisfied. A second situation is shown by the horizontal line $y = L$, where the number of intersection points is three. Once unqualified lines have been identified, they should be removed from the list of horizontal lines. However, simply removing the unqualified lines may distort the uniformity of node distribution. To overcome this problem, the unqualified lines are adjusted by a fraction of L_{avg} such that even intersection points are produced. For this purpose the line $y = M$ is replaced by $y = M + \alpha L_{avg}$, where α is a small number on the order of few percent of L_{avg}. The procedure is illustrated schematically in Figure 15-2. A similar procedure is used for horizontal lines with odd intersection points. Thus, Step 4 provides a set of qualified horizontal lines distributed in a near uniform fashion with a distance of about L_{avg} from each other.

5. Once the intersection points of all the qualified lines have been determined, they are rearranged according to the increasing magnitude of their x-coordinate. For example, the horizontal line $y = H$ has 4 intersection points, with the boundaries identified by points a, b, c, d. The order of these points is rearranged to be b, d, c, and a, i.e., according to increasing x-coordinates.

6. The interior nodes are now distributed along the horizontal lines between every two successive intersection points within the domain. This distribution may be accomplished by a pre-selected distance, i.e., βL_{avg}, where $\beta = 0.5$. For illustration purposes, consider line $y = H$ in Figure 15-2 where the set of points (P_1, P_2) and (P_3, P_4, P_5) have been generated between points b, d, and c, a, respectively.

7. It is possible and, indeed, more likely that some of the nodes to be located would be too close to the boundaries. Such points will distort the near uniformity of node distribution and may be removed. A simple check may be devised to identify such points. If p denotes an interior node and i, k denotes the nodes of edge j, then one may use the following criterion to remove undesirable interior points,

$$(L_{i,k})^2 > (L_{i,p})^2 + (L_{k,p})^2 \tag{15-1}$$

where L designates the length, e.g., $L_{i,k}$ is the length of edge (i, k). Points which satisfy the condition set by (15-1) are removed from the domain.

This description completes the node distribution within the domain. Again, note that the procedure described above is only one technique among many by which node points can be distributed within the domain.

15.3 Domain Triangulation

Among various schemes available for domain triangulation, two techniques are introduced in this section. These schemes are selected because of their relative simplicity. Each scheme has its own advantages and disadvantages, which will be identified as each is introduced.

15.3.1 The Advancing Front Method

Domain triangulation by the Advancing Front Method developed in Ref. [15-1] possesses some important features which include (1) the scheme is simple and straightforward, (2) it is relatively easy to implement for numerical applications, and (3) it can triangularize concave domains without any difficulty or additional effort. It should be noted that many triangulation schemes can only handle convex domains. If a domain is concave, some means must be taken to subdivide the domain into a number of convex regions and subsequently proceed with the triangulation. The advantages of the Advancing Front scheme are, however, accompanied by the following shortcomings of the scheme: (1) the scheme is not as efficient as some of the other triangulation schemes, and (2) control over grid quality is limited. Among factors contributing to grid quality, perhaps the most important is element skewness. For example, in the formation of triangular elements, two possibilities exist which are shown schematically in Figures 15-3a and 15-3b. It is obvious that the elements in Figure 15-3b are of superior quality to that of elements formed in Figure 15-3a, due to their skewness. At this point, the procedure for triangulation by

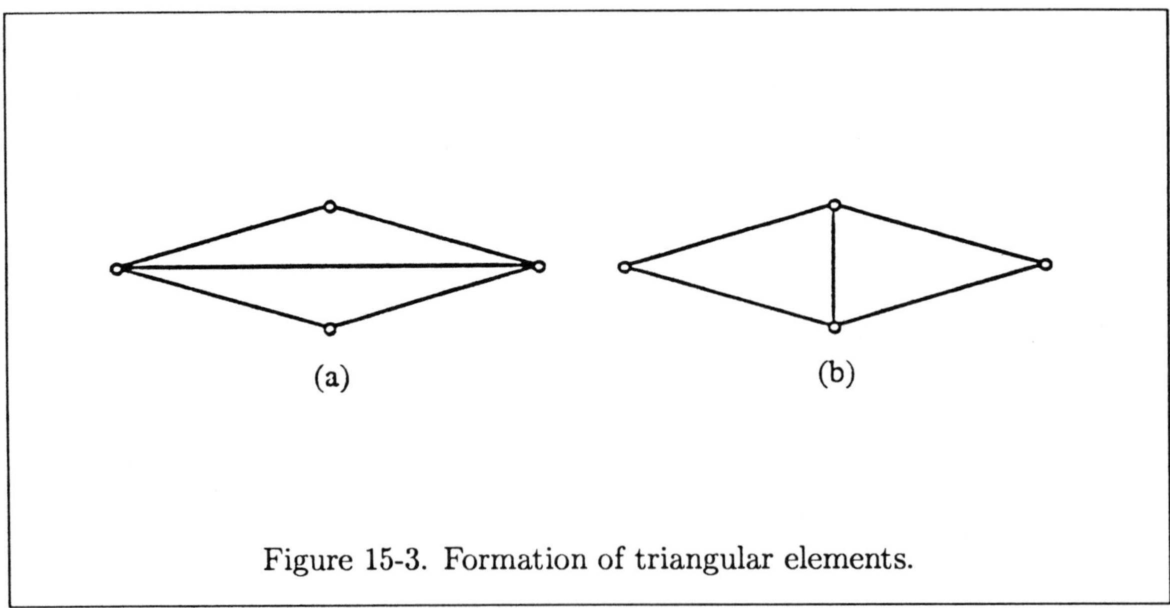

Figure 15-3. Formation of triangular elements.

the Advancing Front Method, as applied to simply-connected domains, is outlined. Subsequently the procedure is extended to multiply-connected domains.

15.3.1.1 Simply-Connected Domain

A simply-connected domain which was defined in Section 9.7.1 is perhaps the simplest domain to discretize. However, it should be noted that when the boundaries of such a domain are highly irregular, discretization will be a challenging task indeed. In order to describe the triangulation procedure by the Advancing Front Method, it is best to accompany it with a simple example. Thus, consider the simple square domain shown in Figure 15-4, which is defined by the boundary points a, d, g, and j.

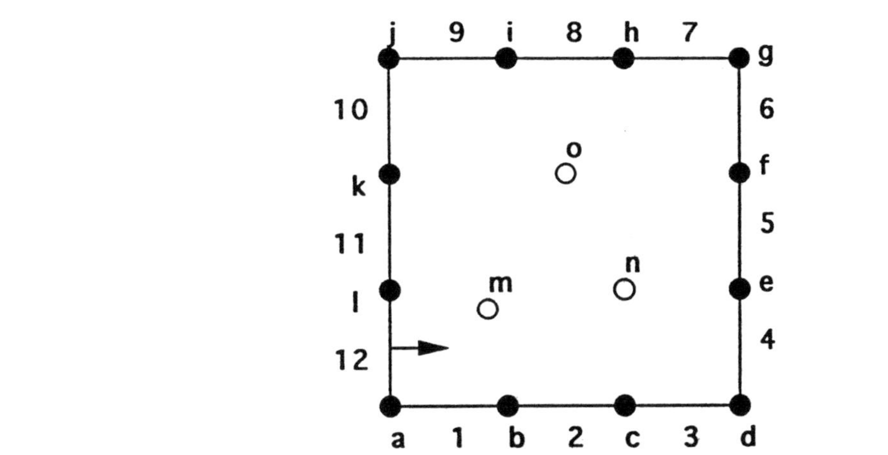

Figure 15-4. Schematic of Step 3 for the simply-connected domain by the Advancing Front Method.

It was previously stated that, before triangulation begins, one must devise a procedure by which the interior domain is nodalized. A scheme to do so was introduced in the previous section. For illustrative purposes, assume three interior points m, n, and o in Figure 15-4 are points which have been distributed within the domain. Recall that these points are referred to as "interior nodes" to distinguish them from the edge nodes, e.g., points a, b, c, $\ldots k$, l, which are located on the boundaries of the domain. The Advancing Front Method proceeds sequentially by the following steps.

1. All the edges along the initial boundary of the domain are numerated in ccw order and saved in an array as E. Recall that an edge is defined as a line segment between two edge nodes. Therefore, the array E is composed of edges $1(a,b)$, $2(b,c)$, $3(c,d)$, $4(d,e)$, $\ldots 12(l,a)$ as shown in Figure 15-4. Edges defined in array E will be used to construct the advancing (or generation) front.

2. All the interior nodes are saved in an array I. Thus for the problem shown in Figure 15-4, array I includes three points m, n, and o.

3. Beginning with the last element in array E which represents the last edge in the boundary, a search is conducted to locate nodes which are on the left-hand side of the edge. This search includes both interior nodes as well as edge nodes. These nodes will be referred to as *qualified nodes.* Among the qualified nodes one needs to select a particular node which will be referred to as the *most suitable* node. Generally, the most suitable node is one which has the minimum norm distance to the two edge nodes among all the qualified nodes. To illustrate this step, recall that the last edge was defined by $12(l,a)$. If the most suitable node is called x, then the criterion for its selection is rewritten as $(L_{xl})^2 + (L_{xa})^2 =$ minimum, where L is the length.

   Now, three nodes, l, a, and x, which are ordered in ccw fashion are used to form a triangle. Referring to Figure 15-5, node m is selected as the most suitable node and, subsequently, the triangle (l,a,m) is formed.

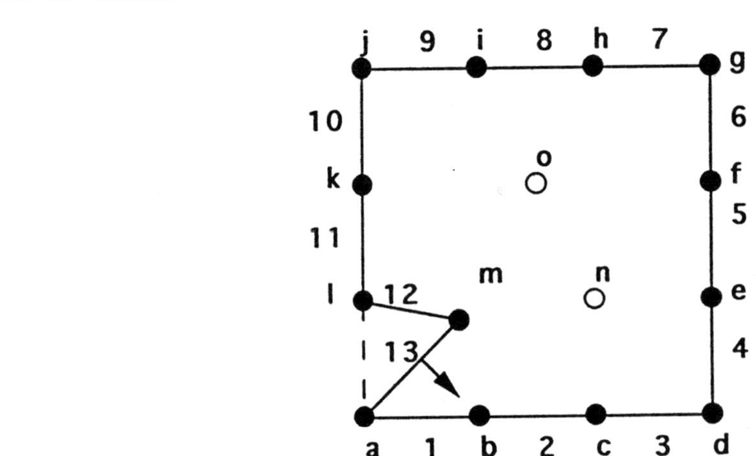

Figure 15-5. Schematic of Step 4 for the simply-connected domain by the Advancing Front Method.

An issue which still needs to be resolved is the procedure for identification of qualified nodes, i.e., how does one judge if a node is located on the left-hand side of a particular edge? Consider, for example, edge 12 in Figure 15-4, and the node m. Define a position vector $\vec{a}$ from point l to point a, and similarly a position vector $\vec{b}$ from point l to point m. If the cross product of vectors $\vec{a}$ and $\vec{b}$ is positive, one concludes that node m is located on the left-hand side of vector $\vec{a}$ which represents edge 12.

4. Now one needs to update the advancing front array E which represents new boundaries as well as the list of interior nodes in array I. It is quite obvious that the edge (l, a) in Figure 15-5 is no longer a part of the advancing front. Therefore, edge $12(l, a)$ must be removed from E, thus reducing the array E to include 11 edges. However, the newly formed edges (l, m) and (m, a) must now be used as a part of the new advancing front and, therefore, they are added to E and registered as edge 12 and edge 13. The list of array E is now arranged as: $1(a, b)$, $2(b, c)$, $3(c, d)$, $4(d, e)$, $\ldots 12(l, m)$, $13(m, a)$. Meanwhile node m is removed from the list of interior nodes I, because it was changed from an interior node to an edge node.

5. Now Steps 3 and 4 as described above are repeated, and the procedure is applied to the last edge in the list of E which, for the example considered, is edge 13. The most suitable node is selected to be point b, and subsequently the triangle (m, a, b) is formed, as shown in Figure 15-6. The list of E is updated by the

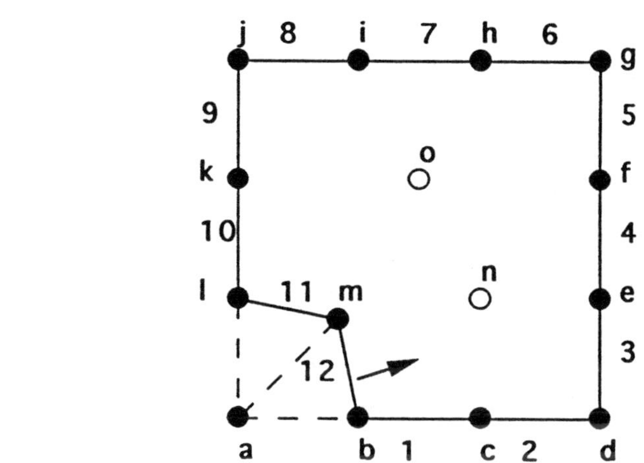

Figure 15-6. Schematic of Step 5 for the simply-connected domain by the Advancing Front Method.

removal the edges (m, a) and (a, b) and the addition of edge (m, b). Therefore, the list of E will be shifted as $1(b, c)$, $2(c, d)$, $3(d, e)$, $\ldots 11(l, m)$, $12(m, b)$. Note that, at this instance, no interior point was used and, therefore, no changes in the list of I will occur.

The following conclusions may be stated at this point, based on Steps 4 and 5. (i) The edge(s) which is/are used in the newly formed triangles that belong to array E is/are removed. (ii) The order of edges in the list of E is reorganized and, subsequently, newly formed edge(s) is/are added.

6. The procedure described in Steps 2 through 5 is repeated until triangle (e, c, d) is formed, as shown in Figure 15-7. This may be considered as a "dead-end" situation for the advancing front, because all three edges $13(e, c)$, $1(c, d)$, and $2(d, e)$, which belonged to E before triangulation must be removed from E. Subsequently, the edge (e, f) is now edge 1, and edge (n, e) is the last edge in E. Thus, the updated "advancing front" array E is: $1(e, f)$, $2(f, g)$, $3(g, h)$, $\ldots 10(n, e)$. Since the edge $10(n, e)$ is now the last edge, it is therefore the edge which is used to continue the triangulation process.

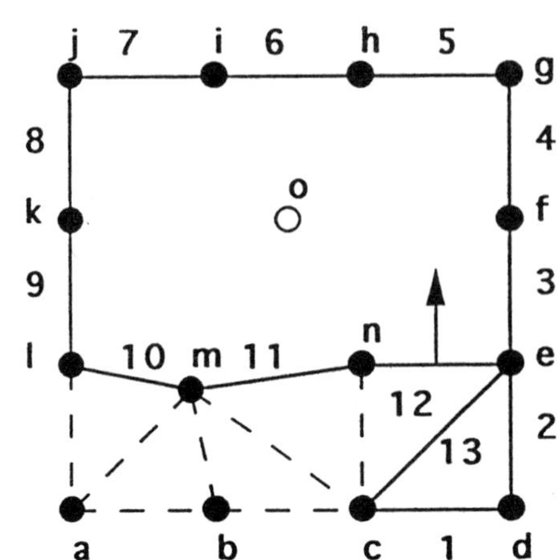

Figure 15-7. Schematic of Step 6 for the simply-connected domain by the Advancing Front Method.

7. The procedures described in the previous steps are continued until all the edges in the list of array E have been removed. As a consequence, the triangulation of the entire domain is completed.

15.3.1.2 Multiply-Connected Domain

For doubly-connected or multiply-connected domains, one or more objects are located within the domain. The advancing front algorithm described in the previous section can be easily extended to multiply-connected domains. However, a point to recognize is that now the advancing front array E is composed of both the outer boundary E_o and the inner boundary E_i. What is important is the organization of the edges in array E. The arrangement of the outer edges must be in ccw order (as for the simply-connected domains), whereas the edges of the inner boundary(ies) should be listed in cw order. In order to clarify this important point, consider the doubly-connected domain shown in Figure 15-8. The advancing front array E is composed of edges of the inner boundary arranged as $1(15, 16)$, $2(16, 17)$,

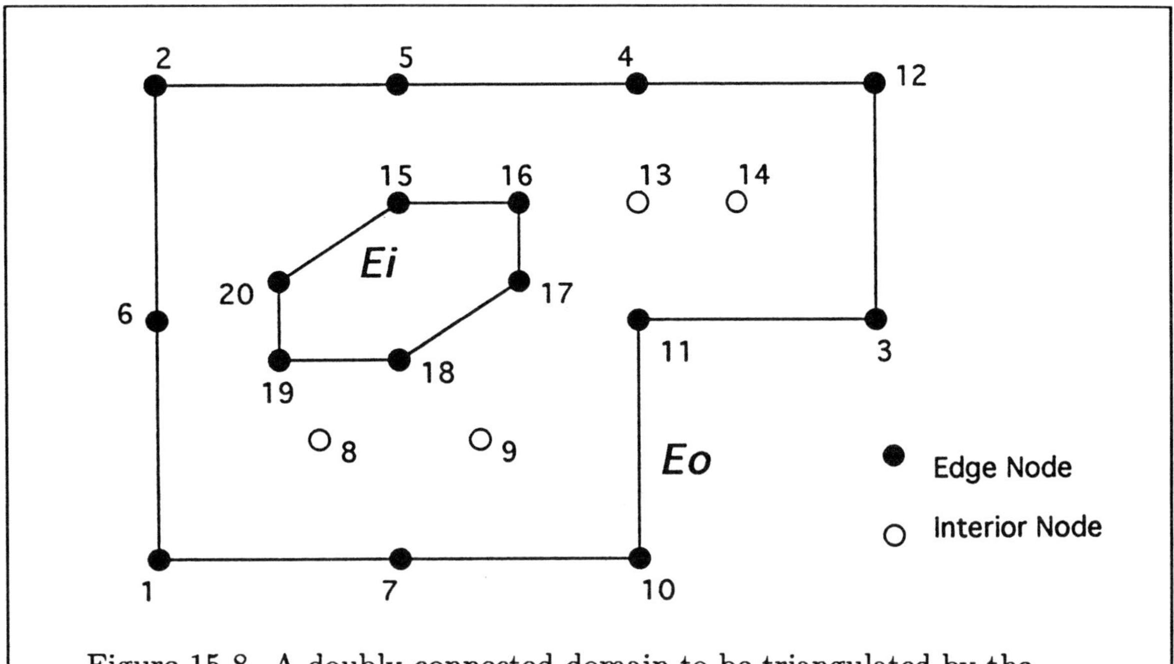

Figure 15-8. A doubly-connected domain to be triangulated by the Advancing Front Method.

$\ldots 6(20,15)$ and edges of the outer boundary arranged as $7(1,7)$, $8(7,10)$, $9(10,11)$, $\ldots 16(6,1)$. Note that the identification of nodes along the boundaries as well as those within the interior of the domain is random. However, it is emphasized again that the arrangement of edges within the advancing front array E must follow the guideline specified above. Now assume a set of points have been distributed within the interior of the domain identified by nodes 8, 9, 13, and 14.

The advancing front algorithm will proceed, as described in the previous section, from the last element of array E, i.e., $16(6,1)$. The result of triangulation is shown in Figure 15-9. A total of 24 triangles is generated. The order of formation of triangles is illustrated by a number within each triangle. Observe that the first dead end situation occurs after triangle 21 is formed. Subsequently, triangulation proceeds from edge $(17,18)$ toward the left.

The application of the scheme to a multiply-connected domain is illustrated in Figure 15-10, which includes two openings within the domain. The domain is first nodalized by the scheme described in Section 15.2 where a total of 35 interior nodes have been generated. Subsequently, the advancing front scheme is used to triangulate the domain. For the example shown, 116 triangles have been formed. For the purpose of clarity, the order of formation of triangles is shown in Figure 15-11.

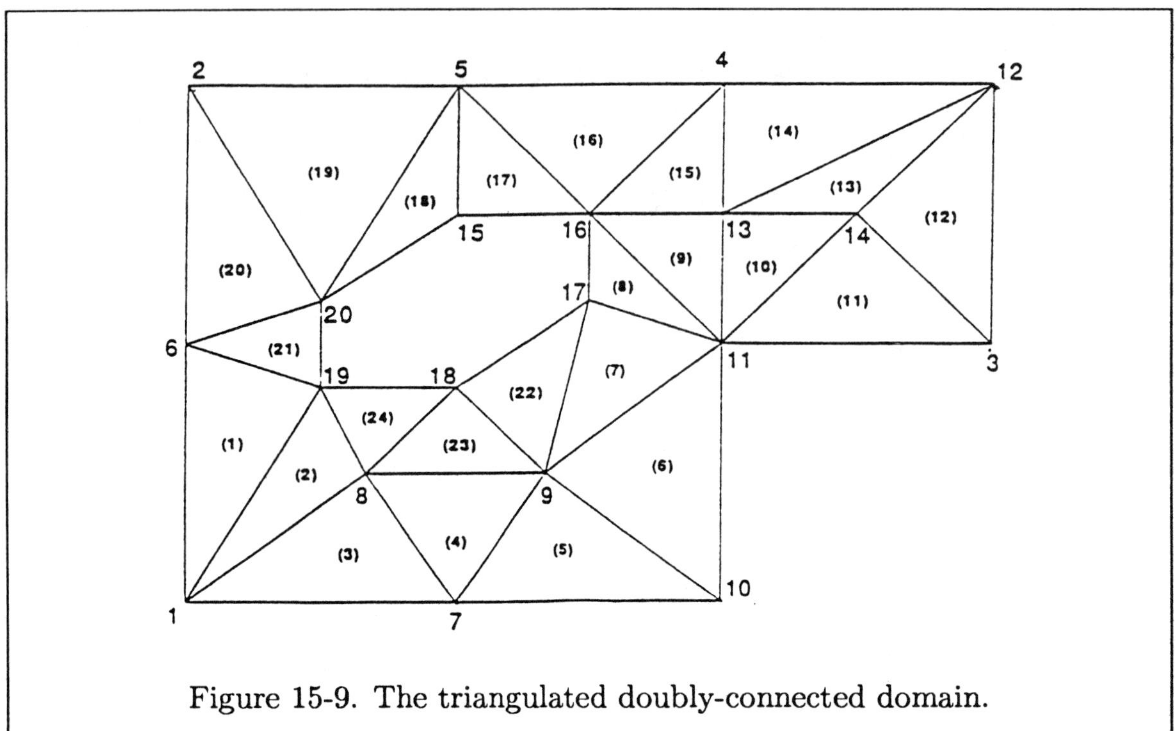

Figure 15-9. The triangulated doubly-connected domain.

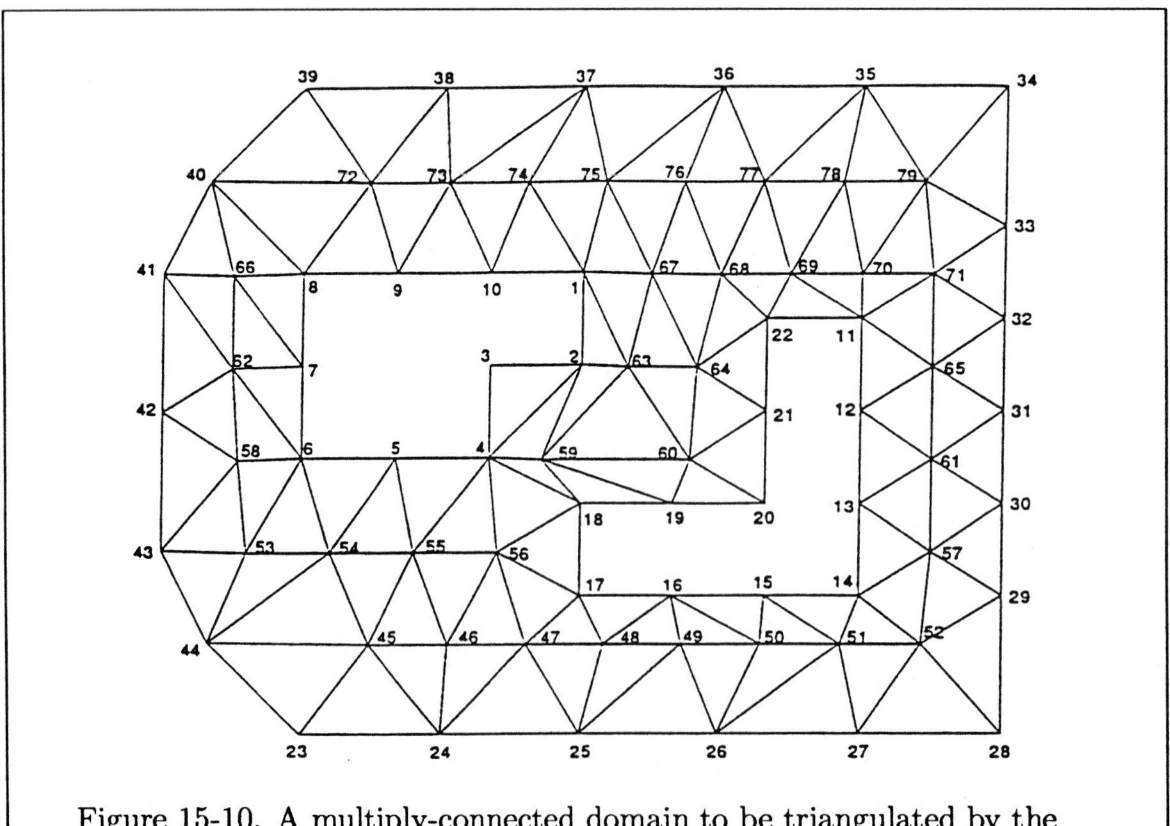

Figure 15-10. A multiply-connected domain to be triangulated by the Advancing Front Method.

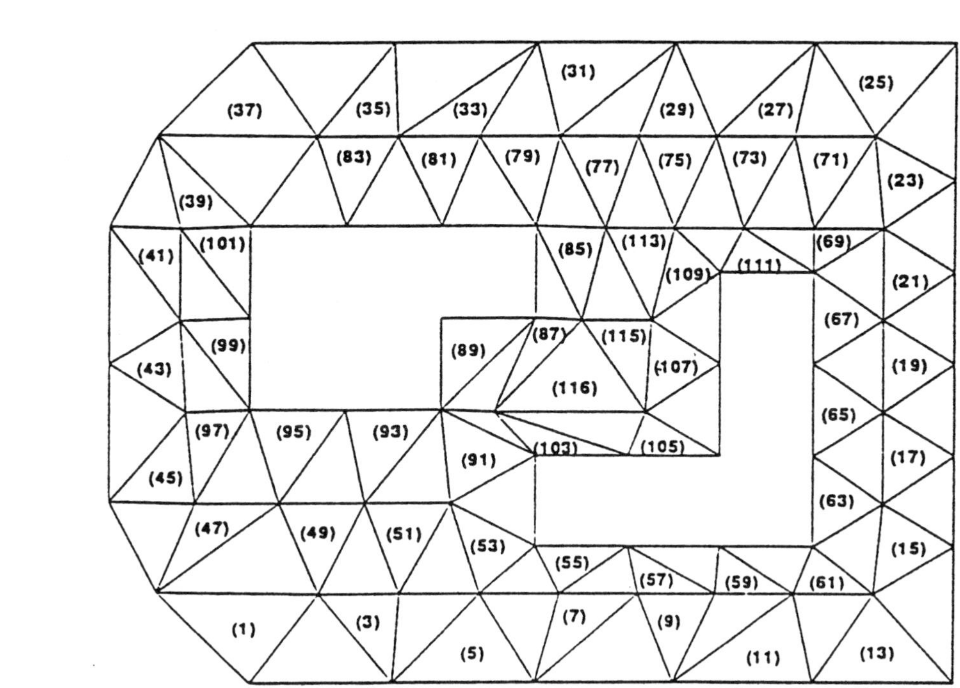

Figure 15-11. The sequence of formation of the triangular elements.

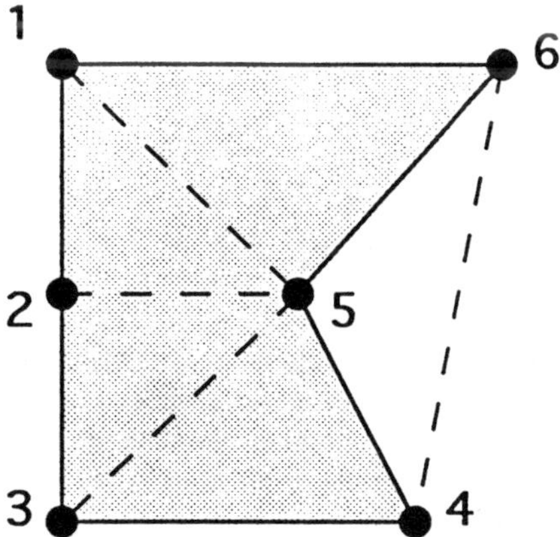

Figure 15-12. A schematic illustration of a false triangle formed in a concave domain by the Delaunay scheme.

15.3.2 The Delaunay Method

This scheme is designed to provide an efficient procedure for connecting a given set of points into an optimum unstructured triangular mesh. The most important aspect of this scheme is its efficiency [15-2], as well as the quality of the generated grid. The disadvantage of the scheme is associated with triangulations of concave domains. For such domains, which occur frequently in practice, triangular elements are generated outside the domain, therefore, violating the boundary faces. To clearly point out this fact, consider the domain shown in Figure 15-12, where a false triangle $(6, 5, 4)$ has been generated. To overcome this difficulty, extra effort is required to identify and subsequently remove those triangles which are generated outside the domain. Various procedures may be used to resolve the difficulty associated with non-convex domains. Further discussions may be found in Reference [15-2].

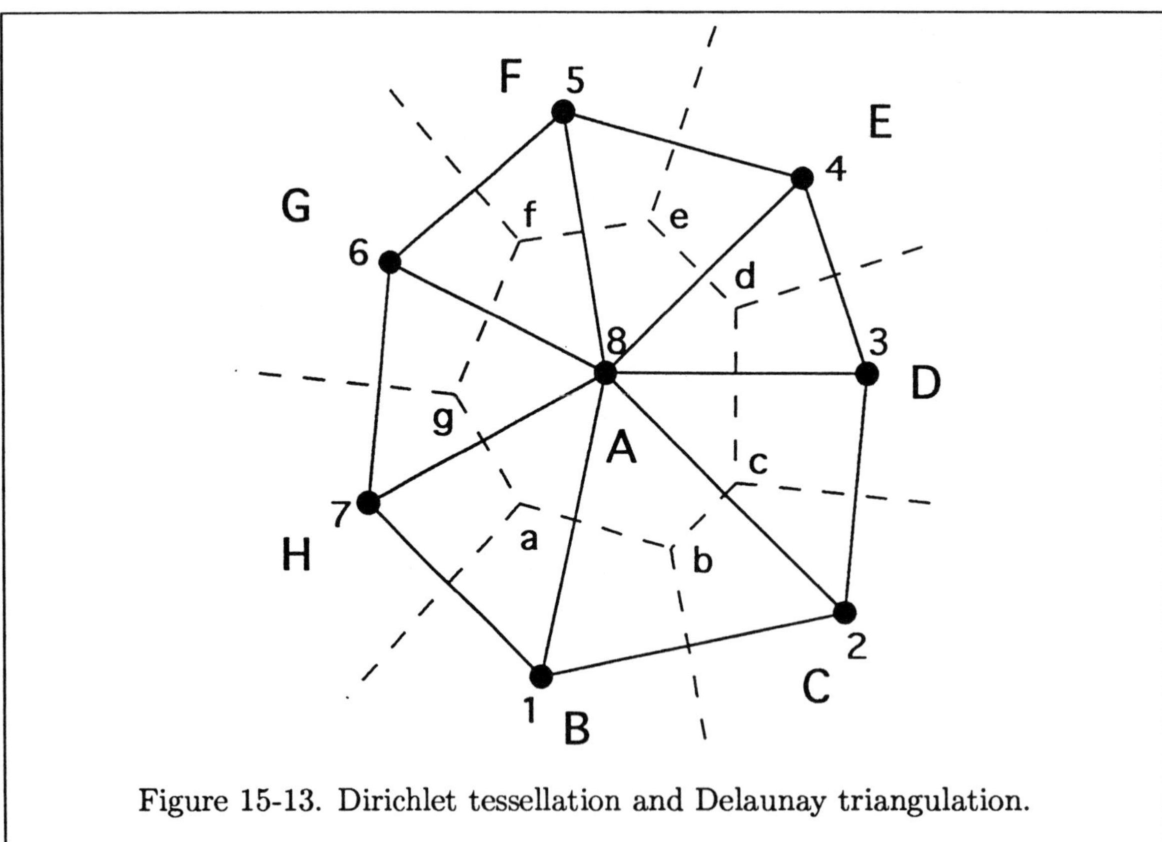

Figure 15-13. Dirichlet tessellation and Delaunay triangulation.

15.3.2.1 Geometrical Description

In order to understand why the Delaunay method generates an optimized triangular grid, some geometric basis of the scheme needs to be explored. In 1850, Dirichlet proposed a method whereby a domain can be decomposed into a set of convex polygons. Each polygon is defined as a "tile" and is associated with a single "generating point." Any point inside a tile is closer to its own generating point than to any other tile's generating point. The tile defined above is also referred to as "Dirichlet tessellation," "Voronoi tessellation," or "Theissen tessellation," [15-3]. The polygon shown in Figure 15-13 defined by points a, b, c, d, e, f, and g is defined as tile A, and the generating point is point 8. The boundaries of the tile are perpendicular bisectors of the lines joining the neighboring generating points. The Delaunay triangulation that corresponds to the Dirichlet tessellation is constructed by connecting generating points of all neighboring tiles. To relate this discussion to Figure 15-13, observe that any point inside tile A should be closer to point 8 (which is the generating point of tile A) than to any of the points 1, 2, 3, 4, 5, 6, or 7 (which are the generating points of the neighboring tiles of A). Furthermore, note that boundary ab of tile A bisects the distance between generating points 1 and 8 and is also perpendicular to line 1-8. Triangles are formed by connecting the generating points.

Before proceeding further, let's explore the concept of locally equiangular. This concept states that for every convex quadrilateral formed by two adjacent triangles, the minimum of the six angles in the two triangles is greater than it would have been if the alternative diagonal had been drawn and the other pair of triangles selected. Schematically, consider points 8, 3, 4, and 5 in Figure 15-13 which needs to be triangularized. Note that if no requirement is imposed, two sets of triangles may be generated. One set includes triangles formed with points $(8,4,5)$ and $(8,3,4)$. The second set is composed of triangles $(8,3,5)$ and $(5,3,4)$. However, imposing the concept of locally equiangular, the first set of triangles, namely $(8,4,5)$ and $(8,3,4)$ is generated. That is because the minimum angle in triangles $(8,4,5)$ and $(8,3,4)$ will be greater than the minimum angle in triangles $(8,3,5)$ and $(5,3,4)$. The algorithm used to generate the locally equiangular triangles is known as *swapping* and will be described in the next section.

15.3.2.2 Outline of the Algorithm

There are a number of algorithms which have been proposed [15-2, 15-4] for the construction of planar Delaunay triangulation. One of the simplest schemes, suggested by Sloan [15-4], is an efficient method for both small and large sets of points and will be explored in this section. An average run time of the algorithm for a domain with N randomly distributed points is $O(N^a)$, where $a \cong 1.06$ for $N < 10,000$.

The details of the scheme are described in this section in several steps followed by an example of a four point triangulation.

1. a. The x- and y-coordinates of all the N points are normalized with respect to the largest length in the domain such that their values are within the range

of 0 to 1, i.e., all the points are in a unit domain. Note that if the coordinate system is originally set such that there are negative x- and/or y-coordinates, it should be shifted so that negative coordinates are replaced.

b. Three points are added to form a supertriangle which completely encompasses all of the N points to be triangulated. The coordinates of the three vertices of the supertriangle can be chosen arbitrarily; however, the vertices of the supertriangle should not be very close to the window enclosing the N points. For example, the values of the vertices [points $N + 1$, $N + 2$, and $N + 3$] can be assigned as $(-100, -100)$, $(100, -100)$ and $(0, 100)$.

c. Create a triangle list array T, where the supertriangle is listed as the first triangle.

2. Partition the domain to be triangulated into approximately $N^{0.25}$ bins. Assign a bin number to each of the N points in the unit domain. Subsequently, all the points are sorted in ascending sequence of bin number. The points are now ordered such that consecutive points are in close proximity of each other. This step is optional. It can increase the efficiency of the algorithm, but is not a necessary requirement.

3. Introduce the first point (from the list of N points) into the supertriangle and generate three triangles by connecting the three vertices of the supertriangle to this point. At the same time, delete the supertriangle from the triangle list T and subsequently add the 3 newly formed triangles into the list of T. The triangles are always defined by their vertices in a ccw order. Note that, at this step, the net increase of triangles in array T is 2.

4. Now the next point (call it point p) is introduced for triangulation. First, one needs to locate an existing triangle (call it triangle x) which encloses point p, and subsequently 3 new triangles can be formed by connecting point p to the vertices of triangle x. Now the original triangle x is deleted from array T and newly formed triangles are added. Thus, the net gain in the total number is 2, i.e., a total of 5 triangles have been formed. In order to accomplish the procedure above, one needs to know a scheme for searching for triangle x. The search begins with the most recently created triangle, i.e., the last triangle in the list of array T. Subsequently, a check is made to see if point p is to the left of all the edges of this triangle. A procedure to do so was described in Step 3 of Section 15.3.1.1. If point p is located to the left of all the three edges, then point p is enclosed by the triangle. However, if point p is located to the right of any edge of the triangle, then the search shifts to the triangle which is adjacent to that edge. The process is repeated until triangle x is found. With this search algorithm, one can avoid the need to search all the triangles in the

list of array T. To complement the discussion above, consider the following example. Assume that the last triangle formed is triangle $14(m, n, o)$ shown in Figure 15-14a. Point p is found to be on the RHS edge mn. Recall the ccw ordering of the triangle, and therefore edge mn (not nm) is used for triangle 14. Now the search moves to the next triangle, that is triangle 13, which shares the same edge with triangle 14. Note that now the edge is nm for triangle 13 in order to satisfy the ccw ordering. A check is made, and it is found that point p is on the RHS of edge mk. Therefore, the search

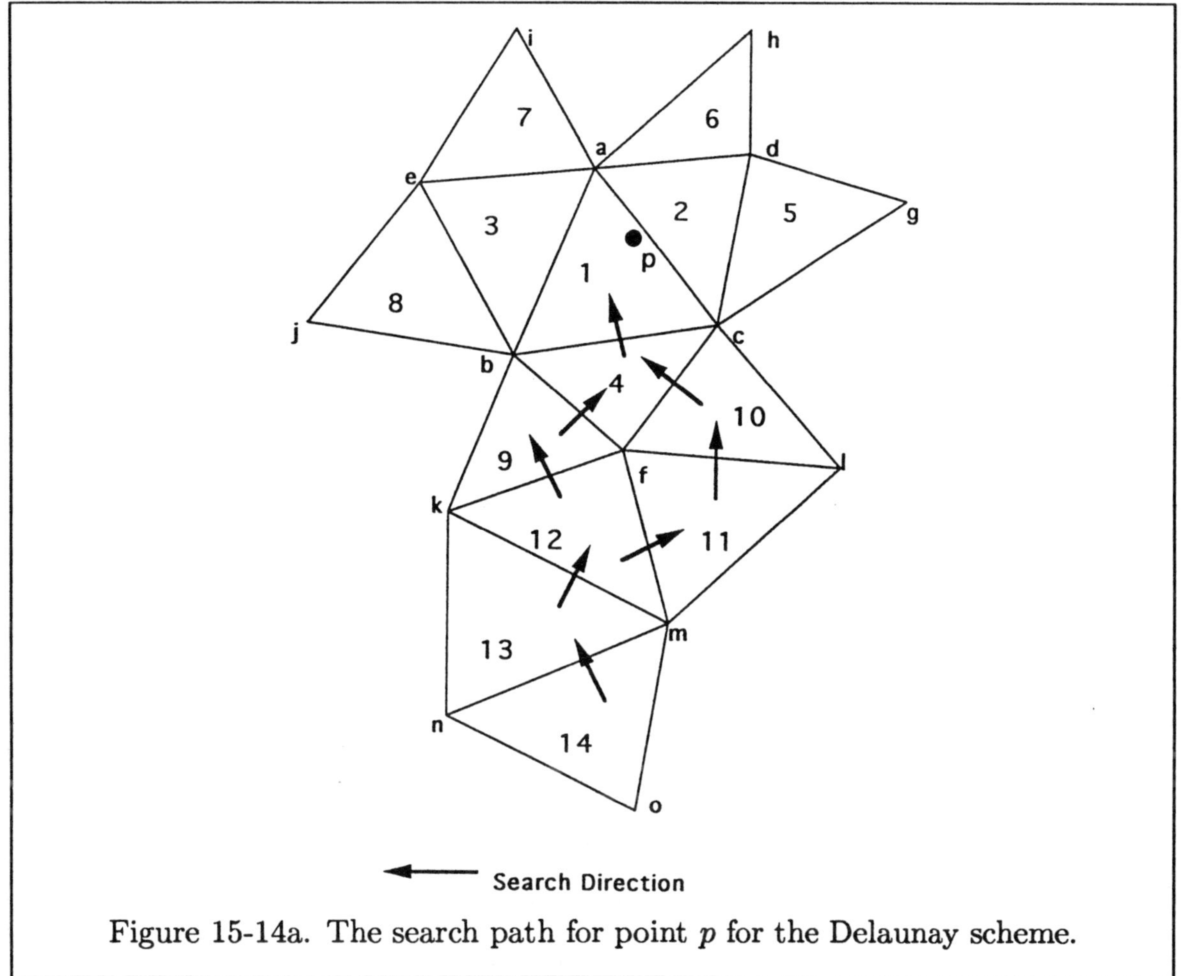

Figure 15-14a. The search path for point p for the Delaunay scheme.

moves to triangle $12(m, f, k)$. At this point, we reach a crossroads in our search. That is point p is located on the RHS of edge mf as well as fk. The issue now is in which direction does one proceed, i.e., triangle $9(k, f, b)$ or triangle $11(f, m, l)$? It turns out that either one will work. For example, if triangle 9 is selected, the search path will be triangles 14, 13, 12, 9, 4, and 1. If triangle 11 is chosen, the search path will be triangles 14, 13, 12, 11, 10, 4, and 1. In either case, the total number of triangles searched would be less than the total number of triangles, which for the example shown is 14. Thus, point p is found to be located within triangle 1. Now, three new triangles are formed and added to the list of array T as $1(a, b, p)$, $15(p, b, c)$, and $16(a, p, c)$, as shown in Figure 15-14b. Observe that the original triangle $1(a, b, c)$ is replaced by triangle $1(a, b, p)$, and the net increase of triangles in list T is 2.

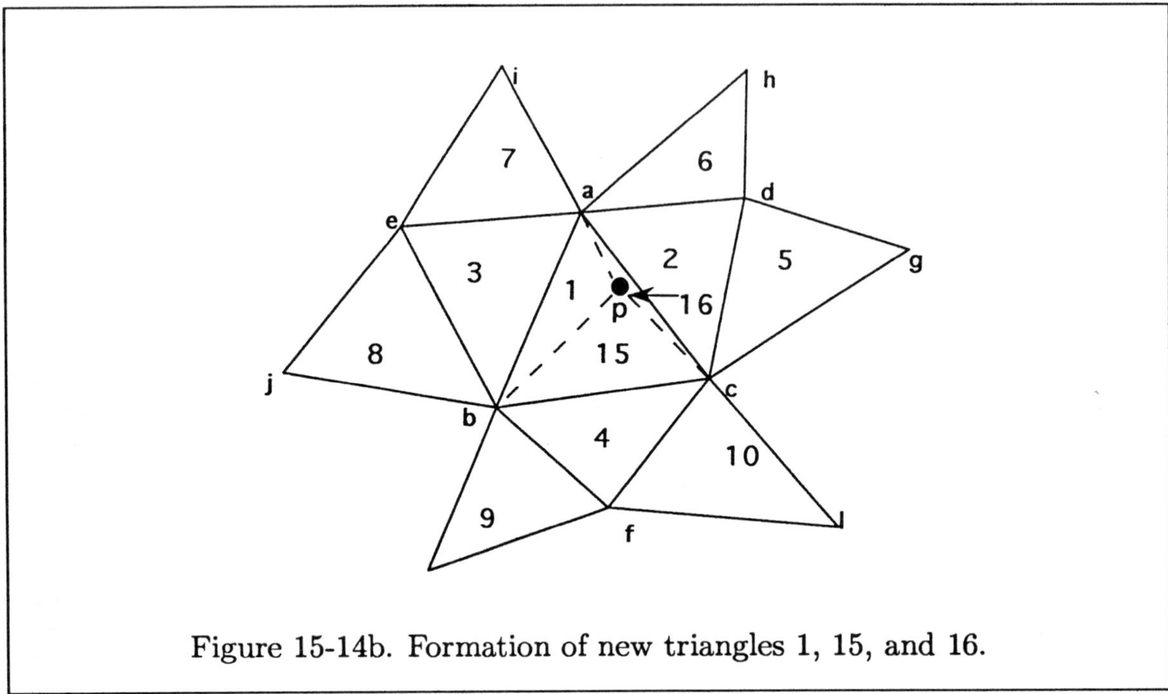

Figure 15-14b. Formation of new triangles 1, 15, and 16.

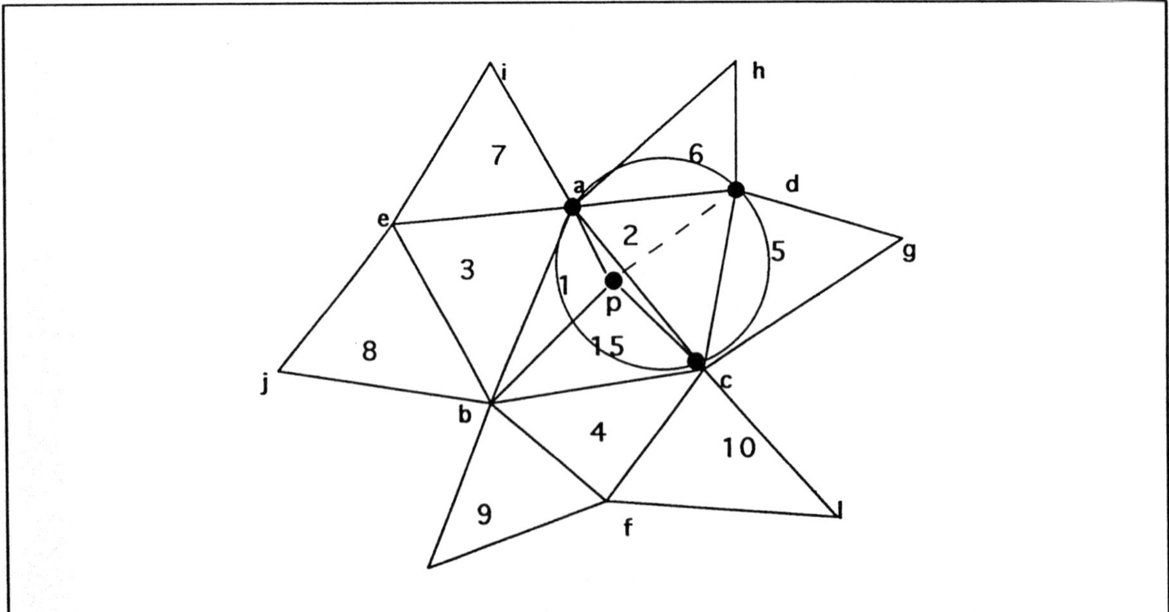

Figure 15-14c. Swapping procedure applied to triangles $2(a, c, d)$ and $16(a, p, c)$.

5. The swapping algorithm is now used to update the existing triangulation to a Delaunay triangulation, i.e., optimization of the grid. All triangles which are adjacent to the edges of the triangle enclosing point p are placed on a last in, first out stack S (a maximum of three triangles are placed on the stack initially, i.e., triangles 2, 3, and 4, as shown in Figure 15-14b). Each triangle is checked, one at a time starting from the last one in the stack, to determine if point p is located within its circumcircle. If that is the case, then the triangle containing point p as a vertex and the adjacent triangle from the convex quadrilateral with its diagonal drawn in the opposite direction must be replaced by the alternative diagonal to preserve the structure of the Delaunay triangulation.
   Once the swap is completed, triangles which are now located to the opposite of point p are added to the stack S. Subsequently, the next triangle is unstacked and the entire process is repeated until the stack is empty. To illustrate the procedure above, consider Figure 15-14c, where the triangles adjacent to the edges opposite to p are triangles $2(a, c, d)$, $3(a, e, b)$ and $4(b, f, c)$. Thus, the list of S includes the triangles 2, 3, and 4. The circumcircle of triangles 2, 3, and 4 are used to determine whether the swapping is necessary. For the example shown in Figure 15-14, more specifically Figures 15-14c through 15-14e, it is concluded that point p is enclosed only by the circumcircle of triangle 2. Therefore, triangles 3 and 4 are removed from list S. Now, triangles 2 and 16 are updated from $2(a, c, d)$ and $16(a, p, c)$ to triangles $2(a, p, d)$ and $16(d, p, c)$, as shown in Figure 15-14d. Subsequent to the swap, triangles $5(d, c, g)$ and $6(a, d, h)$, which are adjacent to the edges opposite to p, are created and added to the list of S. The scheme now proceeds to determine whether a swap between triangles 2, 6 and 16, 5 is necessary. By inspection of Figures 15-14f and 15-14g, it is concluded that no swap is necessary since point p is outside both circumcircles of triangles 5 and 6. Observe that the total number of triangles during the swapping process does not change. That is, the two old triangles are replaced by two new triangles with no net gain.

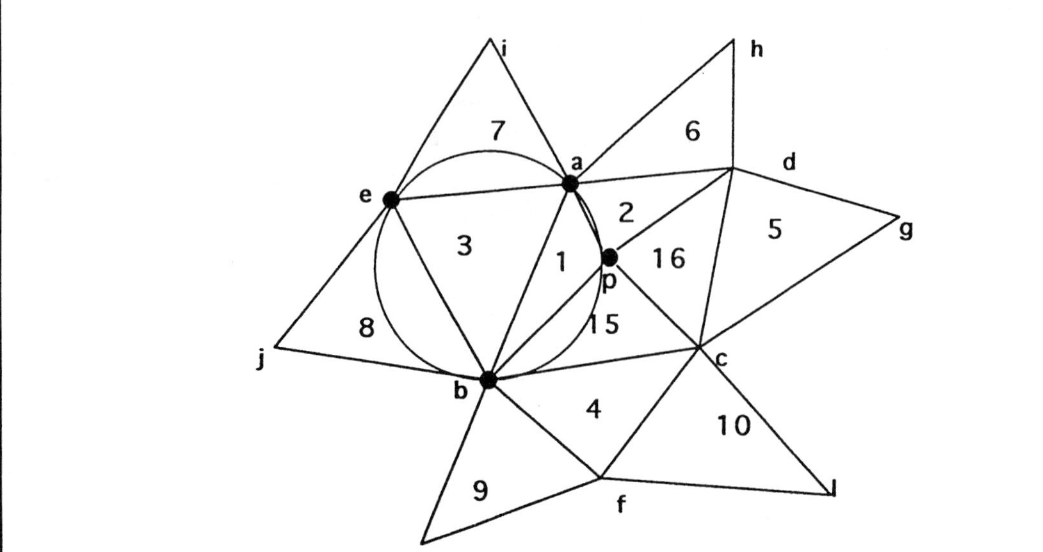

Figure 15-14d. Illustration of swapping procedure for triangles $1(a, b, p)$ and $3(a, e, b)$, which is not required.

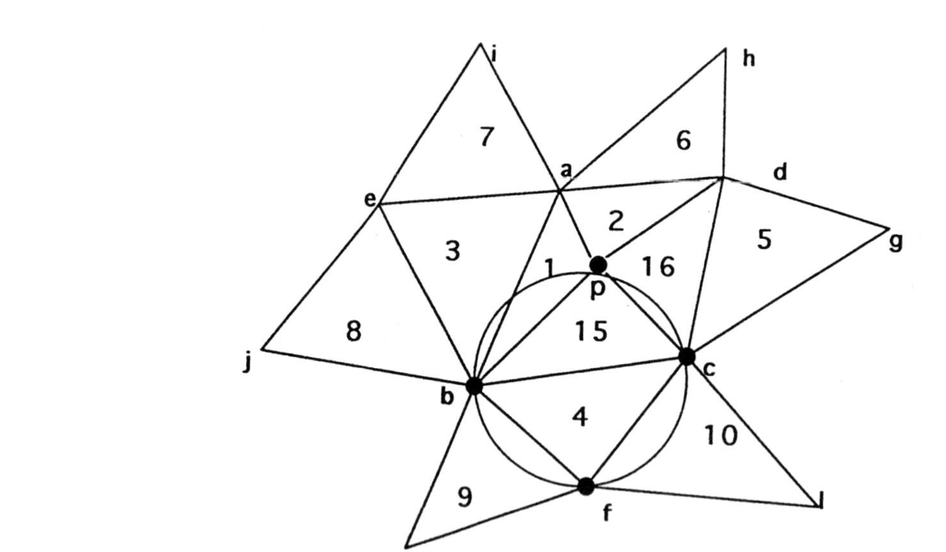

Figure 15-14e. Illustration of swapping procedure for triangles $4(b, f, c)$ and $15(b, c, p)$, which is not required.

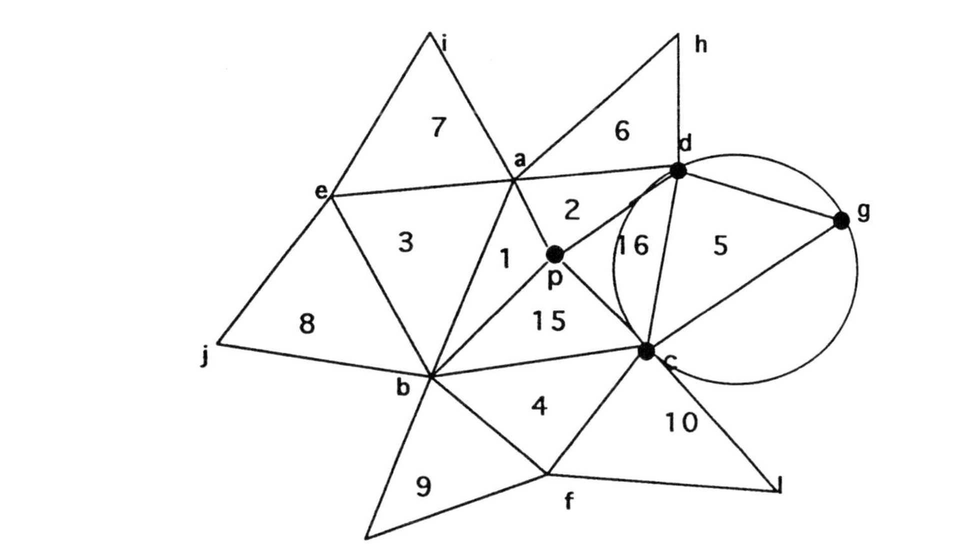

Figure 15-14f. Illustration of swapping procedure for triangles $5(d,c,g)$ and $16(d,p,c)$, which is not required.

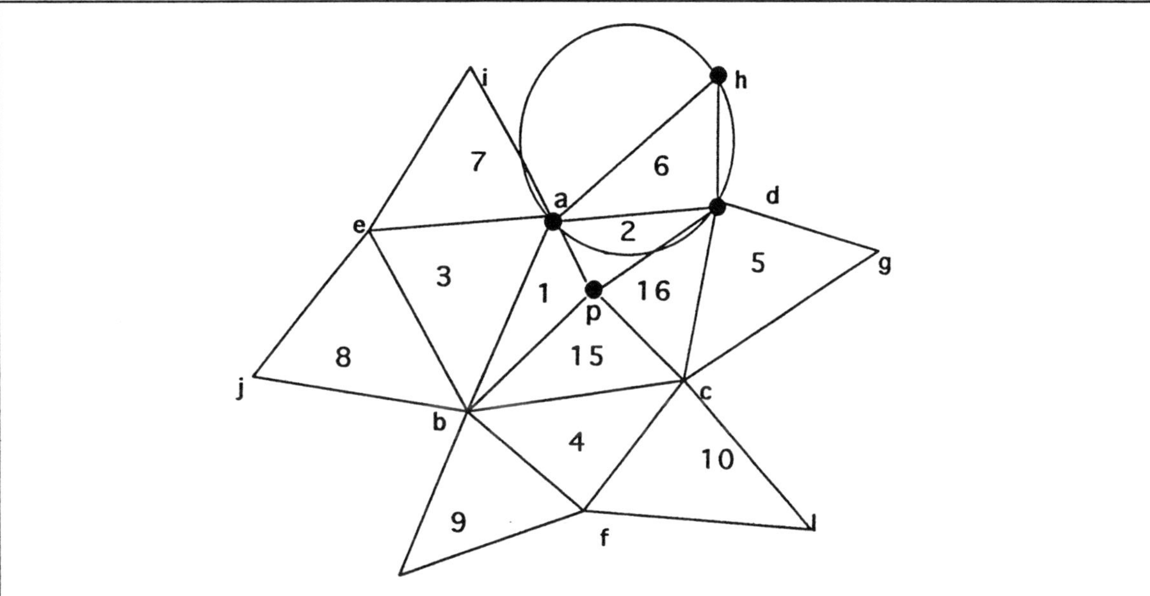

Figure 15-14g. Illustration of swapping procedure for triangles $6(a,d,h)$ and $2(a,p,d)$, which is not required.

At this point, a scheme will be introduced which can be used to determine whether point p is located inside the circumcircle of a triangle. The procedure is described with reference to Figure 15-15. It can be simply stated that if $\alpha+\beta > \pi$, then the point p is located inside the circumcircle of triangle abc, and therefore, a swap is necessary. An equivalent statement to satisfy the criterion above is

$$\sin(\alpha + \beta) < 0$$

which can be expanded as

$$\sin(\alpha + \beta) = \sin\alpha\cos\beta + \cos\alpha\sin\beta < 0$$

and subsequently written in terms of the length of the edges involved as

$$(x_{ac}x_{bc} + y_{ac}y_{bc})(x_{bp}y_{ap} - x_{ap}y_{bp}) < (y_{ac}x_{bc} - x_{ac}y_{bc})(x_{bp}x_{ap} + y_{ap}y_{bp}) \quad (15\text{-}2)$$

Thus, when the criterion previously set by (15-2) is satisfied, it is concluded that point p is located within the circumcircle of triangle abc. Note that the required lengths used in (15-2) are determined by

$$x_{mn} = x_n - x_m$$

and

$$y_{mn} = y_n - y_m$$

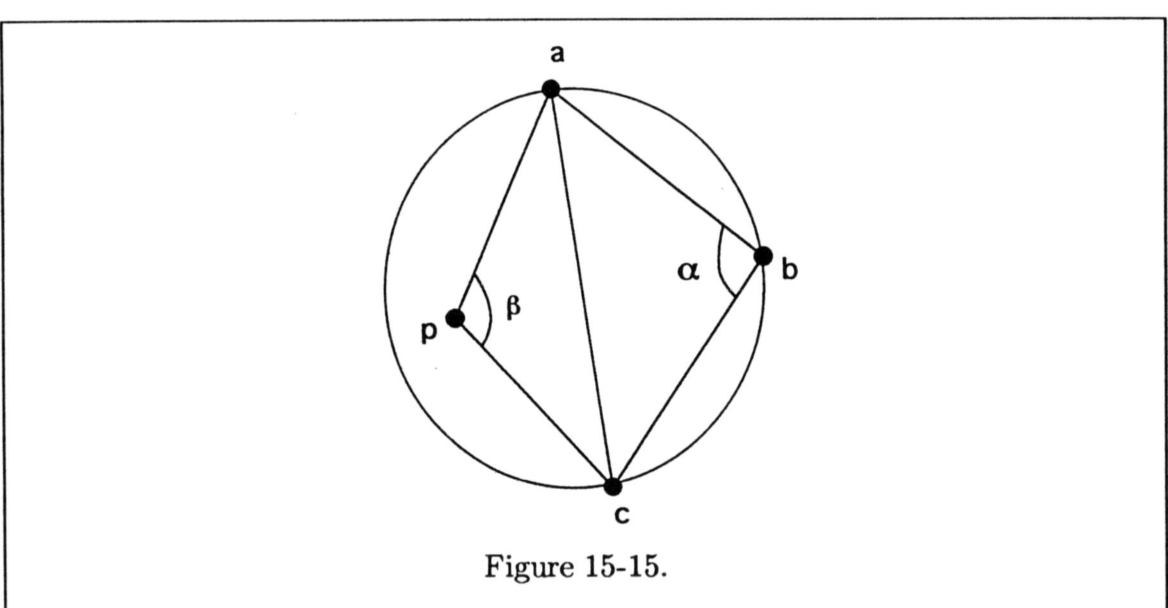

Figure 15-15.

6. Next, a new point is introduced into triangulation process. Therefore, Steps 4 and 5 are repeated. The process continues until all N points are consumed.
7. Finally, all triangles which contain one or more of the vertices of the supertriangle are removed. At the same time, note that any vertex which appears in the deleted triangles, but is not a vertex of a supertriangle, must be on the boundary of the domain.

The various steps outlined above can be reinforced by a simple example proposed in the next section.

15.3.2.3 An Illustrative Example

Consider a domain with four points numbered as points 1, 2, 3, and 4 shown in Figure 15-16a. It is required to triangulate the domain by the Delaunay method. Following the steps outlined in the previous section, the procedure is implemented as follows.

1. Three points identified as points 5, 6, and 7 are added to form the "Supertriangle." Note that, for clarity of the figure, the coordinate of the points are not normalized. In either case, the total number of triangles on the list T is one, i.e., the supertriangle (5,6,7).

2. Since the number of points involved in this example is only four, the sorting of points into bins is not necessary and is skipped.

3. The first point, i.e., point 1, is introduced to form three triangles, which are $a(1,7,5)$, $b(1,5,6)$, and $c(1,6,7)$, shown in Figure 15-16a. The original supertriangle $(5,6,7)$ in the list of T is replaced by $a(1,7,5)$, and the number of triangles is three.

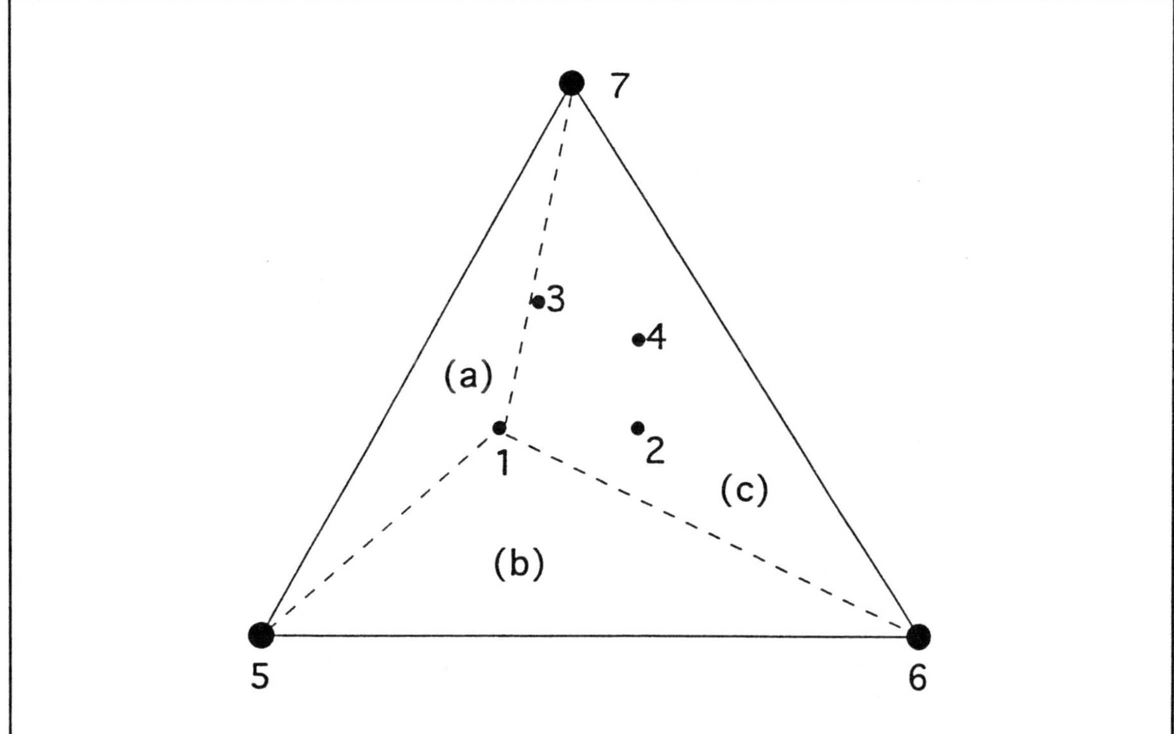

Figure 15-16a. Triangulation of a four-point domain by the Delaunay scheme.

4. Search begins for a triangle that encloses point 2. Recall that the search always starts from the last triangle formed or, equivalently, the last triangle on the list of T. Thus, search begins from triangle c. It happens that, in this example,

point 2 is found to be within this triangle, i.e., it is found in the first try! Now, three new triangles are formed by using point 2 and the vertices of triangle $c(1,6,7)$. The result is shown in Figure 15-16b. Observe that the number of triangles in the list T is now 5.

5. A check is now required to see whether swapping is necessary between triangle pair b,d and a,c. The result indicates that no swapping is required.

6. The search moves now to identify the triangle that encloses point 3. Again, note that since the search always begins from the last triangle formed, the search will start from triangle $e(2,6,7)$. The search path will be triangles e and c, as shown in Figure 15-16b. After point 3 is found to be inside triangle c, three new triangles are formed. The newly formed triangles are $c(3,1,2)$, $f(3,2,7)$ and $g(3,7,1)$, which are shown in Figure 15-16c.

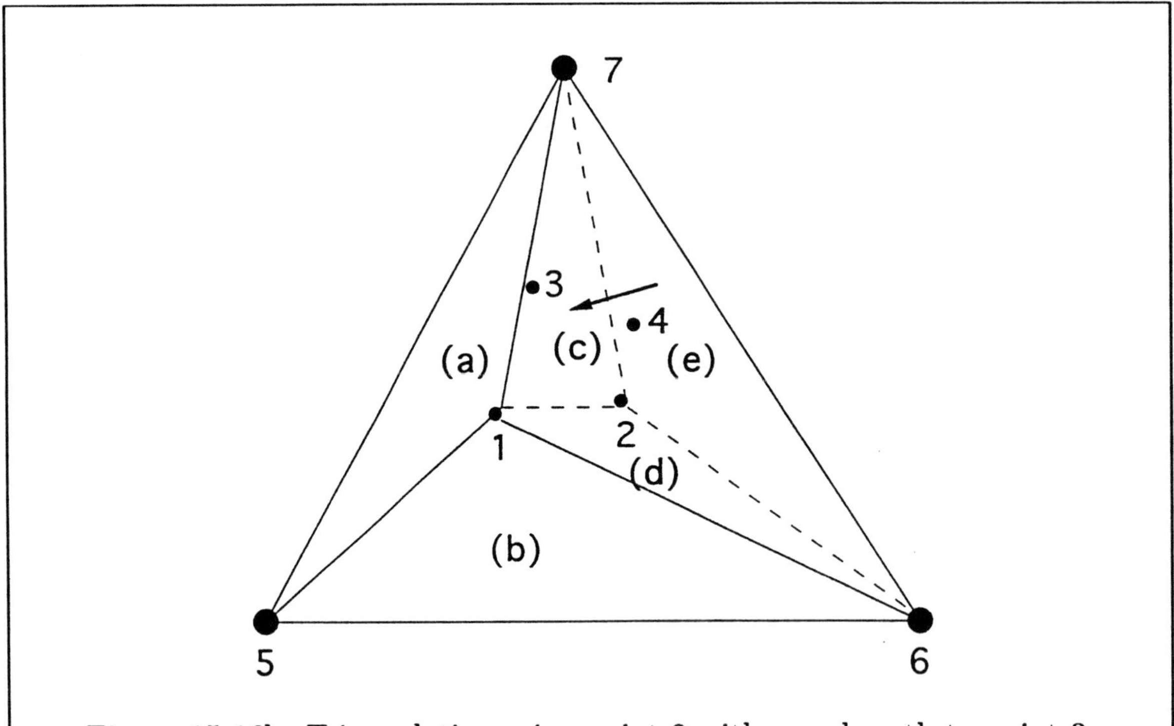

Figure 15-16b. Triangulation using point 2 with search path to point 3.

7. Again, a check is made to see if swapping is necessary between triangle pairs a,g, c,d, and f,e. The results indicate that no swapping is required.

8. The search moves to locate a triangle that encloses point 4. The search starts with the last triangle found, i.e., $g(3,7,1)$. The search path is g, c, f, and e (note that it is also possible to take the path of g, f, and e) as shown in Figure 15-16c. It is found that point 4 is inside the triangle e. Now, three new

triangles are formed as triangles $e(4,6,7)$, $h(4,2,6)$, and $i(4,7,2)$, as shown in Figure 15-16d.

9. A check is made to see whether swapping is necessary between triangle pairs i,f and h,d. The result shows that point 4 will be enclosed by the circumcircle of triangle f and, therefore, swapping is necessary. The two newly formed triangles after swapping are triangles $f(3,4,7)$ and $i(2,4,3)$, which are shown in Figure 15-16e.

10. At this point all four points have been used. Thus, the last step to complete the triangulation is to remove all the triangles that contain the vertices of the supertriangle. Therefore, all triangles which contain any of the points 5, 6, or 7 must be removed. The specific triangles which are removed are triangles a, b, d, e, f, g, and h. Therefore, the result of the triangulation for the four points are two triangles: $(1,2,3)$ and $(3,2,4)$ shown in Figure 15-16f.

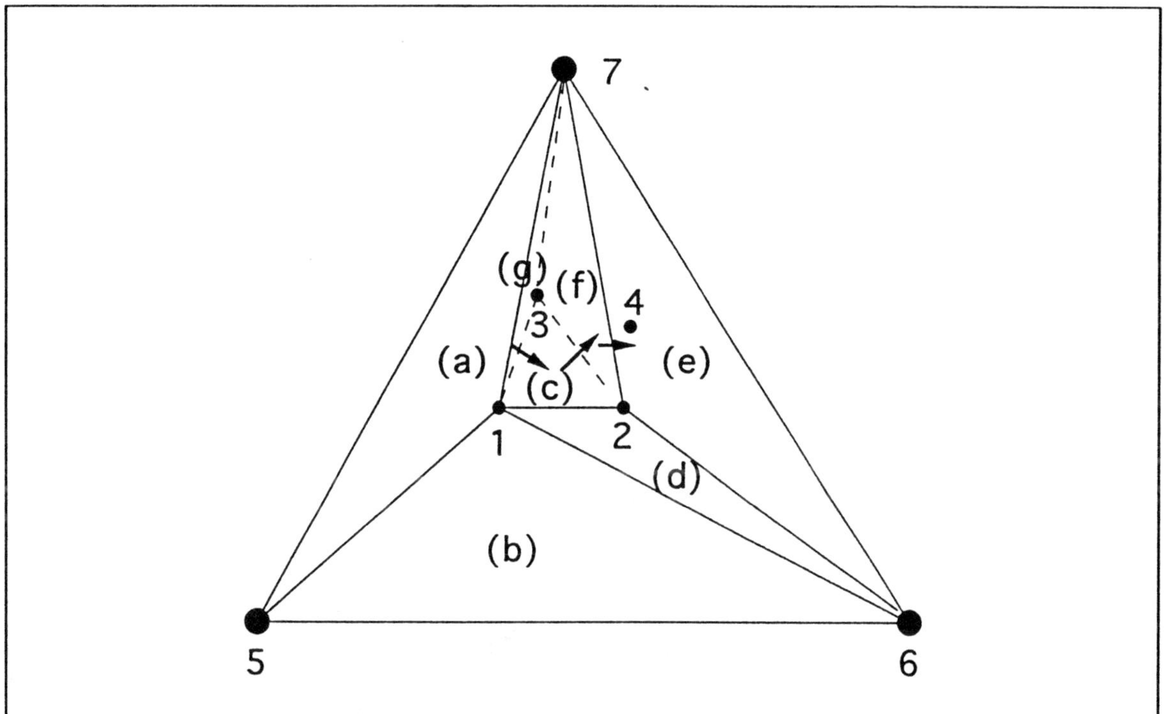

Figure 15-16c. Triangulation using point 3 with search path to point 4.

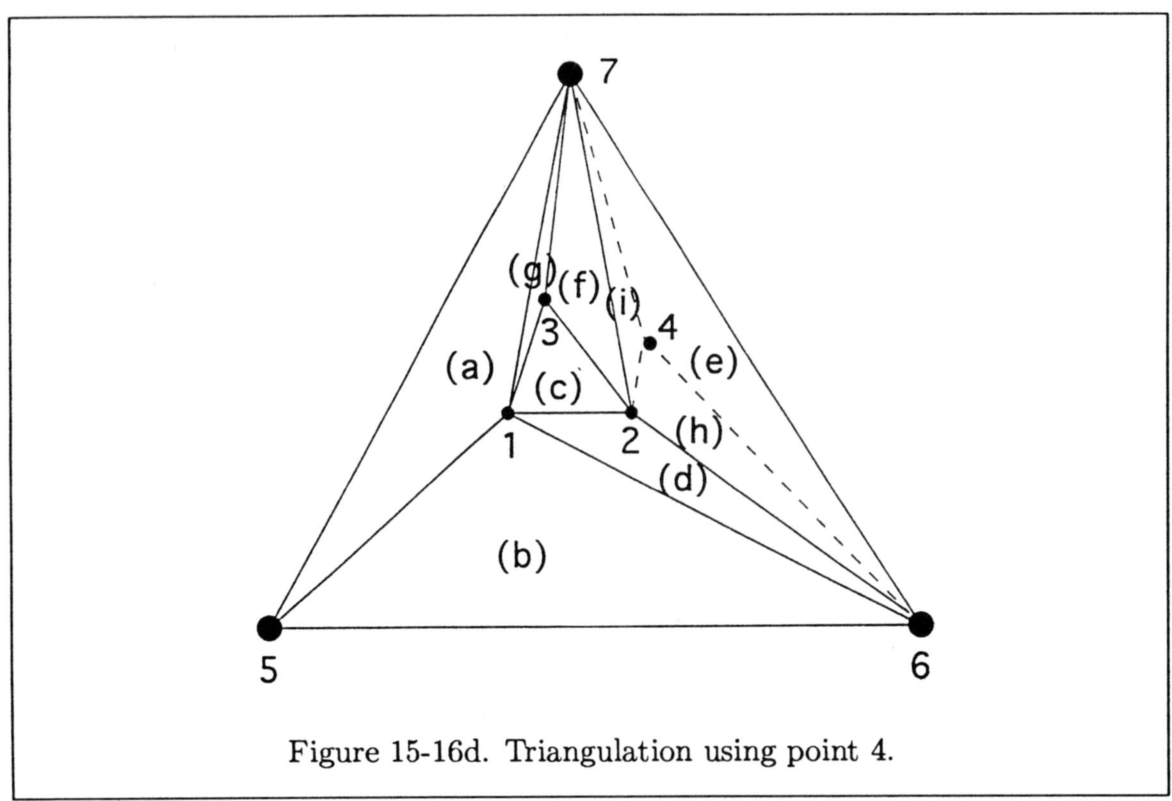

Figure 15-16d. Triangulation using point 4.

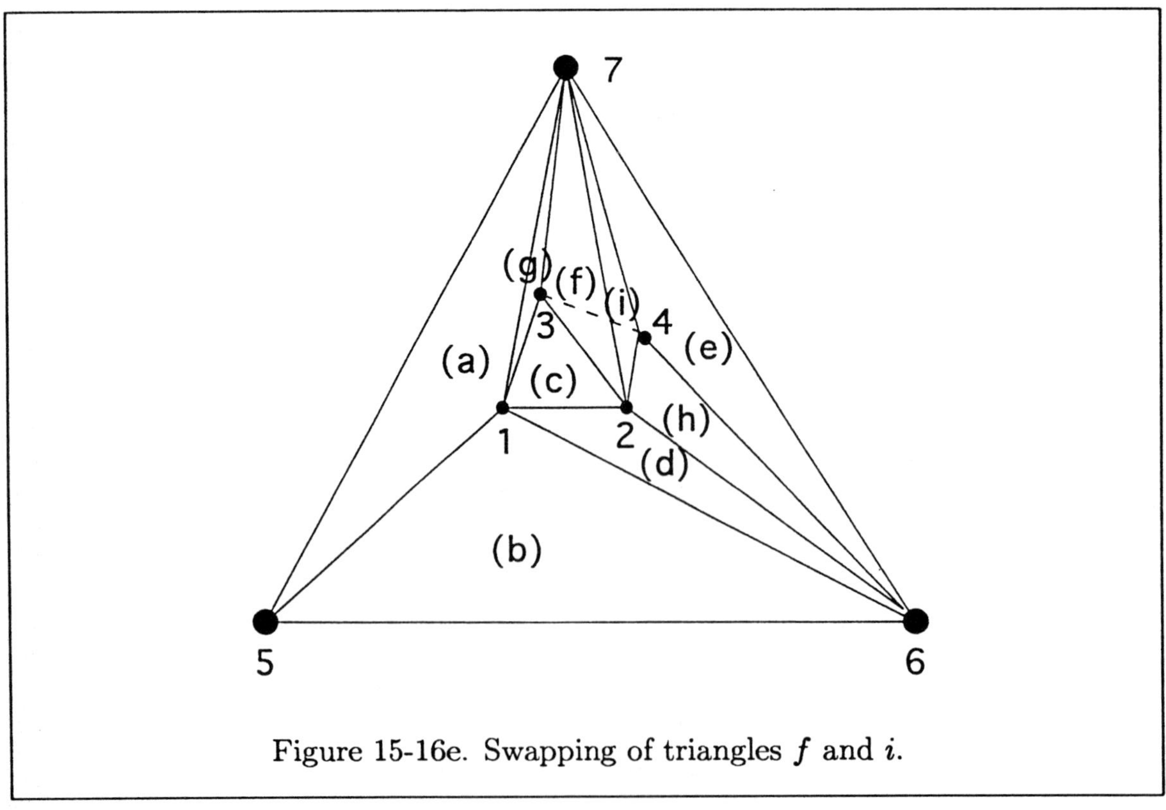

Figure 15-16e. Swapping of triangles *f* and *i*.

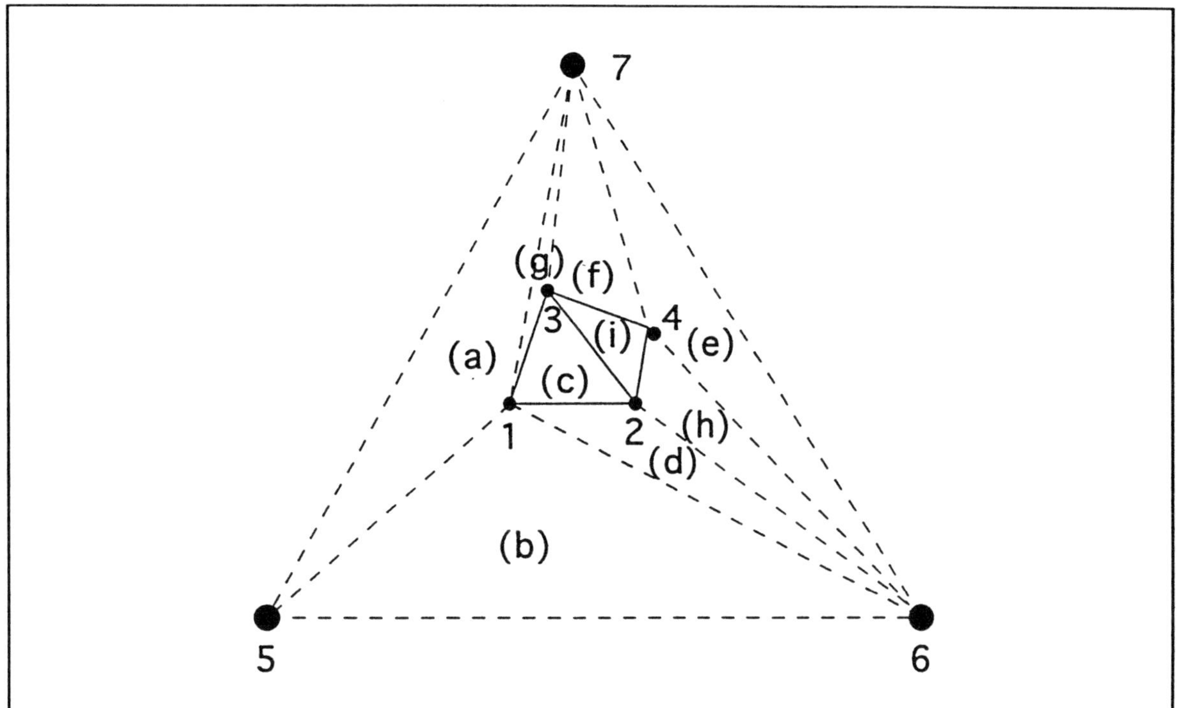

Figure 15-16f. Removal of triangles containing the vertices of the supertriangle.

15.4 Concluding Remarks

Unstructured grids which are used in conjunction with finite volume and finite element schemes were discussed in this chapter at the introductory level. The objectives were accomplished by considering a limited number of methods applied to two-dimensional problems. The specific schemes reviewed are Advancing Front and Delaunay methods. It is hoped that a detailed description of each scheme will facilitate the understanding of other schemes presented in various publications as well as the extension of the schemes to three-dimensions. The procedures introduced in this chapter will be utilized to generate the required grids for the applications of finite volume schemes in the following chapter.

15.5 Problems

15.1 Describe the advantages and disadvantages of unstructured grids. Also propose a scheme whereby a combination of structured and unstructured grids may be used to increase accuracy and efficiency of flowfield computations. For example, such a grid may be used to accurately compute the normal gradients of flow property such as velocity and temperature at the surface.

15.2 Consider a rectangular domain with dimensions of L and H, as shown in Figure P15.2. A circular half-cylinder is placed on the lower boundary at the midpoint. The following geometrical data is provided: $L1 = 2$, $R = 1$, $H = 4$, and $L = 6$. Use incremental distances of $\Delta L = 0.2$, $\Delta H = 0.4$, and an angular increment of $\Delta\theta = 18°$, to distribute grid points on the boundaries. (a) Generate the interior nodes by the scheme described in Section 15.2 and (b) triangularize the domain by using the Advancing Front method.

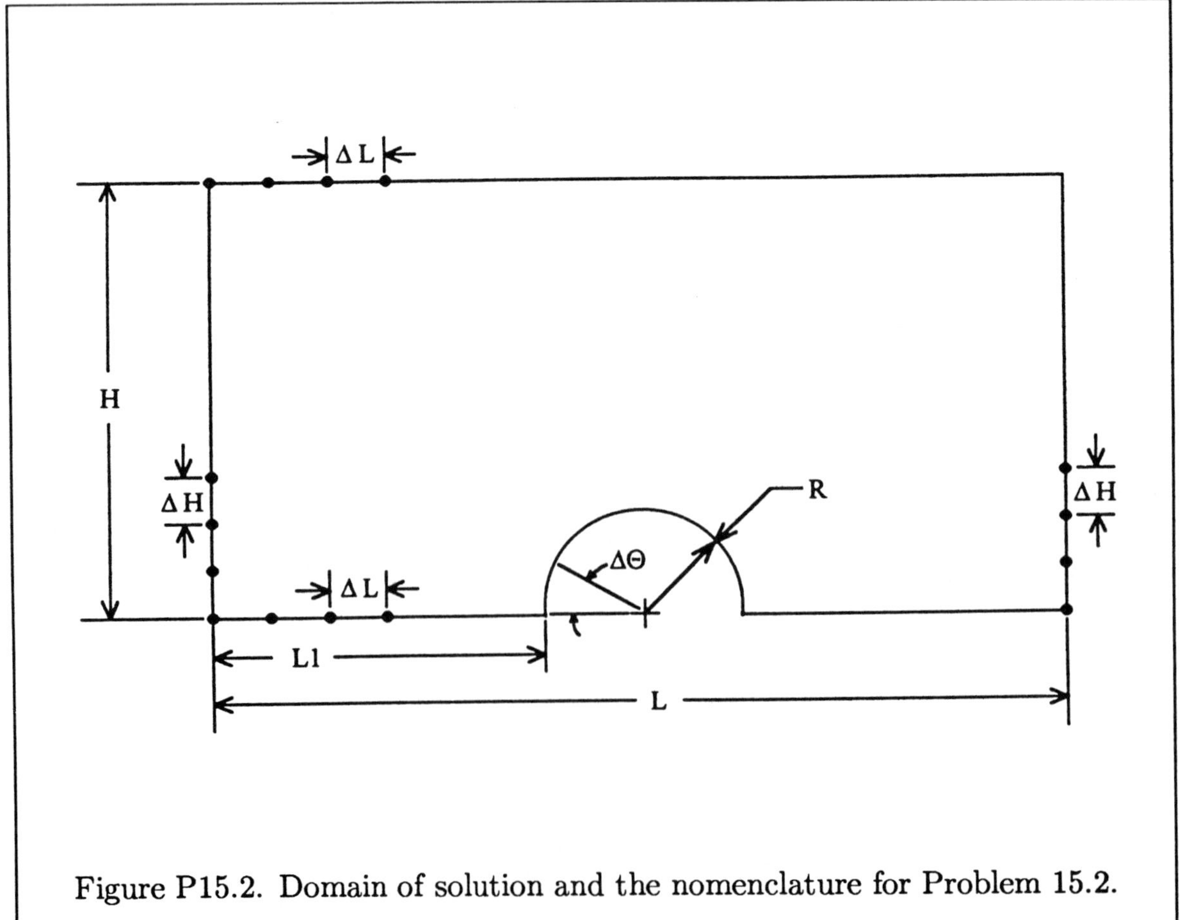

Figure P15.2. Domain of solution and the nomenclature for Problem 15.2.

15.3 Use the Delaunay method to triangularize the following specified nodes.

Node	1	2	3	4	5	6	7	8	9	10	11	12	13
x	0.0	1.0	2.0	2.0	2.0	1.0	0.0	0.0	0.5	1.5	1.0	0.5	1.5
y	0.0	0.0	0.0	1.0	2.0	2.0	2.0	1.0	0.5	0.5	1.0	1.5	1.5

Chapter 16

Finite Volume Method

16.1 Introductory Remarks

The conservation laws of fluid motion may be expressed mathematically in either differential form or integral form. When a numerical scheme is applied to the differential equation, the domain of solution is divided into discrete points, upon which the finite difference equations are solved. On the other hand, when the integral form of the equations is utilized, the domain of solution is divided into small volumes (or areas for a two-dimensional case). Subsequently, the conservation laws in integral form are applied to these elementary volumes. The integral methods include finite volume and finite element methods. In this chapter, the finite volume schemes are introduced.

Before proceeding to the details of the finite volume schemes, it is important to state the differences between the differential and integral methods so that the advantages and disadvantages of each method can be identified. The discussion will be limited to two dimensions, although the conclusions are valid for three dimensions as well.

Recall that the finite difference equations which approximate the partial differential equations are solved within a rectangular domain at equally distanced discrete points. Since the majority of physical domains are irregular in shape, a coordinate transformation from a physical space to a computational space is performed where the computational domain is rectangular. The procedure was described in Chapters 9 and 11, whereas the applications were shown in Chapters 12, 13, and 14. However, even with the coordinate transformation available, domains which are highly irregular would create serious difficulties in accuracy and convergence of the solution. The reason is that the metrics and Jacobian of transformation and the corresponding gradients which are used in the governing equations may include numerical discontinuities if the grid system is not relatively smooth. To overcome this difficulty which is encountered when a complicated domain is involved, one incorporates the use of a zonal (or block) grid system. The zonal grid system discretizes the physical domain into several overlapped subdomains and subsequently solves the finite difference equations on these subdomains. The overlapped regions serve as interfaces between subdomains. The solution of one sub-domain is communicated with other subdomains through these overlapped regions, i.e., these regions are used as boundary conditions for other subdomains. An example of a zonal grid system is shown in Figure 16-1. It is noted that the numerical implementation of

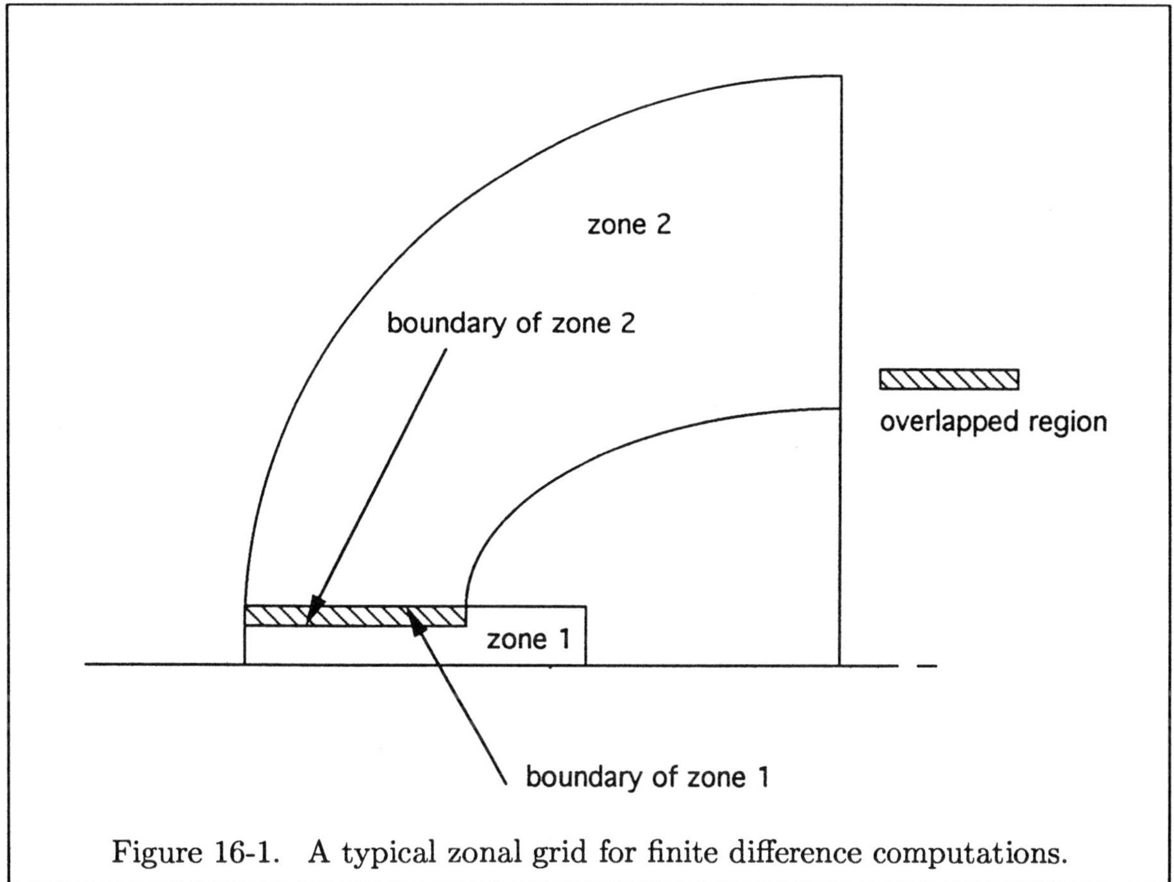

Figure 16-1. A typical zonal grid for finite difference computations.

the zonal grid approach may become cumbersome if one is to develop a general purpose program which can be used for any arbitrary configuration. At this point, it may be concluded that, in general, the finite difference methods possess inherited weaknesses for highly complicated domains. On the other hand, finite volume (or finite element) schemes do not encounter such weakness. That is because the independent variables are integrated directly on the physical domain and, therefore, grid smoothness is no longer an important issue. Thus, the governing equations can be solved if only the domain can be successfully discretized into elements. The geometrical difficulty is now the concern of the grid generation routine and not of the finite volume solver. Furthermore, finite volume schemes do not require a structured grid, as is required of the finite difference schemes; therefore, for most applications, unstructured grids are used. It is also important to emphasize that, since the integral equations are applied directly on the physical domain, a coordinate transformation is no longer required. It is then clear that the finite volume methods have advantages over the finite difference method if the geometry of the domain is complicated. That is, finite volume schemes provide great flexibility, in that a wide range of choices is available for the selection of discrete volumes. However, it should be noted that if the domain can be discretized into a smooth structured grid, the finite difference method would be a better choice due to its efficiency over that of the finite volume or finite element methods.

16.2 General Description of the Finite Volume Method

Finite volume schemes can be generally categorized into two groups; the first group includes the "cell-centered" schemes, and the second group includes the "Nodal Point" schemes. To illustrate each approach and identify the differences between the two, consider the following model equation:

$$\frac{\partial Q}{\partial t} + \frac{\partial E}{\partial x} + \frac{\partial F}{\partial y} = 0 \tag{16-1}$$

Initially, Equation (16-1) is integrated over an element such as quadrilateral mesh *abcd* shown in Figure 16-2a. Thus, one has

$$\int_{abcd} \left(\frac{\partial Q}{\partial t} \right) dxdy = - \int_{abcd} \left(\frac{\partial E}{\partial x} + \frac{\partial F}{\partial y} \right) dxdy \tag{16-2}$$

Subsequently, Green's Theorem is applied to the right-hand side of Equation (16-2). Recall that Green's Theorem converts area integrals to line integrals. Thus, Equation (16-2) is written as

$$\int_{abcd} \left(\frac{\partial Q}{\partial t} \right) dxdy = - \oint_{abcd} (Edy - Fdx) \tag{16-3}$$

Equation (16-3) is used to develop a cell-centered scheme as well as a nodal point scheme in the following sections.

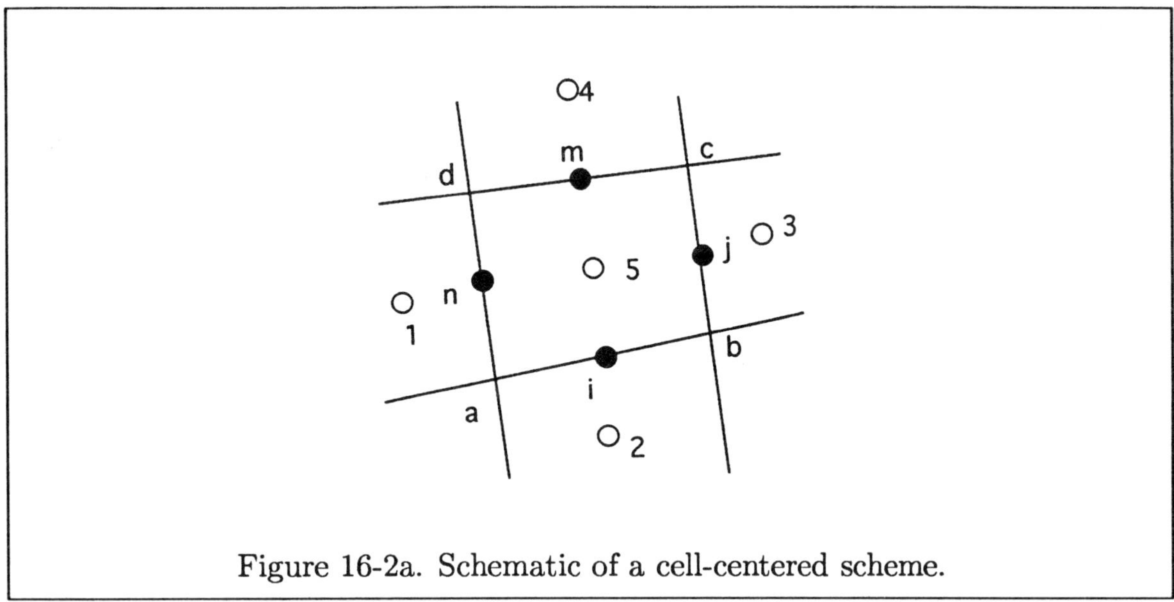

Figure 16-2a. Schematic of a cell-centered scheme.

16.2.1 Cell-Centered Scheme

The integrals in Equation (16-3) are approximated over the element shown in Figure 16-2a. The dependent variable Q is to be solved for at node 5 which is located at the center of the element, and hence the scheme is referred to as a *cell-centered scheme.* The formulation can be expressed in either explicit or implicit forms, depending on how E and F are evaluated. This point is addressed shortly. For now, Equation (16-3) is approximated as

$$\left(\frac{Q_5^{n+1} - Q_5^n}{\Delta t}\right) A_{abcd} = -\left[E_i \Delta y_{ab} + E_j \Delta y_{bc} + E_m \Delta y_{cd} + E_n \Delta y_{da}\right]$$

$$+\left[F_i \Delta x_{ab} + F_j \Delta x_{bc} + F_m \Delta x_{cd} + F_n \Delta x_{da}\right] \qquad (16\text{-}4)$$

where A_{abcd} is the area of the cell and points i, j, m, and n represent the midpoint locations of the edges ab, bc, cd, and da, respectively. Points 1, 2, 3, 4, and 5 are called the control points of the five quadrilaterals. The x and y increments of each edge are determined by the following relations:

$$\Delta x_{ab} = x_b - x_a \ , \quad \Delta x_{bc} = x_c - x_b \ , \quad \Delta x_{cd} = x_d - x_c \ , \quad \Delta x_{da} = x_a - x_d \qquad (16\text{-}5a)$$

$$\Delta y_{ab} = y_b - y_a \ , \quad \Delta y_{bc} = y_c - y_b \ , \quad \Delta y_{cd} = y_d - y_c \ , \quad \Delta y_{da} = y_a - y_d \qquad (16\text{-}5b)$$

The values of functions E and F at the midpoints of the edges can be evaluated by averaging their values from the two control points located on the opposite sides of the corresponding edge. That is,

$$E_i = \frac{1}{2}(E_5^* + E_2^*), \qquad E_j = \frac{1}{2}(E_5^* + E_3^*),$$
$$E_m = \frac{1}{2}(E_5^* + E_4^*), \qquad E_n = \frac{1}{2}(E_5^* + E_1^*), \tag{16-6a}$$

$$F_i = \frac{1}{2}(F_5^* + F_2^*), \qquad F_j = \frac{1}{2}(F_5^* + F_3^*),$$
$$F_m = \frac{1}{2}(F_5^* + F_4^*), \qquad F_n = \frac{1}{2}(F_5^* + F_1^*) \tag{16-6b}$$

In the equations above, if the values of the functions designated by $*$ are evaluated at time level n, then the formulation is an explicit formulation. If the values at $*$ are evaluated at $n+1$ time level, the formulation is an implicit scheme. Furthermore, note that: (1) if the element $abcd$ is rectangular in shape, then $\Delta x_{ab} = -\Delta x_{cd}$, $\Delta x_{bc} = -\Delta x_{da}$, $\Delta y_{ab} = -\Delta y_{cd}$, $\Delta y_{bc} = -\Delta y_{da}$. Therefore, Equation (16-4) is simply equivalent to a finite difference formulation where central difference approximation of the spatial derivatives are used. The proof is required in Problem 16-1. (2) The values of E and F as provided by (16-6a) and (16-6b) may be obtained by other schemes as well. For example, one may use one-sided approximations (such as upwind schemes) to provide the values of E and F. A description of such schemes is presented in Section 16.4.

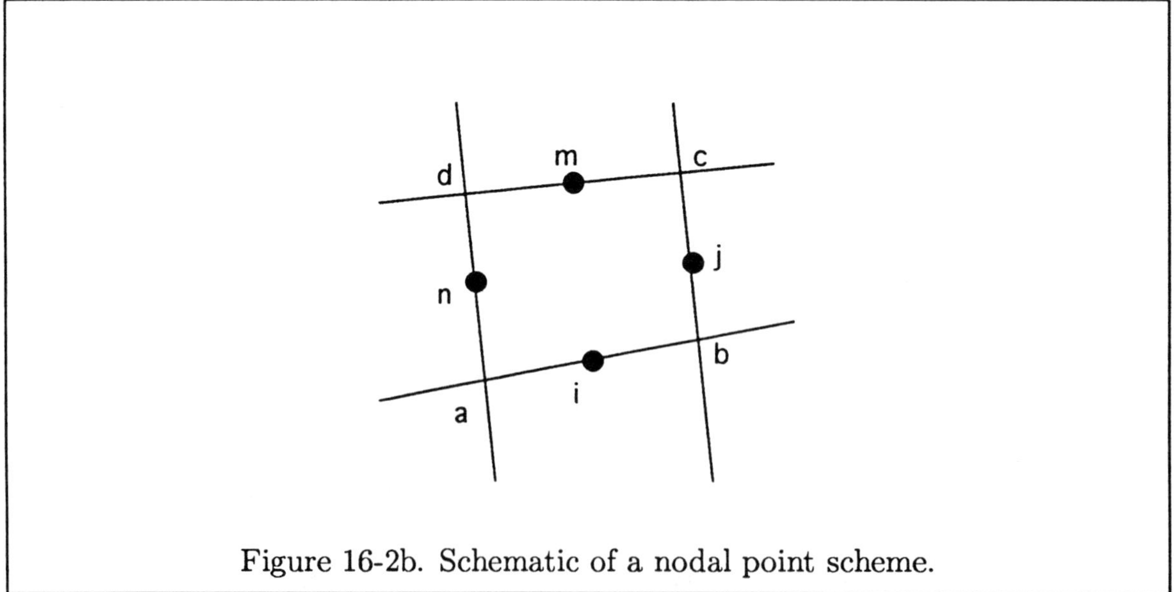

Figure 16-2b. Schematic of a nodal point scheme.

16.2.2 Nodal Point Scheme

In this approach the dependent variable is evaluated at the vertices of the element. Thus, the approximation of the governing equation given by (16-3) over the element shown in Figure 16-2b is as follows:

$$\left[\frac{(Q_a+Q_b+Q_c+Q_d)^{n+1}-(Q_a+Q_b+Q_c+Q_d)^n}{4\Delta t}\right]A_{abcd}=$$

$$-(E_i\Delta y_{ab}+E_j\Delta y_{bc}+E_m\Delta y_{cd}+E_n\Delta y_{da})+(F_i\Delta x_{ab}+F_j\Delta x_{bc}+F_m\Delta x_{cd}+F_n\Delta x_{da}) \quad (16\text{-}7)$$

where

$$E_i=\frac{1}{2}(E_a^*+E_b^*)\,,\quad E_j=\frac{1}{2}(E_b^*+E_c^*)\,,$$
$$E_m=\frac{1}{2}(E_c^*+E_d^*)\,,\quad E_n=\frac{1}{2}(E_d^*+E_a^*)\,, \quad (16\text{-}8a)$$

$$F_i=\frac{1}{2}(F_a^*+F_b^*)\,,\quad F_j=\frac{1}{2}(F_b^*+F_c^*)\,,$$
$$F_m=\frac{1}{2}(F_c^*+F_d^*)\,,\quad F_n=\frac{1}{2}(F_d^*+F_a^*)\,, \quad (16\text{-}8b)$$

Again, the values of E and F can be evaluated at either time levels of n or $n+1$ to provide an explicit scheme or an implicit scheme, respectively.

Our discussion up to this point has been limited to a model equation which contains only time derivatives and convective derivatives, i.e., a first-order partial differential equation. Furthermore, the approximation of the integral equation was performed over a quadrilateral element. Most practical applications include a diffusion term, i.e., the governing partial differential equation is second-order. Thus, the schemes introduced in this section must be extended to a second-order partial differential equation. In addition, to illustrate the application of the scheme to a variety of available elements, instead of a quadrilateral element which was used in this section, a triangular element will be used. Thus, the reader is exposed to the development of the finite volume scheme applied over different types of elements.

An important issue with regard to finite volume schemes, which has not been addressed as yet, is the implementation of the boundary conditions, which will be explored in the next section.

16.3 Two-Dimensional Heat Conduction Equation

To achieve the goals set above, the two-dimensional heat conduction equation is used as the model equation. The particular application is that of Section 3.7. Therefore, an easy comparison between the finite difference solution and the finite volume solution can be made. Obviously, grid generation is the first step in the solution procedure of any scheme. Since one of the goals is to experience the application of the finite volume scheme to triangular elements, the domain of solution is triangularized. For example, consider the domain shown in Figure 16-3, which has been triangularized. The triangular elements can be divided into two groups defined as "boundary triangles" and "interior triangles." A boundary triangle is defined as a triangle with at least two vertices on the boundary of the domain (or, equivalently, at least one edge is coincident with the boundary). The boundary triangles in Figure 16-3 are identified by the shaded triangles. Pre-set boundary conditions must be imposed for the boundary triangles when computations of the temperature at the control points of such triangles is performed. A detailed description of how to impose these requirements is presented in Section 16.3.2. The remaining triangles within the domain belong to the "interior triangle" group. The governing equations can be directly applied to the interior triangles and are described in the next section.

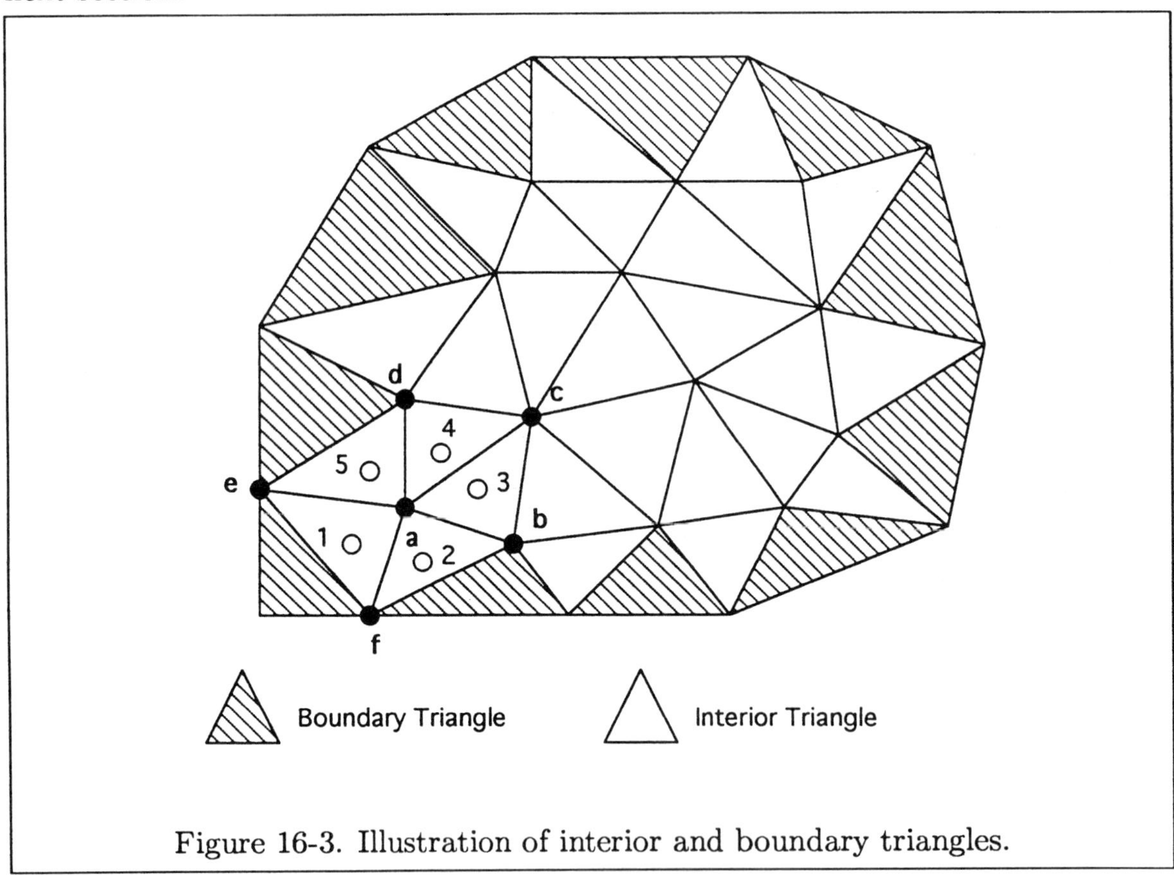

Figure 16-3. Illustration of interior and boundary triangles.

16.3.1 Interior Triangles

Recall the two-dimensional heat conduction equation given by

$$\frac{\partial T}{\partial t} = \alpha \left(\frac{\partial^2 T}{\partial x^2} + \frac{\partial^2 T}{\partial y^2} \right) \tag{16-9}$$

where α is assumed to be a constant and hence, a linear equation. Consider now a triangular element such as that identified by triangle abc of Figure 16-4. Observe

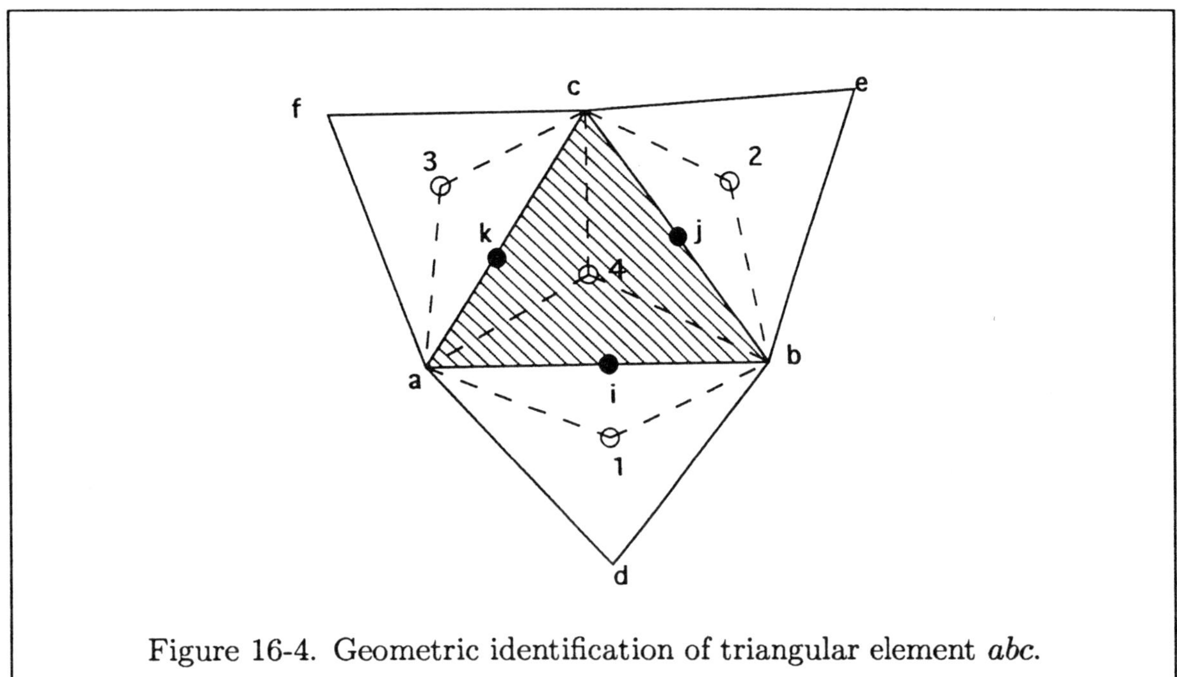

Figure 16-4. Geometric identification of triangular element *abc*.

that the identification of triangles is ordered in a counter-clockwise (ccw) fashion. Integration of Equation (16-9) yields

$$\int_{abc} \left(\frac{\partial T}{\partial t} \right) dx dy = \alpha \int_{abc} \left(\frac{\partial^2 T}{\partial x^2} + \frac{\partial^2 T}{\partial y^2} \right) dx dy \tag{16-10}$$

The approximation of the left integral is simply accomplished by the following expression:

$$\int_{abc} \left(\frac{\partial T}{\partial t} \right) dx dy = \left(\frac{T_4^{n+1} - T_4^n}{\Delta t} \right) A_{abc} \tag{16-11}$$

where T_4 is the value of temperature at the control point 4. Before proceeding with the evaluation of the integrals on the right-hand side of Equation (16-10), let's pose the following question. Where is the exact location of control point 4 (and similarly the location of other control points such as 1, 2, and 3)? A simple scheme to identify its location is to use the centroid (average value of the vertices) of the triangle as the location of the control points. Thus,

$$x_4 = \frac{1}{3}(x_a + x_b + x_c) \tag{16-12a}$$

$$y_4 = \frac{1}{3}(y_a + y_b + y_c) \tag{16-12b}$$

and

$$A_{abc} = \frac{1}{2}(x_b y_c + x_a y_b + x_c y_a - x_b y_a - x_c y_b - x_a y_c) \tag{16-13}$$

What about the initial value of T_4? That is, given the values of temperature at grid points (such as T_a, T_b, and T_c), how does one determine T_4^n, which is required in relation (16-11)? A simple and efficient scheme is to use the weighted average value of the temperature at the three vertices. Therefore,

$$T_4^n = \left(\frac{T_a^n}{L_{4a}} + \frac{T_b^n}{L_{4b}} + \frac{T_c^n}{L_{4c}}\right) \Big/ \left(\frac{1}{L_{4a}} + \frac{1}{L_{4b}} + \frac{1}{L_{4c}}\right) \tag{16-14}$$

where L_{4a}, L_{4b}, and L_{4c} are the distances between control point 4 and the vertices a, b, and c, respectively. The same approach is applied to all other control points of the interior triangles, thus providing their coordinates, the areas, and the values of temperatures at the required time level.

Now return to the right-hand side of Equation (16-10). First, define

$$F = \frac{\partial T}{\partial x} \quad \text{and} \quad G = \frac{\partial T}{\partial y}$$

Therefore,

$$\int_{abc} \left(\frac{\partial^2 T}{\partial x^2} + \frac{\partial^2 T}{\partial y^2}\right) dxdy = \int_{abc} \left(\frac{\partial F}{\partial x} + \frac{\partial G}{\partial y}\right) dxdy = \oint_{abc} (Fdy - Gdx) \tag{16-15}$$

where Green's Theorem has been applied in the last step. Using the triangle *abc*, one may approximate the integral in (16-15) as

$$\begin{aligned} \text{RHS} &= \oint_{abc} (Fdy - Gdx) \\ &= (F_i \Delta y_{ab} + F_j \Delta y_{bc} + F_k \Delta y_{ca}) - (G_i \Delta x_{ab} + G_j \Delta x_{bc} + G_k \Delta x_{ca}) \end{aligned} \tag{16-16}$$

where

$$\Delta x_{ab} = x_b - x_a, \quad \Delta x_{bc} = x_c - x_b, \quad \Delta x_{ca} = x_a - x_c \tag{16-17a}$$

$$\Delta y_{ab} = y_b - y_a, \quad \Delta y_{bc} = y_c - y_b, \quad \Delta y_{ca} = y_a - y_c \tag{16-17b}$$

For an explicit formulation, the unknown T_4 is determined from

$$T_4^{n+1} = T_4^n + \alpha \frac{\Delta t}{A_{abc}} (\text{RHS}) \tag{16-18}$$

What about the computation of functions F and G at points i, j, and k, which are required in Equation (16-16)? They are evaluated by the following relations:

$$\begin{aligned} F_i &= \left(\frac{\partial T}{\partial x}\right)_i = \left[\int_{a1b4} \left(\frac{\partial T}{\partial x}\right) dxdy\right] \Big/ A_{a1b4} \\ &= \Big[(T_a^n + T_1^n)\Delta y_{a1} + (T_1^n + T_b^n)\Delta y_{1b} + (T_b^n + T_4^n)\Delta y_{b4} \\ &\quad + (T_4^n + T_a^n)\Delta y_{4a}\Big] \Big/ (2A_{a1b4}) \end{aligned} \tag{16-19}$$

$$F_j = \left(\frac{\partial T}{\partial x}\right)_j = \left[\int_{b2c4}\left(\frac{\partial T}{\partial x}\right) dxdy\right] \Big/ A_{b2c4}$$

$$= \Big[(T_b^n + T_2^n)\Delta y_{b2} + (T_2^n + T_c^n)\Delta y_{2c} + (T_c^n + T_4^n)\Delta y_{c4}$$

$$+(T_4^n + T_b^n)\Delta y_{4b}\Big] \Big/ (2A_{b2c4}) \tag{16-20}$$

$$F_k = \left(\frac{\partial T}{\partial x}\right)_k = \left[\int_{c3a4}\left(\frac{\partial T}{\partial x}\right) dxdy\right] \Big/ A_{c3a4}$$

$$= \Big[(T_c^n + T_3^n)\Delta y_{c3} + (T_3^n + T_a^n)\Delta y_{3a} + (T_a^n + T_4^n)\Delta y_{a4}$$

$$+(T_4^n + T_c^n)\Delta y_{4c}\Big] \Big/ (2A_{c3a4}) \tag{16-21}$$

where, in Equations (16-19) through (16-21), the Δy increments are computed by

$$\Delta y_{mn} = y_n - y_m$$

The function G is evaluated in a similar fashion according to

$$G_i = \left(\frac{\partial T}{\partial y}\right)_i = \left[\int_{a1b4}\left(\frac{\partial T}{\partial y}\right) dxdy\right] \Big/ A_{a1b4}$$

$$= -\Big[(T_a^n + T_1^n)\Delta x_{a1} + (T_1^n + T_b^n)\Delta x_{1b} + (T_b^n + T_4^n)\Delta x_{b4}$$

$$+ (T_4^n + T_a^n)\Delta x_{4a}\Big] \Big/ (2A_{a1b4}) \tag{16-22}$$

$$G_j = \left(\frac{\partial T}{\partial y}\right)_j = \left[\int_{b2c4}\left(\frac{\partial T}{\partial y}\right) dxdy\right] \Big/ A_{b2c4}$$

$$= -\Big[(T_b^n + T_2^n)\Delta x_{b2} + (T_2^n + T_c^n)\Delta x_{2c} + (T_c^n + T_4^n)\Delta x_{c4}$$

$$+ (T_4^n + T_b^n)\Delta x_{4b}\Big] \Big/ (2A_{b2c4}) \tag{16-23}$$

$$G_k = \left(\frac{\partial T}{\partial x}\right)_k = \left[\int_{c3a4}\left(\frac{\partial T}{\partial x}\right) dxdy\right] \Big/ A_{c3a4}$$

$$= -\Big[(T_c^n + T_3^n)\Delta x_{c3} + (T_3^n + T_a^n)\Delta x_{3a} + (T_a^n + T_4^n)\Delta x_{a4}$$

$$+ (T_4^n + T_c^n)\Delta x_{c4}\Big] \Big/ (2A_{c3a4}) \tag{16-24}$$

Again, the Δx increments are determined by

$$\Delta x_{mn} = x_n - x_m$$

With regard to expressions (16-19) through (16-24), observe the following: (1) In Equations (16-19) through (16-24), the nodal point scheme is used to integrate the expressions for F and G at the three quadrilaterals $a1b4$, $b2c4$, and $c3a4$. (2) The corresponding areas are evaluated according to

$$A_{a1b4} = \frac{1}{3}(A_{abc} + A_{adb})$$

$$A_{b2c4} = \frac{1}{3}(A_{abc} + A_{bec})$$

and

$$A_{c3a4} = \frac{1}{3}(A_{abc} + A_{cfa})$$

Substitution of Equations (16-19) through (16-24) into Equation (16-16) provides the value of RHS which is subsequently used in Equation (16-18) to provide the value of T_4^{n+1}. The procedure described above is applied to all of the interior triangles to provide the temperature values at the control points of interior triangles at the time level of $n+1$. Subsequently, the temperature values are updated at the vertices. This update may be accomplished by averaging of the temperatures of the control points surrounding the vertex. For example, consider vertex a shown in Figure 16-3. Observe that point a is surrounded by control points 1, 2, 3, 4, and 5. Thus, one may approximate

$$T_a^{n+1} = \frac{\sum_{m=1}^{5} \frac{T_m^{n+1}}{L_{ma}}}{\sum_{m=1}^{5} \frac{1}{L_{ma}}} \tag{16-25}$$

where L_{ma} is the distance between point m and the vertex a evaluated by

$$L_{ma} = \left[(x_m - x_a)^2 + (y_m - y_a)^2\right]^{\frac{1}{2}}$$

The next step is to proceed with the boundary triangles, and it is addressed in the following section.

16.3.2 Boundary Triangles

Various types of boundary conditions were defined in Chapter 1 and then implemented into numerical schemes in the subsequent chapters. A review of the physical boundary conditions as applied to the heat conduction equation at this point is helpful. For a Dirichlet-type boundary condition, simply the temperature value, or the temperature distribution along the boundary is specified. For a Neumann-type boundary condition, the heat flux at the boundary is provided. Recall that the heat flux is given by

$$q = -k\frac{\partial T}{\partial n}$$

where k is the thermal conductivity. The third type boundary condition is the mixed type, which for this problem may be expressed as $CT + D(\partial T/\partial n)$. In this section, the implementation of Dirichlet and Neumann types is described. The extension to mixed boundary conditions is straightforward.

16.3.2.1 Dirichlet-Type Boundary Condition

Since the temperature distribution along the boundary is specified for a Dirichlet-type boundary condition, the temperature value at the control points of boundary triangles can be easily determined. For this purpose, the temperature value at the control point is updated using the weighted average scheme, i.e., Equation (16-14).

16.3.2.2 Neumann-Type Boundary Condition

It is apparent that the application of Neumann-type boundary conditions is not as simple as the Dirichlet type. The difficulty is encountered because, in most cases, the locations of nodes are such that normal gradients at the surface cannot be directly defined. Nevertheless, by incorporation of an indirect, iterative scheme, the flux boundary condition can be enforced. To illustrate the difficulty and subsequently the solution procedure, consider a heat distribution given along the boundary ...*idabg*, ...shown in Figure 16-5. As part of the overall solution, it is required to determine the temperature values at points ...i, d, a, b, g, ..., and on the control points 1, 2, 3, 4, ... The solution procedure is developed as follows.

(i) For a typical boundary triangle such as triangle *abc*, define by m the midpoint of boundary edge *ab*. Thus, the coordinates of m can be determined by

$$x_m = \frac{1}{2}(x_a + x_b) \tag{16-26a}$$

and

$$y_m = \frac{1}{2}(y_a + y_b) \tag{16-26b}$$

The value of heat flux is determined by

$$q_m = \frac{1}{2}(q_a + q_b) \tag{16-27}$$

(ii) Generate a line perpendicular to edge *ab* from point m. Identify the line by mm'. Now, determine the intersection of line mm' with the edges of the

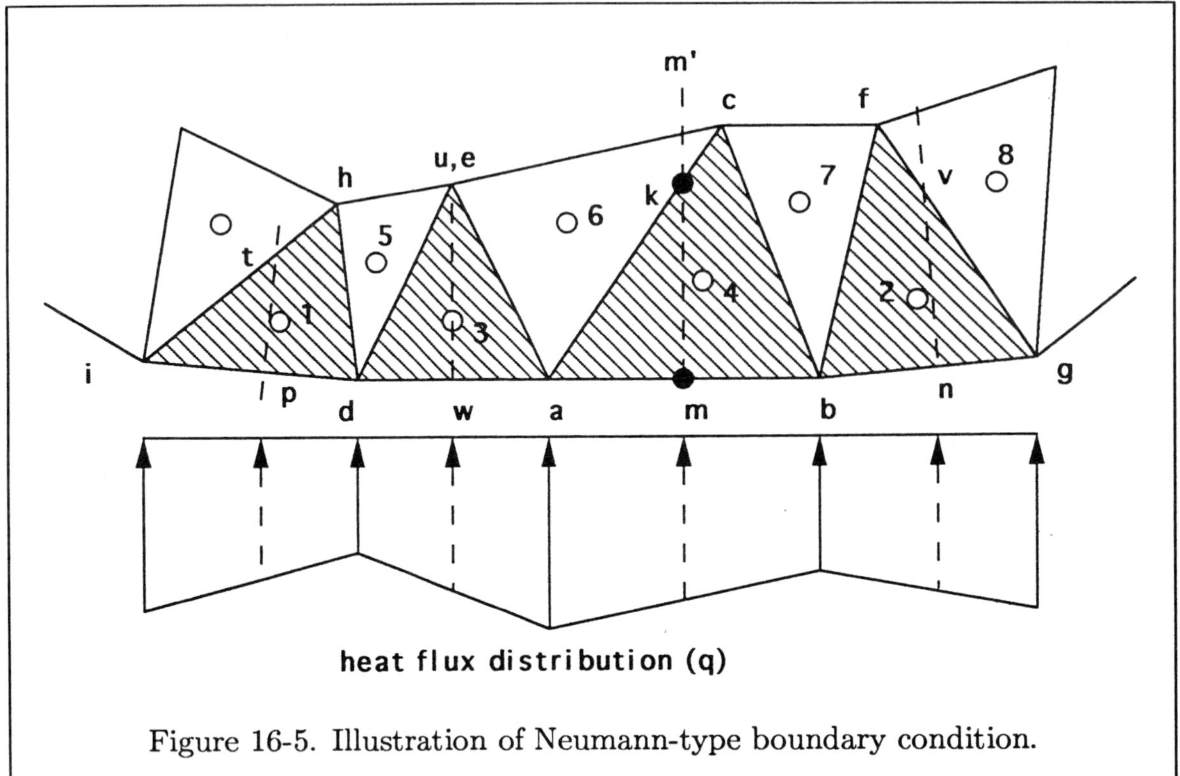

Figure 16-5. Illustration of Neumann-type boundary condition.

triangle, i.e., either edge ac or bc. To generalize, represent either one of the vertices a or b by z. Thus, the edge which intersects line mm' is zc. Call the intersection point k. Define parameter R by

$$R = \frac{L_{kc}}{L_{zc}}$$

where L_{kc} is the distance between point k and vertex c; and similarly, L_{zc} is the distance between point z and vertex c. Now, the coordinates of point k and the temperature value at that point can be determined according to

$$x_k = x_c + R(x_z - x_c) \tag{16-28a}$$

$$y_k = y_c + R(y_z - y_c) \tag{16-28b}$$

$$T_k = T_c + R(T_z - T_c) \tag{16-28c}$$

Recall that z is used to designate either one of the vertices a or b. In order to use Equation (16-28), one needs to know which one of the points a or b must be utilized. The following procedure is used for this purpose. First, the relation between the slopes of the two perpendicular lines km and ab is written as

$$\left(\frac{y_k - y_m}{x_k - x_m}\right)\left(\frac{y_b - y_a}{x_b - x_a}\right) = -1 \tag{16-29}$$

Relations (16-28a) and (16-28b) are substituted into (16-29) and solved for R to provide

$$R = \frac{(x_b - x_a)(x_c - x_m) + (y_b - y_a)(y_c - y_m)}{(x_z - x_c)(x_a - x_b) + (y_z - y_c)(y_a - y_b)} \tag{16-30}$$

Now the coordinate of point a is substituted into (16-30) to provide a value for parameter R; call it R_a. Similarly, one determines R_b by substitution of the coordinate of point b into (16-30). Recall that point k must be located between points z and c and, therefore, R must satisfy the requirement $0 \leq R < 1$. If now $0 \leq R_a < 1$, then the point defined by z is the vertex a. Otherwise $0 \leq R_b < 1$, and point z is assigned to be vertex b. Observe that if $R = 0$, then point k is coincident with vertex c. For the example shown in Figure 16-5, point z would be the vertex a.

(iii) Once point z has been identified, the location of point k is determined by Equations (12-28a) and (12-28b). Thus, the distance between points m and k can be calculated as well. Finally, the temperature value at point m is computed by

$$T_m = T_k - q_m L_{km}/k \tag{16-31}$$

Note that Equation (16-31) is obtained from the heat conduction law, i.e.,

$$q_m = -k\frac{\partial T}{\partial n} = -k\left(\frac{T_m - T_k}{L_{km}}\right)$$

Furthermore, recall that the unit normal to the surface is usually defined as positive in the outward direction of a closed domain. Therefore, for $q_m < 0$, the heat is flowing into the domain. Note that the temperature at point k cannot be treated as a fixed value and must be updated within an iterative loop along with temperatures on the boundaries. Thus, an iterative scheme is required to determine the values of the temperature T_m, T_a, T_b, and T_c. The details of this procedure are described in the following four steps.

1. Calculate the temperatures of all the midpoints, i.e., T_p, T_w, T_m, and T_n in Figure 16-5 by the method described in Steps (i) through (iii).

2. Update the temperature at the vertex by averaging the temperatures of the two neighboring midpoints, e.g.,

$$T_a = (T_m + T_w)/2 \ , \quad T_b = (T_m + T_n)/2 \ , \quad T_d = (T_w + T_p)/2$$

   Note that this simple averaging is based on the assumption that $L_{id} \equiv L_{da} \equiv L_{ab} \equiv L_{bg}$.

3. Update temperatures T_t, T_u, T_k, and T_v of intersection points t, u, k, and v by employing the interpolation method of Equation (16-28c). Note that the temperatures at vertices (T_i, T_d, T_a, T_b, and T_g) are now the updated values as provided from Step 2 and are not their original values.
4. Steps 1 through 3 are repeated until the temperature distribution along the boundary is converged.

(iv) Once the convergence of the temperature distribution on the boundary has been achieved (i.e., T_i, T_d, T_a, T_b, and T_g), the temperature of the control points on the boundary triangles are updated by utilizing the method described by Equation (16-14).

This step completes the solution algorithm for both the interior and boundary elements. Now the scheme is used to solve the heat conduction equation within the rectangular domain described in Section 3.7. Recall that the domain is a 3.5 ft by 3.5 ft rectangular bar with thermal diffusivity of 0.645 ft^2/hr. In addition to the boundary conditions specified in Section 3.7, which are of Dirichlet type and will be referred to as Case (a), a second set of boundary conditions is specified as Case (b). For this case, the boundary conditions along edges $x = 3.5$ and $y = 3.5$ are specified as inflow heat flux and are given by $q(3.5, y) = -10,000\,\text{Btu/hr ft}^2$ and $q(x, 3.5) = -10,000\,\text{Btu/hr ft}^2$.

Solution begins with triangulation of the domain. For this application, the domain is discretized into 2450 triangles, as shown in Figure 16-6. The time step for the explicit formulation given by (16-18) is selected as 0.01 hr. The temperature contours for Case (a) are shown in Figures 16-7 and 16-8, which correspond to time levels of 0.1 and 0.4 hrs, respectively.

The solution is also provided in tabular form in Tables 16-1 and 16-2. These solutions can easily be compared to the solutions provided in Tables 3-7 and 3-8, obtained by a finite difference method. The temperature contours for Case (b) are shown in Figures 16-9 and 16-10 for the time levels of 0.1 and 0.4 hrs, respectively. The temperatures are listed in Tables 16-3 and 16-4 for the corresponding time levels. Note that the values of temperature at the two sides imposed by the heat flux boundary condition increase with time in order to maintain the constant heat flux specified.

An important consideration with regard to the solutions given in Tables 16-3 and 16-4 is as follows. If one evaluates the heat flux based on the computed values of temperature, a value slightly different from that imposed by the boundary condition is obtained! For example, heat flux at point (1.0,3.5) can be calculated according to

$$q = -35\left(\frac{227.4 - 198.99}{3.5 - 3.4}\right) = -9,982\,\text{Btu/hr ft}^2$$

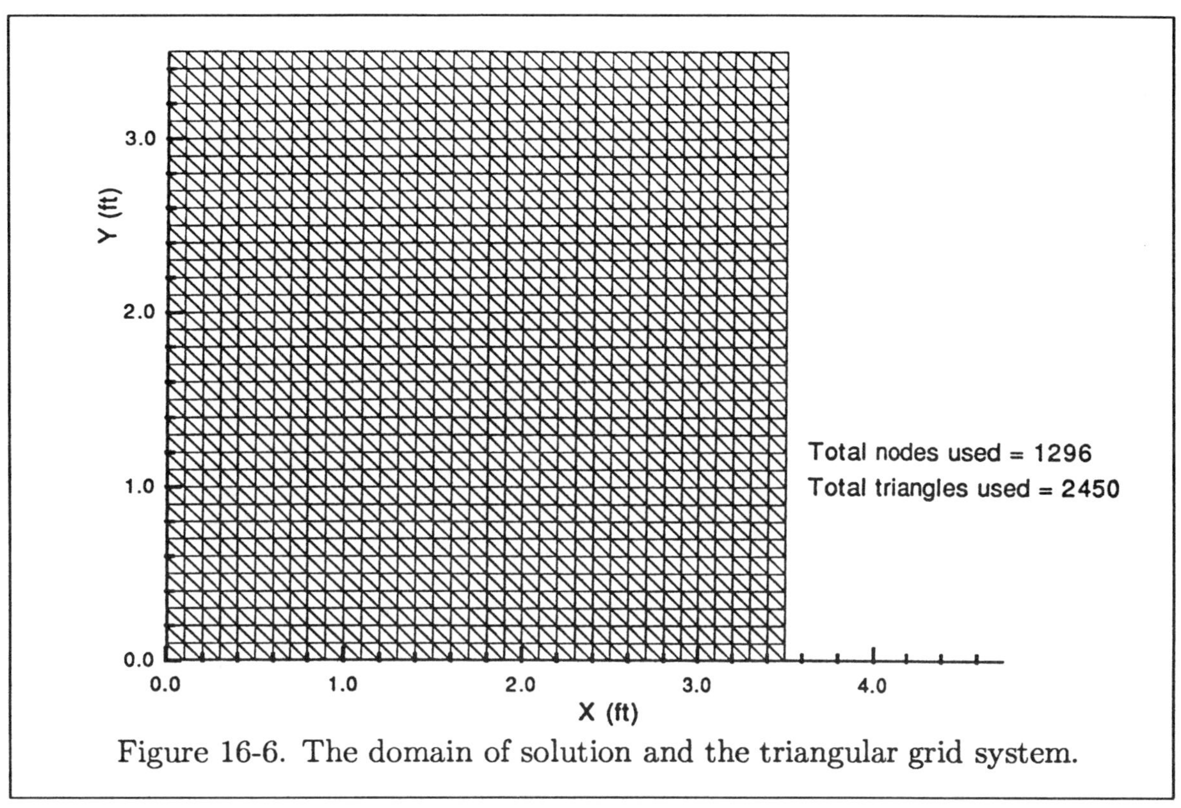

Figure 16-6. The domain of solution and the triangular grid system.

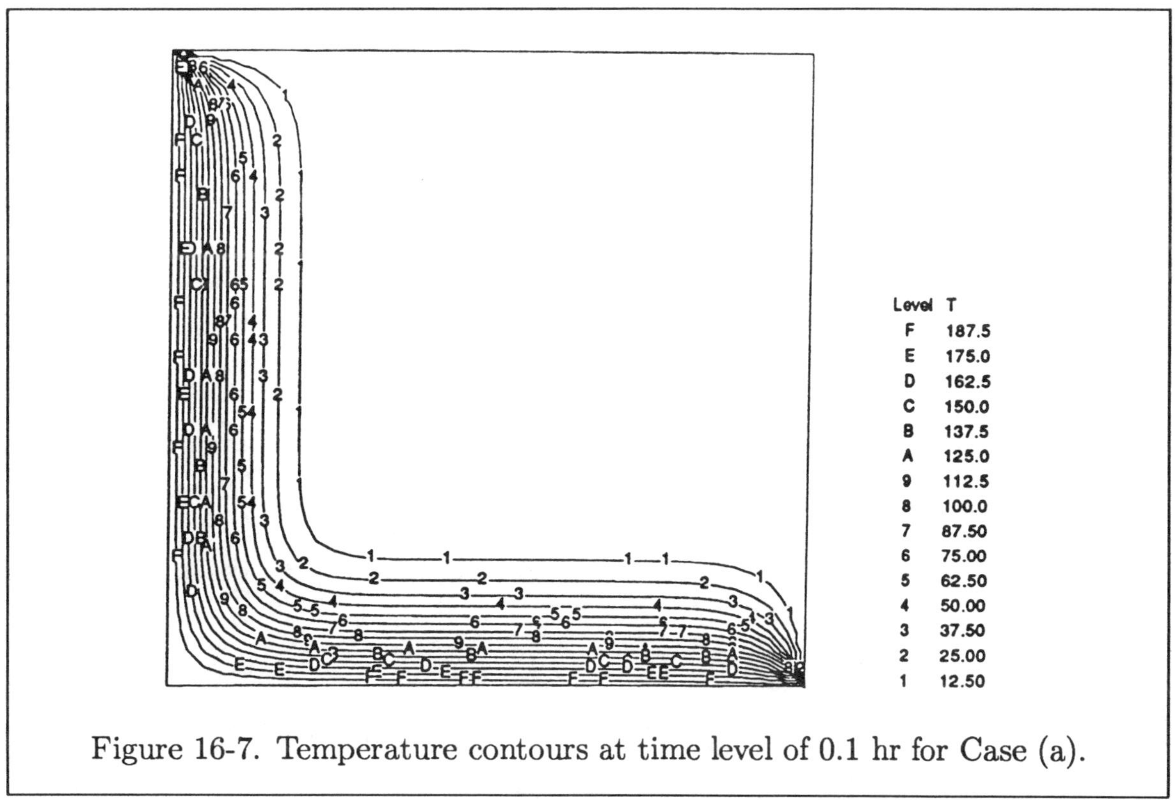

Figure 16-7. Temperature contours at time level of 0.1 hr for Case (a).

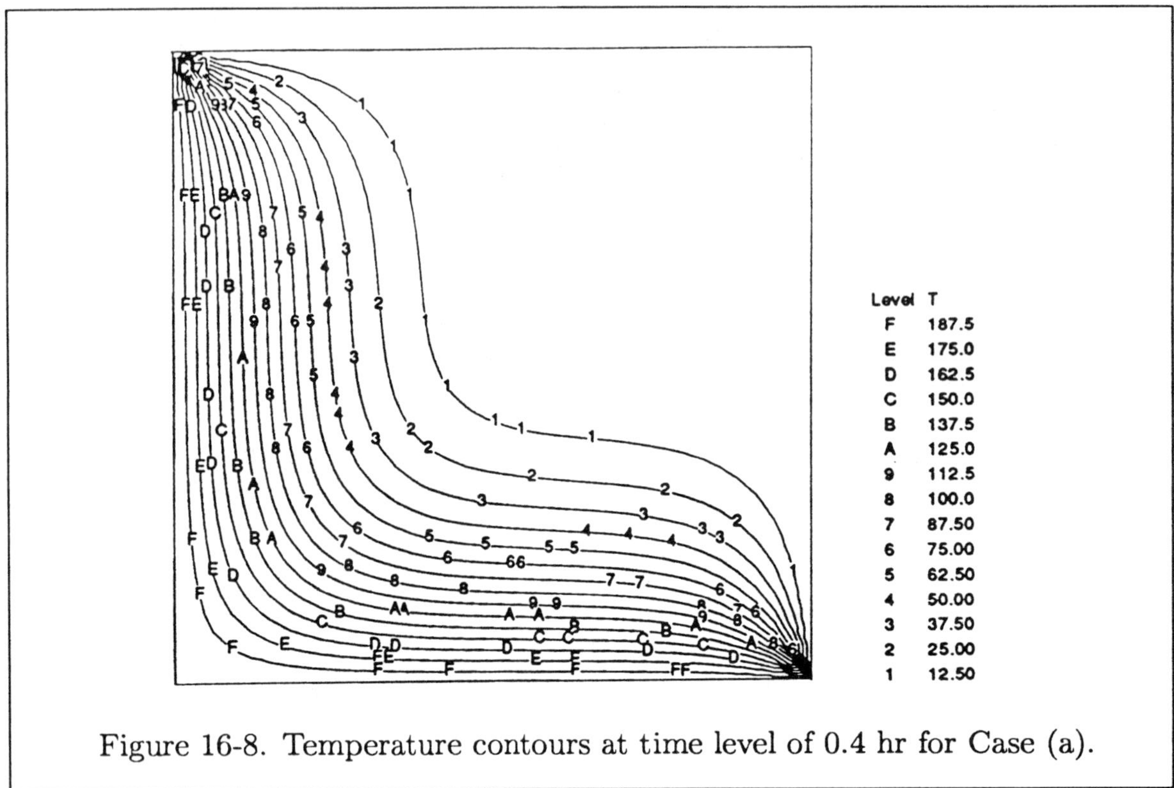

Figure 16-8. Temperature contours at time level of 0.4 hr for Case (a).

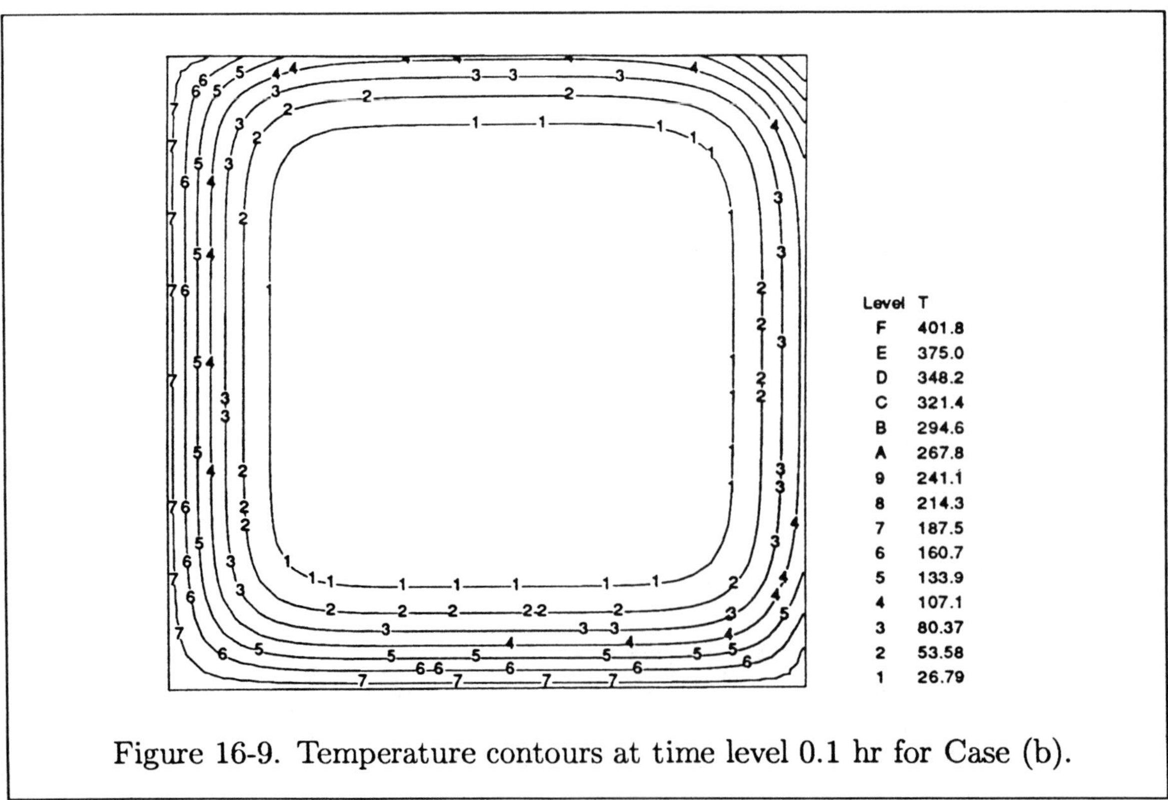

Figure 16-9. Temperature contours at time level 0.1 hr for Case (b).

Recall that the imposed value is $-10,000\,\mathrm{Btu/hr\,ft^2}$. Thus, an error in the order of 0.18% has been introduced. The computation of temperatures at the boundary nodes was performed with only 5 iterations. If the number of iterations is increased, then this error would be reduced.

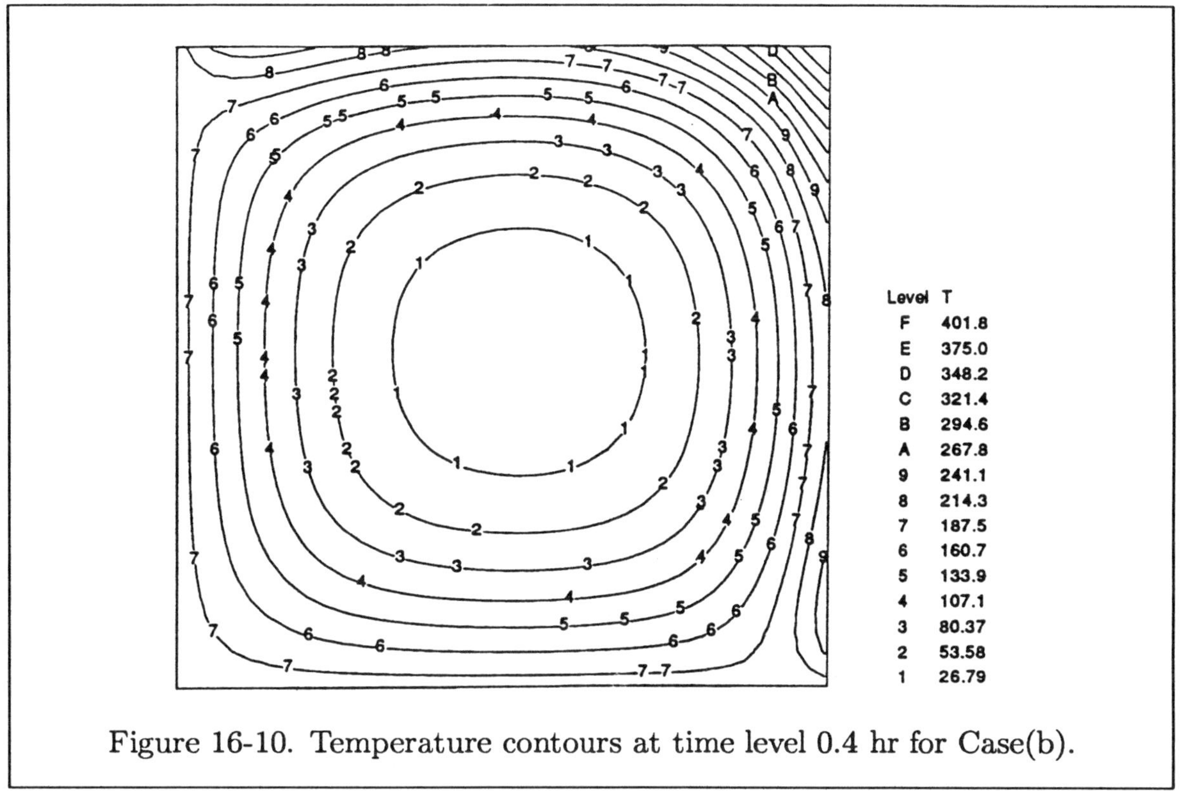

Figure 16-10. Temperature contours at time level 0.4 hr for Case(b).

16.4 Flux Vector Splitting Scheme

Recall that the values of flux E and F used in the finite volume scheme of Section 16.2 were evaluated by averaging the corresponding fluxes at two neighboring control points. Furthermore, for a rectangular element, the resulting formulation is equivalent to the central difference approximation of the finite difference method. It was shown previously in Chapter 6 that a hyperbolic equation is unstable for a formulation with central difference approximation. However, the solution would be stable if the governing equation includes a diffusion term. Indeed, to stabilize the hyperbolic equation approximated by central difference formulation of the convective term, the addition of a damping term would be required. Furthermore, the damping terms are used to reduce oscillations within the domain which may develop in the vicinity of sharp gradients. To avoid the addition of damping terms, one may use the flux vector splitting scheme to formulate the convective term. If the governing equation includes diffusion terms, they are approximated by central difference approximation, as seen in Chapters 13 and 14.

In this section, a flux vector splitting scheme used for a triangular element is explored. The formulation will be developed for the two-dimensional Euler equation given by (12-43), which is repeated here

$$\frac{\partial Q}{\partial t} + \frac{\partial E}{\partial x} + \frac{\partial F}{\partial y} + H = 0 \tag{16-32}$$

where the vector Q and the fluxes E, F, and H are given by (12-44).

16.4.1 Interior Triangles

Equation (16-32) is integrated over the triangular element *abc* shown in Figure 16-11 to provide

$$\int_{abc} \left(\frac{\partial Q}{\partial t} \right) dxdy = - \int_{abc} \left(\frac{\partial E}{\partial x} + \frac{\partial F}{\partial y} + H \right) dxdy \tag{16-33}$$

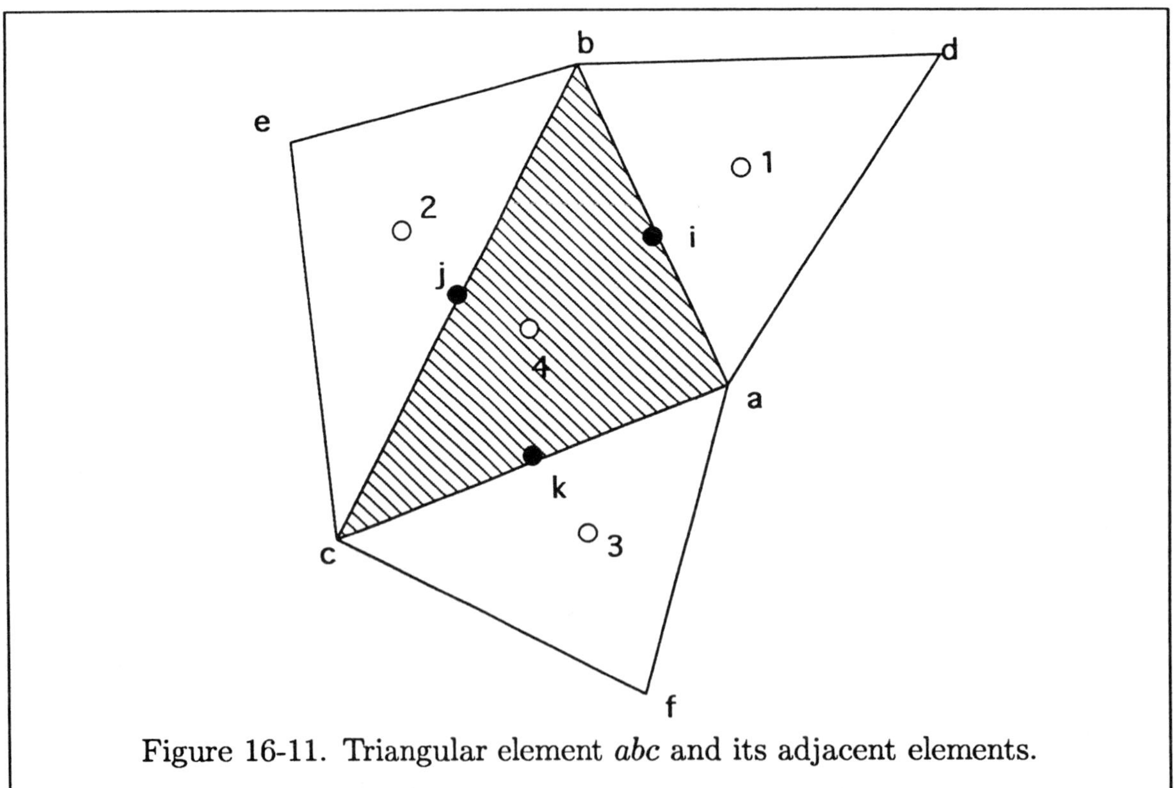

Figure 16-11. Triangular element *abc* and its adjacent elements.

The scheme proceeds with the application of Green's Theorem to the RHS of the equation and subsequently an explicit, cell-centered scheme is employed to approximate (16-33) by

$$\left(\frac{Q_4^{n+1} - Q_4^n}{\Delta t}\right) A_{abc} = -[(E_i \Delta y_{ab} + E_j \Delta y_{bc} + E_k \Delta y_{ca})$$
$$- (F_i \Delta x_{ab} + F_j \Delta x_{bc} + F_k \Delta x_{ca}) + H_4 A_{abc}] \quad (16\text{-}34)$$

Recall that fluxes E and F at points i, j, and k were determined previously by an averaging scheme. The objective now is to use a flux vector splitting scheme to determine these fluxes. The development of the scheme is illustrated by the evaluation of the fluxes E and F at point i located at the midpoint of the edge ab.

Recall that each flux can be split into a positive part and a negative part each associated with the corresponding eigenvalue. Schematically, the fluxes at control point 4 of triangle abc are illustrated in Figure 16-12, which is used as well in subsequent discussions. The specific forms of the fluxes E^+, E^-, F^+, and F^- in the

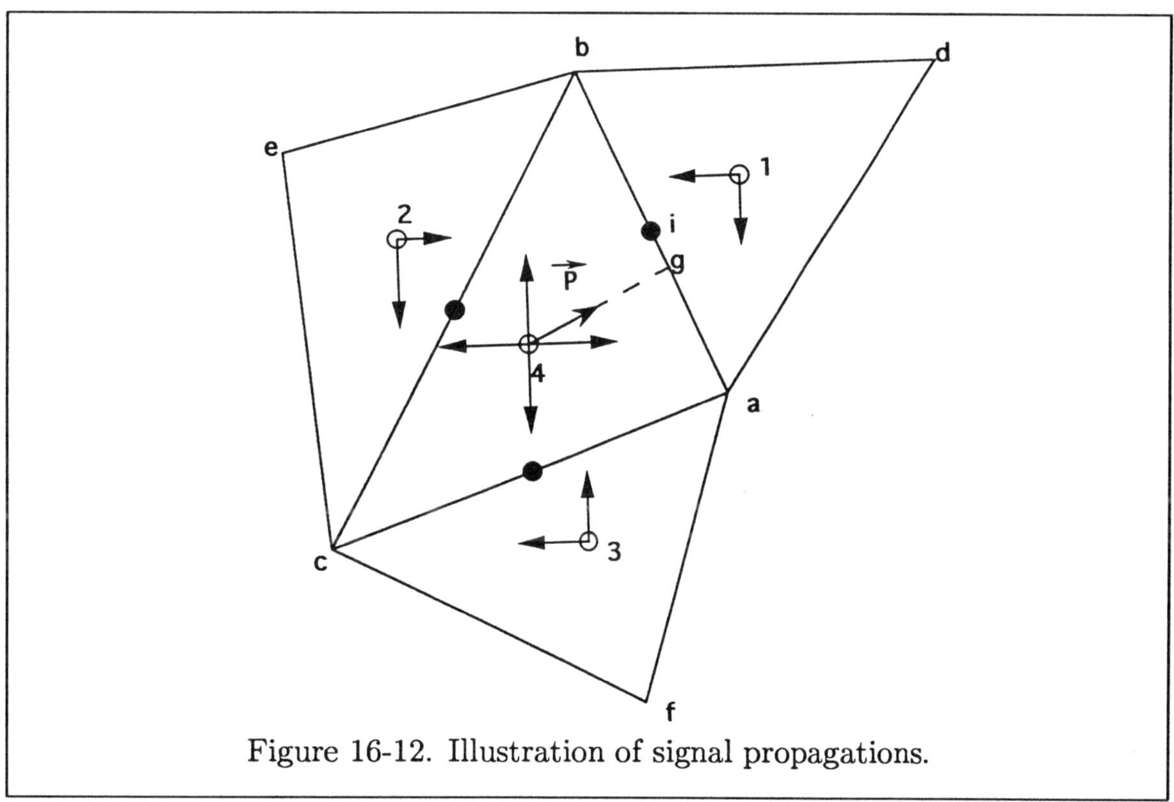

Figure 16-12. Illustration of signal propagations.

transformed computational space were given in Chapter 12. However, since in the finite volume scheme the formulation is applied directly on the physical space, one needs to obtain the associated fluxes in the physical space. The eigenvalues of the Jacobian matrix $A = \dfrac{\partial E}{\partial Q}$ are easily determined to be $\lambda_1 = u$, $\lambda_2 = u$, $\lambda_3 = u + a$, $\lambda_4 = u - a$, where u is the x-component of the velocity and a is the speed of sound.

Observe, as expected, the lack of metrics in the eigenvalues, as compared to those given by (12-58) through (12-61).

The splitting of flux vector E proceeds as follows.

(a) If $u > a$, then all the eigenvalues are positive and therefore,

$$E^+ = E \quad \text{and} \quad E^- = 0$$

(b) If $0 \le u \le a$, then one of the eigenvalues, namely λ_4 is negative, and

$$E^- = (u-a)\left(\frac{\rho}{2\gamma}\right)\begin{bmatrix} 1 \\ u-a \\ v \\ -au + \frac{1}{2}q^2 + \dfrac{a^2}{\gamma-1} \end{bmatrix}$$

where

$$q^2 = u^2 + v^2$$

and

$$E^+ = E - E^-$$

(c) If $-a < u < 0$, then one of the eigenvalues, namely λ_3, is positive, whereas the remaining three are negative. Subsequently,

$$E^+ = (u+a)\left(\frac{\rho}{2\gamma}\right)\begin{bmatrix} 1 \\ u+a \\ v \\ au + \frac{1}{2}q^2 + \dfrac{a^2}{\gamma-1} \end{bmatrix}$$

and

$$E^- = E - E^+$$

(d) If $u \le -a$, then all the eigenvalues are negative and therefore,

$$E^+ = 0$$

and

$$E^- = E$$

The splitting of flux vector E is now concluded. Following a similar procedure, the eigenvalues of Jacobian matrix $B = \dfrac{\partial F}{\partial Q}$ are determined to be $\lambda_1 = v$, $\lambda_2 = v$, $\lambda_3 = v + a$, and $\lambda_4 = v - a$, where v is the y-component of the velocity. The fluxes F^+ and F^- associated with the four possible combinations of λ's are

(a) If $v > a$, then $F^+ = F$ and $F^- = 0$.

(b) If $0 \leq v \leq a$, then

$$F^- = (v - a)\left(\frac{\rho}{2\gamma}\right)\begin{bmatrix} 1 \\ u \\ v - a \\ -av + \frac{1}{2}q^2 + \frac{a^2}{\gamma - 1} \end{bmatrix}$$

and

$$F^+ = F - F^-$$

(c) If $-a < v < 0$, then

$$F^+ = (v + a)\left(\frac{\rho}{2\gamma}\right)\begin{bmatrix} 1 \\ u \\ v + a \\ av + \frac{1}{2}q^2 + \frac{a^2}{\gamma - 1} \end{bmatrix}$$

and

$$F^- = F - F^+$$

(d) If $-a \leq v$, then $F^+ = 0$ and $F^- = F$.

Thus, the fluxes can be decomposed into their positive and negative parts with the exact forms given above. Therefore, at a given point four signals emanate in the four directions. However, observe that only two of the four signals will cross any given edge. For example, only two signals represented by fluxes E_4^+ and F_4^+ will reach edge *ab* from control point 4, which is illustrated in Figure 16-12. Similarly, the signals reaching edge *ab* from control point 1 located on the other side of edge *ab* would be E_1^- and F_1^-. Therefore, the combination of signals reaching point i and contributing to its value are written as

$$E_i = E_4^+ + E_1^-$$

and

$$F_i = F_4^+ + F_1^-$$

A similar argument is extended to edges *bc* and *ca* to provide the following:

$$E_j = E_4^- + E_2^+$$

$$F_j = F_4^+ + F_2^-$$

$$E_k = E_4^+ + E_3^-$$

$$F_k = F_4^- + F_3^+$$

The description of the algorithm above is simply based on graphical inspection, in particular, Figure 16-12. Now, the graphical conclusions are used to develop a numerical scheme. The procedure is described utilizing edge ab as follows.

1. Determine a perpendicular vector from control point 4 to the edge ab and designate it as vector $\vec{P}$. Vector $\vec{P}$ intersects line ab at point g as shown in Figure 16-12. The equation for line ab can be simply expressed as

$$y = mx + n$$

where

$$m = \frac{y_b - y_a}{x_b - x_a}$$

and

$$n = \frac{x_b y_a - y_b x_a}{x_b - x_a}$$

The perpendicular vector $\vec{P}$ is expressed by

$$\vec{P}_4 = (x_g - x_4)\vec{\imath} + (y_g - y_4)\vec{\jmath} = \left[(y_4 - n - mx_4)/(m^2+1)\right](m\vec{\imath} - \vec{\jmath}) = P_{4x}\vec{\imath} + P_{4y}\vec{\jmath}$$

where the coordinate of the intersection point g is

$$x_g = (my_4 + x_4 - mn)/(m^2+1)$$

and

$$y_g = (m^2 y_4 + mx_4 + n)/(m^2+1)$$

Note that if $x_a = x_b$, then

$$\vec{P}_4 = (x_a - x_4)\vec{\imath} = P_{4x}\vec{\imath} + P_{4y}\vec{\jmath}$$

2. The fluxes E and F at the midpoints are determined by inspection of the components of the vector $\vec{P}_4$, which provides the required information on the contribution of each term. The flux vector E at the midpoint i of edge ab is determined as follows.

   (a) If $P_{4x} \geq 0$, then control point 4 contributes E_4^+ and control point 1 will contribute E_1^-. Thus,

   $$E_i = E_4^+ + E_1^-$$

   (b) If $P_{4x} < 0$, then E_4^- is contributed due to control point 4, whereas E_1^+ is contributed due to control point 1. Thus,

   $$E_i = E_4^- + E_1^+$$

The procedure is extended to the flux vector F as well to provide

$$F_i = F_4^+ + F_1^- \qquad \text{for} \quad P_{4y} > 0$$

and

$$F_i = F_4^- + F_1^+ \qquad \text{for} \quad P_{4y} < 0$$

Similarly, the fluxes E and F at point j (midpoint of edge bc) and point k (midpoint of edge ca) are determined. Subsequently, Equation (16-34) is utilized to compute Q_4^{n+1}.

16.4.2 Boundary Triangles

Various types of boundary conditions were identified in the previous chapters. Generally, they are categorized as body surface, symmetry surface, farfield, inflow, and outflow boundaries. To simplify the analysis and implementation of the boundary conditions, the control points of all the boundary triangles are shifted to the middle of the boundary edges, as illustrated in Figure 16-13. The implementation of various boundary conditions is illustrated for an axisymmetric domain of a blunt body in supersonic stream.

1. *Inflow boundary.* When the inflow boundary is set far in the flowfield, the freestream conditions are imposed as the boundary conditions. Thus,

$$p_1 = p_\infty \quad , \quad T_1 = T_\infty \quad , \quad u_1 = u_\infty \quad , \quad \text{and} \quad v_1 = 0$$

2. *Symmetry line.* Along the symmetry line, the y-component of the velocity is zero, whereas the remaining flow properties are approximated by an average value of the properties from the interior points. For example,

$$p_2 = \frac{1}{2}(p_3 + p_4)$$

$$T_2 = \frac{1}{2}(T_3 + T_4)$$

$$u_2 = \frac{1}{2}(u_3 + u_4)$$

and

$$v_2 = 0$$

3. *Body surface.* The no-slip condition for the velocity and the zero-normal gradient for pressure is usually used at the body surface. The temperature may be specified either as a Dirichlet type, where the temperature distribution is provided, or as a Neumann type, where the distribution of the normal gradient is specified.

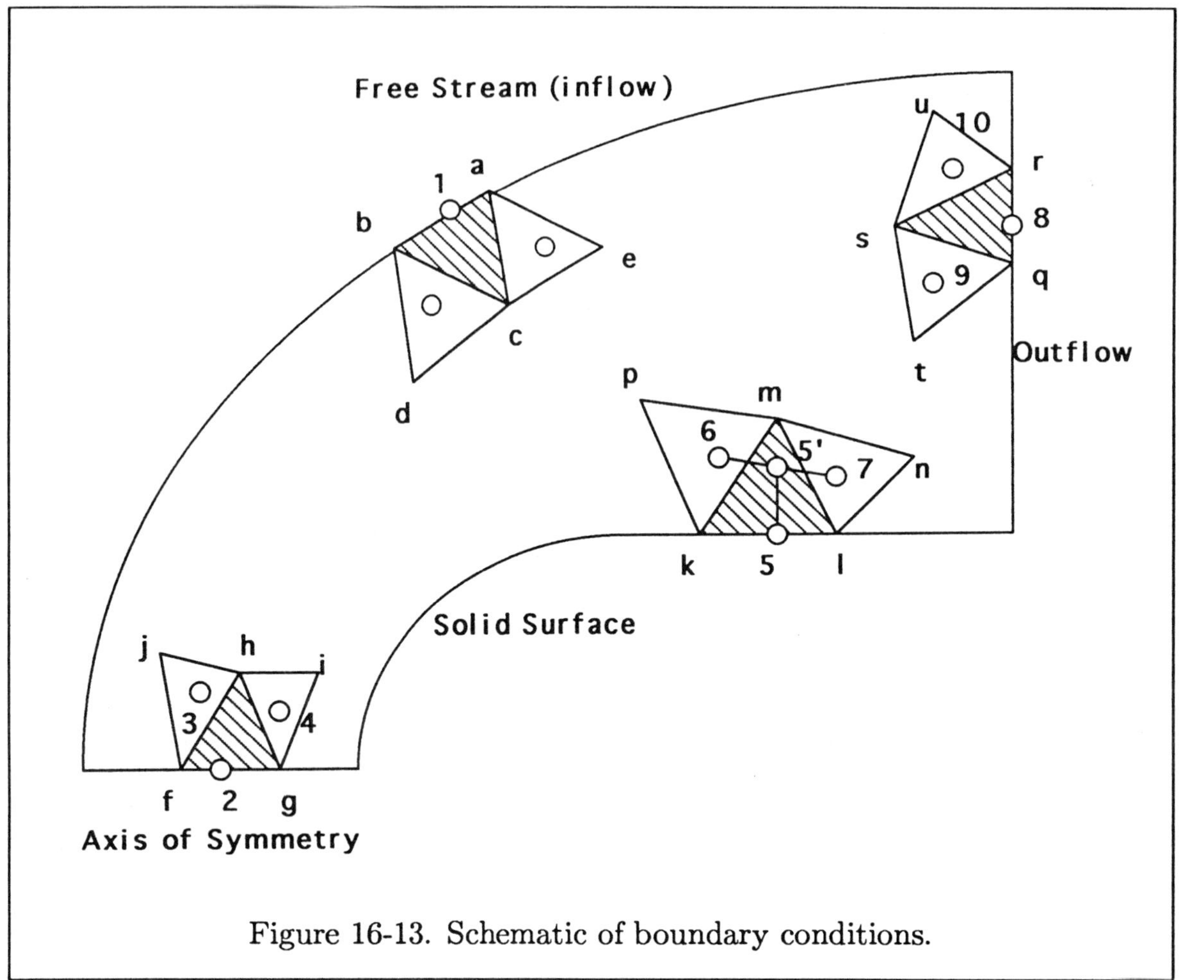

Figure 16-13. Schematic of boundary conditions.

An easy scheme for obtaining an approximate value of the pressure at the surface is as follows. Assume that the midpoint of points 6 and 7 is on a line perpendicular to the surface at point 5. Then, since the pressure gradient at the surface is zero, one may write

$$p_5 = p_{5'} = \frac{1}{2}(p_6 + p_7)$$

Of course, a precise scheme would be that of Section 16.3.2.2 for the implementation of Neumann type boundary conditions. The velocity components are set to zero due to the imposed no-slip condition. Thus, $u_5 = 0$ and $v_5 = 0$. The temperature value is that specified temperature if the boundary condition is Dirichlet type. However, if a Neumann-type boundary condition is imposed, then, depending on the specification, two situations can be identified. (a) If the surface is specified as adiabatic, i.e., $\partial T/\partial n = 0$, then by the same analogy to that of pressure, one may approximate

$$T_5 = \frac{1}{2}(T_6 + T_7)$$

(b) If the heat flux is given, i.e., $\partial T/\partial n$ is specified, then the iterative procedure described in Section 16.3.2.2 is used.

4. *Supersonic outflow.* All the properties at the outflow are determined by extrapolation from the interior domain. A zero-order extrapolation yields:

$$p_8 = \frac{1}{2}(p_9 + p_{10})$$

$$T_8 = \frac{1}{2}(T_9 + T_{10})$$

$$u_8 = \frac{1}{2}(u_9 + u_{10})$$

and

$$v_8 = \frac{1}{2}(v_9 + v_{10})$$

16.5 Concluding Remarks

Fundamental topics in finite volume schemes have been introduced in this chapter. Due to its introductory nature, only a selected number of schemes in two-space dimensions using triangular elements were explored. It is hoped that, at this point, the reader has established a fundamental understanding of the topic. With this background, the review of other finite volume schemes presented elsewhere and extension of the schemes to three dimensions should be facilitated.

16.6 Problems

16.1 Consider the cell-centered formulation described in Section 16.2.1 and illustrated in Figure 16-2a. If the mesh *abcd* has a rectangular shape, where $\Delta x_{ab} = -\Delta x_{cd}$, $\Delta x_{bc} = -\Delta x_{da} = 0$, $\Delta y_{ab} = -\Delta y_{cd} = 0$, and $\Delta y_{bc} = -\Delta y_{da}$, show that the formulation (16-4) can be reduced to a finite difference equation with second-order central difference approximation of the spatial derivatives.

16.2 In Equations (16-19) to (16-21), show that the areas of quadrilaterals are $A_{a1b4} = \frac{1}{3}(A_{abc} + A_{adb})$, $A_{b2c4} = \frac{1}{3}(A_{abc} + A_{bec})$, and $A_{c3a4} = \frac{1}{3}(A_{abc} + A_{cfa})$.

Hint: Use Equations (16-12) and (16-13) to prove that $A_{ab4} = \frac{1}{3}A_{abc}$ and, similarly, $A_{a1b} = \frac{1}{3}A_{adb}$, and so forth.

16.3 Consider the domain of Problem 5.1 which is repeated here for convenience as Figure P16.3. Use the finitc volume scheme described in Section 16.3 to obtain the streamline patterns within the domain. The governing equation to be solved is given by

$$\frac{\partial \psi}{\partial t} = \frac{\partial^2 \psi}{\partial x^2} + \frac{\partial^2 \psi}{\partial y^2}$$

where the time-dependent term is added to represent the iteration, i.e., each time step is equivalent to an iteration step. Recall the analogy between the FTCS scheme applied to a time-dependent parabolic equation and the iterative schemes for the solution of an elliptic equation. Furthermore, note that the time-dependent term $\frac{\partial \psi}{\partial t}$ will approach zero for a large time. Thus, Laplace's equation is recovered.

Select a time step of 0.001 and proceed to a final time of 2.0. That should provide the converged solution. Furthermore, set the initial values of the stream function within the domain to be 50. The appropriate boundary conditions are specified in Figure P16.3.

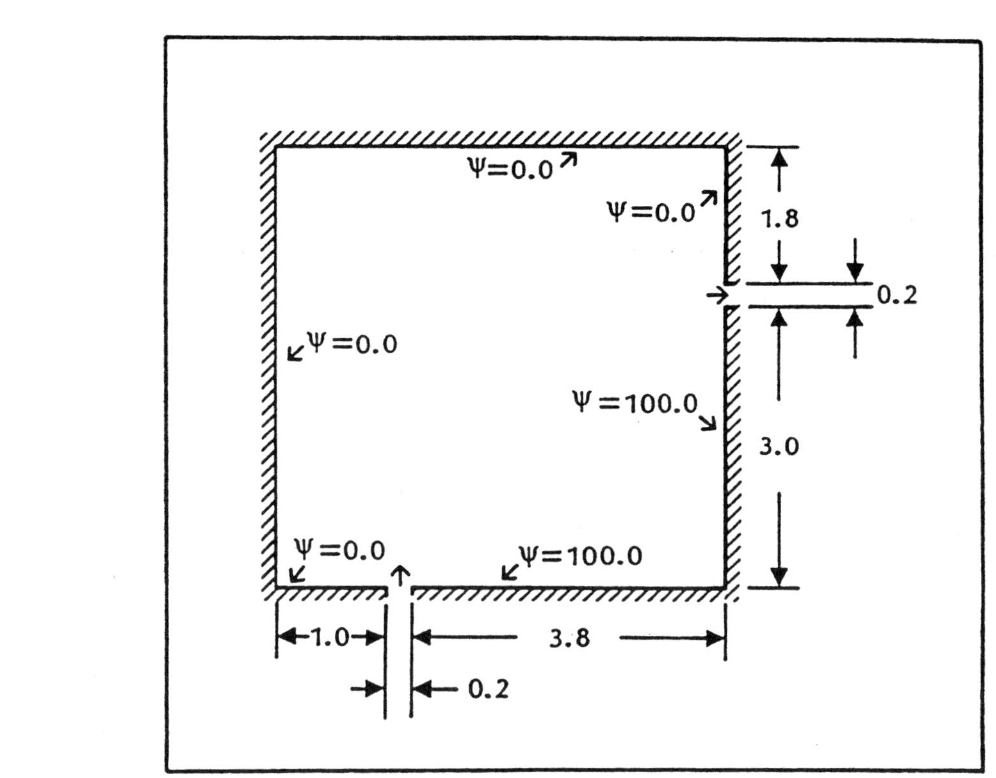

Figure P16.3. The solution domain and the proposed boundary conditions.

16.4 The streamline patterns of an inviscid, incompressible flow within the domain shown in Figure P16.4 are required. The problem requires the solution of Laplace's equation or, equivalently,

$$\frac{\partial \psi}{\partial t} = \frac{\partial^2 \psi}{\partial x^2} + \frac{\partial^2 \psi}{\partial y^2}$$

The concept of using the equation above instead of Laplace's equation is provided in Problem 16.3. Assuming the flow to be uniform at both the inlet and outlet, the following boundary conditions are specified.

1. At the lower surface, $\psi = 0.0$
2. At the inlet, i.e., $x = 0.0$, $\psi = 100y$
3. At the outlet, i.e., $x = 6.0$, $\psi = 100y$
4. At the upper surface, i.e., $y = 4.0$, $\psi = 400.0$

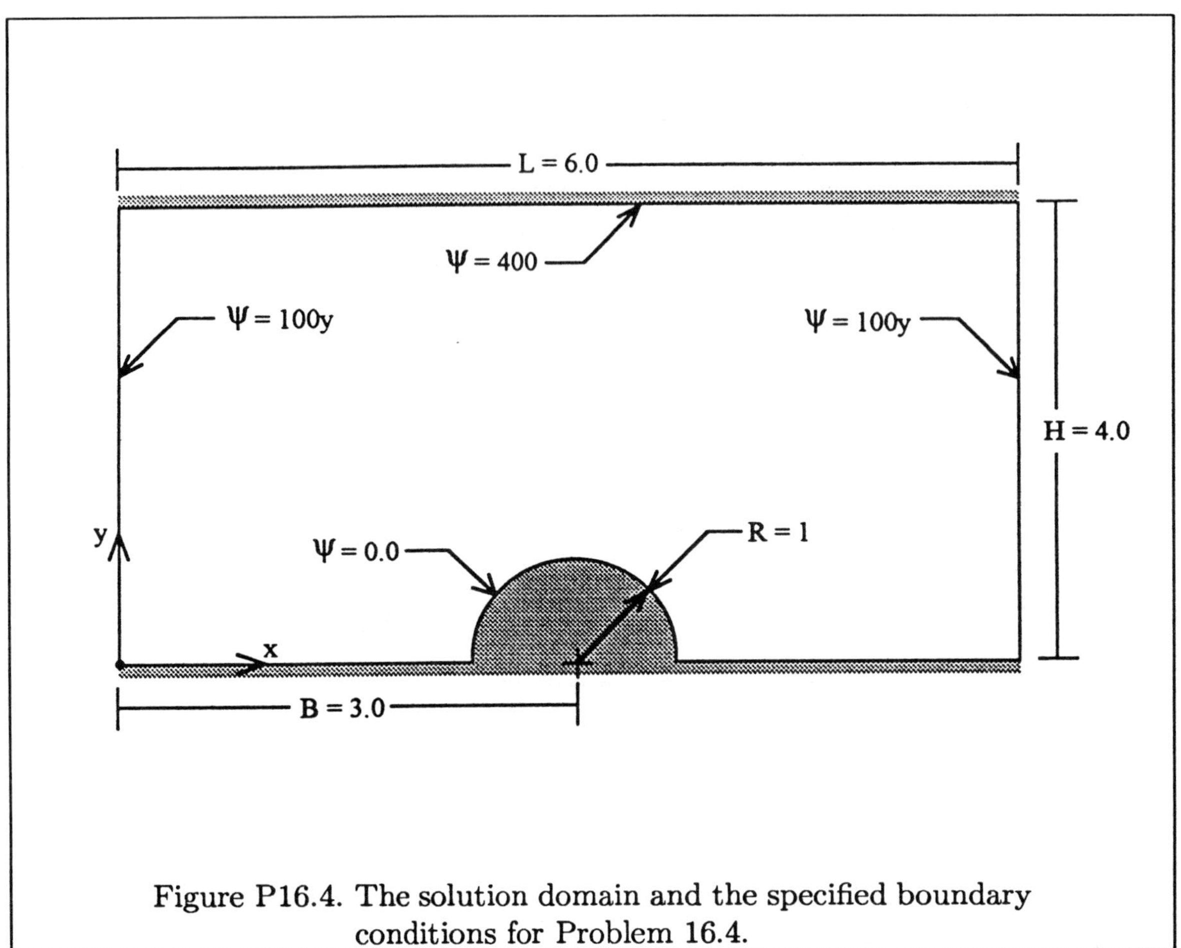

Figure P16.4. The solution domain and the specified boundary conditions for Problem 16.4.

Use a time step of 0.003 and proceed up to a total time of 6.0, which should provide the desired solution. Use the grid system generated in Problem 15.2, and set the initial ψ distribution at 100.0.

Table 16-1. Temperature distribution at $t = 0.1$ hr.

	0.00	0.50	1.00	1.50	2.00	2.50	3.00	3.50
0.00	200.00	200.00	200.00	200.00	200.00	200.00	200.00	200.00
0.10	200.00	169.43	161.60	161.26	161.26	161.24	159.45	0.00
0.20	200.00	136.66	121.83	121.22	121.22	121.18	117.54	0.00
0.30	200.00	107.96	87.10	86.27	86.27	86.21	81.84	0.00
0.40	200.00	84.65	58.97	57.96	57.95	57.90	53.83	0.00
0.50	200.00	67.07	37.81	36.66	36.65	36.60	33.41	0.00
0.60	200.00	54.75	23.01	21.77	21.76	21.73	19.53	0.00
0.70	200.00	46.74	13.41	12.11	12.11	12.08	10.73	0.00
0.80	200.00	41.91	7.64	6.30	6.30	6.28	5.53	0.00
0.90	200.00	39.20	4.42	3.06	3.05	3.05	2.67	0.00
1.00	200.00	37.81	2.75	1.39	1.38	1.38	1.20	0.00
1.10	200.00	37.14	1.96	0.59	0.58	0.58	0.50	0.00
1.20	200.00	36.84	1.60	0.23	0.23	0.23	0.20	0.00
1.30	200.00	36.72	1.46	0.09	0.08	0.08	0.07	0.00
1.40	200.00	36.67	1.41	0.04	0.03	0.03	0.02	0.00
1.50	200.00	36.66	1.39	0.02	0.01	0.01	0.01	0.00
1.60	200.00	36.65	1.38	0.01	0.00	0.00	0.00	0.00
1.70	200.00	36.65	1.38	0.01	0.00	0.00	0.00	0.00
1.80	200.00	36.65	1.38	0.01	0.00	0.00	0.00	0.00
1.90	200.00	36.65	1.38	0.01	0.00	0.00	0.00	0.00
2.00	200.00	36.65	1.38	0.01	0.00	0.00	0.00	0.00
2.10	200.00	36.65	1.38	0.01	0.00	0.00	0.00	0.00
2.20	200.00	36.65	1.38	0.01	0.00	0.00	0.00	0.00
2.30	200.00	36.65	1.38	0.01	0.00	0.00	0.00	0.00
2.40	200.00	36.63	1.38	0.01	0.00	0.00	0.00	0.00
2.50	200.00	36.60	1.38	0.01	0.00	0.00	0.00	0.00
2.60	200.00	36.53	1.37	0.01	0.00	0.00	0.00	0.00
2.70	200.00	36.34	1.36	0.01	0.00	0.00	0.00	0.00
2.80	200.00	35.92	1.34	0.01	0.00	0.00	0.00	0.00
2.90	200.00	35.07	1.29	0.01	0.00	0.00	0.00	0.00
3.00	200.00	33.41	1.20	0.01	0.00	0.00	0.00	0.00
3.10	200.00	30.44	1.06	0.01	0.00	0.00	0.00	0.00
3.20	200.00	25.52	0.86	0.01	0.00	0.00	0.00	0.00
3.30	200.00	18.16	0.58	0.00	0.00	0.00	0.00	0.00
3.40	200.00	8.48	0.26	0.00	0.00	0.00	0.00	0.00
3.50	200.00	0.00	0.00	0.00	0.00	0.00	0.00	0.00

Table 16-2. Temperature distribution at $t = 0.4$ hr.

	0.00	0.50	1.00	1.50	2.00	2.50	3.00	3.50
0.00	200.00	200.00	200.00	200.00	200.00	200.00	200.00	200.00
0.10	200.00	190.45	183.64	180.81	179.97	178.97	172.55	0.00
0.20	200.00	179.61	165.74	160.06	158.37	156.27	142.98	0.00
0.30	200.00	169.11	148.44	140.01	137.52	134.50	116.51	0.00
0.40	200.00	159.17	132.07	121.05	117.82	114.11	93.86	0.00
0.50	200.00	149.94	116.86	103.44	99.55	95.43	74.95	0.00
0.60	200.00	141.51	103.01	87.41	82.95	78.66	59.38	0.00
0.70	200.00	133.98	90.63	73.08	68.13	63.90	46.64	0.00
0.80	200.00	127.37	79.78	60.53	55.17	51.15	36.30	0.00
0.90	200.00	121.67	70.44	49.73	44.05	40.34	27.97	0.00
1.00	200.00	116.86	62.55	40.62	34.68	31.34	21.30	0.00
1.10	200.00	112.88	56.02	33.08	26.94	23.98	16.03	0.00
1.20	200.00	109.63	50.71	26.96	20.67	18.07	11.90	0.00
1.30	200.00	107.04	46.48	22.08	15.68	13.41	8.72	0.00
1.40	200.00	105.01	43.17	18.26	11.79	9.80	6.29	0.00
1.50	200.00	103.44	40.62	15.34	8.81	7.06	4.48	0.00
1.60	200.00	102.24	38.69	13.13	6.57	5.01	3.14	0.00
1.70	200.00	101.33	37.24	11.50	4.92	3.50	2.16	0.00
1.80	200.00	100.62	36.15	10.30	3.73	2.42	1.47	0.00
1.90	200.00	100.05	35.33	9.44	2.88	1.66	0.98	0.00
2.00	200.00	99.55	34.68	8.81	2.28	1.13	0.65	0.00
2.10	200.00	99.05	34.12	8.34	1.87	0.77	0.42	0.00
2.20	200.00	98.48	33.57	7.98	1.58	0.52	0.27	0.00
2.30	200.00	97.76	32.97	7.67	1.38	0.36	0.17	0.00
2.40	200.00	96.79	32.25	7.37	1.24	0.26	0.11	0.00
2.50	200.00	95.43	31.34	7.06	1.13	0.19	0.07	0.00
2.60	200.00	93.54	30.16	6.70	1.03	0.15	0.04	0.00
2.70	200.00	90.88	28.64	6.27	0.94	0.12	0.03	0.00
2.80	200.00	87.17	26.70	5.77	0.85	0.10	0.02	0.00
2.90	200.00	82.02	24.27	5.17	0.75	0.08	0.01	0.00
3.00	200.00	74.95	21.30	4.48	0.65	0.07	0.01	0.00
3.10	200.00	65.38	17.76	3.69	0.53	0.05	0.00	0.00
3.20	200.00	52.69	13.68	2.80	0.40	0.04	0.00	0.00
3.30	200.00	36.55	9.12	1.85	0.26	0.03	0.00	0.00
3.40	200.00	17.38	4.25	0.85	0.12	0.01	0.00	0.00
3.50	200.00	0.00	0.00	0.00	0.00	0.00	0.00	0.00

Table 16-3. Temperature distribution at $t = 0.1$ hr.

	0.00	0.50	1.00	1.50	2.00	2.50	3.00	3.50
0.00	200.00	200.00	200.00	200.00	200.00	200.00	200.00	200.00
0.10	200.00	169.41	161.58	161.24	161.24	161.32	164.20	197.63
0.20	200.00	136.62	121.79	121.19	121.19	121.37	121.67	190.41
0.30	200.00	107.91	87.06	86.23	86.23	86.48	95.63	176.70
0.40	200.00	84.59	58.93	57.92	57.91	58.23	69.49	161.80
0.50	200.00	67.01	37.76	36.62	36.62	36.97	49.60	148.30
0.60	200.00	54.69	22.98	21.74	21.74	22.11	35.50	137.28
0.70	200.00	46.68	13.39	12.09	12.09	12.47	26.18	128.92
0.80	200.00	41.86	7.62	6.29	6.28	6.67	20.43	122.95
0.90	200.00	39.16	4.40	3.05	3.05	3.43	17.12	118.90
1.00	200.00	37.76	2.74	1.38	1.38	1.75	15.34	116.26
1.10	200.00	37.10	1.95	0.59	0.58	0.95	14.43	114.61
1.20	200.00	36.80	1.60	0.23	0.23	0.60	14.00	113.60
1.30	200.00	36.68	1.46	0.09	0.08	0.45	13.80	113.01
1.40	200.00	36.64	1.40	0.04	0.03	0.40	13.71	112.67
1.50	200.00	36.62	1.38	0.02	0.01	0.38	13.67	112.47
1.60	200.00	36.62	1.38	0.01	0.00	0.37	13.66	112.36
1.70	200.00	36.61	1.38	0.01	0.00	0.37	13.65	112.30
1.80	200.00	36.61	1.38	0.01	0.00	0.37	13.65	112.28
1.90	200.00	36.61	1.38	0.01	0.00	0.37	13.65	112.27
2.00	200.00	36.62	1.38	0.01	0.00	0.37	13.65	112.30
2.10	200.00	36.62	1.38	0.01	0.01	0.38	13.66	112.36
2.20	200.00	36.63	1.39	0.03	0.02	0.39	13.68	112.49
2.30	200.00	36.67	1.43	0.06	0.06	0.43	13.74	112.77
2.40	200.00	36.76	1.52	0.15	0.15	0.52	13.89	113.29
2.50	200.00	36.97	1.75	0.38	0.37	0.75	14.21	114.25
2.60	200.00	37.45	2.26	0.88	0.87	1.26	14.89	115.93
2.70	200.00	38.43	3.33	1.92	1.91	2.31	16.22	118.77
2.80	200.00	40.34	5.38	3.93	3.92	4.35	18.70	123.34
2.90	200.00	43.78	9.08	7.56	7.54	8.02	23.05	130.39
3.00	200.00	49.60	15.34	13.67	13.65	14.21	30.22	140.74
3.10	200.00	58.87	25.22	23.34	23.30	24.00	41.38	155.27
3.20	200.00	72.78	39.90	37.69	37.63	38.55	57.78	174.64
3.30	200.00	92.55	60.46	57.78	57.68	58.92	80.63	198.97
3.40	200.00	119.05	87.50	84.21	84.07	85.76	110.63	225.94
3.50	200.00	148.30	116.26	112.47	112.30	114.25	140.74	239.57

Table 16-4. Temperature distribution at $t = 0.4$ hr.

	0.00	0.50	1.00	1.50	2.00	2.50	3.00	3.50
0.00	200.00	200.00	200.00	200.00	200.00	200.00	200.00	200.00
0.10	200.00	190.44	183.64	180.87	180.51	182.37	190.05	225.65
0.20	200.00	179.60	165.75	160.19	159.56	163.58	179.94	244.37
0.30	200.00	169.11	148.45	140.22	139.32	145.40	169.69	252.48
0.40	200.00	159.17	132.08	121.32	120.16	128.13	159.35	254.39
0.50	200.00	149.93	116.88	103.77	102.36	112.02	149.09	252.59
0.60	200.00	141.51	103.03	87.79	86.15	97.26	139.12	248.66
0.70	200.00	133.98	90.66	73.51	71.65	83.99	129.69	243.60
0.80	200.00	127.37	79.81	60.99	58.94	72.29	120.99	238.09
0.90	200.00	121.68	70.48	50.23	48.00	62.16	113.16	232.58
1.00	200.00	116.88	62.62	41.15	38.77	53.56	106.31	227.40
1.10	200.00	112.91	56.12	33.66	31.13	46.41	100.44	222.72
1.20	200.00	109.70	50.86	27.59	24.95	40.60	95.56	218.68
1.30	200.00	107.16	46.71	22.80	20.07	35.97	91.61	215.31
1.40	200.00	105.21	43.52	19.12	16.31	32.41	88.53	212.65
1.50	200.00	103.77	41.15	16.39	13.53	29.77	86.24	210.69
1.60	200.00	102.76	39.51	14.49	11.58	27.94	84.70	209.43
1.70	200.00	102.14	38.49	13.30	10.37	26.83	83.83	208.86
1.80	200.00	101.88	38.03	12.76	9.83	26.38	83.64	209.02
1.90	200.00	101.94	38.12	12.83	9.92	26.57	84.12	209.94
2.00	200.00	102.36	38.77	13.53	10.65	27.43	85.32	211.70
2.10	200.00	103.17	40.02	14.90	12.08	29.03	87.33	214.40
2.20	200.00	104.43	41.97	17.06	14.31	31.46	90.26	218.17
2.30	200.00	106.23	44.76	20.12	17.49	34.88	94.28	223.18
2.40	200.00	108.71	48.55	24.29	21.79	39.48	99.60	229.64
2.50	200.00	112.02	53.56	29.77	27.43	45.49	106.42	237.74
2.60	200.00	116.34	60.04	36.81	34.68	53.17	115.03	247.73
2.70	200.00	121.92	68.25	45.69	43.80	62.81	125.69	259.84
2.80	200.00	129.01	78.49	56.70	55.09	74.70	138.69	274.31
2.90	200.00	137.95	91.07	70.12	68.84	89.14	154.33	291.35
3.00	200.00	149.09	106.31	86.24	85.32	106.42	172.89	311.12
3.10	200.00	162.89	124.48	105.32	104.82	126.82	194.65	333.72
3.20	200.00	179.87	145.87	127.58	127.49	150.57	219.87	359.12
3.30	200.00	200.61	170.68	153.18	153.57	177.83	248.78	386.87
3.40	200.00	225.67	198.99	182.14	183.04	208.61	281.42	414.93
3.50	200.00	252.59	227.40	210.69	211.70	237.74	311.12	428.63

Chapter 17
Turbulent Flow and Turbulence Models

17.1 Introductory Remarks

An important physical phenomenon which may occur in a flow field and which has not been addressed as yet is turbulence. This topic is rather complex, and an attempt to include all aspects of turbulence is beyond the scope of this text. Rather, an attempt is made to briefly describe the features of the turbulent flows and to illustrate how their inclusion modifies the equations of fluid motion. In this regard, only two-dimensional problems are considered.

17.2 Fundamental Concepts

In this section some fundamental concepts about turbulent flows are explored. This explanation is accomplished by considering the classical turbulent boundary layers.

In many high Reynolds number flows, the effect of viscous forces is dominant in the regions adjacent to the surface. This region is known as the boundary layer. From elementary fluid mechanics, it is well known that a boundary layer usually starts with well-behaved streamlines, and the fluid mixing is at microscopic level. This type of boundary layer flow is identified as a laminar boundary layer. Due to conditions imposed by the geometry and flow field, such as surface roughness, surface temperature, surface injection, and pressure gradient, the mixing of the fluid increases and takes place at the macroscopic level; streamlines are no longer well-behaved. This type of flow is known as turbulent flow. There is a transitional region between laminar and turbulent boundary layers known as the *transition region.* The various boundary layer regions are shown in Figure 17-1.

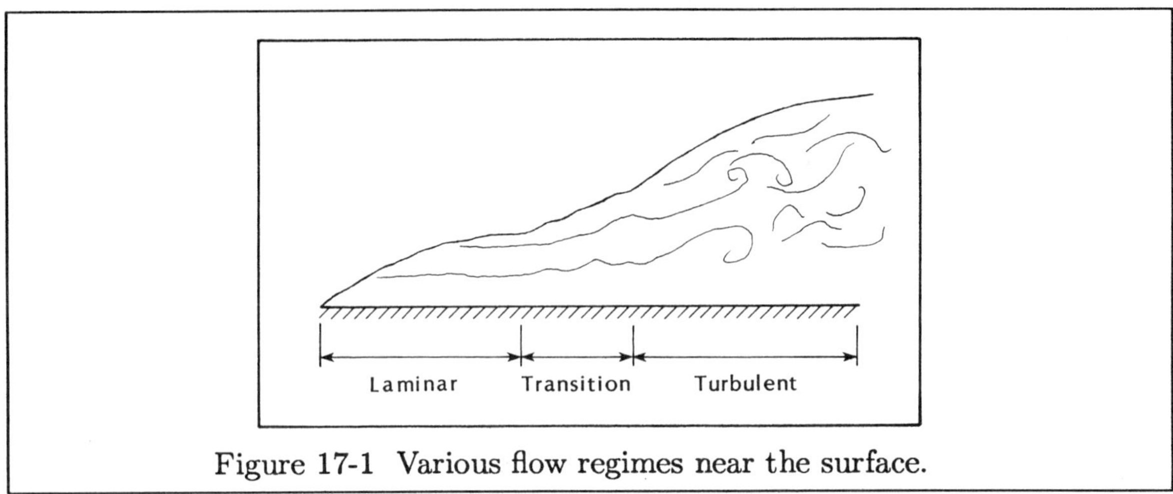

Figure 17-1 Various flow regimes near the surface.

As a result of heavy mixing of the fluid in a turbulent boundary layer and the associated large momentum flux, the velocity profile becomes fuller in a turbulent boundary layer as compared to a laminar boundary layer; i.e., the velocity gradient near the wall for a turbulent boundary layer is larger than the velocity gradient of a laminar boundary layer. Typical velocity profiles for laminar and turbulent boundary layers are shown in Figure 17-2.

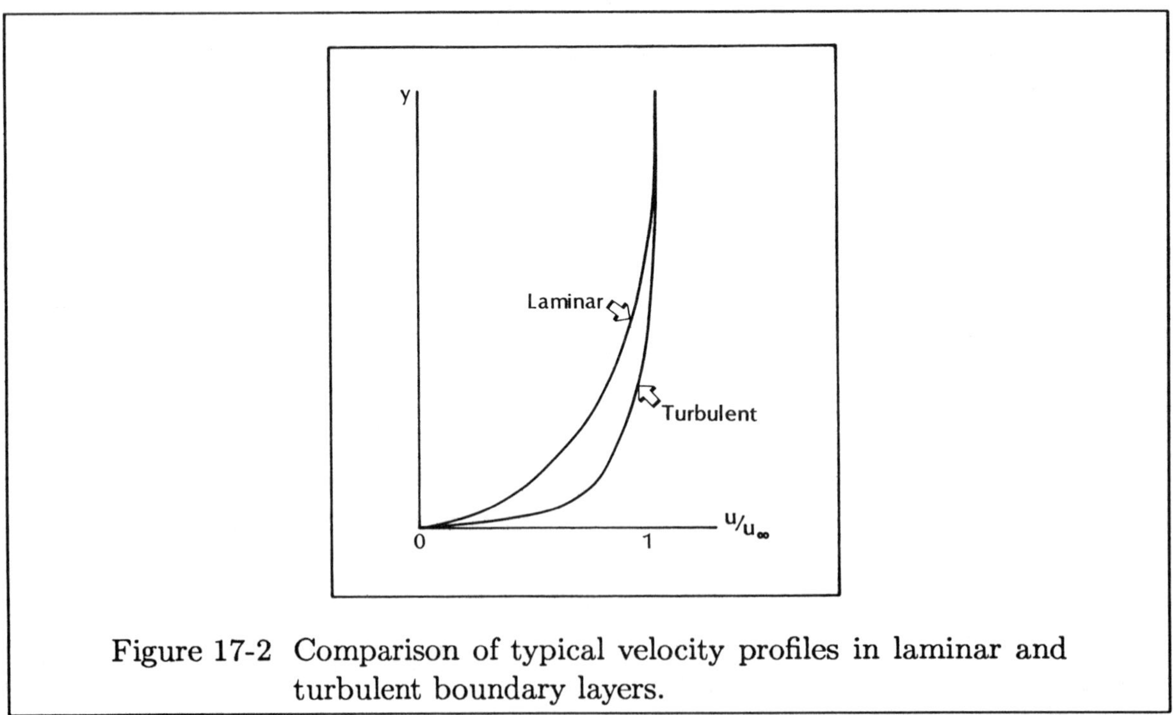

Figure 17-2 Comparison of typical velocity profiles in laminar and turbulent boundary layers.

In addition, boundary layer flows can be classified into velocity boundary layers and thermal boundary layers. For most problems, velocity boundary layer thickness and thermal boundary layer thickness are not identical. Typical velocity and thermal boundary layers are shown in Figure 17-3.

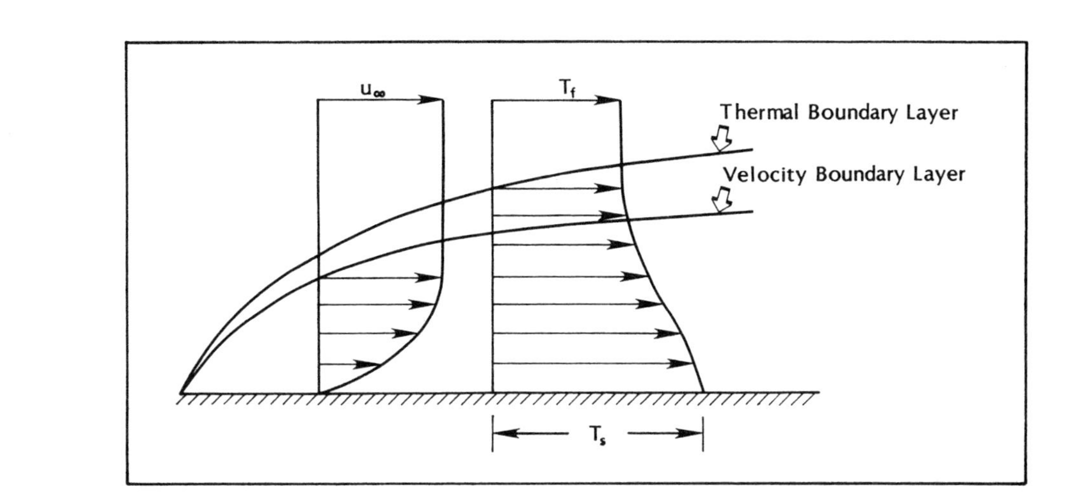

Figure 17-3 A typical comparison of velocity and thermal boundary layers.

Recall that a nondimensional parameter was previously defined as the Prandtl number and expressed as

$$\mathrm{Pr} = \frac{\nu}{\alpha} = \frac{\mu c_p}{k}$$

This parameter represents the ratio of momentum transfer in the flow to that of heat transfer. Therefore, it may represent relative velocity and thermal boundary layer thicknesses. Figure 17-4 illustrates this relation. If δ and δ_t are used to denote

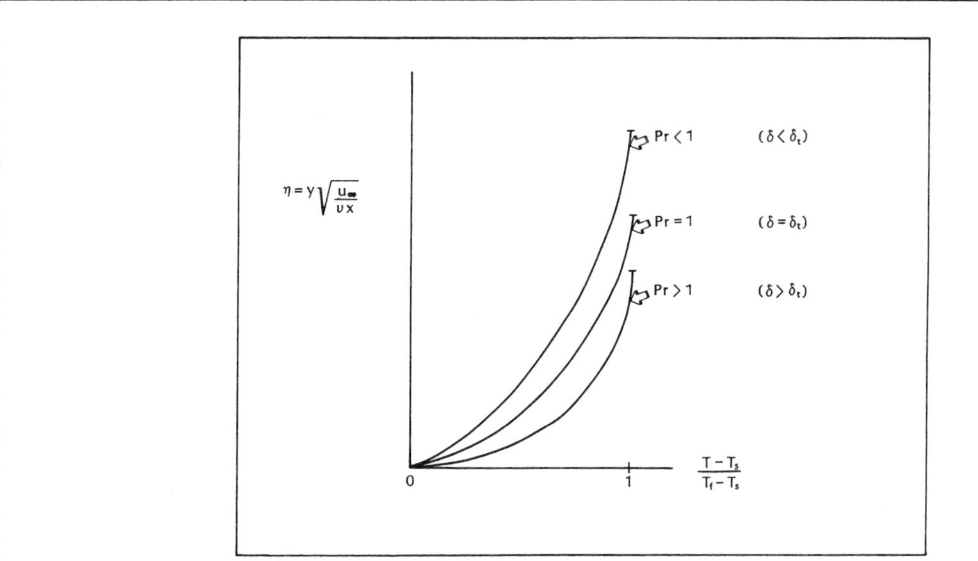

Figure 17-4. The velocity and thermal boundary layer thicknesses for various Prandtl numbers.

the velocity and thermal boundary layer thicknesses, respectively, then,

for	$\mathrm{Pr} < 1$,	$\delta_t > \delta$
	$\mathrm{Pr} = 1$,	$\delta_t = \delta$
and for	$\mathrm{Pr} > 1$,	$\delta_t < \delta$

17.3 Modifications of the Equations of Fluid Motion

In order to include and account for the effect of turbulence in a flow field, the equations of fluid motion are typically modified and amended by turbulence models. Traditionally this modification is accomplished by representing the instantaneous flow quantities by the sum of a mean value (denoted by a bar over the variable) and a time-dependent fluctuating value (denoted by a prime). Mathematically, it is expressed as

$$f = \overline{f} + f' \tag{17-1}$$

where

$$\overline{f} = \frac{1}{\Delta t} \int_{t_o}^{t_o+\Delta t} f dt \tag{17-2}$$

These quantities are shown graphically in Figure 17-5. The time averaging of a fluctuating quantity over a time interval Δt results in

$$\overline{f'} = \frac{1}{\Delta t} \int_{t_o}^{t_o+\Delta t} f' dt = 0 \tag{17-3}$$

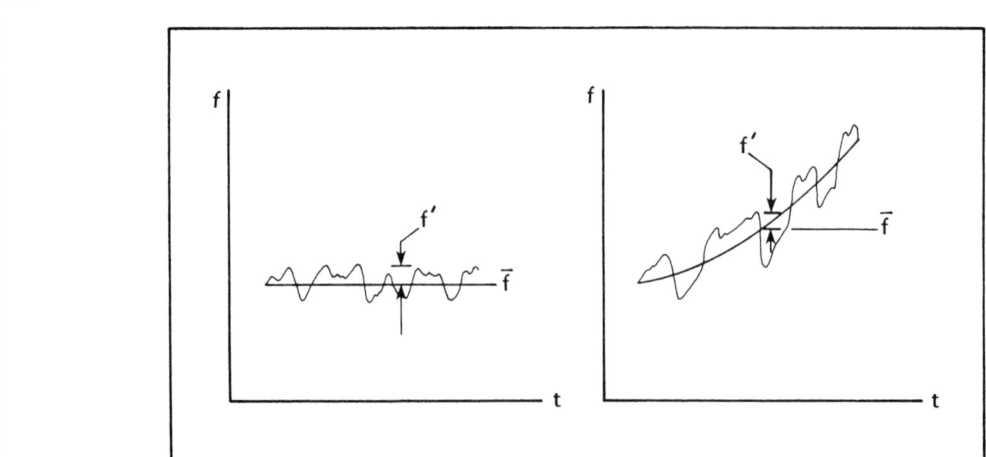

Figure 17-5 Illustration of the average and fluctuating quantities for steady and unsteady flows.

The time interval, Δt, used in the definitions above must be larger than the period of fluctuating quantities but smaller than the time interval associated with the unsteady flow. Thus the time interval is problem dependent, i.e., depends on the geometry and physics of the flow field being investigated. Note that for a steady flow the time-averaged mean value is constant while for an unsteady flow it is a function of time.

In order to carry out the mathematical details, the following averaging rules are applied:

$$\overline{\overline{f}} = \overline{f} \tag{17-4a}$$

$$\overline{f+g} = \overline{f} + \overline{g} \tag{17-4b}$$

$$\overline{\overline{f}g} = \overline{f}\,\overline{g} \tag{17-4c}$$

$$\overline{\frac{\partial f}{\partial x}} = \frac{\partial \overline{f}}{\partial x} \tag{17-4d}$$

$$\overline{f'^2} \neq 0 \tag{17-4e}$$

$$\overline{f'g'} \neq 0 \tag{17-4f}$$

$$\overline{(\overline{f}+f')^2} = \overline{f}^2 + \overline{f'^2} \tag{17-4g}$$

It is important to recognize that the averaged values of the products of fluctuating terms represented by $\overline{f'g'}$ are not, in general, zero. Indeed, these quantities—in particular, those involving the velocity fluctuations—are critically important and represent the cornerstone of turbulent flow effects.

The conservation of mass for a two-dimensional flow is given by the first component of Equation (11-192), which can be expressed in dimensional form as

$$\frac{\partial \rho}{\partial t} + \frac{\partial}{\partial x}(\rho u) + \frac{\partial}{\partial y}(\rho v) + \frac{\alpha}{y}(\rho v) = 0$$

The density and velocity components are replaced by relation (17-1) to provide

$$\frac{\partial}{\partial t}(\overline{\rho} + \rho') + \frac{\partial}{\partial x}\left[(\overline{\rho}+\rho')\,(\overline{u}+u')\right] + \frac{\partial}{\partial y}\left[(\overline{\rho}+\rho')\,(\overline{v}+v')\right]$$

$$+ \frac{\alpha}{y}\left[(\overline{\rho}+\rho')\,(\overline{v}+v')\right] = 0$$

Once this equation is expanded, it is time averaged according to rules set by (17-4) to provide

$$\frac{\partial \overline{\rho}}{\partial t} + \frac{\partial}{\partial x}\left(\overline{\rho}\,\overline{u} + \overline{\rho' u'}\right) + \frac{\partial}{\partial y}\left(\overline{\rho}\,\overline{v} + \overline{\rho' v'}\right) + \frac{\alpha}{y}\left(\overline{\rho}\,\overline{v} + \overline{\rho' v'}\right) = 0 \tag{17-5}$$

The mathematical manipulation for the momentum equation is considered for the x-component, and, subsequently, the results are extended to the y-component. The x-component of the momentum equation in dimensional form is given by the second component of (11-192) as

$$\frac{\partial}{\partial t}(\rho u) + \frac{\partial}{\partial x}(\rho u^2 + p) + \frac{\partial}{\partial y}(\rho u v) + \frac{\alpha}{y}(\rho u v) = \frac{\partial}{\partial x}\left[\mu\left(\frac{4}{3}\frac{\partial u}{\partial x} - \frac{2}{3}\frac{\partial v}{\partial y}\right)\right]$$
$$+ \frac{\partial}{\partial y}\left[\mu\left(\frac{\partial u}{\partial y} + \frac{\partial v}{\partial x}\right)\right] + \frac{\alpha}{y}\left[\mu\left(\frac{\partial u}{\partial y} + \frac{\partial v}{\partial x}\right) - \frac{2}{3}y\frac{\partial}{\partial x}\left(\mu\frac{v}{y}\right)\right]$$

The instantaneous values are replaced by time-averaged mean and fluctuating values to provide

$$\frac{\partial}{\partial t}\left[(\overline{\rho}+\rho')\,(\overline{u}+u')\right] + \frac{\partial}{\partial x}\left[(\overline{\rho}+\rho')\,(\overline{u}+u')^2 + (\overline{p}+p')\right]$$
$$+ \frac{\partial}{\partial y}\left[(\overline{\rho}+\rho')\,(\overline{u}+u')\,(\overline{v}+v')\right] + \frac{\alpha}{y}\left[(\overline{\rho}+\rho')\,(\overline{u}+u')\,(\overline{v}+v')\right] =$$
$$\frac{\partial}{\partial x}\left\{\mu\left[\frac{4}{3}\frac{\partial}{\partial x}(\overline{u}+u') - \frac{2}{3}\frac{\partial}{\partial y}(\overline{v}+v')\right]\right\} + \frac{\partial}{\partial y}\left\{\mu\left[\frac{\partial}{\partial y}(\overline{u}+u') + \frac{\partial}{\partial x}(\overline{v}+v')\right]\right\}$$
$$+ \frac{\alpha}{y}\left\{\mu\left[\frac{\partial}{\partial y}(\overline{u}+u') + \frac{\partial}{\partial x}(\overline{v}+v')\right] - \frac{2}{3}y\frac{\partial}{\partial x}\left[\frac{\mu}{y}(\overline{v}+v')\right]\right\}$$

Since the fluctuating component of viscosity is usually small, it has not been included in this equation. Now the entire equation is time averaged and subsequently rearranged as

$$\frac{\partial}{\partial t}\left(\overline{\rho}\,\overline{u} + \overline{\rho' u'}\right) + \frac{\partial}{\partial x}\left(\overline{\rho}\,\overline{u}^2 + \overline{p} + \overline{\rho}\overline{u'^2} + \overline{\rho' u'^2} + 2\overline{u}\,\overline{\rho' u'}\right)$$
$$\frac{\partial}{\partial y}\left(\overline{\rho}\,\overline{u}\,\overline{v} + \overline{\rho}\overline{u'v'} + \overline{u}\,\overline{\rho' v'} + \overline{v}\,\overline{\rho' u'} + \overline{\rho' u' v'}\right) + \frac{\alpha}{y}\left(\overline{\rho}\,\overline{u}\,\overline{v} + \overline{\rho}\overline{u'v'} + \overline{u}\,\overline{\rho' v'} + \overline{v}\,\overline{\rho' u'} + \overline{\rho' u' v'}\right) =$$
$$\frac{\partial}{\partial x}\left[\mu\left(\frac{4}{3}\frac{\partial \overline{u}}{\partial x} - \frac{2}{3}\frac{\partial \overline{v}}{\partial y}\right)\right] + \frac{\partial}{\partial y}\left[\mu\left(\frac{\partial \overline{u}}{\partial y} + \frac{\partial \overline{v}}{\partial x}\right)\right] + \frac{\alpha}{y}\left[\mu\left(\frac{\partial \overline{u}}{\partial y} + \frac{\partial \overline{v}}{\partial x}\right) - \frac{2}{3}y\frac{\partial}{\partial x}\left(\mu\frac{\overline{v}}{y}\right)\right]$$

Writing this equation in terms of shear stresses, one obtains

$$\frac{\partial}{\partial t}\left(\overline{\rho}\,\overline{u} + \overline{\rho' u'}\right) + \frac{\partial}{\partial x}\left(\overline{\rho}\,\overline{u}^2 + \overline{p} + \overline{u}\,\overline{\rho' u'}\right) + \frac{\partial}{\partial y}\left(\overline{\rho}\,\overline{u}\,\overline{v} + \overline{u}\,\overline{\rho' v'}\right) + \frac{\alpha}{y}\left(\overline{\rho}\,\overline{u}\,\overline{v} + \overline{u}\,\overline{\rho' v'}\right) =$$
$$\frac{\partial}{\partial x}\left(\tau_{xxp} - \overline{\rho}\overline{u'^2} - \overline{\rho' u'^2} - \overline{u}\,\overline{\rho' u'}\right) + \frac{\partial}{\partial y}\left(\tau_{xy} - \overline{\rho}\overline{u'v'} - \overline{\rho' u' v'} - \overline{v}\,\overline{\rho' u'}\right) +$$

$$+\frac{\alpha}{y}\left[\overline{\tau}_{xy}-\frac{2}{3}y\frac{\partial}{\partial x}\left(\mu\frac{\overline{v}}{y}\right)-\overline{\rho}\overline{u'v'}-\overline{v}\overline{\rho'u'}-\overline{\rho'u'v'}\right] \tag{17-6}$$

Similarly, the y-component of the momentum equation becomes

$$\frac{\partial}{\partial t}\left(\overline{\rho}\,\overline{v}+\overline{\rho'v'}\right)+\frac{\partial}{\partial x}\left(\overline{\rho}\,\overline{u}\,\overline{v}+\overline{v}\overline{\rho'v'}\right)+\frac{\partial}{\partial y}\left(\overline{\rho}\,\overline{u}^2+\overline{p}+\overline{v}\overline{\rho'v'}\right)$$

$$+\frac{\alpha}{y}\left(\overline{\rho}\,\overline{v}^2+\overline{v}\overline{\rho'v'}\right)=\frac{\partial}{\partial x}\left(\overline{\tau}_{xy}-\overline{\rho}\overline{u'v'}-\overline{u}\overline{\rho'v'}-\overline{\rho'u'v'}\right)$$

$$+\frac{\partial}{\partial y}\left(\overline{\tau}_{yyp}-\overline{\rho}\overline{v'^2}-\overline{\rho'v'^2}-\overline{v}\overline{\rho'v'}\right)+\frac{\alpha}{y}\left[\overline{\tau}_{yyp}-\overline{\tau}_{\theta\theta}\right.$$

$$\left.-\frac{2}{3}\frac{\mu}{y}\overline{v}-\frac{2}{3}y\frac{\partial}{\partial y}\left(\mu\frac{\overline{v}}{y}\right)-\overline{\rho}\overline{v'^2}-\overline{\rho'v'^2}-\overline{v}\overline{\rho'v'}\right] \tag{17-7}$$

The energy equation given by the fourth component of (11-192) in terms of the total energy per unit mass is expressed as

$$\frac{\partial}{\partial t}(\rho e_t)+\frac{\partial}{\partial x}\left[(\rho e_t+p)\,u\right]+\frac{\partial}{\partial y}\left[(\rho e_t+p)\,v\right]+\frac{\alpha}{y}\left[(\rho e_t+p)\,v\right]=$$

$$+\frac{\partial}{\partial x}\left[u\tau_{xxp}+v\tau_{xy}-q_x\right]+\frac{\partial}{\partial y}\left[u\tau_{xy}+v\tau_{yyp}-q_y\right]$$

$$+\frac{\alpha}{y}\left[u\tau_{xy}+v\tau_{yyp}-q_y-\frac{2}{3}\mu\frac{v^2}{y}-y\frac{\partial}{\partial y}\left(\frac{2}{3}\mu\frac{v^2}{y}\right)-y\frac{\partial}{\partial x}\left(\frac{2}{3}\mu\frac{uv}{y}\right)\right]$$

Note that the total energy is assumed to be composed of internal energy and kinetic energy. The mathematical procedure used to include turbulence is similar to previous manipulations. The final form of the energy equation becomes

$$\frac{\partial}{\partial t}\left[\overline{\rho}\,\overline{e}_t+\frac{1}{2}\overline{\rho}\left(\overline{u'^2}+\overline{v'^2}\right)+\overline{\rho'e'}+\frac{1}{2}\left(\overline{\rho'u'^2}+\overline{\rho'v'^2}\right)+\overline{u}\overline{\rho'u'}+\overline{v}\overline{\rho'v'}\right]$$

$$+\frac{\partial}{\partial x}\left[\overline{u}\left(\overline{\rho}\,\overline{e}_t+\overline{p}\right)+\frac{1}{2}\overline{\rho}\,\overline{u}\left(\overline{u'^2}+\overline{v'^2}\right)+\overline{u}\left(\overline{\rho'e'}+\overline{u}\overline{\rho'u'}+\overline{v}\overline{\rho'v'}\right)\right.$$

$$+\frac{1}{2}\overline{u}\left(\overline{\rho'u'^2}+\overline{\rho'v'^2}\right)+\overline{\rho}\left(\overline{e'u'}+\frac{1}{2}\overline{u'^3}+\frac{1}{2}\overline{u'v'^2}\right)+\overline{\rho}\left(\overline{u}\overline{u'^2}+\overline{v}\overline{u'v'}\right)$$

$$\left.+\overline{e}_t\overline{\rho'u'}+\overline{\rho'e'u'}+\frac{1}{2}\left(\overline{\rho'u'^3}+2\overline{u}\overline{\rho'u'^2}+\overline{\rho'u'v'^2}+2\overline{v}\overline{\rho'u'v'}\right)+\overline{p'u'}\right]$$

$$+\frac{\partial}{\partial y}\left[\overline{v}\left(\overline{\rho}\,\overline{e}_t+\overline{p}\right)+\frac{1}{2}\overline{\rho}\,\overline{v}\left(\overline{u'^2}+\overline{v'^2}\right)+\overline{v}\left(\overline{\rho'e'}+\overline{u}\overline{\rho'u'}+\overline{v}\overline{\rho'v'}\right)+\right.$$

$$+\frac{1}{2}\overline{v}\left(\overline{\rho' u'^2}+\overline{\rho' v'^2}\right)+\overline{\rho}\left(\overline{e'v'}+\frac{1}{2}\overline{u'^2 v'}+\frac{1}{2}\overline{v'^3}\right)+\overline{\rho}\left(\overline{u}\overline{u'v'}+\overline{v}\overline{v'^2}\right)$$

$$+\overline{e}_t\overline{\rho' v'}+\overline{\rho' e' v'}+\frac{1}{2}\left(\overline{\rho' v' u'^2}+2\overline{u}\overline{\rho' u' v'}+\overline{\rho' v'^3}+2\overline{v}\overline{\rho' v'^2}\right)+\overline{p' v'}\Bigg]$$

$$+\frac{\alpha}{y}\Bigg[\overline{v}\left(\overline{\rho}\,\overline{e}_t+\overline{p}\right)+\frac{1}{2}\overline{\rho}\,\overline{v}\left(\overline{u'^2}+\overline{v'^2}\right)+\overline{v}\left(\overline{\rho' e'}+\overline{u}\overline{\rho' u'}+\overline{v}\overline{\rho' v'}\right)$$

$$+\frac{1}{2}\overline{v}\left(\overline{\rho' u'^2}+\overline{\rho' v'^2}\right)+\overline{\rho}\left(\overline{e'v'}+\frac{1}{2}\overline{u'^2 v'}+\frac{1}{2}\overline{v'^3}\right)+\overline{\rho}\left(\overline{u}\overline{u'v'}+\overline{v}\overline{v'^2}\right)$$

$$+\overline{e}_t\overline{\rho' v'}+\overline{\rho' e' v'}+\frac{1}{2}\left(\overline{\rho' v' u'^2}+2\overline{u}\overline{\rho' u' v'}+\overline{\rho' v'^3}+2\overline{v}\overline{\rho' v'^2}\right)+\overline{p' v'}\Bigg]=$$

$$\frac{\partial}{\partial x}\Bigg[\overline{u}\,\overline{\tau}_{xxp}+\overline{v}\,\overline{\tau}_{xy}-k\frac{\partial\overline{T}}{\partial x}+\mu\left(\frac{4}{3}\overline{u'\frac{\partial u'}{\partial x}}-\frac{2}{3}\overline{u'\frac{\partial v'}{\partial y}}\right)$$

$$+\mu\left(-\frac{2}{3}\left(\overline{v'\frac{\partial u'}{\partial x}}+\overline{v'\frac{\partial v'}{\partial y}}\right)+\frac{4}{3}\frac{\overline{v'^2}}{y}\right)\Bigg]$$

$$+\frac{\partial}{\partial y}\Bigg[\overline{u}\,\overline{\tau}_{xy}+\overline{v}\,\overline{\tau}_{yyp}-k\frac{\partial\overline{T}}{\partial y}+\mu\left(-\frac{2}{3}\left(\overline{u'\frac{\partial u'}{\partial x}}+\overline{u'\frac{\partial v'}{\partial y}}\right)+\frac{4}{3}\frac{1}{y}\overline{u'v'}\right)$$

$$+\mu\left(\frac{4}{3}\overline{v'\frac{\partial v'}{\partial y}}-\frac{2}{3}\overline{v'\frac{\partial u'}{\partial x}}\right)\Bigg]$$

$$+\frac{\alpha}{y}\Bigg[\overline{u}\,\overline{\tau}_{xy}+\overline{v}\,\overline{\tau}_{yyp}-k\frac{\partial\overline{T}}{\partial y}+\mu\left(-\frac{2}{3}\left(\overline{u'\frac{\partial u'}{\partial x}}+\overline{u'\frac{\partial v'}{\partial y}}\right)+\frac{4}{3}\frac{1}{y}\overline{u'v'}\right)$$

$$+\mu\left(\frac{4}{3}\overline{v'\frac{\partial v'}{\partial y}}-\frac{2}{3}\overline{v'\frac{\partial u'}{\partial x}}\right)-\frac{2}{3}\frac{\mu}{y}\left(\overline{v}^2\right)-\frac{2}{3}\frac{\mu}{y}\left(\overline{v'^2}\right)$$

$$-y\frac{\partial}{\partial y}\left(\frac{2}{3}\frac{\mu}{y}\left(\overline{v}^2+\overline{v'^2}\right)\right)-y\frac{\partial}{\partial x}\left(\frac{2}{3}\frac{\mu}{y}\left(\overline{u}\,\overline{v}+\overline{u'v'}\right)\right)\Bigg] \qquad (17\text{-}8)$$

Now it is obvious that the number of unknowns in the system of equations composed of (17-5) through (17-8) has increased significantly. Before considering additional relations to close the system, some reduction of terms is introduced. It has been shown experimentally that for flows up to Mach five, turbulence structure remains unchanged and is similar to that of incompressible flows (the so-called Morkovin hypothesis). Therefore, the fluctuating correlations involving density fluctuations are usually neglected. In addition, triple correlations such as $\overline{v'^3}$ are assumed smaller than the double correlations and are neglected as well. Finally,

Equations (17-5) through (17-8) are expressed as

$$\frac{\partial Q}{\partial t}+\frac{\partial Q'}{\partial t}+\frac{\partial E}{\partial x}+\frac{\partial F}{\partial y}+\alpha H=\frac{\partial E_v}{\partial x}+\frac{\partial F_v}{\partial y}+\alpha H_v+\frac{\partial E'}{\partial x}+\frac{\partial F'}{\partial y}+\alpha H' \quad (17\text{-}9)$$

where

$$Q=\begin{bmatrix}\bar\rho\\ \bar\rho\,\bar u\\ \bar\rho\,\bar v\\ \bar\rho\,\bar e_t\end{bmatrix} \quad (17\text{-}10a) \qquad Q'=\begin{bmatrix}0\\0\\0\\ \frac{1}{2}\bar\rho\left(\overline{u'^2}+\overline{v'^2}\right)\end{bmatrix} \quad (17\text{-}10b)$$

$$E=\begin{bmatrix}\bar\rho\,\bar u\\ \bar\rho\,\bar u^2+\bar p\\ \bar\rho\,\bar u\,\bar v\\ (\bar\rho\,\bar e_t+\bar p)\,\bar u\end{bmatrix} \quad (17\text{-}10c) \qquad F=\begin{bmatrix}\bar\rho\,\bar v\\ \bar\rho\,\bar u\,\bar v\\ \bar\rho\,\bar v^2+\bar p\\ (\bar\rho\,\bar e_t+\bar p)\,\bar v\end{bmatrix} \quad (17\text{-}10d)$$

$$H=\frac{1}{y}\begin{bmatrix}\bar\rho\,\bar v\\ \bar\rho\,\bar u\,\bar v\\ \bar\rho\,\bar v^2\\ (\bar\rho\,\bar e_t+\bar p)\,\bar v\end{bmatrix} \quad (17\text{-}10e) \qquad E_v=\begin{bmatrix}0\\ \bar\tau_{xxp}\\ \bar\tau_{xy}\\ \bar u\,\bar\tau_{xxp}+\bar v\,\bar\tau_{xy}-\bar q_x\end{bmatrix} \quad (17\text{-}10f)$$

$$F_v=\begin{bmatrix}0\\ \bar\tau_{xy}\\ \bar\tau_{yyp}\\ \bar u\,\bar\tau_{xy}+\bar v\,\bar\tau_{yyp}-q_y\end{bmatrix} \quad (17\text{-}10g)$$

$$H_v=\frac{1}{y}\begin{bmatrix}0\\ \bar\tau_{xy}-\frac{2}{3}y\frac{\partial}{\partial x}\left(\mu\frac{\bar v}{y}\right)\\ \bar\tau_{yyp}-\bar\tau_{\theta\theta}-\frac{2}{3}\frac{\mu}{y}\bar v-\frac{2}{3}y\frac{\partial}{\partial y}\left(\mu\frac{\bar v}{y}\right)\\ \bar u\,\bar\tau_{xy}+\bar v\,\bar\tau_{yyp}-\bar q_y-\frac{2}{3}\frac{\mu}{y}\left(\bar v^2\right)-y\frac{\partial}{\partial y}\left(\frac{2}{3}\mu\frac{\bar v^2}{y}\right)-y\frac{\partial}{\partial x}\left(\frac{2}{3}\mu\frac{\overline{uv}}{y}\right)\end{bmatrix} \quad (17\text{-}10h)$$

$$E'=\begin{bmatrix}0\\ -\bar\rho\overline{u'^2}\\ -\bar\rho\overline{u'v'}\\ -\frac{1}{2}\bar\rho\,\bar u\left(\overline{u'^2}+\overline{v'^2}\right)-\bar\rho\left(\overline{e'u'}\right)-\bar\rho\,\bar u\left(\overline{u'^2}\right)-\bar\rho\,\bar v\left(\overline{u'v'}\right)-\overline{p'u'}\\ +\mu\left(\frac{4}{3}\overline{u'\frac{\partial u'}{\partial x}}-\frac{2}{3}\overline{u'\frac{\partial v'}{\partial y}}\right)+\mu\left[-\frac{2}{3}\left(\overline{v'\frac{\partial u'}{\partial x}}+\overline{v'\frac{\partial v'}{\partial y}}\right)+\frac{4}{3}\frac{\overline{v'^2}}{y}\right]\end{bmatrix} \quad (17\text{-}11a)$$

$$F' = \begin{bmatrix} 0 \\ -\overline{\rho}\overline{u'v'} \\ -\overline{\rho}\overline{v'^2} \\ -\frac{1}{2}\overline{\rho}\,\overline{v}\left(\overline{u'^2}+\overline{v'^2}\right)-\overline{\rho}\left(\overline{e'v'}\right)-\overline{\rho}\,\overline{u}\left(\overline{u'v'}\right)-\overline{\rho}\,\overline{v}\left(\overline{v'^2}\right)-\overline{p'v'} \\ +\mu\left[-\frac{2}{3}\left(\overline{u'\frac{\partial u'}{\partial x}}+\overline{u'\frac{\partial v'}{\partial y}}\right)+\frac{4}{3}\frac{1}{y}\overline{u'v'}\right]+\mu\left(\frac{4}{3}\overline{v'\frac{\partial v'}{\partial y}}-\frac{2}{3}\overline{v'\frac{\partial u'}{\partial x}}\right) \end{bmatrix} \tag{17-11b}$$

$$H' = \frac{1}{y}\begin{bmatrix} 0 \\ -\overline{\rho}\overline{u'v'} \\ -\overline{\rho}\overline{v'^2} \\ -\frac{1}{2}\overline{\rho}\,\overline{v}\left(\overline{u'^2}+\overline{v'^2}\right)-\overline{\rho}\left(\overline{e'v'}\right)-\overline{\rho}\,\overline{u}\left(\overline{u'v'}\right)-\overline{\rho}\,\overline{v}\left(\overline{v'^2}\right)-\overline{p'v'} \\ +\mu\left[-\frac{2}{3}\left(\overline{u'\frac{\partial u'}{\partial x}}+\overline{u'\frac{\partial v'}{\partial y}}\right)+\frac{4}{3}\frac{1}{y}\overline{u'v'}\right]+\mu\left(\frac{4}{3}\overline{v'\frac{\partial v'}{\partial y}}-\frac{2}{3}\overline{v'\frac{\partial u'}{\partial x}}\right) \\ -\frac{2}{3}\frac{\mu}{y}\left(\overline{v'^2}\right)-y\frac{\partial}{\partial y}\left[\frac{2}{3}\frac{\mu}{y}\left(\overline{v'^2}\right)\right]-y\frac{\partial}{\partial x}\left(\frac{2}{3}\frac{\mu}{y}\overline{u'v'}\right) \end{bmatrix} \tag{17-11c}$$

Additional unknowns appear in the system of equations given by (17-9) as a result of modification of the original system (11-192) to include the effect of turbulence. It is obvious that the solution of Equation (17-9), which includes turbulence quantities, would be a very challenging task. In fact, in practice, the effect of turbulence in most applications is simulated by the introduction of turbulent viscosity and turbulent conductivity, as will be shown in Section 17.4.2. In either case, there are additional unknowns in the Navier-Stokes equations, due to turbulence.

In order to close the system, supplementary relations must be introduced. These relations are known as turbulence models, which in general relate the fluctuating correlations to the mean flow properties by means of empirical constants. Turbulence models vary in degree of sophistication from simple algebraic equations to systems of partial differential equations. Surprisingly enough, in many instances simple algebraic models provide as good a result as sophisticated models, with less computation time. Obviously, the ultimate goal is to simulate and resolve turbulence structure directly; however, due to limitations in the capacity of current computers, the direct simulation approach is still in its infancy stage. With the advancement in computer technology, however, this goal will be accomplished in the future. Research toward this goal is currently underway.

17.4 Reynolds Averaged Navier-Stokes Equations and Turbulence Models

17.4.1 General Remarks

A turbulence model is a semi-empirical equation relating the fluctuating correlation to mean flow variables with various constants provided from experimental investigations. When this equation is expressed as an algebraic equation, it is referred to as the *zero-equation model.* When partial differential equations are used, they are referred to as *one-equation* or *two-equation models*, depending on the number of PDEs utilized. Some models employ ordinary differential equations, in which case they are referred to as half-equation models. Finally, it is possible to write a partial differential equation directly for each of the turbulence correlations such as $\overline{u'v'}$, $\overline{u'^2}$, etc., in which case they compose a system of PDEs known as the Reynolds stress equations.

Before various turbulence models are introduced, the commonly defined zones within a turbulent boundary layer are identified. A very small layer next to the surface is defined as the *laminar sublayer* where viscous shear is dominant. The outer region where turbulent shear (terms such as $\overline{u'v'}$) is dominant is called the *fully turbulent zone.* A region which connects these two zones is called the *buffer zone*, where both viscous shear and turbulent shear are of equal importance.

A typical velocity profile for a turbulent flow over a flat surface, indicating various zones, is shown in Figure 17-6. The nondimensional velocity u^+ and space coordinate y^+ are defined as

$$u^+ = \frac{u}{u_\tau} \tag{17-12}$$

and

$$y^+ = y\frac{u_\tau}{\nu} \tag{17-13}$$

where u_τ is known as the friction velocity given by

$$u_\tau = \left(\frac{\tau_\omega}{\rho}\right)^{\frac{1}{2}} \tag{17-14}$$

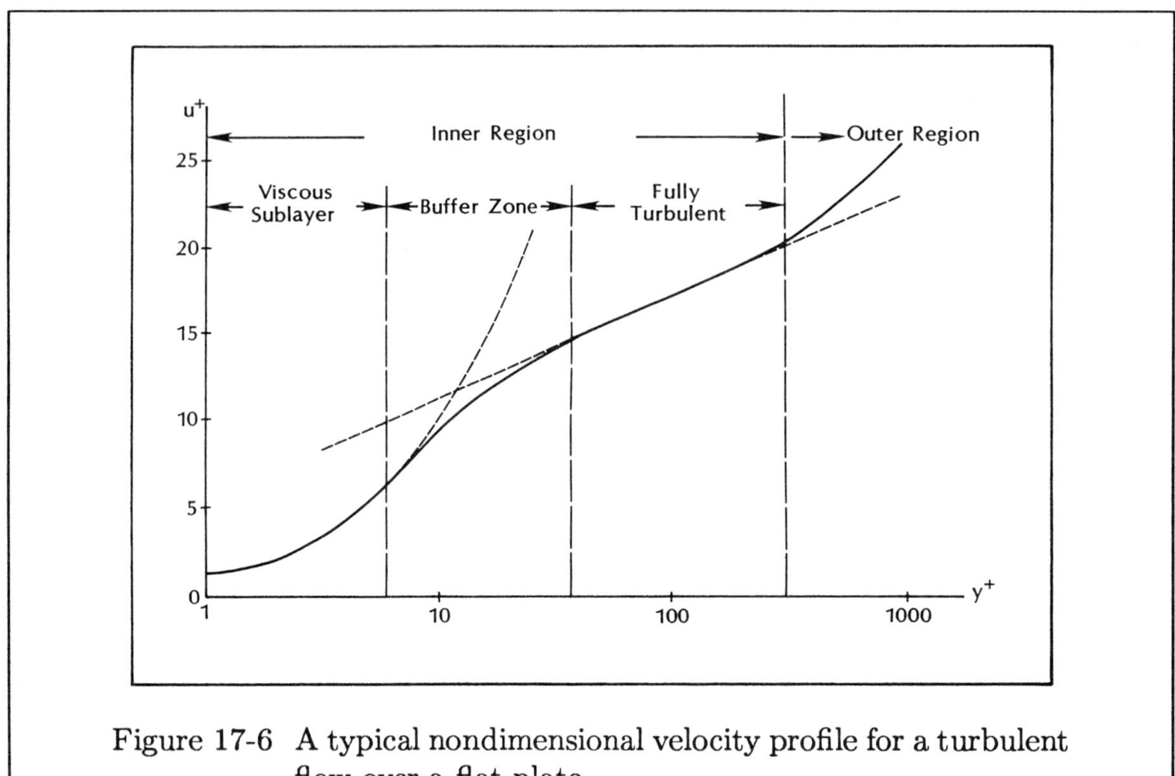

Figure 17-6 A typical nondimensional velocity profile for a turbulent flow over a flat plate.

Another categorization of a turbulent boundary layer is to divide it into inner and outer regions. The inner region includes the laminar sublayer, buffer zone, and part of the fully turbulent zone, as shown in Figure 17-6. It is common to identify the upper limit of the inner region at about y^+ of 400. The remaining part of the turbulent boundary layer is considered to be in the outer region.

17.4.2 Turbulent Shear Stress and Heat Flux

In this section, some of the fundamental concepts introduced by the early investigating pioneers are explored. For the sake of simplicity, the steady incompressible boundary layer equations are considered. The governing equations are expressed as

$$\frac{\partial \overline{u}}{\partial x} + \frac{\partial \overline{v}}{\partial y} = 0 \tag{17-15}$$

$$\overline{u}\frac{\partial \overline{u}}{\partial x} + \overline{v}\frac{\partial \overline{u}}{\partial y} = -\frac{1}{\rho}\frac{\partial \overline{p}}{\partial x} + \nu\frac{\partial^2 \overline{u}}{\partial y^2} - \frac{\partial}{\partial y}\left(\overline{u'v'}\right) \tag{17-16}$$

$$\overline{u}\frac{\partial \overline{T}}{\partial x} + \overline{v}\frac{\partial \overline{T}}{\partial y} = \frac{\nu}{c_p}\left(\frac{\partial \overline{u}}{\partial y}\right)^2 - \frac{1}{c_p}\left(\overline{u'v'}\right)\frac{\partial \overline{u}}{\partial y} + \alpha\frac{\partial^2 \overline{T}}{\partial y^2} - \frac{\partial}{\partial y}\left(\overline{v'T'}\right) \tag{17-17}$$

In the equation above, α is the thermal diffusivity.

It can be shown that the macroscopic momentum exchange due to turbulence is represented by $-\rho\overline{u'v'}$, which is referred to as the *turbulent shear stress* (or *Reynolds stress*), denoted herein by τ_t. It is customary to combine the laminar and turbulent shear stress terms in Equation (17-16) as

$$\nu\frac{\partial^2\overline{u}}{\partial y^2} - \frac{\partial}{\partial y}\left(\overline{u'v'}\right) = \frac{1}{\rho}\frac{\partial}{\partial y}(\tau_l + \tau_t) \tag{17-18}$$

For a laminar flow under the assumptions stated, one may write

$$\tau_l = \mu\frac{\partial\overline{u}}{\partial y} = \rho\nu\frac{\partial\overline{u}}{\partial y} \tag{17-19}$$

In order to express the turbulent shear in a similar form, consider the concept of eddy viscosity, where the turbulent shear stress is related to the gradient of the mean flow velocity. This analogy was introduced by Boussinesq and is referred to as the *Boussinesq assumption.* Thus, one writes

$$\tau_t = \mu_t\frac{\partial\overline{u}}{\partial y} = \rho\nu_t\frac{\partial\overline{u}}{\partial y} \tag{17-20}$$

where μ_t (or ν_t) is known as the turbulent viscosity or eddy viscosity. Hence, Equation (17-16) may be expressed as

$$\overline{u}\frac{\partial\overline{u}}{\partial x} + \overline{v}\frac{\partial\overline{u}}{\partial y} = -\frac{1}{\rho}\frac{\partial\overline{p}}{\partial x} + (\nu + \nu_t)\frac{\partial^2\overline{u}}{\partial^2 y} = -\frac{1}{\rho}\frac{\partial\overline{p}}{\partial y} + \frac{1}{\rho}(\mu + \mu_t)\frac{\partial^2\overline{u}}{\partial^2 y} \tag{17-21}$$

In a similar fashion, turbulent conductivity is defined to combine laminar and turbulent heat fluxes. For a laminar flow, the Fourier heat conduction law is expressed as

$$\left(\frac{q}{A}\right)_l = -k\frac{\partial\overline{T}}{\partial y} = -\rho c_p\alpha\frac{\partial\overline{T}}{\partial y} \tag{17-22}$$

For turbulent heat flux, the following is written,

$$\left(\frac{q}{A}\right)_t = -k_t\frac{\partial\overline{T}}{\partial y} = -\rho c_p\alpha_t\frac{\partial\overline{T}}{\partial y} \tag{17-23}$$

where k_t is the turbulent conductivity and α_t is the turbulent diffusivity. Now, the energy equation may be rearranged as

$$\overline{u}\frac{\partial\overline{T}}{\partial x} + \overline{v}\frac{\partial\overline{T}}{\partial y} = \frac{1}{\rho c_p}(\tau_l + \tau_t)\frac{\partial\overline{u}}{\partial y} - \frac{1}{\rho c_p}\frac{\partial}{\partial y}\left[\left(\frac{q}{A}\right)_l + \left(\frac{q}{A}\right)_t\right] \tag{17-24}$$

or in terms of turbulent parameters,

$$\overline{u}\frac{\partial\overline{T}}{\partial x} + \overline{v}\frac{\partial\overline{T}}{\partial y} = \frac{1}{\rho c_p}(\mu + \mu_t)\left(\frac{\partial\overline{u}}{\partial y}\right)^2 + \frac{1}{\rho c_p}(k + k_t)\frac{\partial^2\overline{T}}{\partial y^2} \tag{17-25}$$

Note that, up to this point, the problem of closure has not been addressed, i.e., two additional unknowns still remain. All that has been accomplished is the rewriting of the equations by introducing the Boussinesq assumption.

It is well known that in a laminar flow the mixing of fluid is at the molecular level, and the viscous stresses and heat fluxes are due to momentum and energy transfer by molecules traveling a distance of mean-free path before collision. A similar concept is extended to turbulent flows, where it is assumed that lumps of fluid travel a finite distance before collision and before losing their identity. The resulting momentum and energy exchange give rise to what was defined as turbulent shear stress and turbulent heat flux. This finite distance is defined as the mixing length, l. This concept was introduced by Prandtl and is known as the Prandtl mixing length. The Prandtl hypothesis is expressed as

$$\tau_t = -\rho\overline{u'v'} = \rho l^2 \left(\frac{\partial \overline{u}}{\partial y}\right)^2 \tag{17-26}$$

Similarly,

$$\left(\frac{q}{A}\right)_t = \rho l_e^2 \left(\frac{\partial \overline{T}}{\partial y}\right)^2 \tag{17-27}$$

In terms of turbulent viscosity and turbulent conductivity, the following may be written:

$$\mu_t = \rho l^2 \left(\frac{\partial \overline{u}}{\partial y}\right) \tag{17-28}$$

$$k_t = \rho c_p l_e^2 \left(\frac{\partial \overline{T}}{\partial y}\right) \tag{17-29}$$

The concept of turbulent viscosity and turbulent conductivity can be easily extended to general flow fields governed by the Navier-Stokes equations. For example, for a 2-D problem,

$$\mu_t = \rho l^2 \left[\left(\frac{\partial \overline{u}}{\partial y}\right)^2 + \left(\frac{\partial \overline{v}}{\partial x}\right)^2\right]^{\frac{1}{2}} \tag{17-30}$$

What has been accomplished up to this point is the introduction of new parameters, and the central issue of closure is not complete as yet since one still has two additional unknowns, now in the form of μ_t and k_t, or, equivalently, l and l_e.

In order to write expressions for mixing lengths, experimental investigations are heavily relied upon; and l and l_e are modeled for the various flow regimes, namely the inner and outer regions.

As previously discussed, these semi-empirical relations are known as turbulence models and may be expressed as algebraic equations (zero-equation), ordinary differential equations (half-equation), or partial differential equations (one-equation or two-equation). In the following sections, some commonly used zero-equation, one-equation, and two-equation turbulence models, as a sample of the wide range of models, are investigated. Several recently published survey papers and comparative analysis papers on turbulence models are excellent references for those interested in various aspects of turbulence models, [17-1] through [17-4].

17.4.3 Reynolds Averaged Navier-Stokes Equations

A simple and practical approach to include the effect of turbulence in the Navier-Stokes equations is simply to replace the molecular viscosity μ in the stress terms by $(\mu + \mu_t)$ and the term (μ/Pr) in the heat conduction term by $(\mu/Pr + \mu_t/Pr_t)$. The parameter Pr_t is called the *turbulent Prandtl number* and is generally taken to be 0.9 for air.

Returning to the nondimensional Navier-Stokes equations in the computational space given by Equation (11-60) and repeated here for convenience, one has

$$\frac{\partial \bar{Q}}{\partial \tau} + \frac{\partial \bar{E}}{\partial \xi} + \frac{\partial \bar{F}}{\partial \eta} + \frac{\partial \bar{G}}{\partial \zeta} = \frac{\partial \bar{E}_v}{\partial \xi} + \frac{\partial \bar{F}_v}{\partial \eta} + \frac{\partial \bar{G}_v}{\partial \zeta} \tag{17-31}$$

where the flux vectors are given by (11-62) through (11-67). Now, to include the effect of turbulence, Equation (17-31) is used along with the following definitions for the viscous stress and heat conduction terms.

$$\begin{aligned} \tau_{xx} &= \frac{1}{Re_\infty}(\mu + \mu_t)\left[\frac{4}{3}(\xi_x u_\xi + \eta_x u_\eta + \zeta_x u_\zeta) - \frac{2}{3}(\xi_y v_\xi + \eta_y v_\eta + \zeta_y v_\zeta)\right. \\ &- \left.\frac{2}{3}(\xi_z w_\xi + \eta_z w_\eta + \zeta_z w_\zeta)\right] \end{aligned} \tag{17-32}$$

$$\begin{aligned} \tau_{yy} &= \frac{1}{Re_\infty}(\mu + \mu_t)\left[\frac{4}{3}(\xi_y v_\xi + \eta_y v_\eta + \zeta_y v_\zeta) - \frac{2}{3}(\xi_x u_\xi + \eta_x u_\eta + \zeta_x u_\zeta)\right. \\ &- \left.\frac{2}{3}(\xi_z w_\xi + \eta_z w_\eta + \zeta_z w_\zeta)\right] \end{aligned} \tag{17-33}$$

$$\begin{aligned} \tau_{zz} &= \frac{1}{Re_\infty}(\mu + \mu_t)\left[\frac{4}{3}(\xi_z w_\xi + \eta_z w_\eta + \zeta_z w_\zeta) - \frac{2}{3}(\xi_x u_\xi + \eta_x u_\eta + \zeta_x u_\zeta)\right. \\ &- \left.\frac{2}{3}(\xi_y v_\xi + \eta_y v_\eta + \zeta_y v_\zeta)\right] \end{aligned} \tag{17-34}$$

$$\tau_{xy} = \tau_{yx} = \frac{1}{Re_\infty}(\mu + \mu_t)(\xi_y u_\xi + \eta_y u_\eta + \zeta_y u_\zeta + \xi_x v_\xi + \eta_x v_\eta + \zeta_x v_\zeta) \tag{17-35}$$

$$\tau_{xz} = \tau_{zx} = \frac{1}{Re_\infty}(\mu + \mu_t)(\xi_z u_\xi + \eta_z u_\eta + \zeta_z u_\zeta + \xi_x w_\xi + \eta_x w_\eta + \zeta_x w_\zeta) \tag{17-36}$$

$$\tau_{yz} = \tau_{zy} = \frac{1}{Re_\infty}(\mu + \mu_t)(\xi_z v_\xi + \eta_z v_\eta + \zeta_z v_\zeta + \xi_y w_\xi + \eta_y w_\eta + \zeta_y w_\zeta) \tag{17-37}$$

$$q_x = -\frac{1}{Re_\infty(\gamma - 1)M_\infty^2}\left(\frac{\mu}{Pr} + \frac{\mu_t}{Pr_t}\right)(\xi_x T_\xi + \eta_x T_\eta + \zeta_x T_\zeta) \tag{17-38}$$

$$q_y = -\frac{1}{Re_\infty(\gamma - 1)M_\infty^2}\left(\frac{\mu}{Pr} + \frac{\mu_t}{Pr_t}\right)(\xi_y T_\xi + \eta_y T_\eta + \zeta_y T_\zeta) \tag{17-39}$$

$$q_z = -\frac{1}{Re_\infty(\gamma - 1)M_\infty^2}\left(\frac{\mu}{Pr} + \frac{\mu_t}{Pr_t}\right)(\xi_z T_\xi + \eta_z T_\eta + \zeta_z T_\zeta) \tag{17-40}$$

where the value of μ_t is provided by a turbulence model.

17.4.4 Zero-Equation Turbulence Models

The zero-equation models are equations wherein the turbulent fluctuating correlations are related to the mean flow field quantities by algebraic relations. The underlying assumption in zero-equation models is that the local rate of production of turbulence and the rate of dissipation of turbulence are approximately equal. Furthermore, they do not include the convection of turbulence. Obviously, this is contrary to the physics of most flow fields, since the past history of the flow must be accounted for. Nevertheless, these models are mathematically simple and their incorporation into a numerical code can be accomplished with relative ease.

Generally, most models employ an inner region/outer region formulation to represent mixing length. A commonly used model utilizes an exponential function (van Driest damping function) for the inner region, whereas the outer region is proportional to the boundary layer thickness. Mathematically they are expressed as

$$l_i = \kappa(1 - e^{-y^+/A^+})y \tag{17-41}$$

and

$$l_o = C_o\delta \tag{17-42}$$

where κ is the von Karman constant (~ 0.41), and A^+ is a parameter which depends on the streamwise pressure gradient. For a zero-pressure gradient flow, it has a value of 26. The constant C_o in Equation (17-42) is usually assigned a value of $0.08 \sim 0.09$, whereas δ is the boundary layer thickness.

Another formulation commonly used for the outer region turbulent viscosity is the Cebeci/Smith model expressed as

$$\nu_t = \alpha \overline{u}_e \delta^* \tag{17-43}$$

where α is usually assigned a value of 0.0168 for flows where Re_θ (momentum thickness Reynolds number) is greater than 5000 and δ^* is the kinetic displacement thickness defined as

$$\delta^* = \int_o^\infty \left(1 - \frac{u}{u_e}\right) dy \quad \textit{for incompressible flow, and} \tag{17-44}$$

$$\delta^* = \int_o^\infty \left(1 - \frac{\rho u}{\rho_e u_e}\right) dy \quad \textit{for compressible flow} \tag{17-45}$$

Recall that the Reynolds number based on momentum thickness is defined as

$$Re_\theta = \frac{\rho_e u_e \theta}{\mu_e} \tag{17-46}$$

where the momentum thickness θ is

$$\theta = \int_o^\infty \frac{u}{u_e}\left(1 - \frac{u}{u_e}\right) dy \quad \textit{for incompressible flow, and} \tag{17-47}$$

$$\theta = \int_o^\infty \frac{\rho u}{\rho_e u_e}\left(1 - \frac{u}{u_e}\right) dy \quad \textit{for compressible flow} \tag{17-48}$$

The algebraic model described above requires information regarding the boundary layer thickness and flow properties at the boundary layer edge. When the Navier-Stokes equations are being solved, it may be a difficult task to determine the boundary layer thickness and the required properties at the edge. That is especially the case when flow separation exists within the domain. However, when it is necessary to determine the extent of the viscous region within the domain, the total enthalpy is usually used.

A turbulence model which is not written in terms of the boundary layer quantities was introduced by Baldwin and Lomax [17-5]. The inner region is approximated by

$$\mu_t = \rho l^2 |\omega| \tag{17-49}$$

where ω is the vorticity defined as

$$\omega = \frac{\partial v}{\partial x} - \frac{\partial u}{\partial y} \tag{17-50}$$

and l is given by (17-41). The nondimensional space coordinate y^+ can be written as

$$y^+ = u_\tau \frac{y}{\nu_w} = \sqrt{\frac{|\tau_w|}{\rho_w}} \frac{\rho_w}{\mu_w} y$$

or

$$y^+ = \sqrt{\rho_w |\tau_w|} \frac{y}{\mu_w}$$

The outer region is approximated by

$$\mu_t = \alpha \bar{\rho} C_{CP} F_{\text{wake}} F_{\text{Kleb}} \tag{17-51}$$

where α is assigned a value of 0.0168 (as in the Cebeci/Smith model) and

$$F_{\text{wake}} = \min \left[y_{\max} G_{\max} \;,\quad C_{\text{wake}} y_{\max} \frac{(\Delta V)^2}{G_{\max}} \right] \tag{17-52}$$

A typical value for C_{wake} is 1.0. In Equation (17-52), the following definitions are employed,

$$G_{\max} = \max \left(\frac{l}{\kappa} |\omega| \right) \tag{17-53}$$

where κ is the von Karman constant, and l, the mixing length, is determined by the van Driest function given by (17-41). The difference between the absolute values of the maximum and minimum velocities within the viscous region is denoted by ΔV. For wall bounded flows, the minimum velocity occurs at the surface where the velocity is zero, then

$$\Delta V = (u^2 + v^2)^{1/2}_{\max}$$

For shear layer flows, ΔV is defined as the difference between the maximum velocity and the velocity at the $y_{\max}$ location, that is,

$$\Delta V = (u^2 + v^2)^{1/2}_{\max} - (u^2 + v^2)^{1/2}_{y_{\max}}$$

F_{Kleb} is the intermittency factor defined as

$$F_{\text{Kleb}} = \left[1 + 5.5 \left(\frac{C_{\text{Kleb}} y}{y_{\max}} \right)^6 \right]^{-1} \tag{17-54}$$

and $y_{\max}$ is the y location where $G_{\max}$ occurs.

The typical values for the Klebanoff constant C_{Kleb} and C_{CP} are respectively 0.3 and 1.6 for zero or mild pressure gradient. However, to include the influence of pressure gradient, the following expressions as suggested in Reference [17-6] may be used:

$$C_{\text{Kleb}} = \frac{2}{3} - \frac{0.01312}{0.1724 - \beta^*} \tag{17-55}$$

where

$$\beta^* = \frac{y_{\max}}{u_\tau} \frac{\partial V}{\partial x}$$

and u_τ is the friction velocity. The velocity gradient in β^* is calculated outside the viscous region. Once the Klebanoff constant is evaluated, C_{CP} is determined from

$$C_{CP} = \frac{3 - 4C_{\text{Kleb}}}{2C_{\text{Kleb}}(2 - 3C_{\text{Kleb}} + C^3_{\text{Kleb}})} \tag{17-56}$$

Finally, the turbulent viscosity distribution across the boundary layer is determined from

$$\mu_t = \min(\mu_{t_i}, \mu_{t_o}) \tag{17-57}$$

To model the eddy diffusivity, α_t, or equivalently the turbulent conductivity, the Reynolds analogy may be used. Recall that the Reynolds analogy assumes a similarity between the momentum transfer and heat transfer. Therefore, a turbulent Prandtl number is defined as

$$\text{Pr}_t = \frac{\nu_t}{\alpha_t} = \frac{\mu_t c_p}{k_t} \tag{17-58}$$

For most flows, it is assumed that the turbulent Prandtl number remains constant across the boundary layer. For air, $\text{Pr}_t = 0.9$. Thus, the turbulent conductivity is determined as

$$k_t = \frac{\mu_t c_p}{\text{Pr}_t} \tag{17-59}$$

where μ_t is provided by the turbulence models just described.

To initiate the computations, a transition location x_{tr} is specified, and the initial value of the turbulent viscosity is set equal to zero everywhere within the domain. Subsequently, in regions where $x > x_{tr}$, the turbulent viscosity is computed from (17-57) and updated after each time step (or iteration).

17.4.5 One-Equation Turbulence Models

As seen in the previous section, zero-equation models employ an algebraic relation for the eddy viscosity. These models specify length and velocity scales in terms of the mean flow, thus implying an equilibrium between mean motion and turbulence.

One-equation models employ a partial differential equation for velocity scale, whereas the length scale is specified algebraically. The velocity scale is typically written in terms of turbulent kinetic energy k defined as

$$k = \frac{1}{2}(\overline{u'^2} + \overline{v'^2} + \overline{w'^2}) \tag{17-60}$$

The turbulent viscosity μ_t is now written as

$$\mu_t = \rho k^{\frac{1}{2}} \ell \tag{17-61}$$

In the following section, two recently developed one-equation turbulence models are reviewed.

17.4.5.1 Baldwin-Barth One-Equation Turbulence Model:

The Baldwin-Barth one-equation model [17-7] is obtained from the standard k-ϵ two-equation model. The two equations are combined to provide a single equation for the turbulence Reynolds number Re_T, where $Re_T = k^2/(\nu\epsilon)$. To account for the near wall regions, the turbulence Reynolds number is split into two parts as $Re_T = \overline{Re_T} f(\overline{Re_T})$, where f is a damping function such that $Re_T \cong \overline{Re_T}$ for large Re_T.

The transport equation for turbulence Reynolds number $\overline{Re_T}$ is written as

$$\begin{aligned}\frac{\partial}{\partial t}(\nu \overline{Re_T}) + \vec{V} \cdot \nabla(\nu \overline{Re_T}) &= \\ \left(\nu + \frac{\nu_t}{\sigma_\epsilon}\right) \nabla^2(\nu \overline{Re_T}) - \frac{1}{\sigma_\epsilon}(\nabla \nu_t) \cdot \nabla(\nu \overline{Re_T}) &+ (c_{\epsilon_2} f_2 - c_{\epsilon_1})\sqrt{\nu \overline{Re_T} P}\end{aligned} \tag{17-62}$$

The turbulence production based on Boussinesq assumption is provided from

$$P = \nu_t \left(\frac{\partial u_i}{\partial x_j} + \frac{\partial u_j}{\partial x_i}\right) \frac{\partial u_i}{\partial x_j} - \frac{2}{3}\nu_t \left(\frac{\partial u_k}{\partial x_k}\right)^2 = \nu_t S^2 \tag{17-63}$$

The kinematic eddy viscosity ν_t is determined from

$$\nu_t = c_\mu (\nu \overline{Re_T}) D_1 D_2 \tag{17-64}$$

where D_1 and D_2 are damping functions which extend the validity of the model to near wall regions and are given by

$$D_1 = 1 - \exp(-y^+/A^+) \tag{17-65}$$

$$D_2 = 1 - \exp(-y^+/A_2^+) \tag{17-66}$$

Various functions and constants appearing in Equation (17-62) are as follows.

$$\frac{1}{\sigma_\epsilon} = (c_{\epsilon_2} - c_{\epsilon_1})\sqrt{c_\mu}/\kappa^2$$

$$\mu_t = \rho\nu_t$$

$$f_2(y^+) = \frac{c_{\epsilon_1}}{c_{\epsilon_2}} + \left(1 - \frac{c_{\epsilon_1}}{c_{\epsilon_2}}\right)\left(\frac{1}{\kappa y^+} + D_1 D_2\right)\left[\sqrt{D_1 D_2} + \right.$$

$$\left.\frac{y^+}{\sqrt{D_1 D_2}}\left(\frac{1}{A^+}\exp(-y^+/A^+)D_2 + \frac{1}{A_2^+}\exp(-y^+/A_2^+)D_1\right)\right] \tag{17-67}$$

$$\kappa = 0.41, \qquad c_{\epsilon_1} = 1.2, \qquad c_{\epsilon_2} = 2.0$$

$$c_\mu = 0.09, \qquad A^+ = 26, \qquad A_2^+ = 10$$

Equation (7-62) can be written as

$$\frac{\partial}{\partial t}(\nu \overline{Re}_T) + \vec{V} \cdot \nabla(\nu \overline{Re}_T) = \left(\nu + \frac{\nu_t}{\sigma_\epsilon}\right)\nabla^2(\nu \overline{Re}_T) - \frac{1}{\sigma_\epsilon}(\nabla \nu_t) \cdot \nabla(\nu \overline{Re}_T)$$
$$+ (c_{\epsilon 2} f_2 - c_{\epsilon 1})\sqrt{c_\mu D_1 D_2} S \nu \overline{Re}_T \tag{17-68}$$

where

$$S^2 = 2\left[\left(\frac{\partial u}{\partial x}\right)^2 + \left(\frac{\partial v}{\partial y}\right)^2 + \frac{\partial u}{\partial y}\frac{\partial v}{\partial x}\right] + \left(\frac{\partial u}{\partial y}\right)^2 + \left(\frac{\partial v}{\partial x}\right)^2 - \frac{2}{3}\left(\frac{\partial u}{\partial x} + \frac{\partial v}{\partial y}\right)^2 \tag{17-69}$$

Due to numerical considerations, Equation (17-68) can be rewritten by manipulation of the diffusion term and with the assumption of constant ν as follows:

$$\frac{\partial}{\partial t}(\nu \overline{Re}_T) + \vec{V} \cdot \nabla(\nu \overline{Re}_T) = 2\left(\nu + \frac{\nu_t}{\sigma_\epsilon}\right)\nabla^2(\nu \overline{Re}_T) - \frac{1}{\sigma_\epsilon}\nabla \cdot \nu_t \nabla(\nu \overline{Re}_T)$$
$$+ (c_{\epsilon 2} f_2 - c_{\epsilon 1})\sqrt{c_\mu D_1 D_2} S \nu \overline{Re}_T \tag{17-70}$$

Note that the unknown in either Equation (17-68) or (17-70) is $\overline{Re}_T$. Once $\overline{Re}_T$ is determined, relation (17-64) is used to determine ν_t.

17.4.5.1.1 Nondimensional form: Since the governing equations to be solved are typically in nondimensional form, the equations for turbulence models are also nondimensionalized. The same nondimensional variables, as defined in Section 11.3, are used, namely,

$$
\begin{aligned}
t^* &= \frac{tu_\infty}{L} \quad , \quad x^* = \frac{x}{L} \quad , \quad y^* = \frac{y}{L} \\
\mu^* &= \frac{\mu}{\mu_\infty} \quad , \quad \rho^* = \frac{\rho}{\rho_\infty} \quad , \quad \nu^* = \frac{\nu}{\nu_\infty} \\
\nu_t^* &= \frac{\nu_t}{\nu_\infty} \quad , \quad k^* = \frac{k}{u_\infty^2} \quad , \quad \epsilon^* = \epsilon\frac{L}{u_\infty^3} \\
Re_T^* &= \frac{{k^*}^2}{\nu^*\epsilon^*}
\end{aligned}
$$

Now, the nondimensional form of the Baldwin-Barth one-equation turbulence model corresponding to (17-68) or (17-70) is given by the following equations where the asterisk has been dropped.

$$
\begin{aligned}
\frac{\partial}{\partial t}(\nu\overline{Re}_T) + \vec{V}\cdot\nabla(\nu\overline{Re}_T) &= \frac{1}{Re_\infty}\left(\nu + \frac{\nu_t}{\sigma_\epsilon}\right)\nabla^2(\nu\overline{Re}_T) - \frac{1}{Re_\infty}\frac{1}{\sigma_\epsilon}(\nabla\nu_t)\cdot\nabla(\nu\overline{Re}_T) \\
&+ (c_{\epsilon 2}f_2 - c_{\epsilon 1})\sqrt{c_\mu D_1 D_2 S\nu\overline{Re}_T}
\end{aligned}
\tag{17-71}
$$

or

$$
\begin{aligned}
\frac{\partial}{\partial t}(\nu\overline{Re}_T) + \vec{V}\cdot\nabla(\nu\overline{Re}_T) &= \frac{2}{Re_\infty}\left(\nu + \frac{\nu_t}{\sigma_\epsilon}\right)\nabla^2(\nu\overline{Re}_T) - \frac{1}{Re_\infty}\frac{1}{\sigma_\epsilon}\nabla\cdot\nu_t\nabla(\nu\overline{Re}_T) \\
&+ (c_{\epsilon 2}f_2 - c_{\epsilon 1})\sqrt{c_\mu D_1 D_2\, S\nu\overline{Re}_T}
\end{aligned}
\tag{17-72}
$$

17.4.5.1.2 Baldwin-Barth turbulence model in computational space: In order to numerically solve either Equation (17-71) or (17-72) for a general domain, a coordinate transformation is performed. This transformation is similar to the transformation of the Navier-Stokes equations given in Chapter 11. The expressions for the Cartesian derivatives are

$$
\frac{\partial}{\partial x} = \xi_x\frac{\partial}{\partial\xi} + \eta_x\frac{\partial}{\partial\eta} \tag{17-73}
$$

$$
\frac{\partial}{\partial y} = \xi_y\frac{\partial}{\partial\xi} + \eta_y\frac{\partial}{\partial\eta} \tag{17-74}
$$

$$
\begin{aligned}
\frac{\partial^2}{\partial x^2} &= \xi_x^2\frac{\partial^2}{\partial\xi^2} + 2\xi_x\eta_x\frac{\partial^2}{\partial\xi\partial\eta} + \eta_x^2\frac{\partial^2}{\partial\eta^2} \\
&+ \xi_x\left[\frac{\partial\xi_x}{\partial\xi}\frac{\partial}{\partial\xi} + \frac{\partial\eta_x}{\partial\xi}\frac{\partial}{\partial\eta}\right] + \eta_x\left[\frac{\partial\xi_x}{\partial\eta}\frac{\partial}{\partial\xi} + \frac{\partial\eta_x}{\partial\eta}\frac{\partial}{\partial\eta}\right]
\end{aligned}
\tag{17-75}
$$

$$\frac{\partial^2}{\partial y^2} = \xi_y^2 \frac{\partial^2}{\partial \xi^2} + 2\xi_y \eta_y \frac{\partial^2}{\partial \xi \partial \eta} + \eta_y^2 \frac{\partial^2}{\partial \eta^2} + \xi_y \left[\frac{\partial \xi_y}{\partial \xi} \frac{\partial}{\partial \xi} + \frac{\partial \eta_y}{\partial \xi} \frac{\partial}{\partial \eta} \right] + \eta_y \left[\frac{\partial \xi_y}{\partial \eta} \frac{\partial}{\partial \xi} + \frac{\partial \eta_y}{\partial \eta} \frac{\partial}{\partial \eta} \right] \tag{17-76}$$

Now, for simplicity, define

$$\nu \overline{Re}_T = R$$

Then, the transformed Baldwin-Barth turbulence model given by Equation (17-71) is written as follows. The details of transformation and mathematical manipulation is provided in Appendix I.

$$\frac{\partial R}{\partial t} = -U \frac{\partial R}{\partial \xi} - V \frac{\partial R}{\partial \eta} + \frac{1}{Re_\infty} \left(\nu + \frac{\nu_t}{\sigma_\epsilon} \right) \left[a_4 \frac{\partial^2 R}{\partial \xi^2} + 2c_5 \frac{\partial^2 R}{\partial \xi \partial \eta} + b_4 \frac{\partial^2 R}{\partial \eta^2} + g_1 \frac{\partial R}{\partial \xi} + g_2 \frac{\partial R}{\partial \eta} \right] - \frac{1}{Re_\infty} \frac{1}{\sigma_\epsilon} \left[a_4 \frac{\partial \nu_t}{\partial \xi} \frac{\partial R}{\partial \xi} + c_5 \left(\frac{\partial \nu_t}{\partial \xi} \frac{\partial R}{\partial \eta} + \frac{\partial \nu_t}{\partial \eta} \frac{\partial R}{\partial \xi} \right) + b_4 \frac{\partial \nu_t}{\partial \eta} \frac{\partial R}{\partial \eta} \right] + (c_{\epsilon 2} f_2 - c_{\epsilon 1}) \sqrt{c_\mu D_1 D_2}\, S R \tag{17-77}$$

where U and V are the contravarient velocity components

$$U = \xi_x u + \xi_y v$$

$$V = \eta_x u + \eta_y v$$

and

$$a_4 = \xi_x^2 + \xi_y^2 \tag{17-78}$$

$$b_4 = \eta_x^2 + \eta_y^2 \tag{17-79}$$

$$c_5 = \xi_x \eta_x + \xi_y \eta_y \tag{17-80}$$

$$g_1 = \xi_x \frac{\partial \xi_x}{\partial \xi} + \eta_x \frac{\partial \xi_x}{\partial \eta} + \xi_y \frac{\partial \xi_y}{\partial \xi} + \eta_y \frac{\partial \xi_y}{\partial \eta} \tag{17-81}$$

$$g_2 = \xi_x \frac{\partial \eta_x}{\partial \xi} + \eta_x \frac{\partial \eta_x}{\partial \eta} + \xi_y \frac{\partial \eta_y}{\partial \xi} + \eta_y \frac{\partial \eta_y}{\partial \eta} \tag{17-82}$$

Furthermore, the expression for S in the last term of Equation (17-77) is

$$S^2 = a_1 \left(\frac{\partial u}{\partial \xi} \right)^2 + b_1 \left(\frac{\partial u}{\partial \eta} \right)^2 + a_2 \left(\frac{\partial v}{\partial \xi} \right)^2 + b_2 \left(\frac{\partial v}{\partial \eta} \right)^2$$

$$+ 2c_1 \left(\frac{\partial u}{\partial \xi} \frac{\partial u}{\partial \eta} \right) + 2c_2 \left(\frac{\partial v}{\partial \xi} \frac{\partial v}{\partial \eta} \right) + 2a_3 \frac{\partial u}{\partial \xi} \frac{\partial v}{\partial \xi}$$

$$+ 2b_3 \frac{\partial u}{\partial \eta} \frac{\partial v}{\partial \eta} + 2c_3 \frac{\partial u}{\partial \xi} \frac{\partial v}{\partial \eta} + 2c_4 \frac{\partial u}{\partial \eta} \frac{\partial v}{\partial \xi} \tag{17-83}$$

where the coefficients a_1, a_2, a_3, b_1, b_2, b_3, c_1, c_2, c_3 and c_4 are given by (11-210a), (11-210b), (11-210c), (11-211a), (11-211b), (11-211c), (11-212a), (11-212b), (11-212c), and (11-212d), respectively. The transformation of Equation (17-72) is carried out in a slightly different fashion, the details of which are given in Appendix I. The transformed equation is given as

$$\begin{aligned} \frac{\partial R}{\partial t} &= -U \frac{\partial R}{\partial \xi} - V \frac{\partial R}{\partial \eta} + \frac{2}{Re_\infty} \left(\nu + \frac{\nu_t}{\sigma_\epsilon} \right) \left[\xi_x \frac{\partial A}{\partial \xi} + \xi_y \frac{\partial B}{\partial \xi} + \eta_x \frac{\partial A}{\partial \eta} + \eta_y \frac{\partial B}{\partial \eta} \right] \\ &\quad - \frac{1}{Re_\infty} \frac{1}{\sigma_\epsilon} \left[\xi_x \frac{\partial \alpha}{\partial \xi} + \xi_y \frac{\partial \beta}{\partial \xi} + \eta_x \frac{\partial \alpha}{\partial \eta} + \eta_y \frac{\partial \beta}{\partial \eta} \right] \\ &\quad + (c_{\epsilon 2} f_2 - c_{\epsilon 1}) \sqrt{c_\mu D_1 D_2} \, SR \end{aligned} \tag{17-84}$$

where

$$A = \frac{\partial R}{\partial x} = \xi_x \frac{\partial R}{\partial \xi} + \eta_x \frac{\partial R}{\partial \eta} \tag{17-85}$$

$$B = \frac{\partial R}{\partial y} = \xi_y \frac{\partial R}{\partial \xi} + \eta_y \frac{\partial R}{\partial \eta} \tag{17-86}$$

$$\alpha = \nu_t \left(\xi_x \frac{\partial R}{\partial \xi} + \eta_x \frac{\partial R}{\partial \eta} \right) \tag{17-87}$$

$$\beta = \nu_t \left(\xi_y \frac{\partial R}{\partial \xi} + \eta_y \frac{\partial R}{\partial \eta} \right) \tag{17-88}$$

In order to simplify the development of finite difference equations, Equation (17-84) is expressed as

$$\frac{\partial R}{\partial t} = M + P \tag{17-89}$$

The term M, which represents the convection and diffusion of turbulence can be written as

$$M = M^1 + M_\xi^2 + M_\eta^2 + M_\xi^3 + M_\eta^3 \tag{17-90}$$

where

$$M^1 = -U \frac{\partial R}{\partial \xi} - V \frac{\partial R}{\partial \eta} \tag{17-91}$$

$$M_\xi^2 = -\frac{1}{Re_\infty}\frac{1}{\sigma_\epsilon}\left(\xi_x\frac{\partial\alpha}{\partial\xi} + \xi_y\frac{\partial\beta}{\partial\xi}\right) \tag{17-92}$$

$$M_\eta^2 = -\frac{1}{Re_\infty}\frac{1}{\sigma_\epsilon}\left(\eta_x\frac{\partial\alpha}{\partial\eta} + \eta_y\frac{\partial\beta}{\partial\eta}\right) \tag{17-93}$$

$$M_\xi^3 = \frac{2}{Re_\infty}\left(\nu + \frac{\nu_t}{\sigma_\epsilon}\right)\left(\xi_x\frac{\partial A}{\partial\xi} + \xi_y\frac{\partial B}{\partial\xi}\right) \tag{17-94}$$

$$M_\eta^3 = \frac{2}{Re_\infty}\left(\nu + \frac{\nu_t}{\sigma_\epsilon}\right)\left(\eta_x\frac{\partial A}{\partial\eta} + \eta_y\frac{\partial B}{\partial\eta}\right) \tag{17-95}$$

and

$$P = (c_{\epsilon 2}f_2 - c_{\epsilon 1})\sqrt{c_\mu D_1 D_2}\,SR$$

The initial and boundary conditions are set according to:

(1) Initial condition: A value of $\overline{Re}_T = (\overline{Re}_T)_\infty < 1$ is specified within the domain to initiate the solution.

(2) Boundary conditions: The value of $\overline{Re}_T$ is set equal to zero at the solid surface, and it is set to its initial value of $(\overline{Re}_T)_\infty$ at the inflow. At the outflow, extrapolation from the interior domain is used to specify $\overline{Re}_T$ at the boundary.

17.4.5.2 Spalart-Allmaras One-Equation Turbulence Model: The Spalart-Allmaras model solves a transport equation for a working variable $\overline{\nu}$ which is related to the eddy viscosity. The governing equation is derived by using empiricism, dimensional analysis, Galilean invariance, and selected dependence on the molecular viscosity [17-8]. The transport equation is expressed as

$$\frac{d\overline{\nu}}{dt} = \frac{1}{\sigma}\left[\nabla\cdot\left((\nu+\overline{\nu})\nabla\overline{\nu}\right) + c_{b2}(\nabla\overline{\nu})^2\right] + c_{b1}\bar{S}\overline{\nu}(1-f_{t2})$$

$$- \left[c_{w1}f_w - \frac{c_{b1}}{\kappa^2}f_{t2}\right]\left[\frac{\overline{\nu}}{d}\right]^2 + f_{t1}(\Delta q)^2 \tag{17-96}$$

where the eddy viscosity is given by

$$\nu_t = \overline{\nu}f_{v1} \tag{17-97}$$

and

$$f_{v1} = \frac{\chi^3}{\chi^3 + c_{v1}^3} \tag{17-98}$$

$$\chi = \frac{\overline{\nu}}{\nu} \tag{17-99}$$

Various functions and constants appearing in Equation (17-96) are defined as

$$\bar{S} = S + \frac{\overline{\nu}}{\kappa^2 d^2} f_{v2} \tag{17-100}$$

where d is the distance to the wall, κ is the von Karman constant, S is the magnitude of the vorticity,

$$S = \left| \frac{\partial v}{\partial x} - \frac{\partial u}{\partial y} \right| \tag{17-101}$$

and

$$f_{v2} = 1 - \frac{\chi}{1 + \chi f_{v1}} \tag{17-102}$$

It should be noted that expression (17-69) has also been used for S. The function f_w is

$$f_w(r) = g \left(\frac{1 + c_{w3}^6}{g^6 + c_{w3}^6} \right)^{\frac{1}{6}} \tag{17-103}$$

where

$$g = r + c_{w2}(r^6 - r) \tag{17-104}$$

and

$$r = \frac{\overline{\nu}}{\bar{S} \kappa^2 d^2} \tag{17-105}$$

Large values of r should be truncated to a value of about 10. The function f_{t2} is given by

$$f_{t2} = c_{t3} \exp(-c_{t4} \chi^2) \tag{17-106}$$

and the trip function f_{t1} is

$$f_{t1} = c_{t1} g_t \exp \left[-c_{t2} \left(\frac{\omega_t}{\Delta q} \right)^2 (d^2 + g_t^2 d_t^2) \right] \tag{17-107}$$

The following are used in Equation (17-107):
d_t: The distance from the field point to the trip which is located on the surface.
ω_t: The wall vorticity at the trip.
Δq: The difference between the velocities at the field point and trip.
g_t: $g_t = \min[1.0, \Delta q / \omega_t \Delta x]$, where Δx is the grid spacing along the wall at the trip.

The constants used in the equations above are

$$\sigma = \frac{2}{3} \qquad c_{b1} = 0.1355 \qquad c_{b2} = 0.622$$

$$c_{w1} = \frac{c_{b1}}{\kappa^2} + (1 + c_{b2})/\sigma \qquad c_{w2} = 0.3 \qquad c_{w3} = 2 \qquad \kappa = 0.41$$

$$c_{v1} = 7.1 \qquad c_{t1} = 1.0 \qquad c_{t2} = 2.0 \qquad c_{t3} = 1.1 \qquad c_{t4} = 2.0$$

The Spalart-Allmaras turbulence model given by (17-96) can also be written as

$$\frac{d\overline{\nu}}{dt} = \left(\frac{1+c_{b2}}{\sigma}\right)\nabla\cdot[(\nu+\overline{\nu})\nabla\overline{\nu}] - \frac{c_{b2}}{\sigma}(\nu+\overline{\nu})\nabla^2\overline{\nu}$$

$$+ c_{b1}(1-f_{t2})\overline{\nu}\overline{S} - \left[c_{w1}f_w - \frac{c_{b1}}{\kappa^2}f_{t2}\right]\left[\frac{\overline{\nu}}{d}\right]^2 + f_{t1}(\Delta q)^2 \quad \text{(17-108a)}$$

The first two terms on the right-hand side of Equation (17-108a) can be expanded and recombined to provide the following equation.

$$\frac{d\overline{\nu}}{dt} = \left(\frac{1+c_{b2}}{\sigma}\right)[\nabla(\nu+\overline{\nu})\cdot\nabla\overline{\nu}] + \frac{1}{\sigma}(\nu+\overline{\nu})\nabla^2\overline{\nu}$$

$$+ c_{b1}(1-f_{t2})\overline{\nu}\overline{S} - \left[c_{w1}f_w - \frac{c_{b1}}{\kappa^2}f_{t2}\right]\left[\frac{\overline{\nu}}{d}\right]^2 + f_{t1}(\Delta q)^2 \quad \text{(17-108b)}$$

Either one of Equations (17-108a) or (17-108b) can be used.

The initial condition for $\bar{\nu}$ is specified to be zero up to a value of $\nu_\infty/10$, that is, $\bar{\nu} = 0 \rightarrow \nu_\infty/10$. The boundary conditions are set according to:

(1) At the inflow, $\bar{\nu} = \bar{\nu}_\infty$,

(2) At the solid surface $\bar{\nu} = 0$, and

(3) At the outflow, extrapolation is used.

17.4.5.2.1 Nondimensional form: Using the same nondimensional variables defined previously, the nondimensionalized Spalart-Allmaras turbulence model is given by the following equation. Again, as in the previous model, the notation of asterisk has been dropped.

$$\frac{d\overline{\nu}}{dt} = \left(\frac{1}{Re_\infty}\right)\left(\frac{1+c_{b2}}{\sigma}\right)\nabla\cdot[(\nu+\overline{\nu})\nabla\overline{\nu}] - \left(\frac{1}{Re_\infty}\right)\left(\frac{c_{b2}}{\sigma}\right)(\nu+\overline{\nu})\nabla^2\overline{\nu}$$

$$+ c_{b1}(1-f_{t2})S\overline{\nu} - \left(\frac{1}{Re_\infty}\right)(c_{w1}f_w)\left(\frac{\overline{\nu}}{d}\right)^2$$

$$+ \left(\frac{1}{Re_\infty}\right)\left(\frac{c_{b1}}{\kappa^2}\right)[(1-f_{t2})f_{v2}+f_{t2}]\left(\frac{\overline{\nu}}{d}\right)^2 + Re_\infty f_{t1}(\Delta q)^2 \quad \text{(17-109a)}$$

which corresponds to Equation (17-108a) and

$$\frac{d\overline{\nu}}{dt} = \left(\frac{1}{Re_\infty}\right)\left(\frac{1+c_{b2}}{\sigma}\right)[\nabla(\nu+\overline{\nu})\cdot\nabla\overline{\nu}] + \left(\frac{1}{Re_\infty}\right)\frac{1}{\sigma}(\nu+\overline{\nu})\nabla^2\overline{\nu}$$

$$+ c_{b1}(1-f_{t2})\overline{\nu}\overline{S} - \left(\frac{1}{Re_\infty}\right)(c_{w1}f_w)\left(\frac{\overline{\nu}}{d}\right)^2$$

$$+ \left(\frac{1}{Re_\infty}\right)\left(\frac{c_{b1}}{\kappa^2}\right)[(1-f_{t2})f_{v2}+f_{t2}]\left(\frac{\overline{\nu}}{d}\right)^2 + Re_\infty f_{t1}(\Delta q)^2 \quad \text{(17-109b)}$$

corresponding to Equation (17-108b).

17.4.5.2.2 Spalart-Allmaras turbulence model in computational space: Using the expressions for Cartesian derivatives given by (17-73) through (17-76), the Spalart-Allmaras turbulence model given by (17-109b) in generalized coordinates is written as

$$\frac{\partial \overline{\nu}}{\partial t} + U\frac{\partial \overline{\nu}}{\partial \xi} + V\frac{\partial \overline{\nu}}{\partial \eta} = \left(\frac{1}{Re_\infty}\right)\left(\frac{1+c_{b2}}{\sigma}\right)\left\{a_4\frac{\partial \overline{\nu}}{\partial \xi}\frac{\partial}{\partial \xi}(\nu+\overline{\nu}) + b_4\frac{\partial \overline{\nu}}{\partial \eta}\frac{\partial}{\partial \eta}(\nu+\overline{\nu})\right.$$
$$\left.+ c_5\left[\frac{\partial \overline{\nu}}{\partial \eta}\frac{\partial}{\partial \xi}(\nu+\overline{\nu}) + \frac{\partial \overline{\nu}}{\partial \xi}\frac{\partial}{\partial \eta}(\nu+\overline{\nu})\right]\right\} + \left(\frac{1}{Re_\infty}\right)\left(\frac{1}{\sigma}\right)(\nu+\overline{\nu})\left[a_4\frac{\partial^2 \overline{\nu}}{\partial \xi^2}\right.$$
$$\left.+ 2c_5\frac{\partial^2 \overline{\nu}}{\partial \xi \partial \eta} + b_4\frac{\partial^2 \overline{\nu}}{\partial \eta^2} + g_1\frac{\partial \overline{\nu}}{\partial \xi} + g_2\frac{\partial \overline{\nu}}{\partial \eta}\right] + c_{b1}(1-f_{t2})S\,\overline{\nu} - \left(\frac{1}{Re_\infty}\right)(c_{w1}f_w)\left(\frac{\overline{\nu}}{d}\right)^2$$
$$+ \left(\frac{1}{Re_\infty}\right)\left(\frac{c_{b1}}{\kappa^2}\right)\left[(1-f_{t2})f_{v2} + f_{t2}\right]\left(\frac{\overline{\nu}}{d}\right)^2 + Re_\infty f_{t1}(\Delta q)^2 \tag{17-110}$$

where

$$S = \left|\xi_y\frac{\partial u}{\partial \xi} + \eta_y\frac{\partial u}{\partial \eta} - \xi_x\frac{\partial v}{\partial \xi} - \eta_x\frac{\partial v}{\partial \eta}\right| \tag{17-111}$$

and, as defined previously,

$$a_4 = \xi_x^2 + \xi_y^2 \tag{17-112}$$

$$b_4 = \eta_x^2 + \eta_y^2 \tag{17-113}$$

$$c_5 = \xi_x\eta_x + \xi_y\eta_y \tag{17-114}$$

$$g_1 = \xi_x\frac{\partial \xi_x}{\partial \xi} + \eta_x\frac{\partial \xi_x}{\partial \eta} + \xi_y\frac{\partial \xi_y}{\partial \xi} + \eta_y\frac{\partial \xi_y}{\partial \eta} \tag{17-115}$$

$$g_2 = \xi_x\frac{\partial \eta_x}{\partial \xi} + \eta_x\frac{\partial \eta_x}{\partial \eta} + \xi_y\frac{\partial \eta_y}{\partial \xi} + \eta_y\frac{\partial \eta_y}{\partial \eta} \tag{17-116}$$

The transformed Spalart-Allmaras turbulence model is given by (17-109a).

$$\frac{\partial \overline{\nu}}{\partial t} + U\frac{\partial \overline{\nu}}{\partial \xi} + V\frac{\partial \overline{\nu}}{\partial \eta} = \left(\frac{1}{Re_\infty}\right)\left(\frac{1+c_{b2}}{\sigma}\right)\left\{\xi_x\frac{\partial \alpha}{\partial \xi} + \eta_x\frac{\partial \alpha}{\partial \eta} + \xi_y\frac{\partial \beta}{\partial \xi} + \eta_y\frac{\partial \beta}{\partial \eta}\right\}$$
$$- \left(\frac{1}{Re_\infty}\right)\frac{c_{b2}}{\sigma}(\nu+\overline{\nu})\left\{\xi_x\frac{\partial A}{\partial \xi} + \eta_x\frac{\partial A}{\partial \eta} + \xi_y\frac{\partial B}{\partial \xi} + \eta_y\frac{\partial B}{\partial \eta}\right\}$$

$$+\left(\frac{1}{Re_\infty}\right)\left(\frac{c_{b1}}{\kappa^2}\right)\left[(1-f_{t2})f_{v2}+f_{t2}\right]\left(\frac{\overline{\nu}}{d}\right)^2-\left(\frac{1}{Re_\infty}\right)(c_{w1}\,f_w)\left(\frac{\overline{\nu}}{d}\right)^2$$

$$+\,c_{b1}(1-f_{t2})S\,\overline{\nu}+Re_\infty\,f_{t1}(\Delta q)^2 \tag{17-117}$$

where

$$A = \frac{\partial\overline{\nu}}{\partial x}=\xi_x\frac{\partial\overline{\nu}}{\partial\xi}+\eta_x\frac{\partial\overline{\nu}}{\partial\eta} \tag{17-118}$$

$$B = \frac{\partial\overline{\nu}}{\partial y}=\xi_y\frac{\partial\overline{\nu}}{\partial\xi}+\eta_y\frac{\partial\overline{\nu}}{\partial\eta} \tag{17-119}$$

$$\alpha = (\nu+\overline{\nu})\left(\xi_x\frac{\partial\overline{\nu}}{\partial\xi}+\eta_x\frac{\partial\overline{\nu}}{\partial\eta}\right) \tag{17-120}$$

$$\beta = (\nu+\overline{\nu})\left(\xi_y\frac{\partial\overline{\nu}}{\partial\xi}+\eta_y\frac{\partial\overline{\nu}}{\partial\eta}\right) \tag{17-121}$$

Equation (17-117) is now written as

$$\frac{\partial\overline{\nu}}{\partial t}=M+P-D+T \tag{17-122}$$

where

(a) The term M represents the convection and diffusion of turbulence and, due to numerical consideration, is decomposed as follows:

$$M=M^1+M^2+M^3 \tag{17-123}$$

Each term in expression (17-123) is defined as

$$M^1 = -\left(U\frac{\partial\overline{\nu}}{\partial\xi}+V\frac{\partial\overline{\nu}}{\partial\eta}\right) \tag{17-124}$$

$$M^2 = \left(\frac{1}{Re_\infty}\right)\left(\frac{1+c_{b2}}{\sigma}\right)\left\{\xi_x\frac{\partial\alpha}{\partial\xi}+\eta_x\frac{\partial\alpha}{\partial\eta}+\xi_y\frac{\partial\beta}{\partial\xi}+\eta_y\frac{\partial\beta}{\partial\eta}\right\} \tag{17-125}$$

$$M^3 = -\left(\frac{1}{Re_\infty}\right)\frac{c_{b2}}{\sigma}(\nu+\overline{\nu})\left\{\xi_x\frac{\partial A}{\partial\xi}+\eta_x\frac{\partial A}{\partial\eta}+\xi_y\frac{\partial B}{\partial\xi}+\eta_y\frac{\partial B}{\partial\eta}\right\} \tag{17-126}$$

The terms M^2 and M^3 can be further regrouped as terms including ξ derivatives and terms involving η derivatives as follow

$$M^2_\xi = \left(\frac{1}{Re_\infty}\right)\left(\frac{1+c_{b2}}{\sigma}\right)\left(\xi_x\frac{\partial\alpha}{\partial\xi}+\xi_y\frac{\partial\beta}{\partial\xi}\right) \tag{17-127}$$

$$M_\eta^2 = \left(\frac{1}{Re_\infty}\right)\left(\frac{1+c_{b2}}{\sigma}\right)\left(\eta_x\frac{\partial\alpha}{\partial\eta} + \eta_y\frac{\partial\beta}{\partial\eta}\right) \tag{17-128}$$

$$M_\xi^3 = -\left(\frac{1}{Re_\infty}\right)\frac{c_{b2}}{\sigma}(\nu+\overline{\nu})\left(\xi_x\frac{\partial A}{\partial\xi} + \xi_y\frac{\partial B}{\partial\xi}\right) \tag{17-129}$$

$$M_\eta^3 = -\left(\frac{1}{Re_\infty}\right)\frac{c_{b2}}{\sigma}(\nu+\overline{\nu})\left(\eta_x\frac{\partial A}{\partial\eta} + \eta_y\frac{\partial B}{\partial\eta}\right) \tag{17-130}$$

(b) The second term in Equation (17-122) represents the production of turbulence and is

$$\begin{aligned} P &= c_{b1}(1-f_{t2})S\overline{\nu} + \left(\frac{1}{Re_\infty}\right)\frac{c_{b1}}{\kappa^2}(1-f_{t2})f_{v2}\left[\frac{\overline{\nu}}{d}\right]^2 \\ &= c_{b1}(1-f_{t2})\overline{\nu}\left[S + \frac{\overline{\nu}}{Re_\infty\kappa^2 d^2}f_{v2}\right] = c_{b1}(1-f_{t2})\overline{\nu}\overline{S} \end{aligned} \tag{17-131}$$

Note that $\overline{S} = S + \frac{1}{Re_\infty}\left(\frac{\overline{\nu}}{\kappa^2 d^2}\right) f_{v2}$

(c) The third term D represents the destruction of turbulence and is given by

$$D = -\left(\frac{1}{Re_\infty}\right)\left(c_{w1}f_w - \frac{c_{b1}}{\kappa^2}f_{t2}\right)\left[\frac{\overline{\nu}}{d}\right]^2 \tag{17-132}$$

(d) And, finally, the last term in Equation (17-122) represents the trip term which causes/ triggers transition from laminar flow to turbulent flow.

$$T = (Re_\infty)f_{t1}(\Delta q)^2 \tag{17-133}$$

17.4.6 Two-Equation Turbulence Models

It was pointed out previously that the convection of turbulence is not modeled in zero-equation models. Therefore, the physical effect of past history of the flow is not included in simple algebraic models. In order to account for this physical effect, a transport equation based on the Navier-Stokes equation may be derived. When one such equation is employed, it is referred to as a one-equation model which was discussed in the previous section. When two transport equations are used, it is known as a *two-equation model.*

Complex flowfields which include massively separated flows, unsteadiness, and flows involving multiple-length scales occur frequently in fluid mechanics applications. In these types of complex flows, the lower order turbulence models, that is, zero-, half-, or one-equation models, become very complicated and often ambiguous. Two-equation models are developed to better represent the physics of turbulence in these types of complex flowfields. In this section, k-ϵ and k-ω two-equation turbulence models are reviewed.

17.4.6.1 k-ε Two-Equation Turbulence Model: A commonly used two-equation turbulence model is the k-ϵ model. The first low Reynolds number k-ϵ model was developed by Jones and Launder [17-9] and has subsequently been modified by several investigators.

The partial differential equations are derived for kinetic energy of turbulence (k), and the dissipation of turbulence (ϵ), where

$$k = \frac{1}{2}\left[\overline{u'^2} + \overline{v'^2} + \overline{w'^2}\right] \tag{17-134}$$

and

$$\epsilon = \nu_t \overline{\left(\frac{\partial u_i'}{\partial x_j}\right)\left(\frac{\partial u_i'}{\partial x_j}\right)} \tag{17-135}$$

The standard k-ϵ two-equation model may be expressed by the turbulent kinetic equation

$$\rho\frac{dk}{dt} = \frac{\partial}{\partial x_j}\left[\left(\mu + \frac{\mu_t}{\sigma_k}\right) + \frac{\partial k}{\partial x_j}\right] + P_k - \rho\epsilon \tag{17-136}$$

and the dissipation rate equation

$$\rho\frac{d\epsilon}{dt} = \frac{\partial}{\partial x_j}\left[\left(\mu + \frac{\mu_t}{\sigma_\epsilon}\right)\frac{\partial \epsilon}{\partial x_j}\right] + c_{\epsilon 1} f_1 P_k \frac{\epsilon}{k} - c_{\epsilon 2} f_2 \rho \frac{\epsilon^2}{k} \tag{17-137}$$

where P_k is the production of turbulence defined as $P_k = \tau_{ij}(\partial u_i/\partial x_j)$.

The turbulent kinetic equation and the dissipation rate equation can be written in an expanded form as

$$\rho\left(\frac{\partial k}{\partial t} + u\frac{\partial k}{\partial x} + v\frac{\partial k}{\partial y}\right) = \frac{\partial}{\partial x}\left[\left(\mu + \frac{\mu_t}{\sigma_k}\right)\frac{\partial k}{\partial x}\right] + \frac{\partial}{\partial y}\left[\left(\mu + \frac{\mu_t}{\sigma_k}\right)\frac{\partial k}{\partial y}\right] + P_k - \rho\epsilon \tag{17-138}$$

and

$$\rho\left(\frac{\partial \epsilon}{\partial t} + u\frac{\partial \epsilon}{\partial x} + v\frac{\partial \epsilon}{\partial y}\right) = \frac{\partial}{\partial x}\left[\left(\mu + \frac{\mu_t}{\sigma_\epsilon}\right)\frac{\partial \epsilon}{\partial x}\right] + \frac{\partial}{\partial y}\left[\left(\mu + \frac{\mu_t}{\sigma_\epsilon}\right)\frac{\partial \epsilon}{\partial y}\right] + c_{\epsilon 1} f_1 P_k \frac{\epsilon}{k} - c_{\epsilon 2} f_2 \rho \frac{\epsilon^2}{k} \tag{17-139}$$

where typical constants are

$$\sigma_k = 1.0 \quad \sigma_\epsilon = 1.3 \quad c_{\epsilon 1} = 1.44 \quad c_{\epsilon 2} = 1.92$$

and the turbulent viscosity is related to ϵ by

$$\mu_t = \rho c_\mu \frac{k^2}{\epsilon} \tag{17-140}$$

where $c_\mu = 0.09$.

As for the zero-equation models, a two-layer approach is incorporated. That is, Equations (17-138) and (17-139) are used in the outer region down to the vicinity of the viscous sublayer. For a typical wall flow, that corresponds to a y^+ range of 30–50. Adjacent to the surface, wall functions such as (17-41) are employed. The values of k and ϵ must be provided at the interface as the inner boundary conditions for Equations (17-138) and (17-139). These values may be obtained by the following: if y_i denotes the interface between the outer region and the viscous sublayer, then

$$k_{y=y_i} = \frac{\tau_i}{\rho c_D^{\frac{3}{2}}} \tag{17-141}$$

where c_D is typically 0.164, and τ_i is the shear stress at y_i location. Similarly,

$$\epsilon_{y=y_i} = \frac{c_D k_{y=y_i}^{\frac{3}{2}}}{l} \tag{17-142}$$

where the mixing length may be evaluated by the simple linear relation for the viscous sublayer given by

$$l = \kappa y \tag{17-143}$$

Once the local values of the kinetic energy of turbulence and dissipation of turbulence have been computed, turbulent viscosity given by (17-140) may be determined. It is noted that correlations such as $\overline{u'^2}$, $\overline{v'^2}$, and $\overline{w'^2}$ can be determined by semi-empirical relations. For example,

$$\overline{u'^2} = 2a_2 k \tag{17-144}$$

$$\overline{v'^2} = 2a_3 k \tag{17-145}$$

$$\overline{w'^2} = 2(1 - a_2 - a_3)k \tag{17-146}$$

where a_2 and a_3 are structural scales usually assigned values of 0.556 and 0.15, respectively.

The initial and boundary conditions must be specified for both k and ϵ. The initial condition for the turbulent kinetic energy k is specified in terms of the freestream turbulence intensity T_i, as follows

$$k_\infty = 1.5(T_i U_\infty)^2$$

where $T_i = 0.0001$ to 10 percent of mean velocity. The turbulent viscosity is specified as

$$\mu_{t_\infty} = (0.1 \rightarrow 100)\mu_\infty$$

and, therefore,

$$\epsilon_\infty = \bar{\rho} c_\mu \frac{k_\infty^2}{\mu_{t_\infty}}$$

The boundary conditions are

(1) At the inflow, the freestream values are specified, that is, $\epsilon = \epsilon_\infty$ and $k = k_\infty$.

(2) At the solid surface, $k = 0$, $\mu_t = 0$, and $\epsilon = \epsilon_w = 0$. For some k-ϵ models, a value for ϵ_w is prescribed.

(3) At the outflow, extrapolation is used.

17.4.6.2 k-ω Two-Equation Turbulence Model:

This two-equation model includes one equation for the turbulent kinetic energy k, and a second equation for the specific turbulent dissipation rate (or turbulent frequency) ω. As in the case of the k-ϵ model, there are several versions of the k-ω model. The k-ω model of Wilcox [17-10] is provided below.

The turbulent kinetic energy equation is given by

$$\bar{\rho}\frac{dk}{dt} = \frac{\partial}{\partial x_j}\left[(\mu + \sigma^* \mu_t)\frac{\partial k}{\partial x_j}\right] + P_k - \beta^* \rho \omega k \tag{17-147}$$

and the specific dissipation rate equation is

$$\bar{\rho}\frac{d\omega}{dt} = \frac{\partial}{\partial x_j}\left[(\mu + \sigma \mu_t)\frac{\partial \omega}{\partial x_j}\right] + \alpha\frac{\omega}{k}P_k - \beta \rho \omega^2 \tag{17-148}$$

The eddy viscosity is determined from

$$\mu_t = \rho\frac{k}{\omega} \tag{17-149}$$

and the auxiliary relations are

$$\epsilon = \beta^* \omega k \tag{17-150}$$

$$\ell = \frac{k^{\frac{1}{2}}}{\omega} \tag{17-151}$$

The constants used in Equations (17-147) and (17-148) are

$$\alpha = \frac{5}{9} \qquad \beta = \frac{3}{40} \qquad \beta^* = \frac{9}{100} \qquad \sigma = \frac{1}{2} \qquad \sigma^* = \frac{1}{2}$$

17.4.6.3 Combined k-ε/k-ω Two-Equation Turbulence Model: The k-ω turbulence model performs well and, in fact, is superior to the k-ϵ model within the laminar sublayer and the logarithmic region of the boundary layer. However, the k-ω model has been shown to be influenced strongly by the specification of freestream value of ω outside the boundary layer. Therefore, the k-ω model does not appear to be an ideal model for applications in the wake region of the boundary layer. On the other hand, the k-ϵ model behaves superior to that of the k-ω model in the outer portion and wake regions of the boundary layer, but inferior in the inner region of the boundary layer.

To include the best features of each model, Menter [17-11] has combined different elements of the k-ϵ and k-ω models to form a new two-equation turbulence model. This model incorporates the k-ω model for the inner region of the boundary layer, and it switches to the k-ϵ model for the outer and wake regions of the boundary layer. Two versions of the model introduced by Menter are referred to as the *baseline (BSL) model* and, a modified version of the BSL model, the *shear-stress transport (SST) model.* It has been shown that the SST model performs well in the prediction of flows with adverse pressure gradient.

The combined k-ω/k-ϵ two-equation model is given by

$$\bar{\rho}\frac{dk}{dt} = \frac{\partial}{\partial x_j}\left[(\mu + \sigma_k \mu_t)\frac{\partial k}{\partial x_j}\right] + P_k - \beta^* \rho\omega k \tag{17-152}$$

and

$$\rho\frac{d\omega}{dt} = \frac{\partial}{\partial x_j}\left[(\mu + \sigma_\omega \mu_t)\frac{\partial \omega}{\partial x_j}\right] + 2(1 - F_1)\rho\sigma_{\omega 2}\frac{1}{\omega}\frac{\partial k}{\partial x_j}\frac{\partial \omega}{\partial x_j} + \frac{\gamma}{\nu_t}P_k - \beta\rho\omega^2 \tag{17-153}$$

where the production of turbulence will be defined by

$$P_k = \tau_{ij}\frac{\partial u_i}{\partial x_j} \tag{17-154}$$

and

$$\tau_{ij} = \mu_t\left(\frac{\partial u_i}{\partial x_j} + \frac{\partial u_j}{\partial x_i} - \frac{2}{3}\delta_{ij}\frac{\partial u_k}{\partial x_k}\right) - \frac{2}{3}\rho k \delta_{ij}$$

The constants appearing in Equations (17-152) and (17-153) are expressed in a general compact form as

$$\phi = F_1\phi_1 + (1 - F_1)\phi_2 \tag{17-155}$$

where ϕ_1 represents the constants associated with the k-ω model, and ϕ_2 represents the constants associated with the k-ϵ model.

The difference between the baseline model and the shear stress transport model is in the definition of turbulent viscosity and the specification of constants.

17.4.6.3.1 Baseline model: The turbulent viscosity for the baseline model is defined by

$$\nu_t = \frac{k}{\omega} \tag{17-156}$$

and the constants for ϕ are set as follows.

The values of the constants for set ϕ_1 are specified according to

$$\sigma_{k1} = 0.5 \ , \qquad \sigma_{\omega 1} = 0.5 \ , \qquad \beta_1 = 0.075$$

$$\beta^* = 0.09 \ , \qquad \kappa = 0.41$$

$$\gamma_1 = \frac{\beta_1}{\beta^*} - \sigma_{\omega 1}\kappa^2/\sqrt{\beta^*}$$

And the constants for ϕ_2 are

$$\sigma_{k2} = 1.0 \ , \qquad \sigma_{\omega 2} = 0.856 \ , \qquad \beta_2 = 0.0828$$

$$\beta^* = 0.09 \ , \qquad \kappa = 0.41$$

$$\gamma_2 = \frac{\beta_2}{\beta^*} - \sigma_{\omega 2}\kappa^2/\sqrt{\beta^*}$$

In addition, the following definitions are used.

$$F_1 = \text{Tanh}\,(\text{arg}_1^4) \tag{17-157}$$

$$\text{arg}_1 = \min\left[\max\left(\frac{\sqrt{k}}{0.09\omega y} \ , \ \frac{500\nu}{\omega y^2}\right) \ , \ \frac{4\rho\sigma_{\omega 2}k}{CD_{k\omega}y^2}\right] \tag{17-158}$$

where y is the distance to the nearest surface, and $CD_{k\omega}$ is the positive portion of the cross-diffusion term

$$CD_{k\omega} = \max\left[2\rho\,\sigma_{\omega 2}\frac{1}{\omega}\frac{\partial k}{\partial x_i}\frac{\partial \omega}{\partial x_i} \ , \ 10^{-20}\right] \tag{17-159}$$

Note that far away from solid surface as $y \to$ large, term arg_1 approaches zero due to $1/y$ and $1/y^2$. Furthermore, the three arguments within arg_1 represent the following:

1) The term $(\sqrt{k}/0.09\omega y)$ is the turbulence length scale divided by y. It is equal to 2.5 in the log layer of the boundary layer and approaches zero at the boundary-layer edge.

2) The term $(500\nu/\omega y^2)$ enforces F_1 to be one in the sublayer region and $1/y$ in the log layer. Recall that ω is proportional to $1/y^2$ near the surface, and it is proportional to $1/y$ in the log region. Therefore, $1/\omega y^2$ will be constant near the surface and will approach zero in the log layer.

3) The third term $(4\rho\sigma_{\omega 2}k/CD_{\omega k}y^2)$ is included to prevent solution dependency on the freestream. It can be shown that, as the boundary-layer edge is approached, arg_1 as well as F_1 become zero and, therefore, the k-ϵ model is utilized in this region.

The following freestream conditions are recommended.

$$\omega_\infty = m\frac{u_\infty}{L} \ , \qquad \nu_{t_\infty} = 10^{-n}\nu_\infty \ , \qquad k_\infty = \nu_{t_\infty}\omega_\infty$$

where

$$1 \leq m \leq 10 \ , \qquad 2 \leq n \leq 5 \ ,$$

and L is a characteristic length of the problem, for example, the length of computational domain. At the solid surface, the boundary condition for ω is set according to

$$\omega = 10\frac{6\nu}{\beta_1(\Delta y_1)^2} \tag{17-160}$$

where Δy_1 is the distance of the first point away from the wall. This boundary condition is specified for a smooth wall, and Δy_1^+ must be less than about 3. At the outflow boundary, extrapolation is used.

17.4.6.3.2 Shear-stress transport model: The turbulent viscosity is now defined as

$$\nu_t = \frac{a_1 k}{\max(a_1\omega \, , \, \Omega F_2)} \tag{17-161}$$

where $a_1 = 0.31$, Ω is the absolute value of vorticity $\Omega = |\partial v/\partial x - \partial u/\partial y|$, and $F_2 = \text{Tanh}\,(\text{arg}_2^2)$ with

$$\text{arg}_2 = \max\left[2\frac{\sqrt{k}}{0.09\omega y} \, , \, \frac{500\nu}{\omega y^2}\right] \tag{17-162}$$

The constants for ϕ_1 and ϕ_2 are identical to the baseline model, except σ_{k1}, where

$$\sigma_{k1} = 0.85$$

17.4.6.3.3 Nondimensional form: The nondimensional variables defined previously are used to express Equations (17-152) and (17-153) in a nondimensional form. Again, the notation of asterisk has been dropped. However, the appearance of the Reynolds number in the equation is a good indication of nondimensional equation. The nondimensional k-ϵ/k-ω two-equation model is written as

$$\rho\frac{dk}{dt} = \rho\left(\frac{\partial k}{\partial t} + u\frac{\partial k}{\partial x} + v\frac{\partial k}{\partial y}\right) = \frac{1}{Re_\infty}\frac{\partial}{\partial x}\left[(\mu + \sigma_k\mu_t)\frac{\partial k}{\partial x}\right]$$

$$+\frac{1}{Re_\infty}\frac{\partial}{\partial y}\left[(\mu+\sigma_k\,\mu_t)\frac{\partial k}{\partial y}\right]+P_k-\beta^*\rho\omega k \tag{17-163}$$

and

$$\rho\frac{d\omega}{dt}=\rho\left(\frac{\partial\omega}{\partial t}+u\frac{\partial\omega}{\partial x}+v\frac{\partial\omega}{\partial y}\right)=\frac{1}{Re_\infty}\frac{\partial}{\partial x}\left[(\mu+\sigma_\omega\,\mu_t)\frac{\partial\omega}{\partial x}\right]$$

$$+\frac{1}{Re_\infty}\frac{\partial}{\partial y}\left[(\mu+\sigma_\omega\,\mu_t)\frac{\partial\omega}{\partial y}\right]+2\rho(1-F_1)\sigma_{\omega 2}\frac{1}{\omega}\left[\frac{\partial k}{\partial x}\frac{\partial\omega}{\partial x}+\frac{\partial k}{\partial y}\frac{\partial\omega}{\partial y}\right]$$

$$+\alpha\frac{\omega}{k}P_k-\beta\rho\omega^2 \tag{17-164}$$

where the production term P_k is determined from

$$P_k=\tau_{ij}\frac{\partial u_i}{\partial x_j}=\left[\mu_t\left(\frac{\partial u_i}{\partial x_j}+\frac{\partial u_j}{\partial x_i}-\frac{2}{3}\delta_{ij}\frac{\partial u_k}{\partial x_k}\right)-\frac{2}{3}\rho k\delta_{ij}\right]\frac{\partial u_i}{\partial x_j} \tag{17-165}$$

17.4.6.3.4 k-ϵ / k-ω Turbulence model in computational space: Using the expressions for the Cartesian derivatives given by (17-73) and (17-74), the k-equation in the computational space becomes

$$\rho\frac{\partial k}{\partial t}=-\rho U\frac{\partial k}{\partial\xi}-\rho V\frac{\partial k}{\partial\eta}+\frac{1}{Re_\infty}\left\{\xi_x\frac{\partial}{\partial\xi}\left[(MUS)\,(KX)\right]+\eta_x\frac{\partial}{\partial\eta}\left[(MUS)\,(KX)\right]+\right.$$

$$\left.\xi_y\frac{\partial}{\partial\xi}\left[(MUS)\,(KY)\right]+\eta_y\frac{\partial}{\partial\eta}\left[(MUS)\,(KY)\right]\right\}+P_k-\beta^*\rho\omega k \tag{17-166}$$

where

$$MUS = \mu+\sigma_k\mu_t \tag{17-167}$$

$$KX = \xi_x\frac{\partial k}{\partial\xi}+\eta_x\frac{\partial k}{\partial\eta} \tag{17-168}$$

$$KY = \xi_y\frac{\partial k}{\partial\xi}+\eta_y\frac{\partial k}{\partial\eta} \tag{17-169}$$

Similarly, the ω-equation is given by

$$\rho\frac{\partial k}{\partial t} = -\rho U\frac{\partial\omega}{\partial\xi}-\rho V\frac{\partial\omega}{\partial\eta}+\frac{1}{Re_\infty}\left\{\xi_x\frac{\partial}{\partial\xi}\left[(MUR)\,(OX)\right]\right.$$

$$\left.+\eta_x\frac{\partial}{\partial\eta}\left[(MUR)\,(OX)\right]+\xi_y\frac{\partial}{\partial\xi}\left[(MUR)\,(OY)\right]+\eta_y\frac{\partial}{\partial\eta}\left[(MUR)\,(OY)\right]\right\}$$

$$+ \left[2\rho(1-F_1)\sigma_{\omega 2}\frac{1}{\omega}\right][(KX)(OX)+(KY)(OY)] - \beta\rho\omega^2 \tag{17-170}$$

where

$$MUR = \mu + \sigma_\omega \mu_t \tag{17-171}$$

$$OX = \xi_x \frac{\partial \omega}{\partial \xi} + \eta_x \frac{\partial \omega}{\partial \eta} \tag{17-172}$$

$$OY = \xi_y \frac{\partial \omega}{\partial \xi} + \eta_y \frac{\partial \omega}{\partial \eta} \tag{17-173}$$

The production term P_k given by (17-154) is first expanded and subsequently transformed to the computational space as follows.

$$P_k = \tau_{ij}\frac{\partial u_i}{\partial x_j} = \tau_{xx}\frac{\partial u}{\partial x} + \tau_{xy}\left(\frac{\partial u}{\partial y} + \frac{\partial v}{\partial x}\right) + \tau_{yy}\frac{\partial v}{\partial y} \tag{17-174}$$

where

$$\tau_{xx} = \left(\frac{1}{Re_\infty}\right)\mu_t\left(\frac{4}{3}\frac{\partial u}{\partial x} - \frac{2}{3}\frac{\partial v}{\partial y}\right) - \frac{2}{3}\rho k \tag{17-175}$$

$$\tau_{xy} = \left(\frac{1}{Re_\infty}\right)\mu_t\left(\frac{\partial u}{\partial y} + \frac{\partial v}{\partial x}\right) \tag{17-176}$$

$$\tau_{yy} = \left(\frac{1}{Re_\infty}\right)\mu_t\left(\frac{4}{3}\frac{\partial v}{\partial y} - \frac{2}{3}\frac{\partial u}{\partial x}\right) - \frac{2}{3}\rho k \tag{17-177}$$

Utilizing the expressions (17-73) and (17-74), the shear stresses defined by (17-175) through (17-177) become

$$\tau_{xx} = \left(\frac{1}{Re_\infty}\right)\mu_t\left[\frac{4}{3}\left(\xi_x\frac{\partial u}{\partial \xi} + \eta_x\frac{\partial u}{\partial \eta}\right) - \frac{2}{3}\left(\xi_y\frac{\partial v}{\partial \xi} + \eta_y\frac{\partial v}{\partial \eta}\right)\right] - \frac{2}{3}\rho k = \tau_{\xi\xi} \tag{17-178}$$

$$\tau_{xy} = \left(\frac{1}{Re_\infty}\right)\mu_t\left[\xi_y\frac{\partial u}{\partial \xi} + \eta_y\frac{\partial u}{\partial \eta} + \xi_x\frac{\partial v}{\partial \xi} + \eta_x\frac{\partial v}{\partial \eta}\right] = \tau_{\xi\eta} \tag{17-179}$$

$$\tau_{yy} = \left(\frac{1}{Re_\infty}\right)\mu_t\left[\frac{4}{3}\left(\xi_y\frac{\partial v}{\partial \xi} + \eta_y\frac{\partial v}{\partial \eta}\right) - \frac{2}{3}\left(\xi_x\frac{\partial u}{\partial \xi} + \eta_x\frac{\partial u}{\partial \eta}\right)\right] - \frac{2}{3}\rho k = \tau_{\eta\eta} \tag{17-180}$$

Now, the production term becomes

$$
\begin{aligned}
P_k &= \tau_{xx}\frac{\partial u}{\partial x} + \tau_{xy}\left(\frac{\partial u}{\partial y} + \frac{\partial v}{\partial x}\right) + \tau_{yy}\frac{\partial u}{\partial y} \\
&= \tau_{\xi\xi}\left(\xi_x\frac{\partial u}{\partial \xi} + \eta_x\frac{\partial u}{\partial \eta}\right) + \tau_{\xi\eta}\left(\xi_y\frac{\partial u}{\partial \xi} + \eta_y\frac{\partial u}{\partial \eta}\right. \\
&\quad \left. + \xi_x\frac{\partial v}{\partial \xi} + \eta_x\frac{\partial v}{\partial \eta}\right) + \tau_{\eta\eta}\left(\xi_y\frac{\partial v}{\partial \xi} + \eta_y\frac{\partial v}{\partial \eta}\right) \\
&= (\xi_x\,\tau_{\xi\xi} + \xi_y\,\tau_{\xi\eta})\frac{\partial u}{\partial \xi} + (\eta_x\,\tau_{\xi\xi} + \eta_y\tau_{\xi\eta})\frac{\partial u}{\partial \eta} \\
&\quad + (\xi_x\,\tau_{\xi\eta} + \xi_y\,\tau_{\eta\eta})\frac{\partial v}{\partial \xi} + (\eta_x\,\tau_{\xi\eta} + \eta_y\,\tau_{\eta\eta})\frac{\partial v}{\partial \eta} \\
&= (P1)\frac{\partial u}{\partial \xi} + (P2)\frac{\partial u}{\partial \eta} + (P3)\frac{\partial v}{\partial \xi} + (P4)\frac{\partial v}{\partial \eta} \qquad \text{(17-181)}
\end{aligned}
$$

where

$$P1 = \xi_x\,\tau_{\xi\xi} + \xi_y\,\tau_{\xi\eta} \qquad \text{(17-182)}$$

$$P2 = \eta_x\,\tau_{\xi\xi} + \eta_y\,\tau_{\xi\eta} \qquad \text{(17-183)}$$

$$P3 = \xi_x\,\tau_{\xi\eta} + \xi_y\,\tau_{\eta\eta} \qquad \text{(17-184)}$$

$$P4 = \eta_x\,\tau_{\xi\eta} + \eta_y\,\tau_{\eta\eta} \qquad \text{(17-185)}$$

17.5 Numerical Considerations

Numerical simulation of turbulent flow may be divided into three categories. The simplest and most practical approach in use at present is the solution of the time-averaged Navier-Stokes equations along with a turbulence model. The second level of sophistication includes the simulation of time-dependent, large-scale eddy motion. In this approach, the effects of smaller scales are included by turbulence models. The third category includes the direct numerical simulation of all important scales of turbulence.

The second and third categories are presently in the infancy stage. One of the difficulties associated with these methods is the limitation of present day computers. However, with the advancement in computer technology, large eddy and full turbulent simulations seem possible for practical problems within the next decade or so.

Currently, the system of equations such as (17-9) along with some turbulence models are solved for engineering applications. The solution schemes for the system of equations given by (17-9) have been discussed in the previous chapters. One important point to emphasize here is grid resolution. In order to accurately resolve the feature of turbulent flow near the surface, a grid point within the laminar sublayer must be included. Generally, this corresponds to a grid point within y^+ of 2.0.

The inclusion of a turbulence model into the system of governing equations may be accomplished either explicitly or implicitly. One may solve the governing equations [such as (17-9)] implicitly and use the turbulence quantities from the previous time-level, i.e., explicitly. On the other hand, the governing equations, along with turbulence equations such as (17-136) and (17-137), may be solved simultaneously, i.e., implicitly. Obviously, a fully implicit formulation will increase the size of the Jacobian matrices and will require major modification of the (laminar) Navier-Stokes code. On the other hand, explicit treatment of turbulence is simple and straightforward, and modification of an existing Navier-Stokes code to include turbulence can be accomplished with ease.

17.6 Finite Difference Formulations

The one-equation and two-equation turbulence models in different forms given in Section 17.4.5 are parabolic equations. Therefore, various numerical methods discussed previously for the solution of parabolic equations can be used to develop solution algorithms.

To illustrate the development of the numerical schemes for the turbulence models, both explicit and implicit formulations are considered. However, the implicit formulations will be limited to the one-equation models. The turbulence models in nondimensional form and in the generalized curvilinear coordinate system are considered to provide the FDEs for general applications.

17.6.1 Baldwin-Barth One-Equation Turbulence Model

A first-order upwind approximation is used for the convective terms, and a second-order central difference approximation is used for the diffusion terms. The finite difference expression for each term is developed separately due to the complexity of the equation. The first-order upwind scheme can be easily extended to a second-order upwind scheme.

17.6.1.1 Implicit Formulation: The Baldwin-Barth one-equation turbulence model written as

$$\frac{\partial R}{\partial t} = M + P \tag{17-186}$$

is used to illustrate the development of an implicit scheme. Equation (17-186) is written as

$$\left(\frac{\partial R}{\partial t}\right)^{n+1} = M^{n+1} + P^{n+1} \tag{17-187}$$

Since the governing equation is nonlinear, a linearization procedure similar to that introduced previously, for example, by relation (11-81), is used to provide the following.

$$M^{n+1} = M^n + \frac{\partial M}{\partial R}\Delta R = M^n + \overline{M}^n \Delta R \tag{17-188}$$

and

$$P^{n+1} = P^n + \frac{\partial P}{\partial R}\Delta R = P^n + \overline{P}^n \Delta R \tag{17-189}$$

Now Equation (17-187) is written as

$$\frac{\Delta R}{\Delta t} - (\overline{M} + \overline{P})\Delta R = (M^n + P^n) \tag{17-190}$$

Recall that the term M can be decomposed into terms involving ξ or η derivatives as defined by (17-91) through (17-95). Thus, M is expressed as $M = M_\xi + M_\eta$. Now, Equation (17-190) is written as

$$\frac{\Delta R}{\Delta t} - (\overline{M}_\xi + \overline{M}_\eta + \overline{P})\,\Delta R = (M_\xi^n + M_\eta^n + P^n) \tag{17-191}$$

or

$$\left[1 - (\overline{M}_\xi + \overline{M}_\eta + \overline{P})\Delta t\right]\Delta R = (M_\xi^n + M_\eta^n + P^n)\Delta t \tag{17-192}$$

and decomposed as

$$\left[1 - \Delta t \overline{M}_\xi\right]\left[1 - \Delta t(\overline{M}_\eta + \overline{P})\right]\Delta R = (M_\xi^n + M_\eta^n + P^n)\Delta t \tag{17-193}$$

Equation (17-193) is solved sequentially be the following equations.

$$(1 - \Delta t\,\overline{M}_\xi)\Delta R^* = (M_\xi^n + M_\eta^n + P^n)\Delta t \tag{17-194}$$

and

$$\left[1 - \Delta t(\overline{M}_\eta + \overline{P})\right]\Delta R = \Delta R^* \tag{17-195}$$

The finite difference expressions for each one of the terms appearing in Equations (17-194) and (17-195) are developed separately.

The convective term M^1 given by (17-91) is approximated by a first-order upwind formulation as follows.

$$\begin{aligned} M^1 &= -U\frac{\partial R}{\partial \xi} - V\frac{\partial R}{\partial \eta} \\ &= -\left\{\frac{1}{2}(U_{i,j} + |U_{i,j}|)\left(\frac{R_{i,j} - R_{i-1,j}}{\Delta\xi}\right)\right. \\ &\quad + \frac{1}{2}(U_{i,j} - |U_{i,j}|)\left(\frac{R_{i+1,j} - R_{i,j}}{\Delta\xi}\right) \\ &\quad + \frac{1}{2}(V_{i,j} + |V_{i,j}|)\left(\frac{R_{i,j} - R_{i,j-1}}{\Delta\eta}\right) \\ &\quad \left. + \frac{1}{2}(V_{i,j} - |V_{i,j}|)\left(\frac{R_{i,j+1} - R_{i,j}}{\Delta\eta}\right)\right\} \end{aligned} \tag{17-196}$$

If a second-order approximation for the convective term is desired, then the formulation is as follows.

$$\begin{aligned} M^1 &= -U\frac{\partial R}{\partial \xi} - V\frac{\partial R}{\partial \eta} \\ &= -\left\{\frac{1}{2}(U_{i,j} + |U_{i,j}|)\left(\frac{3R_{i,j} - 4R_{i-1,j} + R_{i-2,j}}{2\Delta\xi}\right)\right. \\ &\quad + \frac{1}{2}(U_{i,j} - |U_{i,j}|)\left(\frac{-R_{i+2,j} + 4R_{i+1,j} - 3R_{i,j}}{2\Delta\xi}\right) \\ &\quad + \frac{1}{2}(V_{i,j} + |V_{i,j}|)\left(\frac{3R_{i,j} - 4R_{i,j-1} + R_{i,j-2}}{2\Delta\eta}\right) \\ &\quad \left. + \frac{1}{2}(V_{i,j} - |V_{i,j}|)\left(\frac{-R_{i,j+2} + 4R_{i,j+1} - 3R_{i,j}}{2\Delta\eta}\right)\right\} \end{aligned} \tag{17-197}$$

Since all the terms within M^2 given by (17-92) or (17-93) are similar, the development of the finite difference expression will be illustrated in detail for one term only. Subsequently, the conclusion will be extended to the remaining terms. Now, consider the term $\xi_x(\partial\alpha/\partial\xi)$ which is approximated by a second-order central difference expression as follows.

$$\xi_x\frac{\partial \alpha}{\partial \xi} \cong \xi_{x_{i,j}}\frac{\alpha_{i+\frac{1}{2},j} - \alpha_{i-\frac{1}{2},j}}{\Delta\xi} \tag{17-198}$$

Furthermore, recall that from (17-87)

$$\alpha = \nu_t \left(\xi_x \frac{\partial R}{\partial \xi} + \eta_x \frac{\partial R}{\partial \eta} \right) \tag{17-199}$$

The finite difference expression for α can be written as

$$\alpha_{i+\frac{1}{2},j} = \nu_{t_{i+\frac{1}{2},j}} \left[\xi_{x_{i+\frac{1}{2},j}} \left(\frac{R_{i+1,j} - R_{i,j}}{\Delta\xi} \right) + \eta_{x_{i+\frac{1}{2},j}} \frac{R_{i+\frac{1}{2},j+\frac{1}{2}} - R_{i+\frac{1}{2},j-\frac{1}{2}}}{\Delta\eta} \right] \tag{17-200}$$

and

$$\alpha_{i-\frac{1}{2},j} = \nu_{t_{i-\frac{1}{2},j}} \left[\xi_{x_{i-\frac{1}{2},j}} \left(\frac{R_{i,j} - R_{i-1,j}}{\Delta\xi} \right) + \eta_{x_{i-\frac{1}{2},j}} \frac{R_{i-\frac{1}{2},j+\frac{1}{2}} - R_{i-\frac{1}{2},j-\frac{1}{2}}}{\Delta\eta} \right] \tag{17-201}$$

Substitution of (17-200) and (17-201) into (17-198) yields

$$\begin{aligned}
\xi_x \frac{\partial \alpha}{\partial \xi} &= \xi_{x_{i,j}} \frac{1}{\Delta\xi} \left\{ \nu_{t_{i+\frac{1}{2},j}} \left[\xi_{x_{i+\frac{1}{2},j}} \left(\frac{R_{i+1,j} - R_{i,j}}{\Delta\xi} \right) + \eta_{x_{i+\frac{1}{2},j}} \left(\frac{R_{i+\frac{1}{2},j+\frac{1}{2}} - R_{i+\frac{1}{2},j-\frac{1}{2}}}{\Delta\eta} \right) \right] \right. \\
&\left. - \nu_{t_{i-\frac{1}{2},j}} \left[\xi_{x_{i-\frac{1}{2},j}} \left(\frac{R_{i,j} - R_{i-1,j}}{\Delta\xi} \right) + \eta_{x_{i-\frac{1}{2},j}} \left(\frac{R_{i-\frac{1}{2},j+\frac{1}{2}} - R_{i-\frac{1}{2},j-\frac{1}{2}}}{\Delta\eta} \right) \right] \right\}
\end{aligned} \tag{17-202}$$

A similar expression is developed for $\xi_y(\partial\beta/\partial\xi)$ and is combined with $\xi_x(\partial\alpha/\partial\xi)$ to provide an approximation for M_ξ^2 as follows.

$$M_\xi^2 = -\frac{1}{Re_\infty} \frac{1}{\sigma_\epsilon} \left(\xi_x \frac{\partial \alpha}{\partial \xi} + \xi_y \frac{\partial \beta}{\partial \xi} \right) \tag{17-203}$$

where

$$\begin{aligned}
\xi_x \frac{\partial \alpha}{\partial \xi} + \xi_y \frac{\partial \beta}{\partial \xi} &\cong \nu_{t_{i+\frac{1}{2},j}} \left\{ \left(\xi_{x_{i,j}} \xi_{x_{i+\frac{1}{2},j}} + \xi_{y_{i,j}} \xi_{y_{i+\frac{1}{2},j}} \right) \left[\frac{R_{i+1,j} - R_{i,j}}{(\Delta\xi)^2} \right] \right\} \\
&- \nu_{t_{i-\frac{1}{2},j}} \left\{ \left(\xi_{x_{i,j}} \xi_{x_{i-\frac{1}{2},j}} + \xi_{y_{i,j}} \xi_{y_{i-\frac{1}{2},j}} \right) \left[\frac{R_{i,j} - R_{i-1,j}}{(\Delta\xi)^2} \right] \right\} \\
&+ \nu_{t_{i+\frac{1}{2},j}} \left\{ \left(\xi_{x_{i,j}} \eta_{x_{i+\frac{1}{2},j}} + \xi_{y_{i,j}} \eta_{y_{i+\frac{1}{2},j}} \right) \left[\frac{R_{i+\frac{1}{2},j+\frac{1}{2}} - R_{i+\frac{1}{2},j-\frac{1}{2}}}{\Delta\xi\Delta\eta} \right] \right\} \\
&- \nu_{t_{i-\frac{1}{2},j}} \left\{ \left(\xi_{x_{i,j}} \eta_{x_{i-\frac{1}{2},j}} + \xi_{y_{i,j}} \eta_{y_{i-\frac{1}{2},j}} \right) \left[\frac{R_{i-\frac{1}{2},j+\frac{1}{2}} - R_{i-\frac{1}{2},j-\frac{1}{2}}}{\Delta\xi\Delta\eta} \right] \right\}
\end{aligned} \tag{17-204}$$

Similarly for M_η^2,

$$\eta_x \frac{\partial \alpha}{\partial \eta} + \eta_y \frac{\partial \beta}{\partial \eta} \cong \nu_{t_{i,j+\frac{1}{2}}} \left\{ \left(\eta_{x_{i,j}} \eta_{x_{i,j+\frac{1}{2}}} + \eta_{y_{i,j}} \eta_{y_{i,j+\frac{1}{2}}} \right) \left[\frac{R_{i,j+1} - R_{i,j}}{(\Delta\eta)^2} \right] \right\}$$

$$- \nu_{t_{i,j-\frac{1}{2}}} \left\{ \left(\eta_{x_{i,j}} \eta_{x_{i,j-\frac{1}{2}}} + \eta_{y_{i,j}} \eta_{y_{i,j-\frac{1}{2}}} \right) \left[\frac{R_{i,j} - R_{i,j-1}}{(\Delta\eta)^2} \right] \right\}$$

$$+ \nu_{t_{i,j+\frac{1}{2}}} \left\{ \left(\eta_{x_{i,j}} \xi_{x_{i,j+\frac{1}{2}}} + \eta_{y_{i,j}} \xi_{y_{i,j+\frac{1}{2}}} \right) \left[\frac{R_{i+\frac{1}{2},j+\frac{1}{2}} - R_{i-\frac{1}{2},j+\frac{1}{2}}}{\Delta\xi\Delta\eta} \right] \right\}$$

$$- \nu_{t_{i,j-\frac{1}{2}}} \left\{ \left(\eta_{x_{i,j}} \xi_{x_{i,j-\frac{1}{2}}} + \eta_{y_{i,j}} \xi_{y_{i,j-\frac{1}{2}}} \right) \left[\frac{R_{i+\frac{1}{2},j-\frac{1}{2}} - R_{i-\frac{1}{2},j-\frac{1}{2}}}{\Delta\xi\Delta\eta} \right] \right\} \tag{17-205}$$

The terms in M^3 are also similar to that of M^2. Recall that

$$M_\xi^3 = \frac{2}{Re_\infty} \left(\nu + \frac{\nu_t}{\sigma_\epsilon} \right) \left(\xi_x \frac{\partial A}{\partial \xi} + \xi_y \frac{\partial B}{\partial \xi} \right) \tag{17-206}$$

$$M_\eta^3 = \frac{2}{Re_\infty} \left(\nu + \frac{\nu_t}{\sigma_\epsilon} \right) \left(\eta_x \frac{\partial A}{\partial \eta} + \eta_y \frac{\partial B}{\partial \eta} \right) \tag{17-207}$$

and

$$A = \xi_x \frac{\partial R}{\partial \xi} + \eta_x \frac{\partial R}{\partial \eta} \tag{17-208}$$

$$B = \xi_y \frac{\partial R}{\partial \xi} + \eta_y \frac{\partial R}{\partial \eta} \tag{17-209}$$

Now the finite difference approximation for M_ξ^3 following (17-204) can be expressed as

$$M_\xi^3 = \frac{2}{Re_\infty} \left(\nu + \frac{\nu_t}{\sigma_\epsilon} \right)_{i,j} \left\{ \left(\xi_{x_{i,j}} \xi_{x_{i+\frac{1}{2},j}} + \xi_{y_{i,j}} \xi_{y_{i+\frac{1}{2},j}} \right) \left[\frac{R_{i+1,j} - R_{i,j}}{(\Delta\xi)^2} \right] \right.$$

$$- \left(\xi_{x_{i,j}} \xi_{x_{i-\frac{1}{2},j}} + \xi_{y_{i,j}} \xi_{y_{i-\frac{1}{2},j}} \right) \left[\frac{R_{i,j} - R_{i-1,j}}{(\Delta\xi)^2} \right]$$

$$+ \left(\xi_{x_{i,j}} \eta_{x_{i+\frac{1}{2},j}} + \xi_{y_{i,j}} \eta_{y_{i+\frac{1}{2},j}} \right) \left[\frac{R_{i+\frac{1}{2},j+\frac{1}{2}} - R_{i+\frac{1}{2},j-\frac{1}{2}}}{\Delta\xi\Delta\eta} \right]$$

$$\left. - \left(\xi_{x_{i,j}} \eta_{x_{i-\frac{1}{2},j}} + \xi_{y_{i,j}} \eta_{y_{i-\frac{1}{2},j}} \right) \left[\frac{R_{i-\frac{1}{2},j+\frac{1}{2}} - R_{i-\frac{1}{2},j-\frac{1}{2}}}{\Delta\xi\Delta\eta} \right] \right\} \tag{17-210}$$

Similarly, following (17-205)

$$M_\eta^3 = \frac{2}{Re_\infty}\left(\nu + \frac{\nu_t}{\sigma_\epsilon}\right)_{i,j}\left\{(\eta_{x_{i,j}}\eta_{x_{i,j+\frac{1}{2}}} + \eta_{y_{i,j}}\eta_{y_{i,j+\frac{1}{2}}})\left[\frac{R_{i,j+1} - R_{i,j}}{(\Delta\eta)^2}\right]\right.$$

$$- (\eta_{x_{i,j}}\eta_{x_{i,j-\frac{1}{2}}} + \eta_{y_{i,j}}\eta_{y_{i,j-\frac{1}{2}}})\left[\frac{R_{i,j} - R_{i,j-1}}{(\Delta\eta)^2}\right]$$

$$+ (\eta_{x_{i,j}}\xi_{x_{i,j+\frac{1}{2}}} + \eta_{y_{i,j}}\xi_{y_{i,j+\frac{1}{2}}})\left[\frac{R_{i+\frac{1}{2},j+\frac{1}{2}} - R_{i-\frac{1}{2},j+\frac{1}{2}}}{\Delta\xi\Delta\eta}\right]$$

$$\left. - (\eta_{x_{i,j}}\xi_{x_{i,j-\frac{1}{2}}} + \eta_{y_{i,j}}\xi_{y_{i,j-\frac{1}{2}}})\left[\frac{R_{i+\frac{1}{2},j-\frac{1}{2}} - R_{i-\frac{1}{2},j-\frac{1}{2}}}{\Delta\xi\Delta\eta}\right]\right\} \tag{17-211}$$

At this point, the finite difference approximation of the terms in Equations (17-194) and (17-195) have been identified. However, the bar quantities such as $\overline{M}$ require some deliberation.

The procedure for obtaining these quantities is illustrated for $\overline{M}_\xi^1$ in detail, and the remaining terms can be obtained in a similar fashion.

Recall that M_ξ^1 is approximated by (17-196) when first-order approximation is used. Thus,

$$M_\xi^1 = -U\frac{\partial R}{\partial \xi} = -\left\{\frac{1}{2}(U_{i,j} + |U_{i,j}|)\left(\frac{R_{i,j} - R_{i-1,j}}{\Delta\xi}\right)\right.$$

$$\left. + \frac{1}{2}(U_{i,j} - |U_{i,j}|)\left(\frac{R_{i+1,j} - R_{i,j}}{\Delta\xi}\right)\right\} \tag{17-212}$$

Note that, as before, when second-order approximation is used, that is, Equation (17-197), the terms at grid points $i-2$ and $i+2$ are taken to the right-hand side of the equation and treated explicitly. Now from Equation (17-212) expressions for $\overline{M}_\xi^1$ are developed as follow.

$$M_\xi^1\Big|_{i-1,j} = \frac{\partial M_\xi^1}{\partial R}\Big|_{i-1,j} = \left\{\frac{1}{2}(U_{i,j} + |U_{i,j}|)\left(\frac{1}{\Delta\xi}\right)\right\}$$

$$M_\xi^1\Big|_{i,j} = \frac{\partial M_\xi^1}{\partial R}\Big|_{i,j} = -\left\{\frac{1}{2}(U_{i,j} + |U_{i,j}|)\left(\frac{1}{\Delta\xi}\right)\right.$$

$$\left. + \frac{1}{2}(U_{i,j} - |U_{i,j}|)\left(\frac{-1}{\Delta\xi}\right)\right\}$$

$$= -\left\{|U_{i,j}|\left(\frac{1}{\Delta\xi}\right)\right\}$$

$$\overline{M}_\xi^1\Big|_{i+1,j} = \frac{\partial M_\xi^1}{\partial R}\Big|_{i+1,j} = -\left\{\frac{1}{2}(U_{i,j} - |U_{i,j}|)\left(\frac{1}{\Delta\xi}\right)\right\}$$

Now, with all the terms in Equation (17-194) and (17-195) identified, the two tridiagonal systems of equations are solved sequentially by any tridiagonal solver, for example, by the scheme introduced in Appendix B.

17.6.1.2 Explicit Formulation: The Baldwin-Barth turbulence model given by (17-77) is considered to develop an explicit scheme. Again, as in the implicit formulation, an upwind approximation is used for the convective term which could be either first-order or second-order, and a central difference approximation of second-order is used for the diffusion terms. In the following formulation, a first-order upwind scheme is used. However, as seen in the previous section, extension to second-order is simple and straightforward.

$$\begin{aligned}
R^{n+1} &= R^n - \Delta t\left\{\frac{1}{2}(U_{i,j} + |U_{i,j}|)\left(\frac{R_{i,j} - R_{i-1,j}}{\Delta\xi}\right) + \right. \\
&+ \frac{1}{2}(U_{i,j} - |U_{i,j}|)\left(\frac{R_{i+1,j} - R_{i,j}}{\Delta\xi}\right) \\
&+ \frac{1}{2}(V_{i,j} + |V_{i,j}|)\left(\frac{R_{i,j} - R_{i,j-1}}{\Delta\eta}\right) \\
&\left. + \frac{1}{2}(V_{i,j} - |V_{i,j}|)\left(\frac{R_{i,j+1} - R_{i,j}}{\Delta\eta}\right)\right\} \\
&- \Delta t\left(\frac{1}{Re_\infty}\right)\left(\frac{1}{\sigma_\epsilon}\right)\left[\nu_{t_{i+\frac{1}{2},j}}\left\{(\xi_{x_{i,j}}\,\xi_{x_{i+\frac{1}{2},j}} + \xi_{y_{i,j}}\,\xi_{y_{i+\frac{1}{2},j}})\left[\frac{R_{i+1,j} - R_{i,j}}{(\Delta\xi)^2}\right]\right\}\right. \\
&- \nu_{t_{i-\frac{1}{2},j}}\left\{(\xi_{x_{i,j}}\,\xi_{x_{i-\frac{1}{2},j}} + \xi_{y_{i,j}}\,\xi_{y_{i-\frac{1}{2},j}})\left[\frac{R_{i,j} - R_{i-1,j}}{(\Delta\xi)^2}\right]\right\} \\
&+ \nu_{t_{i+\frac{1}{2},j}}\left\{(\xi_{x_{i,j}}\,\eta_{x_{i+\frac{1}{2},j}} + \xi_{y_{i,j}}\,\eta_{y_{i+\frac{1}{2},j}})\left[\frac{R_{i+\frac{1}{2},j+\frac{1}{2}} - R_{i+\frac{1}{2},j-\frac{1}{2}}}{\Delta\xi\Delta\eta}\right]\right\} \\
&- \nu_{t_{i-\frac{1}{2},j}}\left\{(\xi_{x_{i,j}}\,\eta_{x_{i-\frac{1}{2},j}} + \xi_{y_{i,j}}\,\eta_{y_{i-\frac{1}{2},j}})\left[\frac{R_{i-\frac{1}{2},j+\frac{1}{2}} - R_{i-\frac{1}{2},j-\frac{1}{2}}}{\Delta\xi\Delta\eta}\right]\right\} \\
&+ \nu_{t_{i,j+\frac{1}{2}}}\left\{(\eta_{x_{i,j}}\,\eta_{x_{i,j+\frac{1}{2}}} + \eta_{y_{i,j}}\,\eta_{y_{i,j+\frac{1}{2}}})\left[\frac{R_{i,j+1} - R_{i,j}}{(\Delta\eta)^2}\right]\right\}
\end{aligned}$$

$$- \nu_{t_{i,j-\frac{1}{2}}} \left\{ \left(\eta_{x_{i,j}} \eta_{x_{i,j-\frac{1}{2}}} + \eta_{y_{i,j}} \eta_{y_{i,j-\frac{1}{2}}} \right) \left[\frac{R_{i,j} - R_{i,j-1}}{(\Delta\eta)^2} \right] \right\}$$

$$+ \nu_{t_{i,j+\frac{1}{2}}} \left\{ \left(\eta_{x_{i,j}} \xi_{x_{i,j+\frac{1}{2}}} + \eta_{y_{i,j}} \xi_{y_{i,j+\frac{1}{2}}} \right) \left[\frac{R_{i+\frac{1}{2},j+\frac{1}{2}} - R_{i-\frac{1}{2},j+\frac{1}{2}}}{\Delta\xi\Delta\eta} \right] \right\}$$

$$- \nu_{t_{i,j-\frac{1}{2}}} \left\{ \left(\eta_{x_{i,j}} \xi_{x_{i,j-\frac{1}{2}}} + \eta_{y_{i,j}} \xi_{y_{i,j-\frac{1}{2}}} \right) \left[\frac{R_{i+\frac{1}{2},j-\frac{1}{2}} - R_{i-\frac{1}{2},j-\frac{1}{2}}}{\Delta\xi\Delta\eta} \right] \right\} \Bigg]$$

$$+ \Delta t \left(\frac{2}{Re_\infty} \right) \left(\nu + \frac{\nu_t}{\sigma_\epsilon} \right)_{i,j} \Bigg[\left(\xi_{x_{i,j}} \xi_{x_{i+\frac{1}{2},j}} + \xi_{y_{i,j}} \xi_{y_{i+\frac{1}{2},j}} \right) \left[\frac{R_{i+1,j} - R_{i,j}}{(\Delta\xi)^2} \right]$$

$$- \left(\xi_{x_{i,j}} \xi_{x_{i-\frac{1}{2},j}} + \xi_{y_{i,j}} \xi_{y_{i-\frac{1}{2},j}} \right) \left[\frac{R_{i,j} - R_{i-1,j}}{(\Delta\xi)^2} \right]$$

$$+ \left(\xi_{x_{i,j}} \eta_{x_{i+\frac{1}{2},j}} + \xi_{y_{i,j}} \eta_{y_{i+\frac{1}{2},j}} \right) \left[\frac{R_{i+\frac{1}{2},j+\frac{1}{2}} - R_{i+\frac{1}{2},j-\frac{1}{2}}}{\Delta\xi\Delta\eta} \right]$$

$$- \left(\xi_{x_{i,j}} \eta_{x_{i-\frac{1}{2},j}} + \xi_{y_{i,j}} \eta_{y_{i-\frac{1}{2},j}} \right) \left[\frac{R_{i-\frac{1}{2},j+\frac{1}{2}} - R_{i-\frac{1}{2},j-\frac{1}{2}}}{\Delta\xi\Delta\eta} \right]$$

$$+ \left(\eta_{x_{i,j}} \eta_{x_{i,j+\frac{1}{2}}} + \eta_{y_{i,j}} \eta_{y_{i,j+\frac{1}{2}}} \right) \left[\frac{R_{i,j+1} - R_{i,j}}{(\Delta\eta)^2} \right]$$

$$- \left(\eta_{x_{i,j}} \eta_{x_{i,j-\frac{1}{2}}} + \eta_{y_{i,j}} \eta_{y_{i,j-\frac{1}{2}}} \right) \left[\frac{R_{i,j} - R_{i,j-1}}{(\Delta\eta)^2} \right]$$

$$+ \left(\eta_{x_{i,j}} \xi_{x_{i,j+\frac{1}{2}}} + \eta_{y_{i,j}} \xi_{y_{i,j+\frac{1}{2}}} \right) \left[\frac{R_{i+\frac{1}{2},j+\frac{1}{2}} - R_{i-\frac{1}{2},j+\frac{1}{2}}}{\Delta\xi\Delta\eta} \right]$$

$$- \left(\eta_{x_{i,j}} \xi_{x_{i,j-\frac{1}{2}}} + \eta_{y_{i,j}} \xi_{y_{i,j-\frac{1}{2}}} \right) \left[\frac{R_{i+\frac{1}{2},j-\frac{1}{2}} - R_{i-\frac{1}{2},j-\frac{1}{2}}}{\Delta\xi\Delta\eta} \right] \Bigg]$$

$$+ \Delta t \left(c_{\epsilon 2} f_{2_{i,j}} - c_{\epsilon 1} \right) \sqrt{c_\mu D_{1_{i,j}} D_{2_{i,j}}}\, S_{i,j} R_{i,j} \tag{17-213}$$

where

$$S_{i,j}^2 = a_{1_{i,j}} \left(\frac{u_{i+1,j} - u_{i-1,j}}{2\Delta\xi} \right)^2 + b_{1_{i,j}} \left(\frac{u_{i,j+1} - u_{i,j-1}}{2\Delta\eta} \right)^2$$

$$+ a_{2_{i,j}} \left(\frac{v_{i+1,j} - v_{i-1,j}}{2\Delta\xi} \right)^2 + b_{2_{i,j}} \left(\frac{v_{i,j+1} - v_{i,j-1}}{2\Delta\eta} \right)^2$$

$$+ 2c_{1_{i,j}}\left(\frac{u_{i+1,j}-u_{i-1,j}}{2\Delta\xi}\right)\left(\frac{u_{i,j+1}-u_{i,j-1}}{2\Delta\eta}\right)+2c_{2_{i,j}}\left(\frac{v_{i+1,j}-v_{i-1,j}}{2\Delta\xi}\right)\left(\frac{v_{i,j+1}-v_{i,j-1}}{2\Delta\eta}\right)$$

$$+ 2a_{3_{i,j}}\left(\frac{u_{i+1,j}-u_{i-1,j}}{2\Delta\xi}\right)\left(\frac{v_{i+1,j}-v_{i-1,j}}{2\Delta\xi}\right)+2b_{3_{i,j}}\left(\frac{u_{i,j+1}-u_{i,j-1}}{2\Delta\eta}\right)\left(\frac{v_{i,j+1}-v_{i,j-1}}{2\Delta\eta}\right)$$

$$+ 2c_{3_{i,j}}\left(\frac{u_{i+1,j}-u_{i-1,j}}{2\Delta\xi}\right)\left(\frac{v_{i,j+1}-v_{i,j-1}}{2\Delta\eta}\right)+2c_{4_{i,j}}\left(\frac{u_{i,j+1}-u_{i,j-1}}{2\Delta\eta}\right)\left(\frac{v_{i+1,j}-v_{i-1,j}}{2\Delta\xi}\right) \tag{17-214}$$

17.6.2 Spalart-Allmaras One-Equation Turbulence Model

As in the previous section, an upwind approximation is used for the convective terms, and central difference approximation is used for the viscous terms. Since the approach is similar to that of the previous section, the final form of the finite difference expression and/or equations are provided in the following subsections.

17.6.2.1 Implicit Formulation: The convective term can be approximated by either a first-order or a second-order upwind formulation. The first-order approximation yields

$$\begin{aligned} M^1 &= -\left(U\frac{\partial\overline{\nu}}{\partial\xi}+V\frac{\partial\overline{\nu}}{\partial\eta}\right) \\ &= -\left\{\frac{1}{2}(U_{i,j}+|U_{i,j}|)\left(\frac{\overline{\nu}_{i,j}-\overline{\nu}_{i-1,j}}{\Delta\xi}\right)\right. \\ &\quad+\frac{1}{2}(U_{i,j}-|U_{i,j}|)\left(\frac{\overline{\nu}_{i+1,j}-\overline{\nu}_{i,j}}{\Delta\xi}\right) \\ &\quad+\frac{1}{2}(V_{i,j}+|V_{i,j}|)\left(\frac{\overline{\nu}_{i,j}-\overline{\nu}_{i,j-1}}{\Delta\eta}\right) \\ &\quad\left.+\frac{1}{2}(V_{i,j}-|V_{i,j}|)\left(\frac{\overline{\nu}_{i,j+1}-\overline{\nu}_{i,j}}{\Delta\eta}\right)\right\} \end{aligned} \tag{17-215}$$

The second-order approximation provides

$$M^1 = -\left\{\frac{1}{2}(U_{i,j}+|U_{i,j}|)\left(\frac{3\overline{\nu}_{i,j}-4\overline{\nu}_{i-1,j}+\overline{\nu}_{i-2,j}}{2\Delta\xi}\right)\right.$$

$$= +\frac{1}{2}(U_{i,j} - |U_{i,j}|)\left(\frac{-\overline{\nu}_{i+2,j} + 4\overline{\nu}_{i+1,j} - 3\overline{\nu}_{i,j}}{2\Delta\xi}\right)$$

$$+\frac{1}{2}(V_{i,j} + |V_{i,j}|)\left(\frac{3\overline{\nu}_{i,j} - 4\overline{\nu}_{i,j-1} + \overline{\nu}_{i,j-2}}{2\Delta\eta}\right)$$

$$+\frac{1}{2}(V_{i,j} - |V_{i,j}|)\left(\frac{-\overline{\nu}_{i,j+2} + 4\overline{\nu}_{i,j+1} - 3\overline{\nu}_{i,j}}{2\Delta\eta}\right)\Bigg\} \qquad (17\text{-}216)$$

The terms M^2 and M^3 given by (17-127) through (17-130) are approximated following the procedure illustrated in the previous section.

$$M^2_\xi = \left(\frac{1}{Re_\infty}\right)\left(\frac{1+c_{b2}}{\sigma}\right)\left(\xi_x\frac{\partial\alpha}{\partial\xi} + \xi_y\frac{\partial\beta}{\partial\xi}\right)$$

where

$$\xi_x\frac{\partial\alpha}{\partial\xi} + \xi_y\frac{\partial\beta}{\partial\xi} \cong (\nu+\overline{\nu})_{i+\frac{1}{2},j}\left\{(\xi_{x_{i,j}}\xi_{x_{i+\frac{1}{2},j}} + \xi_{y_{i,j}}\xi_{y_{i+\frac{1}{2},j}})\left[\frac{\overline{\nu}_{i+1,j} - \overline{\nu}_{i,j}}{(\Delta\xi)^2}\right]\right\}$$

$$-(\nu+\overline{\nu})_{i-\frac{1}{2},j}\left\{(\xi_{x_{i,j}}\xi_{x_{i-\frac{1}{2},j}} + \xi_{y_{i,j}}\xi_{y_{i-\frac{1}{2},j}})\left[\frac{\overline{\nu}_{i,j} - \overline{\nu}_{i-1,j}}{(\Delta\xi)^2}\right]\right\}$$

$$+(\nu+\overline{\nu})_{i+\frac{1}{2},j}\left\{(\xi_{x_{i,j}}\eta_{x_{i+\frac{1}{2},j}} + \xi_{y_{i,j}}\eta_{y_{i+\frac{1}{2},j}})\left[\frac{\overline{\nu}_{i+\frac{1}{2},j+\frac{1}{2}} - \overline{\nu}_{i+\frac{1}{2},j-\frac{1}{2}}}{\Delta\xi\Delta\eta}\right]\right\}$$

$$-(\nu+\overline{\nu})_{i-\frac{1}{2},j}\left\{(\xi_{x_{i,j}}\eta_{x_{i-\frac{1}{2},j}} + \xi_{y_{i,j}}\eta_{y_{i-\frac{1}{2},j}})\left[\frac{\overline{\nu}_{i-\frac{1}{2},j+\frac{1}{2}} - \overline{\nu}_{i-\frac{1}{2},j-\frac{1}{2}}}{\Delta\xi\Delta\eta}\right]\right\} \qquad (17\text{-}217)$$

$$M^2_\eta = \left(\frac{1}{Re_\infty}\right)\left(\frac{1+c_{b2}}{\sigma}\right)\left(\eta_x\frac{\partial\alpha}{\partial\eta} + \eta_y\frac{\partial\beta}{\partial\eta}\right)$$

where

$$\eta_x\frac{\partial\alpha}{\partial\eta} + \eta_y\frac{\partial\beta}{\partial\eta} \cong (\nu+\overline{\nu})_{i,j+\frac{1}{2}}\left\{(\eta_{x_{i,j}}\eta_{x_{i,j+\frac{1}{2}}} + \eta_{y_{i,j}}\eta_{y_{i,j+\frac{1}{2}}})\left[\frac{\overline{\nu}_{i,j+1} - \overline{\nu}_{i,j}}{(\Delta\eta)^2}\right]\right\}$$

$$-(\nu+\overline{\nu})_{i,j-\frac{1}{2}}\left\{(\eta_{x_{i,j}}\eta_{x_{i,j-\frac{1}{2}}} + \eta_{y_{i,j}}\eta_{y_{i,j-\frac{1}{2}}})\left[\frac{\overline{\nu}_{i,j} - \overline{\nu}_{i,j-1}}{(\Delta\eta)^2}\right]\right\}$$

$$+(\nu+\overline{\nu})_{i,j+\frac{1}{2}}\left\{(\eta_{x_{i,j}}\xi_{x_{i,j+\frac{1}{2}}} + \eta_{y_{i,j}}\xi_{y_{i,j+\frac{1}{2}}})\left[\frac{\overline{\nu}_{i+\frac{1}{2},j+\frac{1}{2}} - \overline{\nu}_{i-\frac{1}{2},j+\frac{1}{2}}}{\Delta\xi\Delta\eta}\right]\right\}$$

$$-(\nu+\overline{\nu})_{i,j-\frac{1}{2}}\left\{(\eta_{x_{i,j}}\xi_{x_{i,j-\frac{1}{2}}} + \eta_{y_{i,j}}\xi_{y_{i,j-\frac{1}{2}}})\left[\frac{\overline{\nu}_{i+\frac{1}{2},j-\frac{1}{2}} - \overline{\nu}_{i-\frac{1}{2},j-\frac{1}{2}}}{\Delta\xi\Delta\eta}\right]\right\} \qquad (17\text{-}218)$$

$$M_\xi^3 = -\left(\frac{1}{Re_\infty}\right)\frac{c_{b2}}{\sigma}(\nu+\overline{\nu})\left(\xi_x\frac{\partial A}{\partial \xi}+\xi_y\frac{\partial B}{\partial \xi}\right)$$

where

$$-(\nu+\overline{\nu})\left(\xi_x\frac{\partial A}{\partial \xi}+\xi_y\frac{\partial B}{\partial \xi}\right)\cong$$

$$-(\nu+\overline{\nu})_{i,j}\left\{(\xi_{x_{i,j}}\,\xi_{x_{i+\frac{1}{2},j}}+\xi_{y_{i,j}}\,\xi_{y_{i+\frac{1}{2},j}})\left[\frac{\overline{\nu}_{i+1,j}-\overline{\nu}_{i,j}}{(\Delta\xi)^2}\right]\right\}$$

$$+(\nu+\overline{\nu})_{i,j}\left\{(\xi_{x_{i,j}}\,\xi_{x_{i-\frac{1}{2},j}}+\xi_{y_{i,j}}\,\xi_{y_{i-\frac{1}{2},j}})\left[\frac{\overline{\nu}_{i,j}-\overline{\nu}_{i-1,j}}{(\Delta\xi)^2}\right]\right\}$$

$$-(\nu+\overline{\nu})_{i,j}\left\{(\xi_{x_{i,j}}\,\eta_{x_{i+\frac{1}{2},j}}+\xi_{y_{i,j}}\,\eta_{y_{i+\frac{1}{2},j}})\left[\frac{\overline{\nu}_{i+\frac{1}{2},j+\frac{1}{2}}-\overline{\nu}_{i+\frac{1}{2},j-\frac{1}{2}}}{\Delta\xi\Delta\eta}\right]\right\}$$

$$+(\nu+\overline{\nu})_{i,j}\left\{(\xi_{x_{i,j}}\,\eta_{x_{i-\frac{1}{2},j}}+\xi_{y_{i,j}}\,\eta_{y_{i-\frac{1}{2},j}})\left[\frac{\overline{\nu}_{i-\frac{1}{2},j+\frac{1}{2}}-\overline{\nu}_{i-\frac{1}{2},j-\frac{1}{2}}}{\Delta\xi\Delta\eta}\right]\right\} \quad (17\text{-}219)$$

$$M_\eta^3 = -\left(\frac{1}{Re_\infty}\right)\frac{c_{b2}}{\sigma}(\nu+\overline{\nu})\left(\eta_x\frac{\partial A}{\partial \eta}+\eta_y\frac{\partial B}{\partial \eta}\right)$$

where

$$-(\nu+\overline{\nu})\left(\eta_x\frac{\partial A}{\partial \eta}+\eta_y\frac{\partial B}{\partial \eta}\right)\cong$$

$$-(\nu+\overline{\nu})_{i,j}\left\{(\eta_{x_{i,j}}\eta_{x_{i,j+\frac{1}{2}}}+\eta_{y_{i,j}}\eta_{y_{i,j+\frac{1}{2}}})\left[\frac{\overline{\nu}_{i,j+1}-\overline{\nu}_{i,j}}{(\Delta\eta)^2}\right]\right\}$$

$$+(\nu+\overline{\nu})_{i,j}\left\{(\eta_{x_{i,j}}\eta_{x_{i,j-\frac{1}{2}}}+\eta_{y_{i,j}}\eta_{y_{i,j-\frac{1}{2}}})\left[\frac{\overline{\nu}_{i,j}-\overline{\nu}_{i,j-1}}{(\Delta\eta)^2}\right]\right\}$$

$$-(\nu+\overline{\nu})_{i,j}\left\{(\eta_{x_{i,j}}\xi_{x_{i,j+\frac{1}{2}}}+\eta_{y_{i,j}}\xi_{y_{i,j+\frac{1}{2}}})\left[\frac{\overline{\nu}_{i+\frac{1}{2},j+\frac{1}{2}}-\overline{\nu}_{i-\frac{1}{2},j+\frac{1}{2}}}{\Delta\xi\Delta\eta}\right]\right\}$$

$$+(\nu+\overline{\nu})_{i,j}\left\{(\eta_{x_{i,j}}\xi_{x_{i,j-\frac{1}{2}}}+\eta_{y_{i,j}}\xi_{y_{i,j-\frac{1}{2}}})\left[\frac{\overline{\nu}_{i+\frac{1}{2},j-\frac{1}{2}}-\overline{\nu}_{i-\frac{1}{2},j-\frac{1}{2}}}{\Delta\xi\Delta\eta}\right]\right\} \quad (17\text{-}220)$$

The production term in this model is simulated by

$$P = c_{b_1}\overline{S}\overline{\nu}(1-f_{t2})$$

where $c_{b_1} = 0.1355$ is a constant, and

$$\overline{S} = S + \frac{\overline{\nu}}{Re_\infty k^2 d^2} f_{v2} \quad , \quad \kappa = 0.41$$

$$f_{v2} = 1 - \frac{\chi}{1 + \chi f_{v1}} \quad , \quad \chi = \frac{\overline{\nu}}{\nu}$$

$$f_{v1} = \frac{\chi^3}{\chi^3 + c_{v1}^3} \quad , \quad c_{v1} = 7.1$$

$$f_{t2} = c_{t3} \exp(-c_{t4}\chi^2) \quad , \quad c_{t4} = 2.0$$

At location (i, j),

$$P_{i,j} = c_{b1}\overline{S}_{i,j}\overline{\nu}_{i,j}(1 - f_{t2_{i,j}}) \tag{17-221}$$

$$\overline{S}_{i,j} = S_{i,j} + \frac{\overline{\nu}_{i,j}}{Re_\infty k^2 d_{i,j}^2} f_{v2_{i,j}} \tag{17-222}$$

$$\begin{aligned} S_{i,j} &= \left| \frac{\partial v}{\partial x} - \frac{\partial u}{\partial y} \right|_{i,j} \\ &= \left| \xi_x \frac{\partial v}{\partial \xi} + \eta_x \frac{\partial v}{\partial \eta} - \xi_y \frac{\partial u}{\partial \xi} - \eta_y \frac{\partial u}{\partial \eta} \right|_{i,j} \\ &= \left| \xi_{x_{i,j}} \frac{v_{i+1,j} - v_{i-1,j}}{2\Delta\xi} + \eta_{x_{i,j}} \frac{v_{i,j+1} - v_{i,j-1}}{2\Delta\eta} \right. \\ &\quad \left. - \xi_{y_{i,j}} \frac{u_{i+1,j} - u_{i-1,j}}{2\Delta\xi} - \eta_{y_{i,j}} \frac{u_{i,j+1} - u_{i,j-1}}{2\Delta\eta} \right| \end{aligned} \tag{17-223}$$

$$\chi_{i,j} = \frac{\overline{\nu}_{i,j}}{\nu_{i,j}} \tag{17-224}$$

$$f_{t2_{i,j}} = c_{t_3} \exp(-c_{t_4}\chi_{i,j}^2) \tag{17-225}$$

$$f_{v2_{i,j}} = 1 - \frac{\chi_{i,j}}{1 + f_{v1_{i,j}}} \tag{17-226}$$

$$f_{v1_{i,j}} = \frac{\chi_{i,j}^3}{\chi_{i,j}^3 + c_{v1}^3} \tag{17-227}$$

The destruction term is given by

$$D = -\frac{1}{Re_\infty}\left(c_{w_1} f_w - \frac{c_{b1}}{k^2} f_{t2}\right)\left[\frac{\overline{\nu}}{d}\right]^2$$

$$f_w = g\left(\frac{1+c_{w3}^6}{g^6+c_{w3}^6}\right)^{\frac{1}{6}}$$

$$g = r + c_{w_2}(r^6 - r)$$

$$r = \frac{1}{Re_\infty}\frac{\overline{\nu}}{\overline{S}\kappa^2 d^2}$$

$$\overline{S} = S + \frac{\overline{\nu}}{Re_\infty \kappa^2 d^2} f_{v_2}$$

Thus,

$$D_{i,j} = -\frac{1}{Re_\infty}\left(c_{w_1} f_{w_{i,j}} - \frac{c_{b_1}}{\kappa^2} f_{t2_{i,j}}\right)\left(\frac{\overline{\nu}_{i,j}}{d_{i,j}}\right)^2 \qquad (17\text{-}228)$$

$$f_{w_{i,j}} = g_{i,j}\left(\frac{1+c_{w3}^6}{g_{i,j}^6+c_{w3}^6}\right)^{\frac{1}{6}} \qquad (17\text{-}229)$$

$$g_{i,j} = r_{i,j} + c_{w_2}(r_{i,j}^6 - r_{i,j}) \qquad (17\text{-}230)$$

$$r_{i,j} = \frac{1}{Re_\infty}\frac{\overline{\nu}_{i,j}}{\overline{S}_{i,j}\kappa^2 d_{i,j}^2} \qquad (17\text{-}231)$$

The trip term is given by

$$T = Re_\infty f_{t1}(\Delta q)^2$$

where

$$f_{t1} = c_{t1} g_t \exp\left[-c_{t2}\frac{\omega_t^2}{(\Delta q)^2}\left(d^2 + g_t^2 d_t^2\right)\right]$$

and

$$g_t = \min\left(0.1\,,\, \frac{\Delta q}{\omega_t(\Delta x)_t}\right)$$

Δq: Is the difference between the velocities at the field point and at the trip point at the wall. Since the velocity at the wall is zero (no suction), Δq is simply the velocity at the field point.

ω_t: Is the wall vorticity at the trip.

Δx_t: Is the grid spacing along the wall at the trip.

d_t: Is the distance from the field point to the trip point.

d: Is the nearest distance from the field point to the wall.

At this point, all the finite difference expressions are in place to develop a numerical scheme. The ADI formulation, which was introduced in Chapter 3, will be used for this purpose. Recall that the governing equation given by (17-122) is

$$\frac{\partial \overline{\nu}}{\partial t} = M + P - D + T \tag{17-232}$$

Furthermore, Equation (17-232) is nonlinear due to the terms in M, P, and D. Thus, before a numerical scheme is considered, Equation (17-232) must be linearized. The linearization procedure developed in Chapter 6, and similarly developed in Chapter 11, and given by Relation (11-81), is used to linearize Equation (17-232). The general expression is

$$E^{n+1} = E^n + A\Delta u \tag{17-233}$$

where

$$A = \frac{\partial E}{\partial u}$$

Now, to develop an implicit scheme, Equation (17-232) is applied at time level $n+1$. Therefore,

$$\left(\frac{\partial \overline{\nu}}{\partial t}\right)^{n+1} = M^{n+1} + P^{n+1} - D^{n+1} + T^n \tag{17-234}$$

where the trip term is treated as a source term evaluated at time level n. Now the linearization (17-233) provides

$$M^{n+1} = M^n + \frac{\partial M}{\partial \overline{\nu}}\Delta\overline{\nu} = M^n + \overline{M}\Delta\overline{\nu} \tag{17-235}$$

$$P^{n+1} = P^n + \frac{\partial P}{\partial \overline{\nu}}\Delta\overline{\nu} = P^n + \overline{P}\Delta\overline{\nu} \tag{17-236}$$

$$D^{n+1} = D^n + \frac{\partial D}{D\overline{\nu}}\Delta\overline{\nu} = D^n + \overline{D}\Delta\overline{\nu} \tag{17-237}$$

Now Equation (17-234) can be written as

$$\frac{\Delta\overline{\nu}}{\Delta t} - (\overline{M} + \overline{P} - \overline{D})\Delta\overline{\nu} = M^n + P^n - D^n + T^n \tag{17-238}$$

or

$$\left[1 - (\overline{M} + \overline{P} - \overline{D})\Delta t\right]\Delta\overline{\nu} = (M^n + P^n - D^n + T^n)\Delta t \tag{17-239}$$

Recall that, in the development of finite difference expressions for M, it was decomposed as M_ξ and M_η. Therefore, the Equation (17-238) can be written as

$$\left[1 - (\overline{M}_\xi + \overline{M}_\eta + \overline{P} - \overline{D})\Delta t\right] \Delta\overline{\nu} = (M^n + P^n - D^n + T^n)\Delta t$$

Now this equation is factored as

$$\left[1 - \Delta t\overline{M}_\xi\right] \left[1 - \Delta t\left(\overline{M}_\eta + \overline{P} - \overline{D}\right)\right] \Delta\overline{\nu} = \left(M_\xi^n + M_\eta^n + P^n - D^n + T^n\right) \Delta t$$

and solved sequentially with the following equations

$$(1 - \Delta t\,\overline{M}_\xi)\,\Delta\overline{\nu}^* = RHS \tag{17-240}$$

$$\left[1 - \Delta t\,(\overline{M}_\eta + \overline{P} - \overline{D})\right] \Delta\overline{\nu} = \Delta\overline{\nu}^* \tag{17-241}$$

As seen previously, each one of these equations, once applied to all of the grid points, will result in a tridiagonal system. For example, Equation (17-240) will yield

$$(1 - \Delta t\,\overline{M}_\xi)_{i-1,j}\Delta\overline{\nu}^*_{i-1,j} + (1 - \Delta t\,\overline{M}_\xi)_{i,j}\Delta\overline{\nu}^*_{i,j} + (1 - \Delta t\,\overline{M}_\xi)_{i+1,j}\Delta\overline{\nu}^*_{i+1,j} = RHS_{i,j} \tag{17-242}$$

The finite difference expressions for the terms in $RHS_{i,j}$ of Equation (17-242) have already been identified. The "bar" quantities are determined in a similar fashion as that of Section 17.6.1.1. For example, the results for $\overline{M}_\xi^1$ are as follows

$$\overline{M}_\xi^1\Big|_{i+1,j} = -\frac{1}{2}(U_{i,j} - |U_{i,j}|)\left(\frac{1}{\Delta\xi}\right)$$

$$\overline{M}_\xi^1\Big|_{i,j} = -\left\{|U_{i,j}|\left(\frac{1}{\Delta\xi}\right)\right\}$$

$$\overline{M}_\xi^1\Big|_{i-1,j} = \frac{1}{2}(U_{i,j} + |U_{i,j}|)\left(\frac{1}{\Delta\xi}\right)$$

Similar expressions can be obtained for the remaining terms.

Once the tridiagonal system of equations given by (17-242) is solved, the right-hand side of (17-241) is known, and, subsequently, it can be solved. The system of tridiagonal equations given by (17-241) or (17-242) can be solved by any tridiagonal solver such as the scheme discussed in Appendix B.

17.6.2.2 Explicit Formulation: The explicit formulation of Equation (17-110) is similar to that of (17-213) and is given by

$$\overline{\nu}_{i,j}^{n+1} = \overline{\nu}_{i,j}^{n} - \Delta t\left\{\frac{1}{2}(U_{i,j}+|U_{i,j}|)\left(\frac{\overline{\nu}_{i,j}-\overline{\nu}_{i-1,j}}{\Delta\xi}\right)+\frac{1}{2}(U_{i,j}-|U_{i,j}|)\left(\frac{\overline{\nu}_{i+1,j}-\overline{\nu}_{i,j}}{\Delta\xi}\right)\right.$$
$$\left.+\frac{1}{2}(V_{i,j}+|V_{i,j}|)\left(\frac{\overline{\nu}_{i,j}-\overline{\nu}_{i,j-1}}{\Delta\eta}\right)+\frac{1}{2}(V_{i,j}-|V_{i,j}|)\left(\frac{\overline{\nu}_{i,j+1}-\overline{\nu}_{i,j}}{\Delta\eta}\right)\right\}$$

$$+\Delta t\left(\frac{1}{Re_\infty}\right)\left(\frac{1+c_{b2}}{\sigma}\right)\left[(\nu+\overline{\nu})_{i+\frac{1}{2},j}\left\{(\xi_{x_{i,j}}\xi_{x_{i+\frac{1}{2},j}}+\xi_{y_{i,j}}\xi_{y_{i+\frac{1}{2},j}})\left[\frac{\overline{\nu}_{i+1,j}-\overline{\nu}_{i,j}}{(\Delta\xi)^2}\right]\right\}\right.$$
$$-(\nu+\overline{\nu})_{i-\frac{1}{2},j}\left\{(\xi_{x_{i,j}}\xi_{x_{i-\frac{1}{2},j}}+\xi_{y_{i,j}}\xi_{y_{i-\frac{1}{2},j}})\left[\frac{\overline{\nu}_{i,j}-\overline{\nu}_{i-1,j}}{(\Delta\xi)^2}\right]\right\}$$
$$+(\nu+\overline{\nu})_{i+\frac{1}{2},j}\left\{(\xi_{x_{i,j}}\eta_{x_{i+\frac{1}{2},j}}+\xi_{y_{i,j}}\eta_{y_{i+\frac{1}{2},j}})\left[\frac{\overline{\nu}_{i+\frac{1}{2},j+\frac{1}{2}}-\overline{\nu}_{i+\frac{1}{2},j-\frac{1}{2}}}{\Delta\xi\Delta\eta}\right]\right\}$$
$$-(\nu+\overline{\nu})_{i-\frac{1}{2},j}\left\{(\xi_{x_{i,j}}\eta_{x_{i-\frac{1}{2},j}}+\xi_{y_{i,j}}\eta_{y_{i-\frac{1}{2},j}})\left[\frac{\overline{\nu}_{i-\frac{1}{2},j+\frac{1}{2}}-\overline{\nu}_{i-\frac{1}{2},j-\frac{1}{2}}}{\Delta\xi\Delta\eta}\right]\right\}$$
$$(\nu+\overline{\nu})_{i,j+\frac{1}{2}}\left\{(\eta_{x_{i,j}}\eta_{x_{i,j+\frac{1}{2}}}+\xi_{y_{i,j}}\eta_{y_{i,j+\frac{1}{2}}})\left[\frac{\overline{\nu}_{i,j+1}-\overline{\nu}_{i,j}}{(\Delta\eta)^2}\right]\right\}$$
$$-(\nu+\overline{\nu})_{i,j-\frac{1}{2}}\left\{(\eta_{x_{i,j}}\eta_{x_{i,j-\frac{1}{2}}}+\eta_{y_{i,j}}\eta_{y_{i,j-\frac{1}{2}}})\left[\frac{\overline{\nu}_{i,j}-\overline{\nu}_{i,j-1}}{(\Delta\eta)^2}\right]\right\}$$
$$+(\nu+\overline{\nu})_{i,j+\frac{1}{2}}\left\{(\eta_{x_{i,j}}\xi_{x_{i,j+\frac{1}{2}}}+\eta_{y_{i,j}}\xi_{y_{i,j+\frac{1}{2}}})\left[\frac{\overline{\nu}_{i+\frac{1}{2},j+\frac{1}{2}}-\overline{\nu}_{i-\frac{1}{2},j+\frac{1}{2}}}{\Delta\xi\Delta\eta}\right]\right\}$$
$$\left.-(\nu+\overline{\nu})_{i,j-\frac{1}{2}}\left\{(\eta_{x_{i,j}}\xi_{x_{i,j-\frac{1}{2}}}+\eta_{y_{i,j}}\xi_{y_{i,j-\frac{1}{2}}})\left[\frac{\overline{\nu}_{i+\frac{1}{2},j-\frac{1}{2}}-\overline{\nu}_{i-\frac{1}{2},j-\frac{1}{2}}}{\Delta\xi\Delta\eta}\right]\right\}\right]$$
$$+\Delta t\left(\frac{1}{Re_\infty}\right)\left(\frac{c_{b2}}{\sigma}\right)\left[-(\nu+\overline{\nu})_{i,j}\left\{(\xi_{x_{i,j}}\xi_{x_{i+\frac{1}{2},j}}+\xi_{y_{i,j}}\xi_{y_{i+\frac{1}{2},j}})\left[\frac{\overline{\nu}_{i+1,j}-\overline{\nu}_{i,j}}{(\Delta\xi)^2}\right]\right\}\right.$$
$$+(\nu+\overline{\nu})_{i,j}\left\{(\xi_{x_{i,j}}\xi_{x_{i-\frac{1}{2},j}}+\xi_{y_{i,j}}\xi_{y_{i-\frac{1}{2},j}})\left[\frac{\overline{\nu}_{i,j}-\overline{\nu}_{i-1,j}}{(\Delta\xi)^2}\right]\right\}$$

$$- (\nu + \bar{\nu})_{i,j} \left\{ (\xi_{x_{i,j}} \eta_{x_{i+\frac{1}{2},j}} + \xi_{y_{i,j}} \eta_{y_{i+\frac{1}{2},j}}) \left[\frac{\bar{\nu}_{i+\frac{1}{2},j+\frac{1}{2}} - \bar{\nu}_{i+\frac{1}{2},j-\frac{1}{2}}}{\Delta\xi\Delta\eta} \right] \right\}$$

$$+ (\nu + \bar{\nu})_{i,j} \left\{ (\xi_{x_{i,j}} \eta_{x_{i-\frac{1}{2},j}} + \xi_{y_{i,j}} \eta_{y_{i-\frac{1}{2},j}}) \left[\frac{\bar{\nu}_{i-\frac{1}{2},j+\frac{1}{2}} - \bar{\nu}_{i-\frac{1}{2},j-\frac{1}{2}}}{\Delta\xi\Delta\eta} \right] \right\}$$

$$- (\nu + \bar{\nu})_{i,j+\frac{1}{2}} \left\{ (\eta_{x_{i,j}} \eta_{x_{i,j+\frac{1}{2}}} + \xi_{y_{i,j}} \eta_{y_{i,j+\frac{1}{2}}}) \left[\frac{\bar{\nu}_{i,j+1} - \bar{\nu}_{i,j}}{(\Delta\eta)^2} \right] \right\}$$

$$+ (\nu + \bar{\nu})_{i,j} \left\{ (\eta_{x_{i,j}} \eta_{x_{i,j-\frac{1}{2}}} + \eta_{y_{i,j}} \eta_{y_{i,j-\frac{1}{2}}}) \left[\frac{\bar{\nu}_{i,j} - \bar{\nu}_{i,j-1}}{(\Delta\eta)^2} \right] \right\}$$

$$- (\nu + \bar{\nu})_{i,j} \left\{ (\eta_{x_{i,j}} \xi_{x_{i,j+\frac{1}{2}}} + \eta_{y_{i,j}} \xi_{y_{i,j+\frac{1}{2}}}) \left[\frac{\bar{\nu}_{i+\frac{1}{2},j+\frac{1}{2}} - \bar{\nu}_{i-\frac{1}{2},j+\frac{1}{2}}}{\Delta\xi\Delta\eta} \right] \right\}$$

$$+ (\nu + \bar{\nu})_{i,j} \left\{ (\eta_{x_{i,j}} \xi_{x_{i,j-\frac{1}{2}}} + \eta_{y_{i,j}} \xi_{y_{i,j-\frac{1}{2}}}) \left[\frac{\bar{\nu}_{i+\frac{1}{2},j-\frac{1}{2}} - \bar{\nu}_{i-\frac{1}{2},j-\frac{1}{2}}}{\Delta\xi\Delta\eta} \right] \right\} \Bigg]$$

$$+ \Delta t \Bigg\{ c_{b_1} \overline{S}_{i,j} \bar{\nu}_{i,j} (1 - f_{t2_{i,j}})$$

$$- \frac{1}{Re_\infty} \left(c_{w1} f_{w_{i,j}} - \frac{c_{b_1}}{\kappa^2} f_{t2_{i,j}} \right) \left(\frac{\nu_{i,j}}{d_{i,j}} \right)^2 + Re_\infty f_{t1} (\Delta q_{i,j})^2 \Bigg\} \qquad (17\text{-}243)$$

17.6.3 *k*-ε / *k*-ω Two-Equation Turbulence Model

Similar to the one-equation models, the convective terms are approximated by either a first-order or a second-order upwind scheme. Second-order central difference approximation is used for the remaining terms either at the grid points or at midpoints. If central difference approximation is applied at the points next to the boundaries, a difficulty will be encountered due to the appearance of points in the finite difference equation which are outside the domain. At these points, one-sided approximation can be used to eliminate the difficulty. When approximation at the midpoints is used, this difficulty does not occur. Both formulations are provided in this section. In the following formulations, the first-order upwind scheme is used. Furthermore, since the many terms in the k-equation and in the ω-equation are similar, only the finite difference equation for the k-equation is provided in detail. Subsequently, the extra term in the ω-equation is investigated. The explicit finite difference equation for the k-equation is written as

$$\begin{aligned}
k_{i,j}^{n+1} &= k_{i,j}^{n} - \frac{\Delta t}{\rho_{i,j}} \left[\left\{ \frac{1}{2} \left(U_{i,j} + |U_{i,j}| \right) \frac{(\rho k)_{i,j}^{n} - (\rho k)_{i-1,j}^{n}}{\Delta \xi} \right. \right. \\
&+ \frac{1}{2} \left(U_{i,j} - |U_{i,j}| \right) \frac{(\rho k)_{i+1,j}^{n} - (\rho k)_{i,j}^{n}}{\Delta \xi} \\
&+ \frac{1}{2} \left(V_{i,j} + |V_{i,j}| \right) \frac{(\rho k)_{i,j}^{n} - (\rho k)_{i,j-1}^{n}}{\Delta \eta} \\
&+ \frac{1}{2} \left(V_{i,j} - |V_{i,j}| \right) \frac{(\rho k)_{i,j+1}^{n} - (\rho k)_{i,j}^{n}}{\Delta \eta} \\
&+ (CU1)_{i,j} \frac{u_{i+1,j} - u_{i-1,j}}{2\Delta \xi} + (CU2)_{i,j} \frac{u_{i,j+1} - u_{i,j-1}}{2\Delta \eta} \\
&+ (CV1)_{i,j} \frac{v_{i+1,j} - v_{i-1,j}}{2\Delta \xi} + (CV2)_{i,j} \frac{v_{i,j+1} - v_{i,j-1}}{2\Delta \eta} \\
&+ \frac{1}{Re_\infty} (P_k)_{i,j} - \beta^* \rho_{i,j} \omega_{i,j} k_{i,j} \\
&+ \xi_{x_{i,j}} \frac{1}{Re_\infty} \left(\frac{1}{2\Delta \xi} \right) \left\{ [(MUS)(KX)]_{i+1,j} - [(MUS)(KX)]_{i-1,j} \right\} \\
&+ \eta_{x_{i,j}} \frac{1}{Re_\infty} \left(\frac{1}{2\Delta \eta} \right) \left\{ [(MUS)(KX)]_{i,j+1} - [(MUS)(KX)]_{i,j-1} \right\} \\
&+ \xi_{y_{i,j}} \frac{1}{Re_\infty} \left(\frac{1}{2\Delta \xi} \right) \left\{ [(MUS)(KY)]_{i+1,j} - [(MUS)(KY)]_{i-1,j} \right\} \\
&\left. + \eta_{y_{i,j}} \frac{1}{Re_\infty} \left(\frac{1}{2\Delta \eta} \right) \left\{ [(MUS)(KY)]_{i,j+1} - [(MUS)(KY)]_{i,j-1} \right\} \right]
\end{aligned} \tag{17-244}$$

To illustrate the midpoint approximation, consider the first four terms on the right-hand side of Equation (17-166). Keep in mind that the convective terms are approximated by upwind scheme, as before. The four terms of (17-166) under consideration are

$$\frac{1}{Re_\infty} \xi_x \frac{\partial}{\partial \xi} \left[(MUS) \left(\xi_x \frac{\partial k}{\partial \xi} + \eta_x \frac{\partial k}{\partial \eta} \right) \right]$$

$$+ \frac{1}{Re_\infty} \eta_x \frac{\partial}{\partial \eta} \left[(MUS) \left(\xi_x \frac{\partial k}{\partial \xi} + \eta_x \frac{\partial k}{\partial \eta} \right) \right]$$

$$+ \frac{1}{Re_\infty} \xi_y \frac{\partial}{\partial \xi} \left[(MUS) \left(\xi_y \frac{\partial k}{\partial \xi} + \eta_y \frac{\partial k}{\partial \eta} \right) \right]$$

$$+ \frac{1}{Re_\infty} \eta_y \frac{\partial}{\partial \eta} \left[(MUS) \left(\xi_y \frac{\partial k}{\partial \xi} + \eta_y \frac{\partial k}{\partial \eta} \right) \right]$$

$$= \frac{1}{Re_\infty} \left\{ \xi_x \frac{\partial A}{\partial \xi} + \xi_y \frac{\partial B}{\partial \xi} \right\} + \frac{1}{Re_\infty} \left\{ \eta_x \frac{\partial A}{\partial \eta} + \eta_y \frac{\partial B}{\partial \eta} \right\} \quad (17\text{-}245)$$

where

$$A = (MUS) \left(\xi_x \frac{\partial k}{\partial \xi} + \eta_x \frac{\partial k}{\partial \eta} \right)$$

and

$$B = (MUS) \left(\xi_y \frac{\partial k}{\partial \xi} + \eta_y \frac{\partial k}{\partial \eta} \right)$$

Now consider the finite difference expression for $(\xi_x \partial A / \partial \xi)$ where central difference approximation is used.

$$\left(\xi_x \frac{\partial A}{\partial \xi} \right)_{i,j} = \xi_{x_{i,j}} \frac{A_{i+\frac{1}{2},j} - A_{i-\frac{1}{2},j}}{\Delta \xi} \quad (17\text{-}246)$$

where

$$A_{i+\frac{1}{2},j} = (MUS)_{i+\frac{1}{2},j} \left(\xi_x \frac{\partial k}{\partial \xi} + \eta_x \frac{\partial k}{\partial \eta} \right)_{i+\frac{1}{2},j}$$

$$= (MUS)_{i+\frac{1}{2},j} \left\{ \xi_{x_{i+\frac{1}{2},j}} \frac{k_{i+1,j} - k_{i,j}}{\Delta \xi} + \eta_{x_{i+\frac{1}{2},j}} \frac{k_{i+\frac{1}{2},j+\frac{1}{2}} - k_{i+\frac{1}{2},j-\frac{1}{2}}}{\Delta \eta} \right\} \quad (17\text{-}247)$$

and

$$A_{i-\frac{1}{2},j} = (MUS)_{i-\frac{1}{2},j} \left\{ \xi_{x_{i-\frac{1}{2},j}} \frac{k_{i,j} - k_{i-1,j}}{\Delta \xi} + \eta_{x_{i-\frac{1}{2},j}} \frac{k_{i-\frac{1}{2},j+\frac{1}{2}} - k_{i-\frac{1}{2},j-\frac{1}{2}}}{\Delta \eta} \right\} \quad (17\text{-}248)$$

Similarly,

$$B_{i+\frac{1}{2},j} = (MUS)_{i+\frac{1}{2},j} \left\{ \xi_{y_{i+\frac{1}{2},j}} \frac{k_{i+1,j} - k_{i,j}}{\Delta \xi} + \eta_{y_{i+\frac{1}{2},j}} \frac{k_{i+\frac{1}{2},j+\frac{1}{2}} - k_{i+\frac{1}{2},j-\frac{1}{2}}}{\Delta \eta} \right\} \quad (17\text{-}249)$$

$$B_{i-\frac{1}{2},j} = (MUS)_{i-\frac{1}{2},j} \left\{ \xi_{y_{i-\frac{1}{2},j}} \frac{k_{i,j} - k_{i-1,j}}{\Delta \xi} + \eta_{y_{i-\frac{1}{2},j}} \frac{k_{i-\frac{1}{2},j+\frac{1}{2}} - k_{i-\frac{1}{2},j-\frac{1}{2}}}{\Delta \eta} \right\} \quad (17\text{-}250)$$

Now the first term of expression (17-244) becomes

$$\frac{1}{Re_\infty}\left\{\xi_x\frac{\partial A}{\partial \xi}+\xi_y\frac{\partial B}{\partial \xi}\right\} =$$

$$\frac{1}{Re_\infty}\left\{(MUS)_{i+\frac{1}{2}j}\frac{\xi_{x_{i,j}}\xi_{x_{i+\frac{1}{2}j}}}{(\Delta\xi)^2}(k_{i+1,j}-k_{i,j})\right.$$

$$+(MUS)_{i+\frac{1}{2}j}\frac{\xi_{x_{i,j}}\eta_{x_{i+\frac{1}{2}j}}}{\Delta\xi\Delta\eta}(k_{i+\frac{1}{2}j+\frac{1}{2}}-k_{i+\frac{1}{2}j-\frac{1}{2}})$$

$$-(MUS)_{i-\frac{1}{2}j}\frac{\xi_{x_{i,j}}\xi_{x_{i-\frac{1}{2}j}}}{(\Delta\xi)^2}(k_{i,j}-k_{i-1,j})$$

$$-(MUS)_{i-\frac{1}{2}j}\frac{\xi_{x_{i,j}}\eta_{x_{i-\frac{1}{2}j}}}{\Delta\xi\Delta\eta}(k_{i-\frac{1}{2}j+\frac{1}{2}}-k_{i-\frac{1}{2}j-\frac{1}{2}})$$

$$+(MUS)_{i+\frac{1}{2}j}\frac{\xi_{y_{i,j}}\xi_{y_{i+\frac{1}{2}j}}}{(\Delta\xi)^2}(k_{i+1,j}-k_{i,j})$$

$$+(MUS)_{i+\frac{1}{2}j}\frac{\xi_{y_{i,j}}\eta_{y_{i+\frac{1}{2}j}}}{\Delta\xi\Delta\eta}(k_{i+\frac{1}{2}j+\frac{1}{2}}-k_{i+\frac{1}{2}j-\frac{1}{2}})$$

$$-(MUS)_{i-\frac{1}{2}j}\frac{\xi_{y_{i,j}}\xi_{y_{i-\frac{1}{2}j}}}{(\Delta\xi)^2}(k_{i,j}-k_{i-1,j})$$

$$\left.-(MUS)_{i-\frac{1}{2}j}\frac{\xi_{y_{i,j}}\eta_{y_{i-\frac{1}{2}j}}}{\Delta\xi\Delta\eta}(k_{i-\frac{1}{2}j+\frac{1}{2}}-k_{i-\frac{1}{2}j-\frac{1}{2}})\right\}$$

$$=\frac{1}{Re_\infty}\left\{(MUS)_{i+\frac{1}{2}j}\frac{1}{(\Delta\xi)^2}\left(\xi_{x_{i,j}}\xi_{x_{i+\frac{1}{2}j}}+\xi_{y_{i,j}}\xi_{y_{i+\frac{1}{2}j}}\right)(k_{i+1,j}-k_{i,j})\right.$$

$$-(MUS)_{i-\frac{1}{2},j}\frac{1}{(\Delta\xi)^2}\left(\xi_{x_{i,j}}\xi_{x_{i-\frac{1}{2}j}}+\xi_{y_{i,j}}\xi_{y_{i-\frac{1}{2}j}}\right)(k_{i,j}-k_{i-1,j})$$

$$+(MUS)_{i+\frac{1}{2},j}\frac{1}{\Delta\xi\Delta\eta}\left(\xi_{x_{i,j}}\eta_{x_{i+\frac{1}{2}j}}+\xi_{y_{i,j}}\eta_{y_{i+\frac{1}{2}j}}\right)\left(k_{i+\frac{1}{2}j+\frac{1}{2}}-k_{i+\frac{1}{2}j-\frac{1}{2}}\right)$$

$$- (MUS)_{i-\frac{1}{2},j} \frac{1}{\Delta\xi\Delta\eta} \left(\xi_{x_{i,j}} \eta_{x_{i-\frac{1}{2},j}} + \xi_{y_{i,j}} \eta_{y_{i-\frac{1}{2},j}}\right) \left(k_{i-\frac{1}{2},j+\frac{1}{2}} - k_{i-\frac{1}{2},j-\frac{1}{2}}\right)\Bigg\} \quad (17\text{-}251)$$

The midpoint values are determined according to

$$\begin{aligned}
(\;)_{i\pm\frac{1}{2},j} &= \frac{1}{2}[(\;)_{i,j} + (\;)_{i\pm1,j}] \\
(\;)_{i\pm\frac{1}{2},j+\frac{1}{2}} &= \frac{1}{4}[(\;)_{i,j} + (\;)_{i\pm1,j} + (\;)_{i\pm1,j+1} + (\;)_{i,j+1}] \\
(\;)_{i\pm\frac{1}{2},j-\frac{1}{2}} &= \frac{1}{4}[(\;)_{i,j} + (\;)_{i\pm1,j} + (\;)_{i,j-1} + (\;)_{i\pm1,j-1}]
\end{aligned}$$

For example,

$$k_{i+\frac{1}{2},j+\frac{1}{2}} = \frac{1}{4}(k_{i,j} + k_{i+1,j} + k_{i+1,j+1} + k_{i,j+1})$$

Similarly, the second term of (17-245) which is given by

$$\frac{1}{Re_\infty}\left\{\eta_x \frac{\partial A}{\partial \eta} + \eta_y \frac{\partial B}{\partial \eta}\right\}$$

is approximated by

$$\begin{aligned}
&\frac{1}{Re_\infty}\Bigg\{(MUS)_{i,j+\frac{1}{2}} \frac{1}{(\Delta\eta)^2} \left(\eta_{x_{i,j}} \eta_{x_{i,j+\frac{1}{2}}} + \eta_{y_{i,j}} \eta_{y_{i,j+\frac{1}{2}}}\right) \left(k_{i,j+\frac{1}{2}} - k_{i,j}\right) \\
&\quad - (MUS)_{i,j-\frac{1}{2}} \frac{1}{(\Delta\eta)^2} \left(\eta_{x_{i,j}} \eta_{x_{i,j-\frac{1}{2}}} + \eta_{y_{i,j}} \eta_{y_{i,j-\frac{1}{2}}}\right) (k_{i,j} - k_{i,j-1}) \\
&\quad + (MUS)_{i,j+\frac{1}{2}} \frac{1}{\Delta\xi\Delta\eta} \left(\eta_{x_{i,j}} \xi_{x_{i,j+\frac{1}{2}}} + \eta_{y_{i,j}} \xi_{y_{i,j+\frac{1}{2}}}\right) \left(k_{i+\frac{1}{2},j+\frac{1}{2}} - k_{i-\frac{1}{2},j+\frac{1}{2}}\right) \\
&\quad - (MUS)_{i,j-\frac{1}{2}} \frac{1}{\Delta\xi\Delta\eta} \left(\eta_{x_{i,j}} \xi_{x_{i,j-\frac{1}{2}}} + \eta_{y_{i,j}} \xi_{y_{i,j-\frac{1}{2}}}\right) \left(k_{i+\frac{1}{2},j-\frac{1}{2}} - k_{i-\frac{1}{2},j-\frac{1}{2}}\right)\Bigg\} \quad (17\text{-}252)
\end{aligned}$$

Finally, the finite difference equation for the k-equation becomes,

$$\begin{aligned}
k_{i,j}^{n+1} &= k_{i,j}^n - \frac{\Delta t}{\rho_{i,j}}\Bigg[\Bigg\{\frac{1}{2}(U_{i,j} + |U_{i,j}|)\frac{(\rho k)_{i,j}^n - (\rho k)_{i-1,j}^n}{\Delta\xi} \\
&\quad + \frac{1}{2}(U_{i,j} - |U_{i,j}|)\frac{(\rho k)_{i+1,j}^n - (\rho k)_{i,j}^n}{\Delta\xi} + \frac{1}{2}(V_{i,j} + |V_{i,j}|)\frac{(\rho k)_{i,j}^n - (\rho k)_{i,j-1}^n}{\Delta\eta}
\end{aligned}$$

$$+\frac{1}{2}(V_{i,j}-|V_{i,j}|)\frac{(\rho k)^n_{i,j+1}-(\rho k)^n_{i,j}}{\Delta\eta}\Bigg\}+\frac{1}{Re_\infty}(P_k)_{i,j}-\beta^*\rho_{i,j}\omega_{i,j}k_{i,j}$$

$$+\frac{1}{Re_\infty}\Bigg\{(MUS)_{i+\frac{1}{2}j}\frac{1}{(\Delta\xi)^2}\left(\xi_{x_{i,j}}\xi_{x_{i+\frac{1}{2}j}}+\xi_{y_{i,j}}\xi_{y_{i+\frac{1}{2}j}}\right)(k_{i+1,j}-k_{i,j})$$

$$-(MUS)_{i-\frac{1}{2}j}\frac{1}{(\Delta\xi)^2}\left(\xi_{x_{i,j}}\xi_{x_{i-\frac{1}{2}j}}+\xi_{y_{i,j}}\xi_{y_{i-\frac{1}{2}j}}\right)(k_{i,j}-k_{i-1,j})$$

$$+(MUS)_{i+\frac{1}{2}j}\frac{1}{\Delta\xi\Delta\eta}\left(\xi_{x_{i,j}}\eta_{x_{i+\frac{1}{2}j}}+\xi_{y_{i,j}}\eta_{y_{i+\frac{1}{2}j}}\right)$$

$$\Big[(0.25)\,(k_{i,j+1}+k_{i+1,j+1}-k_{i,j-1}-k_{i+1,j-1}\Big]$$

$$-(MUS)_{i-\frac{1}{2}j}\frac{1}{\Delta\xi\Delta\eta}\left(\xi_{x_{i,j}}\eta_{x_{i-\frac{1}{2}j}}+\xi_{y_{i,j}}\eta_{y_{i-\frac{1}{2}j}}\right)$$

$$\Big[(0.25)(k_{i,j+1}+k_{i-1,j+1}-k_{i,j-1}-k_{i-1,j-1})\Big]$$

$$+(MUS)_{i,j+\frac{1}{2}}\frac{1}{(\Delta\eta)^2}\left(\eta_{x_{i,j}}\eta_{x_{i,j+\frac{1}{2}}}+\eta_{y_{i,j}}\eta_{y_{i,j+\frac{1}{2}}}\right)(k_{i,j+1}-k_{i,j})$$

$$-(MUS)_{i,j-\frac{1}{2}}\frac{1}{(\Delta\eta)^2}\left(\eta_{x_{i,j}}\eta_{x_{i,j-\frac{1}{2}}}+\eta_{y_{i,j}}\eta_{y_{i,j-\frac{1}{2}}}\right)(k_{i,j}-k_{i,j-1})$$

$$+(MUS)_{i,j+\frac{1}{2}}\frac{1}{\Delta\xi\Delta\eta}\left(\eta_{x_{i,j}}\xi_{x_{i,j+\frac{1}{2}}}+\eta_{y_{i,j}}\xi_{y_{i,j+\frac{1}{2}}}\right)$$

$$\Big[(0.25)\,(k_{i+1,j}+k_{i+1,j+1}-k_{i-1,j}-k_{i-1,j+1})\Big]-$$

$$-(MUS)_{i,j-\frac{1}{2}}\frac{1}{\Delta\xi\Delta\eta}\left(\eta_{x_{i,j}}\xi_{x_{i,j-\frac{1}{2}}}+\eta_{y_{i,j}}\xi_{y_{i,j-\frac{1}{2}}}\right)$$

$$\Big[(0.25)\,(k_{i+1,j}+k_{i+1,j-1}-k_{i-1,j}-k_{i-1,j-1}\Big] \qquad (17\text{-}253)$$

The ω-equation has similar terms as the k-equation and, therefore, the FDE just investigated can be used for the ω-equation by replacing k with ω. However, recall that the ω-equation has an extra term which will be investigated at this point.

The last term of the ω-equation given by (17-153) is

$$\left[2\rho(1-F_1)\sigma_{\omega 2}\frac{1}{\omega}\right]\left[\left(\xi_x\frac{\partial k}{\partial \xi}+\eta_x\frac{\partial k}{\partial \eta}\right)\left(\xi_x\frac{\partial \omega}{\partial \xi}+\eta_x\frac{\partial \omega}{\partial \eta}\right)\right.$$

$$\left.+\left(\xi_y\frac{\partial k}{\partial \xi}+\eta_y\frac{\partial k}{\partial \eta}\right)\left(\xi_y\frac{\partial \omega}{\partial \xi}+\eta_y\frac{\partial \omega}{\partial \eta}\right)\right]$$

The second bracket can be rearranged as

$$(\xi_x^2+\xi_y^2)\frac{\partial k}{\partial \xi}\frac{\partial \omega}{\partial \xi}+(\xi_x\eta_x+\xi_y\eta_y)\frac{\partial k}{\partial \xi}\frac{\partial \omega}{\partial \eta}$$

$$+(\xi_x\eta_x+\xi_y\eta_y)\frac{\partial k}{\partial \eta}\frac{\partial \omega}{\partial \xi}+(\eta_x^2+\eta_y^2)\frac{\partial k}{\partial \eta}\frac{\partial \omega}{\partial \eta}$$

$$=a_4\frac{\partial k}{\partial \xi}\frac{\partial \omega}{\partial \xi}+c_5\left(\frac{\partial k}{\partial \xi}\frac{\partial \omega}{\partial \eta}+\frac{\partial k}{\partial \eta}\frac{\partial \omega}{\partial \xi}\right)+b_4\frac{\partial k}{\partial \eta}\frac{\partial \omega}{\partial \eta}$$

The finite difference for the expression above is written as

$$a_4\frac{(k_{i+1,j}k_{i-1,j})\,(\omega_{i+1,j}-\omega_{i-1,j})}{4(\Delta\xi)^2}$$

$$+\,c_{5_{i,j}}\frac{(k_{i+1,j}-k_{i-1,j})\,(\omega_{i,j+1}-\omega_{i,j-1})+(k_{i,j+1}-k_{i,j-1})\,(\omega_{i+1,j}-\omega_{i-1,j})}{4(\Delta\xi)\,(\Delta\eta)}$$

$$+\,b_{4_{i,j}}\frac{(k_{i,j+1}-k_{i,j-1})\,(\omega_{i,j+1}-\omega_{i,j-1})}{4(\Delta\eta)^2}$$

17.7 Concluding Remarks

Turbulence is a very complex physical phenomenon. Direct simulation of the governing equations with small scale turbulence for a general configuration at high Reynolds numbers is not currently possible. The available approach for practical applications, at the present, uses various types of turbulence models which relate the average effect of turbulence to mean flow. A tremendous amount of work is still required in turbulence modeling. Currently, a universally accepted turbulence model which can be used in diverse flow regimes does not exist. Some models may perform reasonably well in certain flow regimes but result in inaccurate predictions under different situations. Therefore, turbulence models must be carefully investigated for their range of applicability before being utilized.

An attempt has been made in this chapter to introduce some fundamental concepts of turbulent flow, turbulence models, and limited number of solution schemes. It is hoped that this attempt will be beneficial as a first step toward the inclusion of turbulence into the governing equations of fluid motion. An in-depth review of turbulence may be found in various texts such as [17-12] through [17-16].

Chapter 18

An Introduction to High Temperature Gases

18.1 Introductory Remarks

For high speed flows, particularly those in the hypersonic regime, the assumption of calorically perfect gas imposed in the previous equations is no longer valid. That is due to the high temperatures associated with such flow fields. As a consequence of high temperatures, molecules will dissociate and may ionize. Therefore, the effect of chemistry must be accounted for if a reasonable computation is to be carried out. To address numerous issues related to chemically reacting gases, the fundamental concepts are initially explored. Subsequently, a procedure to include the chemistry effect in the equations of motion is introduced. In this regard, only inviscid equations of fluid motion are considered to illustrate the procedure.

18.2 Fundamental Concepts

The underlying assumption of calorically perfect gas results in constant specific heats and, hence, a constant ratio of specific heats, γ. For example, a value of $\gamma = 1.4$ is used for air. In addition, the internal energy of the system is expressed solely as a function of temperature. Indeed, it is assumed that the internal energy is composed of translational and rotational modes of energies only. It is based on these assumptions that the previous set of equations was developed. For a chemically reacting flow, the energy may be a function of temperature as well as pressure. The ratio of specific heats, γ, is no longer constant and, in general, a new definition for γ in the form of the ratio of enthalpy to internal energy is introduced.

In the next few subsections, various definitions from physical as well as mathematical points of view are explored.

18.2.1 Real Gas and Perfect Gas

A molecule possesses a force field due to the electromagnetic actions of electrons and the nuclei. When a domain composed of many molecules is considered, the force field associated with each molecule affects other molecules in that it may act as a repulsive force if molecules are very close to each other or as an attractive force if they are relatively far apart. Under normal conditions, such as atmospheric, the average distance between molecules of air is about 10 molecular diameters, resulting in weak attraction force. Now consider a fixed region and introduce more and more molecules into this fixed region. As a result, the molecules are more compact. This translates into conditions where the pressure is extremely high and/or temperature is very low. Under this condition the intermolecular forces become important and the gas is defined as a real gas. On the other hand, when the intermolecular forces are negligible, the gas is defined as a perfect gas. For the majority of problems in aerodynamics, the assumption of perfect gas is a valid one and is utilized extensively. From an application point of view, the major difference between a real gas and a perfect gas is the use of the equation of state. For a perfect gas, the equation of state $p = \rho RT$ is employed, whereas for a real gas, the van der Waals equation of state expressed as

$$(p + a\rho^2)\,(\frac{1}{\rho} - b) = RT$$

is usually employed. Note that in the equation above a and b are gas-dependent constants. An important point to clarify at this time is the consideration of chemistry. Whether the flow under study is chemically reacting or not has nothing to do with the assumption of perfect gas or real gas. Indeed, the equation of state for a perfect gas is used extensively for chemically reacting gases. Such a chemically reacting flow is considered a mixture of perfect gases. In this regard, the following equation of state for a species s holds

$$p_s = \rho_s R_s T \tag{18-1}$$

where p_s is the partial pressure contributed by species s; ρ_s is the partial density contributed by species s; and R_s is the gas constant for species s defined as $R_s = \frac{\mathcal{R}}{MW_s}$ where $\mathcal{R}$ is the universal gas constant; MW_s is the molecular weight of species s; and T is the temperature. Modification to Equation (18-1) to include real gas consideration may be accomplished by introduction of a so-called compressibility factor Z, such that

$$p_s = Z\rho_s R_s T \tag{18-2}$$

where the compressibility factor Z is usually given as a function of reduced pressure and temperature, e.g., see Reference [18-1]. Thus, when the compressibility factor is about one, the perfect gas equation of state may be employed.

18.2.2 Partial Pressure

Consider a gas mixture composed of various species. For simplicity, assume air composed of 80% nitrogen and 20% oxygen within a fixed region at a pressure of 10 atm. Now consider the region with exactly the same number of nitrogen molecules, i.e., the oxygen molecules have been extracted. The measured pressure is now 8.0 atm. Similarly, when the domain includes the original number of oxygen molecules, the measured pressure is at 2.0 atm. By definition, these pressures are called *partial pressures.* Thus, partial pressure of a species is formally defined as the pressure within a domain if the species s is the only matter within the region. Mathematically, the pressure of a mixture within a domain is written as the sum of the partial pressures, i.e.,

$$p = \sum_{s=1}^{n} p_s \tag{18-3}$$

This relation is known as the Dalton's law of partial pressures. In terms of Equation (18-1), it may be written that

$$p = \sum_{s=1}^{n} \rho_s R_s T = \sum_{s=1}^{n} \rho_s \frac{\mathcal{R}}{MW_s} T \tag{18-4}$$

Now, define the mass fraction of species s as

$$C_s = \frac{\rho_s}{\rho}$$

Substitution into (18-4) yields

$$p = \sum_{s=1}^{n} \rho C_s \frac{\mathcal{R}}{MW_s} T = \rho T \sum_{s=1}^{n} \left(C_s \frac{\mathcal{R}}{MW_s} \right) = \rho R T \tag{18-5}$$

Note that the gas constant for the mixture, R, is defined as

$$R = \sum_{s=1}^{n} C_s R_s$$

18.2.3 Frozen Flow

When the chemical reaction rates within the flow field are extremely slow, such that fluid particles moving within the domain do not experience any change in the chemical composition, it is referred to as a frozen flow. In order to better understand various categories of chemically reacting flows, define two time scales: one associated with the fluid motion and one associated with the chemical reactions. Assume t_f seconds are required for a fluid particle to travel the length of a domain designated by some characteristic length L with a velocity u. Hence, $t_f \approx L/u$. Now denote the time required for chemical reactions to take place by t_c. Then, for a flow field where $t_c \gg t_f$, the flow is assumed to be frozen.

From a physical point of view, recall that the chemical reactions occur due to molecular collisions. Relating the number of molecular collisions to the chemical time scales, it is apparent that for a frozen flow the molecular collisions are sufficiently few such that no chemical processes take place.

18.2.4 Equilibrium Flow

For flow fields where the chemical reaction rates are extremely high, the reactions take place instantaneously. Thus, reactions are completed before the fluid has a chance to move downstream. Such a flow is called *equilibrium* or, more precisely, *chemical equilibrium.* Therefore, for an equilibrium flow, $t_f \gg t_c$. For a flow in an equilibrium state, the specific heats are functions of both pressure and temperature. Therefore, the ratio of specific heats is no longer constant and becomes a function of temperature and pressure as well. The gas constant is also a variable due to changes in the molecular weight of the mixture.

18.2.5 Nonequilibrium Flow

The frozen and equilibrium flows just defined represent two extreme conditions. In reality, chemical reactions occur as particles are moving within the domain. Therefore, in situations where the flow cannot be classified as either frozen or equilibrium, it is referred to as nonequilibrium. For nonequilibrium flows, the perfect gas equation of state still holds, except the gas constant is now a variable because the molecular weight of the mixture is changing.

It is important to realize that, within a flow domain, equilibrium flow may be established in a certain region while in some other region, flow is in a nonequilibrium state. A computer code which incorporates a nonequilibrium model should be able to compute the equilibrium state as well. This point will be illustrated shortly.

18.2.6 Various Modes of Energy

In order to describe the various forms of energy, consider a diatomic molecule in motion within a domain. The energy associated with the translational motion of its center of gravity is called the translational energy. Since the molecules may also rotate about orthogonal axes in space, it also possesses rotational energy. In addition, the atoms forming the molecule are vibrating. The associated energy is known as the vibrational energy. Other modes, such as bending and twisting, may be activated but are usually assumed negligible. Finally the electrons are in motion; therefore, they possess kinetic energy due to orbital motion about the nucleus, and a potential energy, established by an electromagnetic field.

From a computational point of view, what is essential is the change in the value of energy rather than its absolute value. Thus, various modes of energy are measured with respect to a reference datum, usually selected as the lowest allowable energy, theoretically at a temperature of zero degrees absolute. This reference datum is known as the *zero point energy level*. Now, the internal energy for a molecule may be written based on the classification of various forms of energies as

$$\epsilon = \epsilon_t + \epsilon_r + \epsilon_v + \epsilon_e + \epsilon_o$$

where ϵ_t, ϵ_r, ϵ_v, ϵ_e, and ϵ_o represent translational, rotational, vibrational, electronic, and zero point energies, respectively. Note that the energy for an atom is composed of translational, electronic, and zero point energies only.

The various energies defined above are now expressed mathematically for a diatomic molecule in thermal equilibrium flow

$$\epsilon_{t,s} = \frac{3}{2}\frac{\mathcal{R}}{MW_s}T \tag{18-6}$$

$$\epsilon_{r,s} = \frac{2}{2}\frac{\mathcal{R}}{MW_s}T \tag{18-7}$$

$$\epsilon_{v,s} = \frac{\mathcal{R}}{MW_s}\frac{\theta_s}{(e^{\theta_s/T}-1)} \tag{18-8}$$

where θ_s is the characteristic temperature.

A closed form relation such as (18-6) does not exist for electronic energy. For problems where the temperature is less than 8000 K, the electronic energy can be ignored. That may effect at most about 1% of the total energy. Finally, the values of zero point energies are given in tables for various gases. For example, Reference [18-2] provides such tables for air. The zero point energy, $\epsilon_{o,s}$, is usually expressed in terms of $h_{o,s}$, called the *heat of formation*. The heat of formation, in general, is given as a relative value, whereas $\epsilon_{o,s}$ is an absolute value. Note that for flows within the subsonic and supersonic range, the vibrational energy within the internal energy of the system is small and is usually ignored. A benchmark value at which to include the vibrational energy of air for a pressure of one atmosphere is at temperatures of about 800 K and above.

Examination of Equations (18-6) through (18-8) reveals that one temperature is used for various modes of energies, in which case it is referred to as a one temperature model. Sophisticated models based on multi-temperature models are currently under investigation. One such model is a two temperature model in which a translational/rotational temperature T is used in relations (18-6) and (18-7) for the translational and rotational energies, and a vibrational temperature T_v is used in (18-8), where $T \neq T_v$.

Finally, incorporating relations (18-6) through (18-8), the internal energy of a species s may be expressed for an atom as

$$\epsilon_s = \frac{3}{2}\left(\frac{\mathcal{R}}{MW_s}\right)T + h_{o,s} \tag{18-9}$$

and for a molecule as

$$\epsilon_s = \frac{5}{2}\left(\frac{\mathcal{R}}{MW_s}\right)T + \frac{\mathcal{R}}{MW_s}\theta_s / \left[\exp(\theta_s/T) - 1\right] + h_{o,s} \tag{18-10}$$

18.2.7 Reaction Rates

Consider a domain composed of molecules at standard atmospheric conditions. When the temperature of the domain is increased, it is accompanied by an increase in various energy modes. As a result, the collision rate between molecules is increased as well. Now, some of the increased energy is absorbed into breaking the atomic bonds between molecules. This breakdown of molecules to atoms is known as *dissociation*. Next, consider the reverse situation where the energy of atoms is released to bond atoms into the formation of molecules. This process is called *recombination*. For simplicity, consider a diatomic molecule such as oxygen. The definitions above may be expressed mathematically as

$$O_2 + \text{absorbed energy} \rightarrow 2O \qquad \text{for dissociation}$$

$$2O - \text{released energy} \rightarrow O_2 \qquad \text{for recombination}$$

Note that the energy provided for the reactions above is related to the collision between molecules and atoms. With this physical concept in mind, it is customary to express the relations above as

$$O_2 + M \underset{b}{\overset{f}{\rightleftharpoons}} 2O + M \tag{18-11}$$

where M is considered as a nonreacting particle. Dissociation is represented by *forward reaction* denoted in (18-11) by f, and recombination is called *backward reaction*, denoted by b. The reactions in (18-11) occur at specific rates which are functions of temperature and composition.

In order to generalize relation (18-11) to n reacting species, define the concentration of species s per unit volume of the mixture as

$$[X_s] = \frac{\rho C_s}{MW_s}$$

Now, Equation (18-11) is generalized as

$$\sum_{s=1}^{n} a_s[X_s] \underset{K_b}{\overset{K_f}{\rightleftharpoons}} \sum_{s=1}^{n} b_s[X_s] \tag{18-12}$$

where a_s and b_s are the stoichiometric mole numbers of the reactants and products of species s, respectively. The forward reaction rate and backward reaction rate are represented by K_f and K_b. An empirically determined expression for the forward reaction rate K_f may be written as

$$K_f = CT^n \exp\left[-E/KT\right] \tag{18-13}$$

where K is the Boltzmann constant and C, n, and E are constants depending on each dissociating molecule. These constants are given in Appendix J for a five-species model. Instead of introducing a relation for the backward reaction rate directly, an equilibrium constant is first introduced as

$$K_e = \frac{K_f}{K_b}$$

A typical relation for the equilibrium constant is expressed as

$$K_e = \exp(A_1 + A_2 \ell n Z + A_3 Z + A_4 Z^2 + A_5 Z^3) \tag{18-14}$$

where $Z = 10000/T$, and thc coefficients A_1 through A_5 are provided in Appendix J. Now, the backward reaction rate may be determined as

$$K_b = \frac{K_f}{K_e}$$

At this point, consider the net rate of formation of $[X_s]$. From Equation (18-12), one may write

$$\left.\frac{d[X_s]}{dt}\right|_{\text{net}} = \left.\frac{d[X_s]}{dt}\right|_f + \left.\frac{d[X_s]}{dt}\right|_b \tag{18-15}$$

where the forward rate of formation of X_s is

$$\left.\frac{d[X_s]}{dt}\right|_f = (b_s - a_s) K_f \Pi_s \left[X_s\right]^{a_s}$$

and the backward rate of formation of X_s is

$$\left.\frac{d[X_s]}{dt}\right|_b = -(b_s - a_s) K_b \Pi_s \left[X_s\right]^{b_s}$$

The symbol Π is used to represent the product. Note that, by definition, when a chemical equilibrium state exists, the rate of formation of each species is zero; mathematically,

$$\left.\frac{d[X_s]}{dt}\right|_{\text{net}} = 0$$

Obviously, for chemically nonequilibrium flows the net rate of formation is not zero, and relations such as (18-15) will form a set of equations which have to be solved for the concentrations of each species.

In order to establish a benchmark value with regard to molecular dissociation, consider air at atmospheric conditions. The onset of dissociation and ionization is shown in Figure 18-1. This figure illustrates the range of dissociation of N_2, O_2 and the onset temperature for ionization.

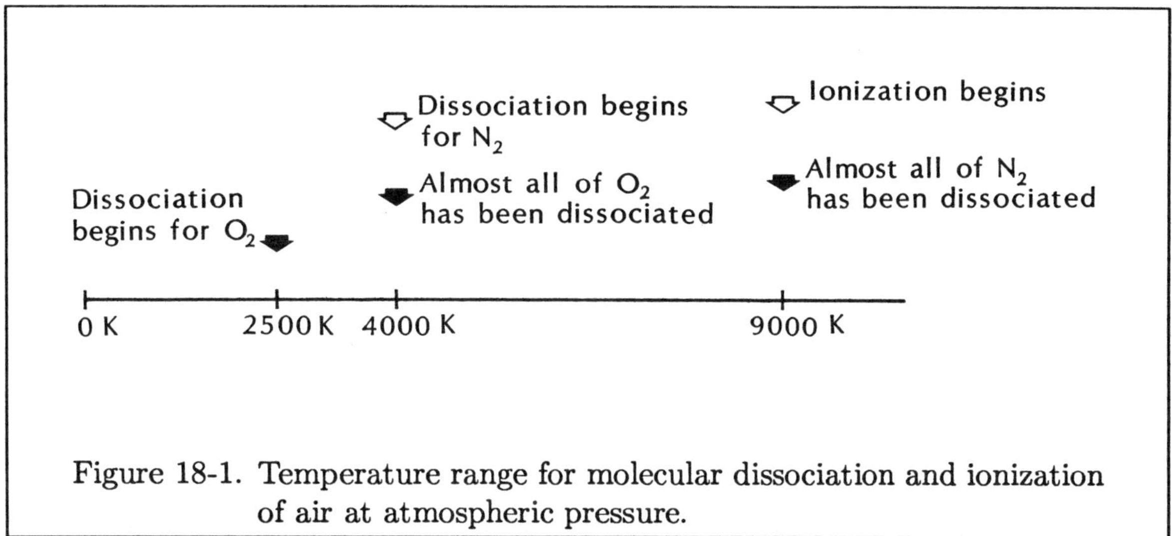

Figure 18-1. Temperature range for molecular dissociation and ionization of air at atmospheric pressure.

It is emphasized that these values correspond to a pressure of one atmosphere and that an increase in pressure will increase the onset temperatures for dissociations and ionizations.

18.2.8 Five-Species Model

In order to illustrate the nonequilibrium effect, a five-species model for air is considered. In this case, the chemical reactions of interest would be

$$O_2 + M \rightleftharpoons 2O + M$$

$$N_2 + M \rightleftharpoons 2N + M$$

$$NO + M \rightleftharpoons N + O + M$$

$$NO + O \rightleftharpoons N + O_2$$

$$O + N_2 \rightleftharpoons N + NO$$

where M represents a nonreacting particle and can be any one of the five reactants, i.e., O_2, N_2, NO, O, or N. The selection of this simple model limits the range of applicability to 8000 K. Recall that for temperatures above 8000 K, ionization must be considered as well; for example,

$$N + O \rightleftharpoons NO^+ + e^-$$

For the application to follow, the five-species model is used.

18.3 Quasi One-Dimensional Flow/Equilibrium Chemistry

For a wide range of applications, the utilization of an equilibrium chemistry model will provide an accurate solution. That is particularly true for high density flows where a sufficient number of molecular collisions take place and the reaction rates are high. For example, a flight regime below 50 km can be adequately modeled by an equilibrium model. In general, specification of two thermodynamic state variables will provide the remaining thermodynamic properties.

The general approach to include an equilibrium chemistry effect is to consider the governing equations for partial pressures of the species. The relevant equations form a system of nonlinear algebraic equations. A typical system is composed of the Dalton's law of partial pressures, equilibrium constants (provided by statistical mechanics as a function of temperature), and equations for conservation of each nuclei. To solve this system of equations, two thermodynamic states, namely pressure and temperature, are required. Once the partial pressures of species are determined, the mass fraction of species may be evaluated. Subsequently, the density, enthalpy, and internal energy of the mixture are computed.

Since many applications in gas dynamics involve air, various sets of tables or graphical plots have been generated by the system of equations just described. Therefore, they may be used to facilitate the computation of equilibrium air. In this regard, two procedures are available. In one procedure, the tabular values are input to a program, along with an interpolation routine. With any two thermodynamic states specified, the remaining variables can be determined by interpolation. In a second approach, the tabular data are curve fit using polynomials. A commonly used scheme in this category is given in Reference [18-3]. This approach is adapted for the example to follow and, therefore, a brief explanation is provided.

As stated earlier, two thermodynamic variables are required to compute the remaining variables. Recall that the equation of fluid motion considered in previous chapters provides velocity, ρ and e_t, from which internal energy can be determined as

$$e = e_t - \frac{1}{2}(u^2 + v^2 + w^2)$$

Hence, for the time being, consider the two given thermodynamic variables to be the density and internal energy. First, an effective γ denoted herein by $\hat{\gamma}$ is computed from the following relation

$$\begin{aligned} \hat{\gamma} &= a_1 + a_2Y_1 + a_3Z_1 + a_4Y_1Z_1 + a_5Y_1^2 + a_6Z_1^2 + a_7Y_1Z_1^2 + a_8Z_1^3 \\ &+ \frac{a_9 + a_{10}Y_1 + a_{11}Z_1 + a_{12}Y_1Z_1}{1 + \exp[(a_{13} + a_{14}Y_1)(Z_1 + a_{15}Y_1 + a_{16})]} \end{aligned} \tag{18-16}$$

where $Y_1 = \log(\rho/1.292)$ and $Z_1 = \log(e/78408.4)$. The corresponding units for the variables in the relations above are: (N/m^2) for pressure, (Kg/m^3) for the density, and (m^2/sec^2) for the internal energy. Once $\hat{\gamma}$ has been determined, the equation of state is used to compute the pressure, i.e.,

$$p = \rho e(\hat{\gamma} - 1)$$

Now, the temperature (in units of Kelvin) is evaluated from the following relation

$$\begin{aligned} \log\left(\frac{T}{151.78}\right) &= b_1 + b_2Y_2 + b_3Z_2 + b_4Y_2Z_2 + b_5Y_2^2 + b_6Z_2^2 + b_7Y_2^2Z_2 + b_8Y_2Z_2^2 \\ &+ \frac{b_9 + b_{10}Y_2 + b_{11}Z_2 + b_{12}Y_2Z_2 + b_{13}Z_2^2}{1 + \exp[(b_{14}Y_2 + b_{15})(Z_2 + b_{16})]} \end{aligned} \tag{18-17}$$

where $Y_2 = \log(\rho/1.225)$, $X_2 = \log(p/1.0314 \times 10^5)$, and $Z_2 = X_2 - Y_2$. The coefficients appearing in relations (18-16) and (18-17) are provided in Reference [18-3], which should be consulted for an in-depth explanation.

Implementation of the equilibrium model into the equation of fluid motion is simple and straightforward. The computation of the flow field at the first time step uses a specified γ, usually 1.4. The solution of the gas dynamic equations provides the values of ρ and e. Now the equilibrium model is used to determine a new $\hat{\gamma}$. The computation for the next time level is carried out with the newly computed value of $\hat{\gamma}$. The procedure continues until the solution converges.

18.4 Quasi One-Dimensional Flow/Nonequilibrium Chemistry

At a high altitude flight regime, the nonequilibrium chemistry model must be used in order to adequately simulate the high speed flow field. This is due to insufficient molecular collisions and the resulting low reaction rates. Mathematical modeling of the nonequilibrium chemistry is similar to that of equilibrium, except now a set of partial differential equations must be solved to provide the necessary mass fractions of various species. It is therefore obvious that the solution procedure for nonequilibrium flows is more difficult than for equilibrium flows. Furthermore, since a system of PDEs is now being solved, the computation time is drastically higher, typically by a factor of about three. In order to illustrate the effects of chemical nonequilibrium, the simple quasi one-dimensional Euler equation is used in the following discussion.

18.4.1 Species Continuity Equation

In conjunction with the quasi one-dimensional Euler equation introduced in Chapter 12, the nonequilibrium species continuity equation is written as

$$\frac{\partial SQ_c}{\partial t_c} + \frac{\partial E_c}{\partial x} + SW = 0 \tag{18-18}$$

where, as before, $S = S(x)$ is the cross-sectional area, and the vectors Q_c, E_c, and W are defined as

$$Q_c = \begin{bmatrix} \rho C_1 \\ \rho C_2 \\ \rho C_3 \\ \rho C_4 \\ \rho C_5 \end{bmatrix} \qquad E_c = S \begin{bmatrix} \rho u C_1 \\ \rho u C_2 \\ \rho u C_3 \\ \rho u C_4 \\ \rho u C_5 \end{bmatrix} \qquad W = - \begin{bmatrix} \dot{w}_1 \\ \dot{w}_2 \\ \dot{w}_3 \\ \dot{w}_4 \\ \dot{w}_5 \end{bmatrix}$$

The subscript number used in the mass fraction, C, and mass production rate, $\dot{w}$, represent reactants as

1 for O_2

2 for N_2

3 for NO

4 for O

5 for N

The components of vector W represent the mass production rate of each species where

$$\dot{w}_s = MW_s \frac{d[X_s]}{dt} \tag{18-19}$$

The rate of formation for the five-species are provided in Appendix J.

The conservation of mass requires that $\sum_{s=1}^{5} \dot{w}_s = 0$, i.e., total mass of the system is conserved. Therefore, the five-species continuity equations must sum to the global continuity equation.

18.4.2 Coupling Schemes

There are two schemes upon which the Euler equation given by (12-8) and the species continuity equation (18-18) are related. One scheme is the so-called fully coupled approach where the three gas dynamic equations and four-species continuity equations are expressed as one vector equation. Note that only four-species continuity equations are used since a global continuity equation is included in the gas dynamic equation. Hence, the coupled system is expressed as

$$\frac{\partial}{\partial t}(SQ) + \frac{\partial E}{\partial x} + H = 0 \tag{18-20}$$

where

$$Q = \begin{bmatrix} \rho \\ \rho u \\ \rho e_t \\ \rho C_1 \\ \rho C_2 \\ \rho C_3 \\ \rho C_4 \end{bmatrix} \quad E = S \begin{bmatrix} \rho u \\ \rho u^2 + p \\ (\rho e_t + p)u \\ \rho u C_1 \\ \rho u C_2 \\ \rho u C_3 \\ \rho u C_4 \end{bmatrix} \quad , \quad H = - \begin{bmatrix} 0 \\ -\frac{dS}{dx}p \\ 0 \\ S\dot{w}_1 \\ S\dot{w}_2 \\ S\dot{w}_3 \\ S\dot{w}_4 \end{bmatrix}$$

Equation (18-20) may be solved by the flux-vector splitting scheme described in Chapter 12. Obviously, some additional mathematical manipulation is required. For example, now the Jacobian matrices for this system must be derived. A second scheme is known as the *loosely coupled method.* In this approach, the communication between the gas dynamic equation and the species continuity equation is performed by defining a thermodynamic property $\hat{\gamma}$ such that

$$\hat{\gamma} = \frac{h}{e} = \frac{p}{\rho e} + 1$$

Note that $\hat{\gamma}$ is a variable which depends on temperature and species mass fraction.

Each approach has its own merit. For the loosely coupled system, previously developed computer programs for ideal gas can be easily modified to include chemistry effects. In addition, the Jacobian matrices are 3×3 and 5×5 for the gas dynamic and species continuity equations, respectively. A disadvantage of the scheme is stability requirement in that it is more restrictive than the fully coupled scheme. For illustrative purposes, the loosely coupled approach is adapted.

18.4.3 Numerical Procedure for the Loosely Coupled Scheme

The quasi one-dimensional Euler equation and the numerical scheme introduced previously in Chapter 12 are used in conjunction with the five-species continuity equation in a loosely coupled fashion. The numerical scheme proceeds along the following steps:

(1) Euler equations are solved for the unknown ΔQs based on the original formulation described in Chapter 12. This solution is designated by an asterisk to distinguish it from the solution obtained after the chemistry adjustment. Hence, the solution of Euler equations yields:

$$\rho^* = \rho^n + \Delta\rho$$

$$(\rho u)^* = (\rho u)^n + \Delta(\rho u)$$

$$(\rho e_t)^* = (\rho e_t)^n + \Delta(\rho e_t)$$

Once the three primary unknowns ρ^*, u^*, and e_t^* are computed, the remaining variables such as T^*, p^*, etc. can be determined. Note that the mass fractions $(C_s)^n$, $\hat{\gamma}^n$, and the gas constant of the mixture R^n are held constant during this step.

(2) With the values of ρ^*, T^*, and C_s^n known, the mass production rate of each species $\dot{w}_s$ is determined from Equation (18-19). Subsequently, the unknowns $\Delta(\rho C_s)$ in the species continuity equations are computed. Now the intermediate partial densities are evaluated by

$$(\rho C_s)^* = \rho^* C_s^n + \Delta(\rho C_s)$$

Subsequently, the species mass fraction and the gas constant of the mixture are updated according to

$$C_s^{n+1} = \frac{(\rho C_s)^*}{\sum_{s=1}^{5}(\rho C_s)^*}$$

$$R^{n+1} = \sum_{s=1}^{5} \frac{C_s^{n+1}\mathcal{R}}{MW_s}$$

(3) The pressure p^*, flux $(\rho u)^*$, and total enthalpy h_t^* are considered as invariant variables during the chemistry step (2). Therefore $p^* = p^{n+1}$, $(\rho u)^* = (\rho u)^{n+1}$, and $h_t^* = h_t^{n+1}$. Now, an equation for temperature is developed in order to evaluate its value at the $n+1$ time level. For this purpose, recall that

$$h_t^{n+1} = e^{n+1} + \frac{1}{2}\left[(u^{n+1})^2\right] + (RT)^{n+1} \tag{18-21}$$

Furthermore

$$e^{n+1} = \sum_{s=1}^{5} C_s^{n+1}\epsilon_s \tag{18-22}$$

and

$$u^{n+1} = \frac{(\rho u)^{n+1}}{\rho^{n+1}} = \frac{(\rho u)^*}{p^{n+1}/(RT)^{n+1}} = \frac{(\rho u)^*(RT)^{n+1}}{p^*} \tag{18-23}$$

Note that ϵ_s is a function of T^{n+1} and is defined by Equations (18-9) and (18-10) for atoms and molecules, respectively. Now, Equations (18-22) and (18-23) are substituted into Equation (18-21) to yield

$$h_t^{n+1} = \sum_{s=1}^{5} C_s^{n+1}\epsilon_s + \frac{1}{2}\left\{\left[\frac{(\rho u)^*}{p^*}\right]^2\right\}\left[(RT)^{n+1}\right]^2 + (RT)^{n+1} = h_t^* \tag{18-24}$$

Equation (18-24) is solved by the Newton-Raphson method for the unknown T^{n+1}.

(4) At this step, the updated value of T^{n+1} is used to recompute all other properties according to

$$\rho^{n+1} = \frac{p^*}{(RT)^{n+1}}$$

$$u^{n+1} = \frac{(\rho u)^*}{\rho^{n+1}}$$

$$e_t^{n+1} = h_t^* - \frac{p^*}{\rho^{n+1}}$$

$$e^{n+1} = e_t^{n+1} - \frac{1}{2}(u^{n+1})^2$$

and

$$\hat{\gamma}^{n+1} = \frac{p^*}{\rho^{n+1}e^{n+1}} + 1$$

(5) Once all the flow properties at $n+1$ time level are updated, the solution is ready to proceed to the next time level, i.e., $n+2$. Thus, steps (1) through (4) are repeated for each time step until a converged solution is reached. A typical convergence criterion may be specified as

$$\text{CONV} = \sum_{i=1}^{i=IM} \frac{|r_i^{n+1} - r_i^n|}{r_i^n}$$

where r is a property, such as temperature. Once CONV is less than CONVMAX, a prescribed value, the solution has converged.

18.5 Applications

In order to illustrate the effect of chemistry on the flow field, two examples are presented. As a first problem, the quasi one-dimensional nozzle flow introduced in Chapter 12 is considered. Subsequently, a two-dimensional axisymmetric configuration is used to demonstrate the differences in the flow fields due to the implementation of different chemistry models.

18.5.1 Quasi One-Dimensional Flow

The diverging nozzle with cross-sectional area given by Equation (12-37) is used to define the domain of interest. For the current application, the nozzle entrance is located at $x = 1.2$ cm, whereas the nozzle exit is at 8.0 cm, providing an exit to inlet area ratio of 1.65. Twenty-one equally spaced grid points are distributed along the nozzle. The inflow boundary conditions are set by specification of the inlet Mach number, temperature, and pressure. The outflow boundary condition is set by specification of pressure so as to position a normal shock at $x = 4.0$ cm. In order to demonstrate the differences in the computed flow fields by equilibrium and nonequilibrium models, inlet pressures of 26500 N/m^2 and 79.8 N/m^2 (corresponding to altitudes of 10 km and 50 km) are used. The inlet temperature is set to a fixed value of 230.44 K, which is about the average value of temperatures at the two altitudes. In addition, to evaluate the effect of the inlet velocities, two conditions are specified. In one case the inlet flow is supersonic with a Mach number of 4, whereas in the second case the inlet flow is hypersonic with a Mach number of 20. Computations based on the ideal gas model are performed as well, for comparison purposes.

In Figures 18-2a and 18-2b, the temperature distributions for the inlet Mach number of 4 for altitudes of 10 and 50 km are presented. Since the post-shock temperature rise is still relatively low, i.e., below 1000 K, the chemistry effect is minimal. Indeed, all three models provide results within a few percent.

For the second case, the inlet Mach number is 20. Now the post-shock temperature, computed by the ideal gas model, is about 18500 K. Obviously, before the temperature of the flow reaches such a high value, chemical reactions will occur, thus absorbing some of the energy and consequently reducing the temperature. Indeed, the actual post-shock temperature for an altitude of 10 km is about 8200 K. At this low altitude, the density is relatively high and, therefore, the chemically-reacting flow is in equilibrium. As demonstrated in Figure 18-3a, the solution obtained by the nonequilibrium model is shown to be identical to that of the equilibrium model.

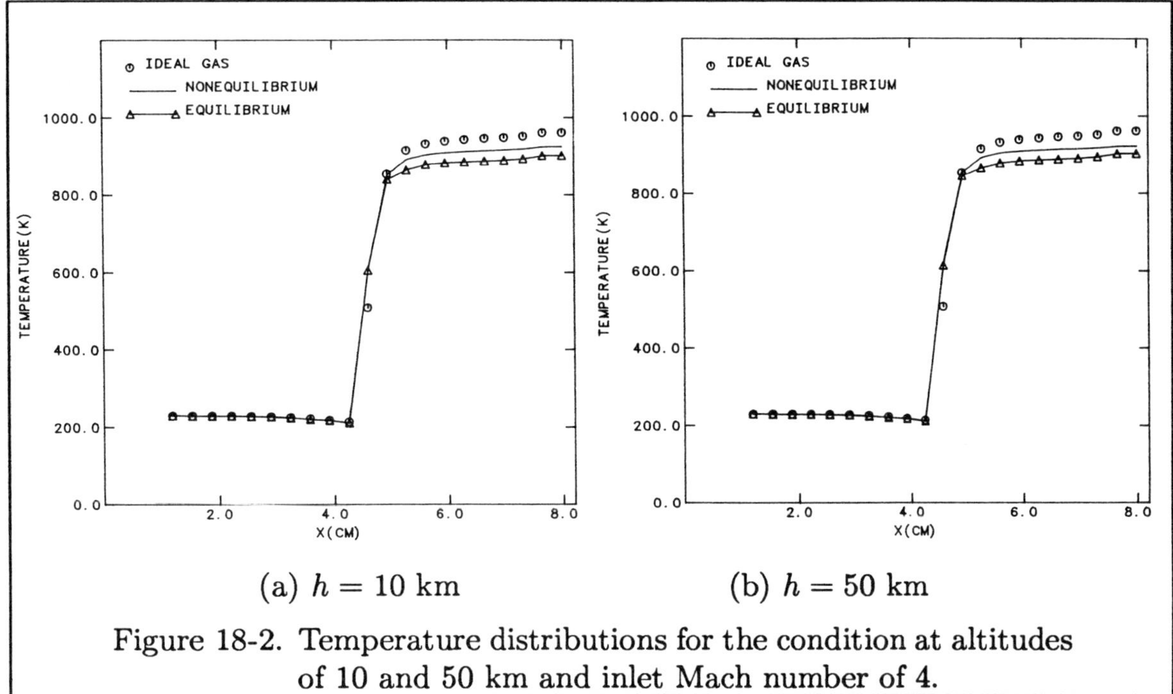

(a) $h = 10$ km (b) $h = 50$ km

Figure 18-2. Temperature distributions for the condition at altitudes of 10 and 50 km and inlet Mach number of 4.

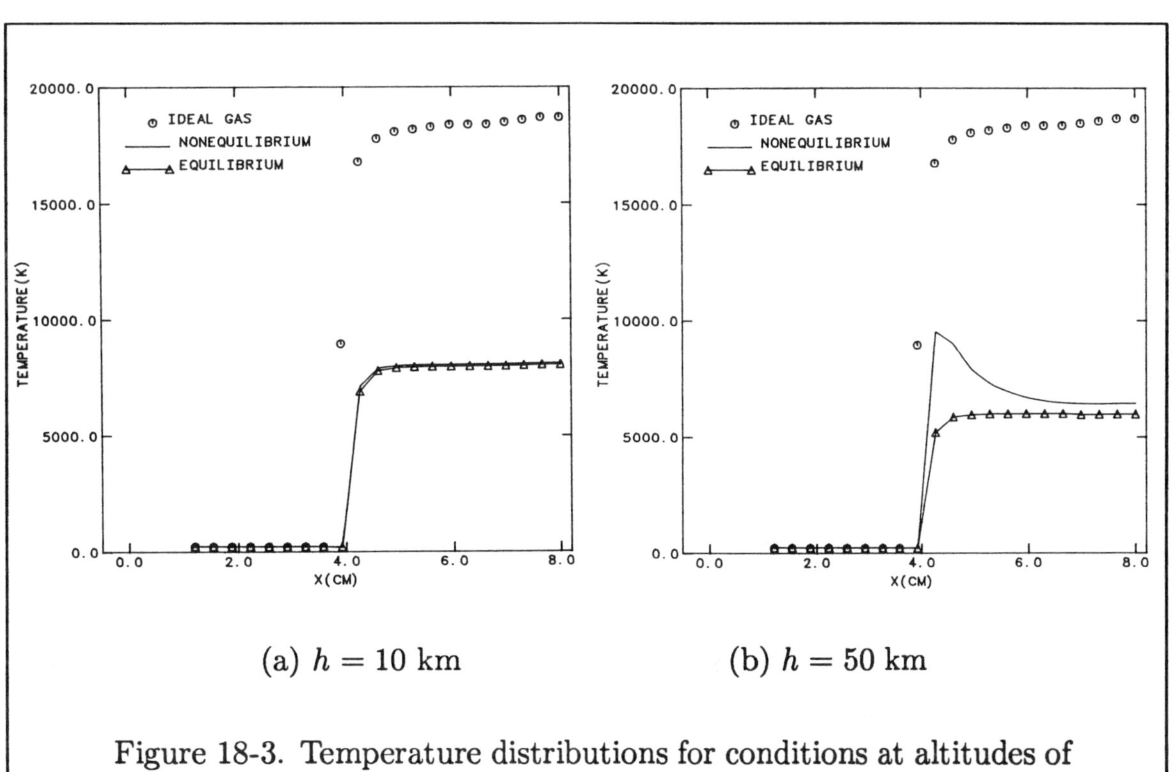

(a) $h = 10$ km (b) $h = 50$ km

Figure 18-3. Temperature distributions for conditions at altitudes of 10 and 50 km and inlet Mach number of 20.

At higher altitudes, the density is low and, therefore, nonequilibrium effects become important. The temperature distributions at an altitude of 50 km, with an inlet Mach number of 20, are shown in Figure 18-3b. In contrast to Figure 18-2b, now a remarkable difference between the solutions obtained by the equilibrium and nonequilibrium models exists. In particular, there is a temperature peak just downstream of the shock. This phenomenon indicates that nonequilibrium conditions exist in that region and, in fact, chemical equilibrium has not been achieved. When the altitude is increased further, the peak's wave length will increase, indicating a larger region of nonequilibrium flow. This behavior is shown in Figure 18-4.

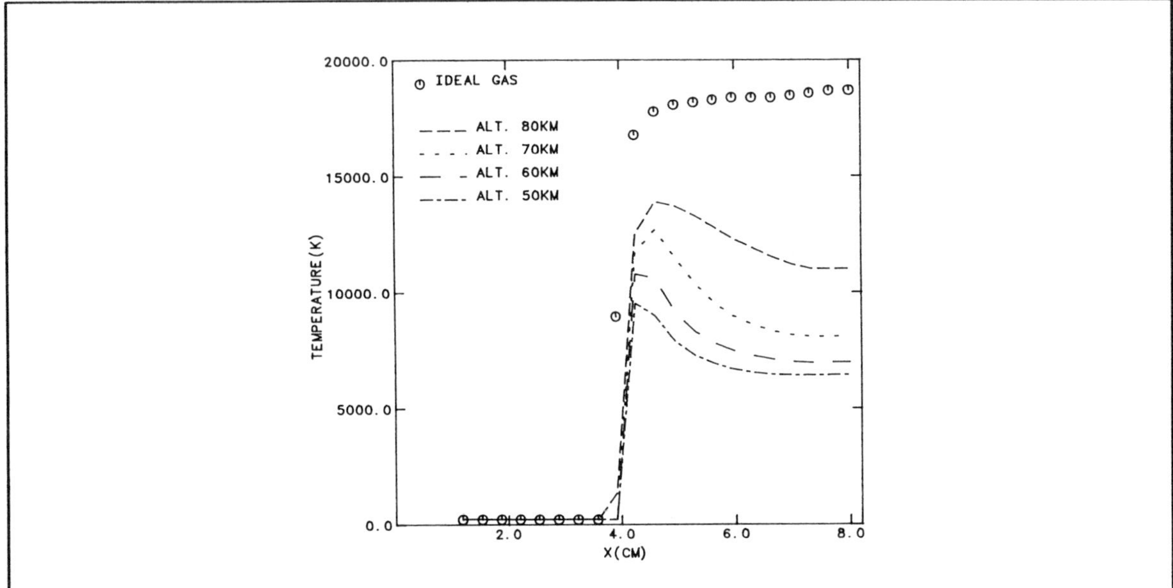

Figure 18-4. Comparison of the temperature distributions for conditions at various altitudes and inlet Mach number of 20.

18.5.2 Two-Dimensional Axisymmetric Flow

In this example, an axisymmetric blunt body at zero degree angle of attack is considered. The governing equations of gas dynamics given by Equation (12-124), along with an ideal gas, equilibrium chemistry model, and nonequilibrium chemistry model are solved within the domain of interest. The freestream Mach number, pressure, and temperature are specified as 18, 1197 N/m^2, and 226.5 K, respectively. Contours of constant temperature obtained by the ideal gas model, equilibrium, and nonequilibrium chemistry models are shown in Figures 18-5, 18-6, and 18-7, respectively. Note the remarkable difference in the shock stand-off distances.

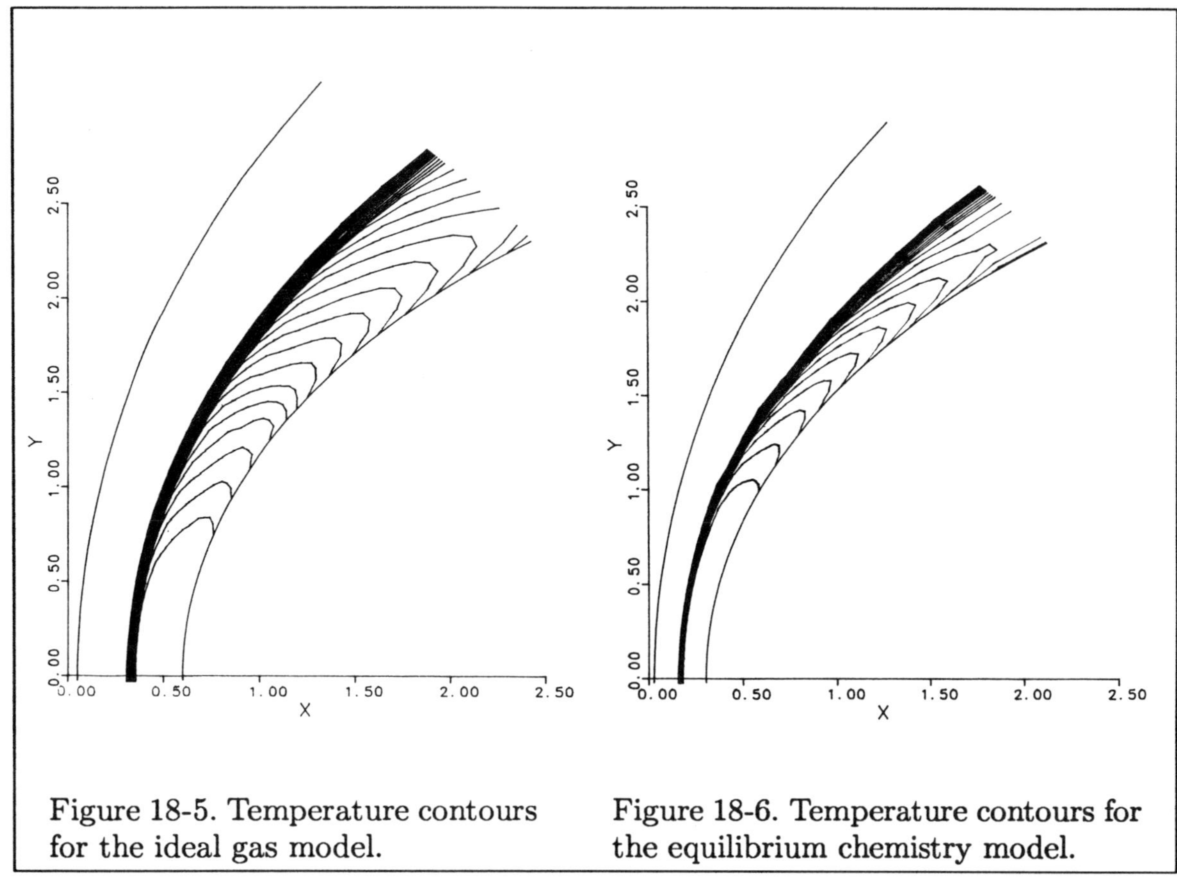

Figure 18-5. Temperature contours for the ideal gas model.

Figure 18-6. Temperature contours for the equilibrium chemistry model.

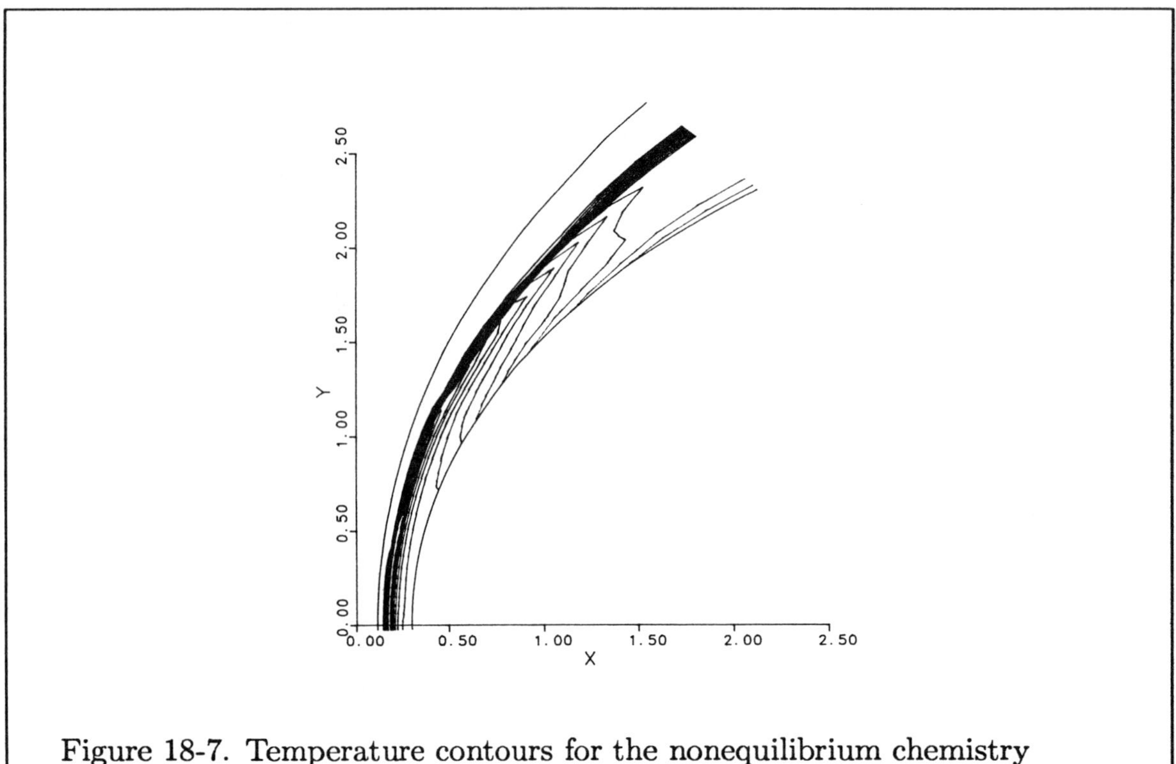

Figure 18-7. Temperature contours for the nonequilibrium chemistry model.

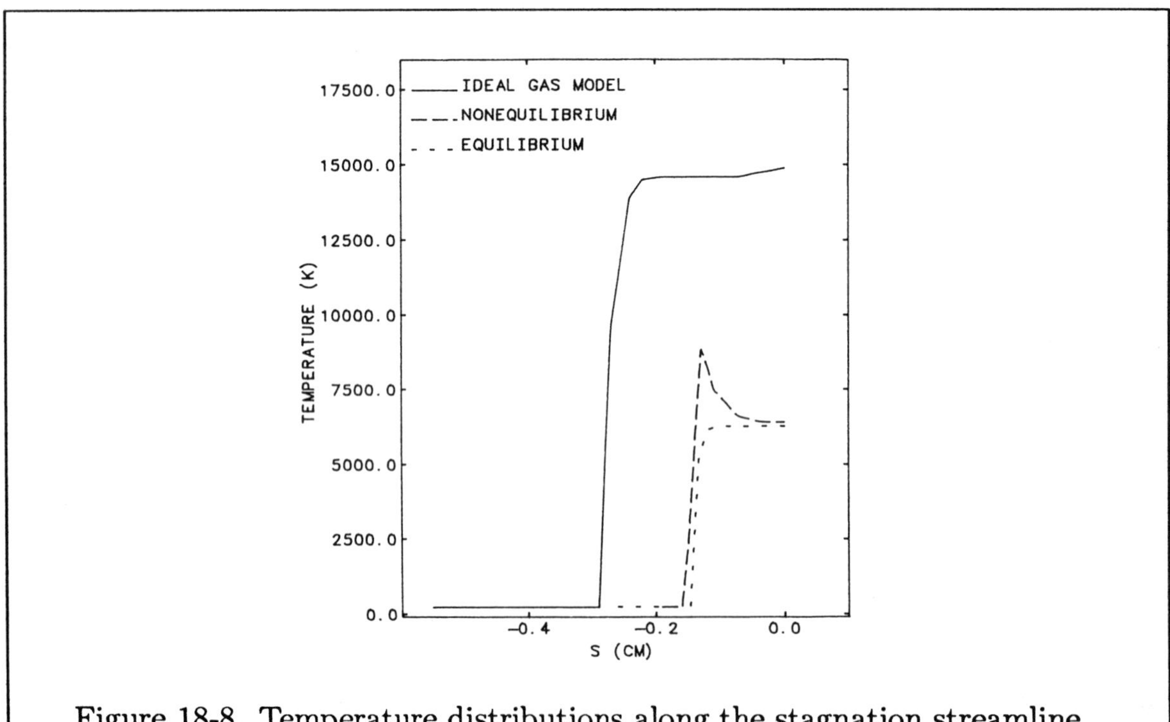

Figure 18-8. Temperature distributions along the stagnation streamline by various models.

Since the post-shock density values obtained by utilizing chemistry models are higher than the values predicted by the ideal gas model, the shock stand-off distance is smaller for chemically reacting flows. The temperature distribution along the stagnation streamline is illustrated in Figure 18-8. Three facts which have already been discussed are notable in this figure. First, the shock stand-off distance obtained by chemistry models is about 2/3 of that obtained from the ideal gas model. Second, the peak in temperature illustrates the nonequilibrium flow region. Third, the stagnation temperature obtained by chemistry models is less than half of the value obtained by the ideal gas model.

18.6 Concluding Remarks

In this chapter, some fundamental concepts and definitions of high temperature gases were introduced. The majority of equations used in high temperature gases are provided by statistical mechanics. No attempt has been made to explore the origin or details of the equations utilized in this chapter. Instead, the primary objective was to show how to implement chemistry models into the gas dynamics equations and observe the differences in the solutions. For in-depth discussions of chemically reacting flows, the classical text of Vincenti and Kruger [18-4] and a recently published text by Anderson [18-5] are strongly recommended.

APPENDIX G: An Introduction to Theory of Characteristics: Euler Equations

G.1 Introductory Remarks

The Euler equations used for the solution of inviscid flowfields is composed of a system of first-order, nonlinear coupled equations. The system of equations may be written in various forms in terms of either conservative variables, primitive variables, or characteristic variables. In this appendix, the concept of characteristics is extended to the Euler equations. For simplicity, the mathematical manipulation is applied to the one-dimensional Euler equations as a first step and, subsequently, the results for the two-dimensional case are provided.

G.2 One-Dimensional Euler Equations

Consider the one-dimensional Euler equations in conservative form given by

$$\frac{\partial \rho}{\partial t} + \frac{\partial}{\partial x}(\rho u) = 0$$

$$\frac{\partial}{\partial t}(\rho u) + \frac{\partial}{\partial x}(\rho u^2 + p) = 0$$

$$\frac{\partial}{\partial t}(\rho e_t) + \frac{\partial}{\partial x}[(\rho e_t + p)\, u] = 0$$

which, in a vector form, may be expressed as

$$\frac{\partial Q}{\partial t} + \frac{\partial E}{\partial x} = 0 \qquad \text{(G-1)}$$

where

$$Q = \begin{bmatrix} \rho \\ \rho u \\ \rho e_t \end{bmatrix} \qquad E = \begin{bmatrix} \rho u \\ \rho u^2 + p \\ (\rho e_t + p) u \end{bmatrix}$$

Linearization of Equation (G-1) yields

$$\frac{\partial Q}{\partial t} + A \frac{\partial Q}{\partial x} = 0 \tag{G-2}$$

where the Jacobian matrix $A = \partial E / \partial Q$ is given by

$$A = \begin{bmatrix} 0 & 1 & 0 \\ \frac{1}{2}(\gamma - 3) u^2 & -(\gamma - 3)\, u & \gamma - 1 \\ -\gamma u e_t + (\gamma - 1)\, u^3 & \gamma e_t - \frac{3}{2}(\gamma - 1)\, u^2 & \gamma u \end{bmatrix}$$

The Euler equations can also be written in terms of the primitive variables ρ, u, and p by the following equations:

$$\frac{\partial \rho}{\partial t} + u \frac{\partial \rho}{\partial x} + \rho \frac{\partial u}{\partial x} = 0$$

$$\frac{\partial u}{\partial t} + u \frac{\partial u}{\partial x} + \frac{1}{\rho} \frac{\partial p}{\partial x} = 0$$

$$\frac{\partial p}{\partial t} + u \frac{\partial p}{\partial x} + \rho a^2 \frac{\partial u}{\partial x} = 0$$

The system of equations is written in a vector form as

$$\frac{\partial Q'}{\partial t} + A' \frac{\partial Q'}{\partial x} = 0 \tag{G-3}$$

where

$$Q' = \begin{bmatrix} \rho \\ u \\ p \end{bmatrix} \quad \text{and} \quad A' = \begin{bmatrix} u & \rho & 0 \\ 0 & u & \frac{1}{\rho} \\ 0 & \rho a^2 & u \end{bmatrix}$$

Obviously, since Equations (G-2) and (G-3) represent the same physical laws, one may relate the two systems by a similarity transformation. Indeed, the coefficient matrix of Equation (G-2) may be diagonalized under this transformation. It will be shown shortly that the elements of the diagonal matrix are the eigenvalues of matrix A.

Proceeding with similarity transformation, Equation (G-2) can be rewritten as

$$\frac{\partial Q}{\partial Q'}\frac{\partial Q'}{\partial t} + A\frac{\partial Q}{\partial Q'}\frac{\partial Q'}{\partial x} = 0 \tag{G-4}$$

Define

$$\frac{\partial Q}{\partial Q'} = M$$

Then Equation (G-4) may be expressed as

$$\frac{\partial Q'}{\partial t} + M^{-1}AM\frac{\partial Q'}{\partial x} = 0 \tag{G-5}$$

Comparison of Equations (G-3) and (G-5) yields

$$A' = M^{-1}AM$$

where

$$M = \frac{\partial Q}{\partial Q'} = \begin{bmatrix} 1 & 0 & 0 \\ u & \rho & 0 \\ \frac{1}{2}u^2 & \rho u & \frac{1}{\gamma - 1} \end{bmatrix}$$

and

$$M^{-1} = \frac{\partial Q'}{\partial Q} = \begin{bmatrix} 1 & 0 & 0 \\ -\frac{u}{\rho} & \frac{1}{\rho} & 0 \\ \frac{1}{2}(\gamma - 1)\,u^2 & -(\gamma - 1)\,u & (\gamma - 1) \end{bmatrix}$$

The eigenvalues of matrix A or matrix A' can easily be determined to be $\lambda_1 = u$, $\lambda_2 = u + a$, and $\lambda_3 = u - a$. As expected, matrices A and A' have the same eigenvalues because of the similarity transformation. Now, define a diagonal matrix D, composed of the eigenvalues of A, i.e.,

$$D = \begin{bmatrix} u & 0 & 0 \\ 0 & u + a & 0 \\ 0 & 0 & u - a \end{bmatrix} \tag{G-6}$$

The significance of matrix D will be shown shortly.

At this point, a simple procedure is introduced by which the characteristic variables may be identified. Define a vector for the characteristic variables by Q''. Recall Equation (G-3) given by

$$\frac{\partial Q'}{\partial t} + A' \frac{\partial Q'}{\partial x} = 0$$

Rewrite the equation as

$$\frac{\partial Q'}{\partial Q''} \frac{\partial Q''}{\partial t} + A' \frac{\partial Q'}{\partial Q''} \frac{\partial Q''}{\partial x} = 0 \tag{G-7}$$

Define the matrix

$$R = \frac{\partial Q'}{\partial Q''} \tag{G-8}$$

Then Equation (G-7) can be expressed as

$$\frac{\partial Q''}{\partial t} + R^{-1} A' R \frac{\partial Q''}{\partial x} = 0$$

where it can be shown that

$$R^{-1} A' R = D$$

or

$$\frac{\partial Q''}{\partial t} + D \frac{\partial Q''}{\partial x} = 0 \tag{G-9}$$

Several points need to be explored at this time. First, the coefficient matrix D is a diagonal matrix which was defined by (G-6). Second, the matrix defined by (G-8) is obtained by the right eigenvectors of A'. Similarly, R^{-1} can be formed by the three left eigenvectors of A', which is determined to be

$$R^{-1} = \begin{bmatrix} \alpha & 0 & -\dfrac{\alpha}{a^2} \\ 0 & \beta & \dfrac{\beta}{\rho a} \\ 0 & \delta & -\dfrac{\delta}{\rho a} \end{bmatrix}$$

where α, β, and δ are arbitrary normalization coefficients. For example, the following values are commonly used:

$$\alpha = 1, \quad \beta = \frac{1}{\sqrt{2}}, \quad \text{and} \quad \delta = -\frac{1}{\sqrt{2}}$$

With coefficients specified above, the matrix R and its inverse are written as

$$R = \begin{bmatrix} 1 & \frac{1}{\sqrt{2}}\frac{\rho}{a} & \frac{1}{\sqrt{2}}\frac{\rho}{a} \\ 0 & \frac{1}{\sqrt{2}} & -\frac{1}{\sqrt{2}} \\ 0 & \frac{1}{\sqrt{2}}\rho a & \frac{1}{\sqrt{2}}\rho a \end{bmatrix} \tag{G-10}$$

and

$$R^{-1} = \begin{bmatrix} 1 & 0 & -\frac{1}{a^2} \\ 0 & \frac{1}{\sqrt{2}} & \frac{1}{\sqrt{2}}\frac{1}{\rho a} \\ 0 & -\frac{1}{\sqrt{2}} & \frac{1}{\sqrt{2}}\frac{1}{\rho a} \end{bmatrix} \tag{G-11}$$

Now, one may proceed to determine the characteristic variables from (G-8) as follows,

$$Q'' = \int R^{-1} dQ'$$

or

$$Q'' = \int \begin{bmatrix} 1 & 0 & -\frac{1}{a^2} \\ 0 & \frac{1}{\sqrt{2}} & \frac{1}{\sqrt{2}}\frac{1}{\rho a} \\ 0 & -\frac{1}{\sqrt{2}} & \frac{1}{\sqrt{2}}\frac{1}{\rho a} \end{bmatrix} \begin{bmatrix} d\rho \\ du \\ dp \end{bmatrix} = \begin{bmatrix} \int \left(d\rho - \frac{dp}{a^2}\right) \\ \frac{1}{\sqrt{2}} \int \left(du + \frac{dp}{\rho a}\right) \\ -\frac{1}{\sqrt{2}} \int \left(du - \frac{dp}{\rho a}\right) \end{bmatrix} = \begin{bmatrix} \int \left(d\rho - \frac{dp}{a^2}\right) \\ \frac{1}{\sqrt{2}} \left(u + \frac{2a}{\gamma - 1}\right) \\ -\frac{1}{\sqrt{2}} \left(u - \frac{2a}{\gamma - 1}\right) \end{bmatrix} \tag{G-12}$$

The integration above is performed by the implementation of relations (A-34a) and (A-34b) of Appendix A. Recall that the expressions

$$\left(u + \frac{2a}{\gamma - 1}\right) \quad \text{and} \quad \left(u - \frac{2a}{\gamma - 1}\right)$$

previously given by (A-35a) and (A-35b) are called the Riemann invariants.

To clearly identify a typical characteristic form of the one-dimensional Euler equation, expand the vector equation (G-9) and rewrite it as

$$\frac{\partial \rho}{\partial t} - \frac{1}{a^2}\frac{\partial p}{\partial t} + u\frac{\partial \rho}{\partial x} - \frac{u}{a^2}\frac{\partial p}{\partial x} = 0 \tag{G-13}$$

$$\frac{\partial}{\partial t}\left(u + \frac{2a}{\gamma - 1}\right) + (u + a)\frac{\partial}{\partial x}\left(u + \frac{2a}{\gamma - 1}\right) = 0 \tag{G-14}$$

$$\frac{\partial}{\partial t}\left(u - \frac{2a}{\gamma - 1}\right) + (u - a)\frac{\partial}{\partial x}\left(u - \frac{2a}{\gamma - 1}\right) = 0 \tag{G-15}$$

Equations (G-13) through (G-15) represent the characteristic form of the Euler equations, i.e., governing equations are expressed in terms of the characteristic variables defined by the components of vector Q''.

The characteristic equations along which the characteristic variables propagate are given by

$$\frac{dx}{dt} = u \tag{G-16}$$

$$\frac{dx}{dt} = u + a \tag{G-17}$$

$$\frac{dx}{dt} = u - a \tag{G-18}$$

Schematically, the characteristic lines are shown in Figure (G-1).

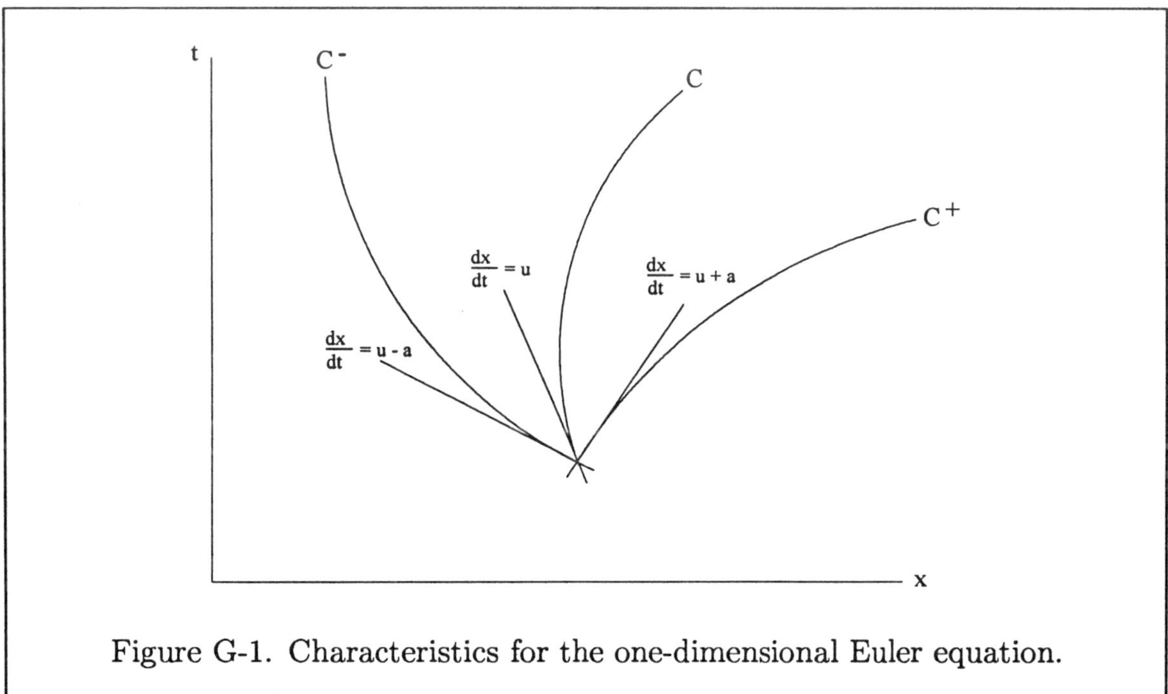

Figure G-1. Characteristics for the one-dimensional Euler equation.

To reemphasize, recall that the quantities represented by the characteristic variables propagate along the characteristic lines. For example, the characteristic quantity $[u + 2a/(\gamma - 1)]$ is propagated along the C^+ characteristic line with the speed of $u + a$. Similarly, the characteristic quantities $(d\rho - dp/a^2)$ and $[u - 2a/(\gamma - 1)]$ are propagated along characteristics C and C^- with speeds of u and $u - a$, respectively. Observe that the C characteristics are coincident with the stream lines of the flow.

G.3 Extension to Two Dimensions

Recall the governing equations for an inviscid flow in the computational space given by Equation (12-126), where for a planar two-dimensional flow it is reduced to

$$\frac{\partial \bar{Q}}{\partial \tau} + \frac{\partial \bar{E}}{\partial \xi} + \frac{\partial \bar{F}}{\partial \eta} = 0 \tag{G-19}$$

Linearization with the help of expressions (12-132) and (12-133) provides

$$\frac{\partial \bar{Q}}{\partial \tau} + A\frac{\partial \bar{Q}}{\partial \xi} + B\frac{\partial \bar{Q}}{\partial \eta} = 0 \tag{G-20}$$

where the matrices A and B are given by (12-136) and (12-137), respectively. Now consider the similarity transformation

$$\frac{\partial \bar{Q}}{\partial \bar{Q}'}\frac{\partial \bar{Q}'}{\partial \tau} + A\frac{\partial \bar{Q}}{\partial \bar{Q}'}\frac{\partial \bar{Q}'}{\partial \xi} + B\frac{\partial \bar{Q}}{\partial \bar{Q}'}\frac{\partial \bar{Q}'}{\partial \eta} = 0 \tag{G-21}$$

where $\bar{Q}'$ represents the vector of primitive variables as given by

$$\bar{Q}' = \frac{1}{J}\begin{bmatrix} \rho \\ u \\ v \\ p \end{bmatrix}$$

and, as previously defined,

$$\frac{\partial \bar{Q}}{\partial \bar{Q}'} = M$$

Therefore, Equation (G-21) is written as

$$M\frac{\partial \bar{Q}'}{\partial \tau} + AM\frac{\partial \bar{Q}'}{\partial \xi} + BM\frac{\partial \bar{Q}'}{\partial \eta} = 0$$

or

$$\frac{\partial \bar{Q}'}{\partial \tau} + M^{-1}AM\frac{\partial \bar{Q}'}{\partial \xi} + M^{-1}BM\frac{\partial \bar{Q}'}{\partial \eta} = 0 \tag{G-22}$$

Rewrite Equation (G-22) as

$$\frac{\partial \bar{Q}'}{\partial \tau} + A'\frac{\partial \bar{Q}'}{\partial \xi} + B'\frac{\partial \bar{Q}'}{\partial \eta} = 0 \tag{G-23}$$

where

$$A' = M^{-1} A M \tag{G-24a}$$

$$B' = M^{-1} B M \tag{G-24b}$$

and

$$M = \frac{\partial \bar{Q}}{\partial \bar{Q}'} = \begin{bmatrix} 1 & 0 & 0 & 0 \\ u & \rho & 0 & 0 \\ v & 0 & \rho & 0 \\ \frac{1}{2}(u^2+v^2) & \rho u & \rho v & \frac{1}{(\gamma-1)} \end{bmatrix} \tag{G-25}$$

and

$$M^{-1} = \frac{\partial \bar{Q}'}{\partial \bar{Q}} = \begin{bmatrix} 1 & 0 & 0 & 0 \\ -\frac{u}{\rho} & \frac{1}{\rho} & 0 & 0 \\ -\frac{v}{\rho} & 0 & \frac{1}{\rho} & 0 \\ \frac{1}{2}(\gamma-1)(u^2+v^2) & -(\gamma-1)u & -(\gamma-1)v & (\gamma-1) \end{bmatrix} \tag{G-26}$$

The eigenvalues of A' and B' are determined to be

$$\lambda_{\xi 1} = \lambda_{\xi 2} = \xi_x u + \xi_y v \tag{G-27a}$$

$$\lambda_{\xi 3} = \xi_x u + \xi_y v + a\sqrt{\xi_x^2 + \xi_y^2} \tag{G-27b}$$

$$\lambda_{\xi 4} = \xi_x u + \xi_y v - a\sqrt{\xi_x^2 + \xi_y^2} \tag{G-27c}$$

and

$$\lambda_{\eta 1} = \lambda_{\eta 2} = \eta_x u + \eta_y v \tag{G-28a}$$

$$\lambda_{\eta 3} = \eta_x u + \eta_y v + a\sqrt{\eta_x^2 + \eta_y^2} \tag{G-28b}$$

$$\lambda_{\eta 4} = \eta_x u + \eta_y v - a\sqrt{\eta_x^2 + \eta_y^2} \tag{G-28c}$$

which are identical to the eigenvalues of A given by relations (12-138) through (12-141), and the eigenvalues of B given by Equations (12-144) through (12-147).

To proceed with the determination of characteristic variables, consider Equation (G-23) and introduce the characteristic variables vector associated with ξ, $\bar{Q}''_\xi$ as follows.

$$\frac{\partial \bar{Q}'}{\partial \bar{Q}''_\xi}\frac{\partial \bar{Q}''_\xi}{\partial \tau} + A'\frac{\partial \bar{Q}'}{\partial \bar{Q}''_\xi}\frac{\partial \bar{Q}''_\xi}{\partial \xi} + B'\frac{\partial \bar{Q}'}{\partial \eta} = 0 \tag{G-29}$$

which is equivalent to

$$\frac{\partial \bar{Q}''_\xi}{\partial \tau} + D_\xi \frac{\partial \bar{Q}''_\xi}{\partial \xi} + R_\xi^{-1} B \frac{\partial \bar{Q}'}{\partial \eta} = 0 \tag{G-30}$$

where

$$D_\xi = \begin{bmatrix} \lambda_{\xi 1} & & & \\ & \lambda_{\xi 2} & & \\ & & \lambda_{\xi 3} & \\ & & & \lambda_{\xi 4} \end{bmatrix} \tag{G-31}$$

Furthermore, recall that

$$D_\xi = R_\xi^{-1} A' R_\xi \tag{G-32}$$

where R_ξ is the matrix of right eigenvectors of A', and is given by

$$R_\xi = \begin{bmatrix} 1 & 0 & \frac{1}{\sqrt{2}}\frac{\rho}{a} & \frac{1}{\sqrt{2}}\frac{\rho}{a} \\ 0 & K_{\xi y} & \frac{1}{\sqrt{2}}K_{\xi x} & -\frac{1}{\sqrt{2}}K_{\xi x} \\ 0 & -K_{\xi x} & \frac{1}{\sqrt{2}}K_{\xi y} & -\frac{1}{\sqrt{2}}K_{\xi y} \\ 0 & 0 & \frac{1}{\sqrt{2}}\rho a & \frac{1}{\sqrt{2}}\rho a \end{bmatrix} \tag{G-33}$$

where

$$K_{\xi x} = \frac{\xi_x}{(\xi_x^2 + \xi_y^2)^{\frac{1}{2}}} \tag{G-34}$$

$$K_{\xi y} = \frac{\xi_y}{(\xi_x^2 + \xi_y^2)^{\frac{1}{2}}} \tag{G-35}$$

The inverse of R_ξ which is the matrix of left eigenvectors of A' is

$$R_\xi^{-1} = \begin{bmatrix} 1 & 0 & 0 & -\frac{1}{a^2} \\ 0 & K_{\xi y} & -K_{\xi x} & 0 \\ 0 & \frac{1}{\sqrt{2}}K_{\xi x} & \frac{1}{\sqrt{2}}K_{\xi y} & \frac{1}{\sqrt{2}}\frac{1}{\rho a} \\ 0 & -\frac{1}{\sqrt{2}}K_{\xi x} & -\frac{1}{\sqrt{2}}K_{\xi y} & \frac{1}{\sqrt{2}}\frac{1}{\rho a} \end{bmatrix} \tag{G-36}$$

The characteristic variables associated with A' are now determined as follows,

$$\bar{Q}'' = \int R_\xi^{-1} d\bar{Q}' \tag{G-37}$$

$$\bar{Q}''_\xi = \int \begin{bmatrix} 1 & 0 & 0 & -\frac{1}{a^2} \\ 0 & K_{\xi y} & -K_{\xi x} & 0 \\ 0 & \frac{1}{\sqrt{2}}K_{\xi x} & \frac{1}{\sqrt{2}}K_{\xi y} & \frac{1}{\sqrt{2}}\frac{1}{\rho a} \\ 0 & -\frac{1}{\sqrt{2}}K_{\xi x} & -\frac{1}{\sqrt{2}}K_{\xi y} & \frac{1}{\sqrt{2}}\frac{1}{\rho a} \end{bmatrix} \begin{bmatrix} d\rho \\ du \\ dv \\ dp \end{bmatrix}$$

$$= \begin{bmatrix} \int \left(d\rho - \frac{dp}{a^2} \right) \\ \int \left[(\xi_y du - \xi_x dv)/(\xi_x^2 + \xi_y^2)^{\frac{1}{2}} \right] \\ \frac{1}{\sqrt{2}} \int \left[(\xi_x du + \xi_y dv)/(\xi_x^2 + \xi_y^2)^{\frac{1}{2}} + \frac{dp}{\rho a} \right] \\ -\frac{1}{\sqrt{2}} \int \left[(\xi_x du + \xi_y dv)/(\xi_x^2 + \xi_y^2)^{\frac{1}{2}} - \frac{dp}{\rho a} \right] \end{bmatrix} \tag{G-38}$$

To proceed with the integrations, the following relations are to be used. First, recall the following relation

$$\int \frac{dp}{\rho a} = \frac{2a}{\gamma - 1} \tag{G-39}$$

Second, the velocity components normal and tangent to the lines of constant ξ can be shown to be

$$V_{n\xi} = \frac{\xi_x u + \xi_y v}{(\xi_x^2 + \xi_y^2)^{\frac{1}{2}}} \tag{G-40}$$

and

$$V_{t\xi} = \frac{\xi_y u - \xi_x v}{(\xi_x^2 + \xi_y^2)^{\frac{1}{2}}} \tag{G-41}$$

The mathematical details to obtain relations (G-40) and (G-41) are provided in Section G.4. Return to relation (G-38) and substitute expressions (G-39) through (G-41) to obtain

$$\bar{Q}''_\xi = \begin{bmatrix} \int \left(d\rho - \frac{dp}{a^2}\right) \\ \int dV_{t\xi} \\ \frac{1}{\sqrt{2}} \int \left(dV_{n\xi} + \frac{dp}{\rho a}\right) \\ -\frac{1}{\sqrt{2}} \int \left(dV_{n\xi} - \frac{dp}{\rho a}\right) \end{bmatrix} = \begin{bmatrix} \int \left(d\rho - \frac{dp}{a^2}\right) \\ V_{t\xi} \\ \frac{1}{\sqrt{2}} \left(V_{n\xi} + \frac{2a}{\gamma - 1}\right) \\ -\frac{1}{\sqrt{2}} \left(V_{n\xi} - \frac{2a}{\gamma - 1}\right) \end{bmatrix} \tag{G-42}$$

The corresponding characteristic equations are

$$\frac{d\xi}{dt} = V_{n\xi} \tag{G-43a}$$

$$\frac{d\xi}{dt} = V_{n\xi} \tag{G-43b}$$

$$\frac{d\xi}{dt} = V_{n\xi} + a \tag{G-43c}$$

$$\frac{d\xi}{dt} = V_{n\xi} - a \tag{G-43d}$$

The significance of Equations (G-42) and (G-43) is reemphasized as follows. The characteristic quantities defined in (G-42) are propagated with the speeds of $V_{n\xi}$, $V_{n\xi} + a$, and $V_{n\xi} - a$ along the characteristic lines defined by Equations (G-43). Schematically, the characteristic lines in the $\xi - t$ plane are shown in Figure G-2.

Similarly, the characteristic variables associated with η, $\bar{Q}''_\eta$ are determined as follow.

First characteristic variable vector $\bar{Q}''_\eta$ is introduced, and Equation (G-23) is written as

$$\frac{\partial \bar{Q}'}{\partial \bar{Q}''_\eta} \frac{\partial \bar{Q}''_\eta}{\partial \tau} + A' \frac{\partial \bar{Q}'}{\partial \xi} + B' \frac{\partial \bar{Q}'}{\partial \bar{Q}''_\eta} \frac{\partial \bar{Q}''_\eta}{\partial \eta} = 0 \tag{G-44}$$

or

$$\frac{\partial Q''_\eta}{\partial \tau} + R_\eta^{-1} A' \frac{\partial \bar{Q}'}{\partial \xi} + D_\eta \frac{\partial \bar{Q}''}{\partial \eta} = 0 \tag{G-45}$$

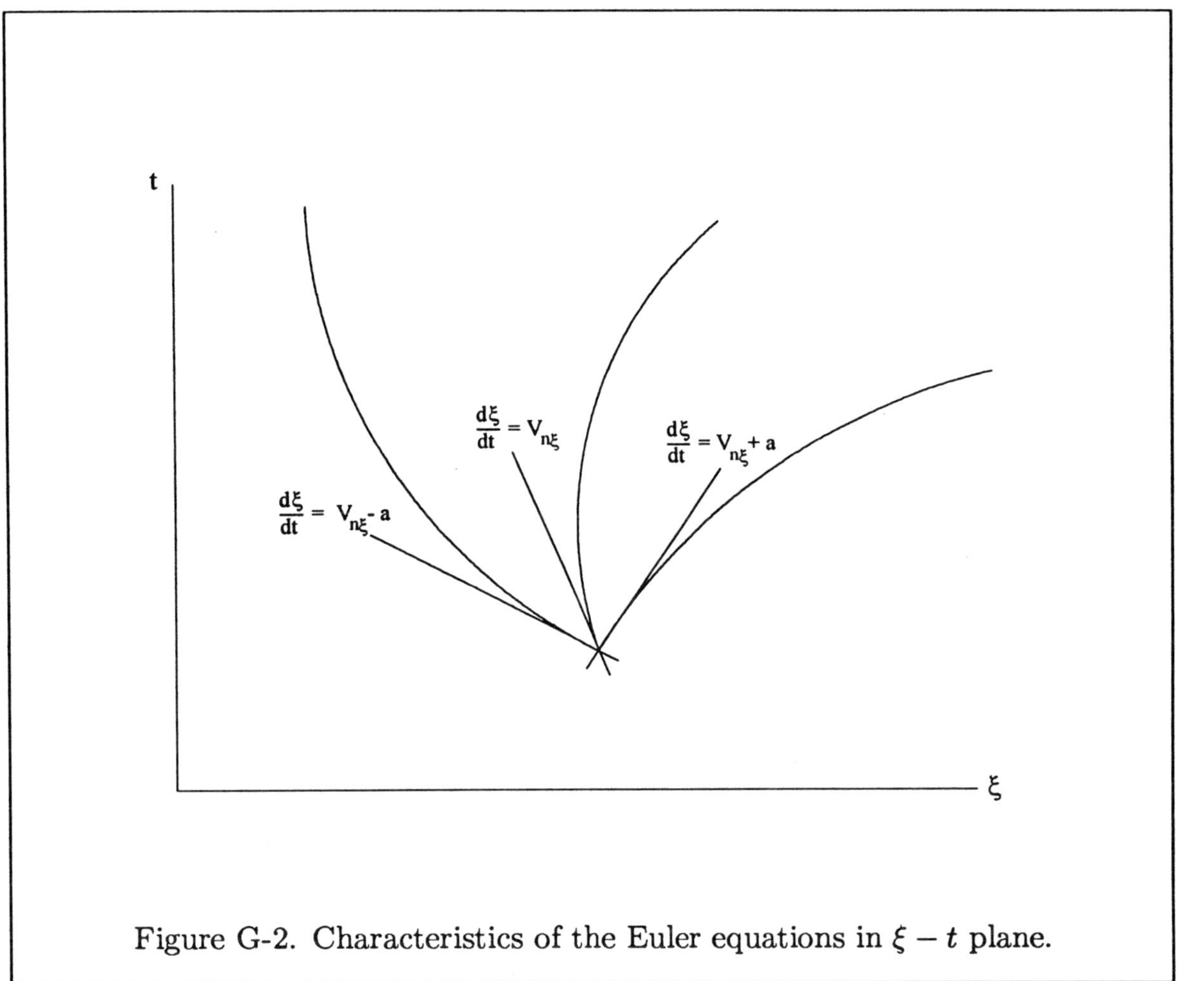

Figure G-2. Characteristics of the Euler equations in $\xi - t$ plane.

where

$$D_n = \begin{bmatrix} \lambda_{\eta_1} & & & \\ & \lambda_{\eta_2} & & \\ & & \lambda_{\eta_3} & \\ & & & \lambda_{\eta_4} \end{bmatrix} \tag{G-46}$$

The matrices R_η and R_η^{-1} have similar forms as R_ξ and R_ξ^{-1}, except $K_{\xi x}$ and $K_{\xi y}$ are replaced by $K_{\eta x}$ and $K_{\eta y}$ defined by

$$K_{\eta x} = \frac{\eta_x}{(\eta_x^2 + \eta_y^2)^{\frac{1}{2}}} \tag{G-47}$$

and

$$K_{\eta y} = \frac{\eta_y}{(\eta_x^2 + \eta_y^2)^{\frac{1}{2}}} \tag{G-48}$$

The characteristic variables associated with B' are now determined as follows.

$$\bar{Q}''_\eta = \int R_\eta^{-1} d\bar{Q}' \tag{G-49}$$

or

$$\bar{Q}''_\eta = \int \begin{bmatrix} 1 & 0 & 0 & -\frac{1}{a^2} \\ 0 & K_{\eta y} & -K_{\eta x} & 0 \\ 0 & \frac{1}{\sqrt{2}} K_{\eta x} & \frac{1}{\sqrt{2}} K_{\eta y} & \frac{1}{\sqrt{2}} \frac{1}{\rho a} \\ 0 & -\frac{1}{\sqrt{2}} K_{\eta x} & -\frac{1}{\sqrt{2}} K_{\eta y} & \frac{1}{\sqrt{2}} \frac{1}{\rho a} \end{bmatrix} \begin{bmatrix} d\rho \\ du \\ dv \\ dp \end{bmatrix}$$

$$= \int \begin{bmatrix} \int \left(d\rho - \frac{dp}{a^2} \right) \\ \int \left[(\eta_y du - \eta_x dv) \,/\, \left(\eta_x^2 + \eta_y^2 \right)^{\frac{1}{2}} \right] \\ \frac{1}{\sqrt{2}} \int \left[(\eta_x du + \eta_y dv) \,/\, \left(\eta_x^2 + \eta_y^2 \right)^{\frac{1}{2}} + \frac{dp}{\rho a} \right] \\ -\frac{1}{\sqrt{2}} \left[(\eta_x du + \eta_y dv) / (\eta_x^2 + \eta_y^2)^{\frac{1}{2}} - \frac{dp}{\rho a} \right] \end{bmatrix} \tag{G-50}$$

which is written as

$$\overline{Q''}_\eta = \begin{bmatrix} \int \left(d\rho - \frac{dp}{a^2} \right) \\ V_{t\eta} \\ \frac{1}{\sqrt{2}} \left(V_{n\eta} + \frac{2a}{\gamma - 1} \right) \\ -\frac{1}{\sqrt{2}} \left(V_{n\eta} - \frac{2a}{\gamma - 1} \right) \end{bmatrix} \tag{G-51}$$

where

$$V_{t\eta} = \frac{\eta_y u - \eta_x v}{\left(\eta_x^2 + \eta_y^2 \right)^{\frac{1}{2}}} \tag{G-52}$$

and

$$V_{n\eta} = \frac{\eta_x u + \eta_y v}{\left(\eta_x^2 + \eta_y^2 \right)^{\frac{1}{2}}} \tag{G-53}$$

The associated characteristic lines are given by

$$\frac{d\eta}{dt} = V_{n\eta} \tag{G-54a}$$

$$\frac{d\eta}{dt} = V_{n\eta} \tag{G-54b}$$

$$\frac{d\eta}{dt} = V_{n\eta} + a \tag{G-54c}$$

$$\frac{d\eta}{dt} = V_{n\eta} - a \tag{G-54d}$$

G.4 Velocity Components

The components of the velocity normal and tangent to the lines of constant ξ and η, which were used in the previous sections, are derived in this section. First, consider lines of constant ξ and designate the normal and tangential components of velocity by $V_{n\xi}$ and $V_{t\xi}$, which are shown schematically in Figure (G-3). The velocity vector and unit normal to the lines of constant ξ are given by

$$\vec{V} = u\vec{i} + v\vec{j}$$

and

$$\vec{n}_\xi = \frac{\nabla\xi}{|\nabla\xi|} = \frac{\xi_x\vec{i} + \xi_y\vec{j}}{\left(\xi_x^2 + \xi_y^2\right)^{\frac{1}{2}}}$$

Now, the normal and tangential components are determined as follows,

$$V_{n\xi} = \vec{V} \cdot \vec{n}_\xi = \frac{\xi_x u + \xi_y v}{\left(\xi_x^2 + \xi_y^2\right)^{\frac{1}{2}}} \tag{G-55}$$

and

$$V_{t\xi} = \left[\vec{V} \cdot \vec{V} - V_{n\xi}^2\right]^{\frac{1}{2}} = \left[\left(u^2 + v^2\right) - \frac{\left(\xi_x u + \xi_y v\right)^2}{\xi_x^2 + \xi_y^2}\right]^{\frac{1}{2}}$$

or

$$V_{t\xi} = \frac{\xi_y u - \xi_x v}{\left(\xi_x^2 + \xi_y^2\right)^{\frac{1}{2}}} \tag{G-56}$$

Similarly, for the lines of constant η, one has

$$\vec{n}_\eta = \frac{\eta_x\vec{i} + \eta_y\vec{j}}{\left(\eta_x^2 + \eta_y^2\right)^{\frac{1}{2}}}$$

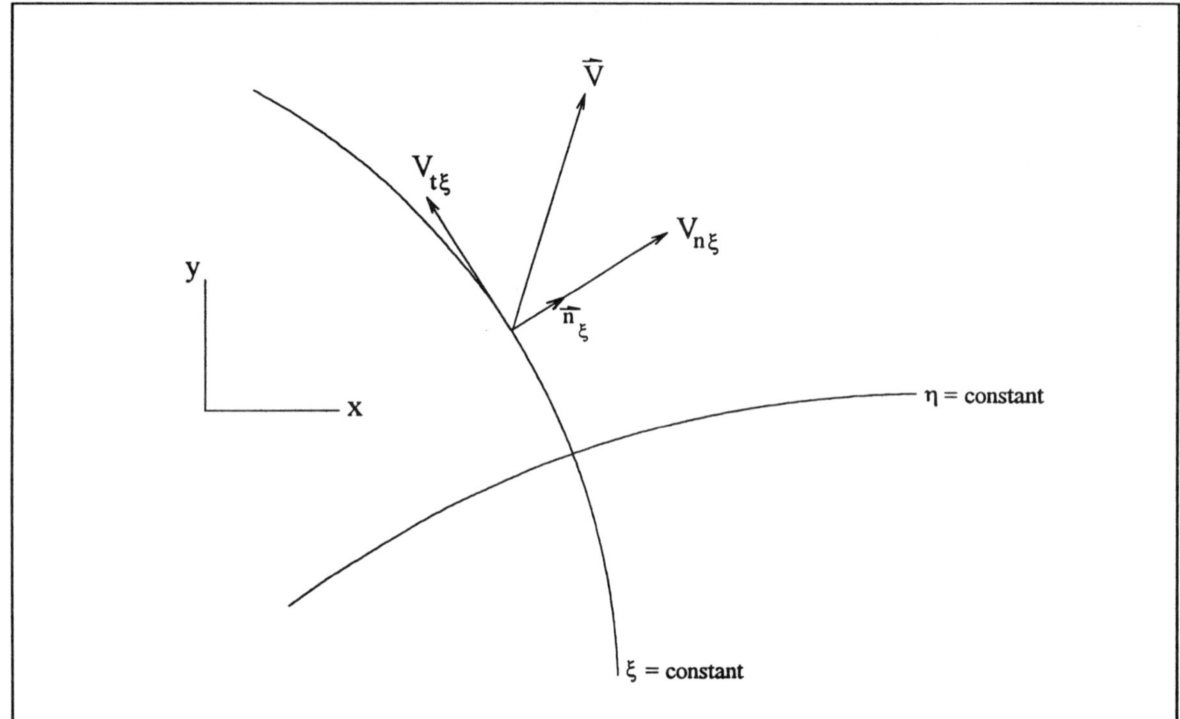

Figure G-3. Illustration of the normal and tangential velocity components.

Therefore,

$$V_{n\eta} = \frac{\eta_x u + \eta_y v}{\left(\eta_x^2 + \eta_y^2\right)^{\frac{1}{2}}} \tag{G-57}$$

and

$$V_{t\eta} = \frac{\eta_y u - \eta_x v}{\left(\eta_x^2 + \eta_y^2\right)^{\frac{1}{2}}} \tag{G-58}$$

G.5 Specification of Boundary Conditions

In this section, the specification of the boundary conditions associated with characteristic variables is reviewed. The numerous physical and/or numerical boundary conditions which must be specified or determined at the boundaries are provided in the appropriate sections of the text.

Before proceeding, consider a brief review of the previous topics related to characteristics. For simplicity, consider a first-order hyperbolic system given by

$$\frac{\partial Q'}{\partial t} + A'\frac{\partial Q'}{\partial x} = 0 \tag{G-59}$$

Since the system is hyperbolic, the $(n \times n)$ matrix A' possesses n real eigenvalues and associated eigenvectors. In the previous section, it was shown that there exists a similarity transformation by which A' is diagonalized, resulting in

$$\frac{\partial Q''}{\partial t} + D\frac{\partial Q''}{\partial x} = 0 \tag{G-60}$$

where

$$D = R^{-1}A'R \tag{G-61}$$

It is emphasized again that the eigenvalues of D are real. Assume D to possess m positive eigenvalues and, therefore, $(n - m)$ negative eigenvalues. Furthermore, assume that the spatial boundaries are located at $x = 0$ and $x = L$. Since there exists m positive eigenvalues along which information is transmitted into the domain at $x = 0$, m boundary conditions at $x = 0$ are required. Similarly, the $(n-m)$ set of data is propagated along the characteristic lines into the domain at $x = L$ location, thus requiring specification of $(n-m)$ boundary conditions at $x = L$. The remaining boundary conditions are usually extrapolated from the interior solution.

To illustrate specification of a typical set of boundary conditions, consider the following simple example. A one-dimensional inviscid flow enters a nozzle with subsonic speed, and it exits subsonically. It was previously shown that the eigenvalues of the system are u, $u + a$, and $u - a$. Therefore, two of the eigenvalues, namely u and $u + a$, are positive at the inflow transmitting information into the domain, thus requiring specification of two boundary conditions at the inflow. Similarly, at the outflow, one characteristic enters the domain from the exterior of the domain along which information is transmitted to the outflow boundary. Therefore, only one boundary condition can be specified at the outflow. The remaining variables at the boundaries must be extrapolated from the interior domain. To proceed with this example, assume that the freestream conditions at the inflow are provided and designate them by "∞." Thus, the following requirements at the inflow for the characteristic variables are specified as

$$R^+ = u_\infty + \frac{2a_\infty}{\gamma - 1} = u + \frac{2a}{\gamma - 1} \tag{G-62}$$

$$R^- = u_e - \frac{2a_e}{\gamma - 1} = u - \frac{2a}{\gamma - 1} \tag{G-63}$$

where e designates the extrapolated value. Now, from relations (G-62) and (G-63), one has

$$u = \frac{1}{2}(R^- + R^+) \tag{G-64}$$

and

$$a^2 = \frac{\gamma - 1}{4}(R^+ - R^-) \tag{G-65}$$

Subsequently, density and pressure may be updated according to

$$\rho = \left(\frac{\gamma}{a^2} \frac{p_\infty}{\rho_\infty^\gamma} \right)^{\frac{1}{\gamma - 1}} \tag{G-66}$$

and

$$p = a^2 \frac{\rho}{\gamma} \tag{G-67}$$

Note that one encounters numerous forms by which boundary conditions may be applied. Just as in the governing equations, some approximations may be required in order to implement the specified boundary conditions. The specification of boundary conditions in the example shown is typical.

APPENDIX H:
Computation of Pressure at the Body Surface

The components of the momentum equation for an axisymmetric flow may be expressed as:

$$\frac{\partial}{\partial t}(\rho u) + \frac{\partial}{\partial x}(\rho u^2 + p) + \frac{\partial}{\partial y}(\rho uv) + \frac{1}{y}\alpha(\rho uv) \quad = \quad 0 \tag{H-1}$$

$$\frac{\partial}{\partial t}(\rho v) + \frac{\partial}{\partial x}(\rho uv) + \frac{\partial}{\partial y}(\rho v^2 + p) + \frac{1}{y}\alpha(\rho v^2) \quad = \quad 0 \tag{H-2}$$

By definition

$$\mathring{m} = \rho \vec{V} \cdot \hat{n} \tag{H-3}$$

and is equal to zero at the surface for a nonporous surface. Now recall that

$$\begin{aligned} \hat{n} &= \frac{\nabla \eta}{|\nabla \eta|} \\ \vec{V} &= (u\hat{\imath} + v\hat{\jmath}) \end{aligned}$$

and

$$\nabla \eta = \eta_x \hat{\imath} + \eta_y \hat{\jmath} = J(-y_\xi \hat{\imath} + x_\xi \hat{\jmath})$$

Substitution of these relations into Equation (H-3) yields:

$$\rho v x_\xi - \rho u y_\xi = 0 \tag{H-4}$$

A time derivative of (H-4) provides

$$x_\xi \frac{\partial}{\partial t}(\rho v) - y_\xi \frac{\partial}{\partial t}(\rho u) = 0 \tag{H-5}$$

Note that the grid system has been assumed to be independent of time. Now the following mathematical manipulation is performed. Equation (H-1) is multiplied by y_ξ and subtracted from the product of x_ξ times Equation (H-2). The result is

$$y_\xi \frac{\partial}{\partial x}(\rho u^2 + p) + y_\xi \frac{\partial}{\partial y}(\rho uv) - x_\xi \frac{\partial}{\partial x}(\rho uv) - x_\xi \frac{\partial}{\partial y}(\rho v^2 + p) = 0 \qquad \text{(H-6)}$$

This equation may be rearranged as (with the help of Equation (H-4))

$$y_\xi \left(\rho u \frac{\partial u}{\partial x} + \rho v \frac{\partial u}{\partial y} + \frac{\partial p}{\partial x} \right) - x_\xi \left(\rho u \frac{\partial v}{\partial x} + \rho v \frac{\partial v}{\partial y} + \frac{\partial p}{\partial y} \right) = 0 \qquad \text{(H-7)}$$

Rewrite this equation as

$$y_\xi \left(\rho u \frac{\partial u}{\partial x} + \rho v \frac{\partial u}{\partial y} \right) - x_\xi \left(\rho u \frac{\partial v}{\partial x} + \rho v \frac{\partial v}{\partial y} \right) = -y_\xi \frac{\partial p}{\partial x} + x_\xi \frac{\partial p}{\partial y}$$

Using Equations (9-4) and (9-5), one obtains

$$\begin{aligned} & y_\xi \left[\rho u (\xi_x u_\xi + \eta_x u_\eta) + \rho v (\xi_y u_\xi + \eta_y u_\eta) \right] \\ & \quad - x_\xi \left[\rho u (\xi_x v_\xi + \eta_x v_\eta) + \rho v (\xi_y v_\xi + \eta_y v_\eta) \right] = \\ & \qquad - y_\xi (\xi_x p_\xi + \eta_x p_\eta) + x_\xi (\xi_y p_\xi + \eta_y p_\eta) \end{aligned}$$

Now equations for the metrics (e.g., Equations (9-14) through (9-17)) and Equation (H-4), i.e., $u y_\xi - v x_\xi = 0$, are employed to provide

$$\begin{aligned} & J\rho \left[(-u_\xi y_\xi + x_\xi v_\xi)(x_\eta v - u y_\eta) \right] = \\ & \qquad (-\xi_x y_\xi + x_\xi \xi_y) p_\xi + (-y_\xi \eta_x + x_\xi \eta_y) p_\eta \end{aligned}$$

Resubstitution of relations for the metrics, i.e., Equations (9-14) through (9-17), yields

$$\begin{aligned} & J\rho \left[-u_\xi \left(-\frac{\eta_x}{J} \right) + v_\xi \left(\frac{\eta_y}{J} \right) \right] \left[\left(-\frac{\xi_y}{J} \right) v - \left(\frac{\xi_x}{J} \right) u \right] \\ & \quad = \left[-\xi_x \left(-\frac{\eta_x}{J} \right) + \xi_y \left(\frac{\eta_y}{J} \right) \right] p_\xi + \left[- \left(-\frac{\eta_x}{J} \right) (\eta_x) + \left(\frac{\eta_y}{J} \right) (\eta_y) \right] p_\eta \end{aligned}$$

which may be simplified to

$$-\frac{\rho}{J}(u\xi_x + v\xi_y)(u_\xi \eta_x + v_\xi \eta_y) = \frac{1}{J}(\xi_x \eta_x + \xi_y \eta_y) p_\xi + \frac{1}{J}(\eta_x^2 + \eta_y^2) p_\eta \qquad \text{(H-8)}$$

This equation may be expressed in a conservative form by the addition of some "zero" terms. The procedure closely follows that of Chapter 11. The LHS of the equation is considered first.

The lefthand side of Equation (H-8), in terms of the contravariant velocity U, is (excluding the minus sign which will be included at the final result)

$$\text{LHS} = \frac{\rho U}{J}(u_\xi \eta_x + v_\xi \eta_y)$$

A zero term is added to provide

$$\text{LHS} = \frac{\rho U}{J}(u_\xi \eta_x + v_\xi \eta_y) + V\left[\left(\frac{\rho U}{J}\right)_\xi + \left(\frac{\rho V}{J}\right)_\eta\right]$$

Note that the added term is zero because V is zero at the surface. Thus,

$$\begin{aligned}
\text{LHS} &= \frac{\rho U}{J}(u_\xi \eta_x + v_\xi \eta_y) + (u\eta_x + v\eta_y)\left[\left(\frac{\rho U}{J}\right)_\xi + \left(\frac{\rho V}{J}\right)_\eta\right] \\
&= u\eta_x\left[\left(\frac{\rho U}{J}\right)_\xi + \left(\frac{\rho V}{J}\right)_\eta\right] + \frac{\rho U}{J}(u_\xi \eta_x + v_\xi \eta_y) \\
&\quad + v\eta_y\left[\left(\frac{\rho U}{J}\right)_\xi + \left(\frac{\rho V}{J}\right)_\eta\right]
\end{aligned}$$

Now the zero terms $(\rho V/J)\, u_\eta$ and $(\rho V/J)\, v_\eta$ are added to yield:

$$\begin{aligned}
\text{LHS} &= \eta_x\left[u\left(\frac{\rho U}{J}\right)_\xi + u_\xi\left(\frac{\rho U}{J}\right)\right] + \eta_y\left[v\left(\frac{\rho U}{J}\right)_\xi + v_\xi\left(\frac{\rho U}{J}\right)\right] \\
&\quad + \eta_x\left[u\left(\frac{\rho V}{J}\right)_\eta + u_\eta\left(\frac{\rho V}{J}\right)\right] + \eta_y\left[v\left(\frac{\rho V}{J}\right)_\eta + v_\eta\left(\frac{\rho V}{J}\right)\right] \\
&= \eta_x\left[\frac{\rho u U}{J}\right]_\xi + \eta_y\left[\frac{\rho v U}{J}\right]_\xi + \eta_x\left[\frac{\rho u V}{J}\right]_\eta + \eta_y\left[\frac{\rho v V}{J}\right]_\eta
\end{aligned} \tag{H-9}$$

Now, the RHS is modified. Recall that

$$\text{RHS} = \frac{1}{J}(\xi_x \eta_x + \xi_y \eta_y)p_\xi + \frac{1}{J}(\eta_x^2 + \eta_y^2)p_\eta$$

Add the following zero term

$$p\left\{\eta_x\left[\left(\frac{\xi_x}{J}\right)_\xi + \left(\frac{\eta_x}{J}\right)_\eta\right] + \eta_y\left[\left(\frac{\xi_y}{J}\right)_\xi + \left(\frac{\eta_y}{J}\right)_\eta\right]\right\} = 0$$

Hence,

$$\text{RHS} = \eta_x\left[\left(\frac{\xi_x}{J}\right)_\xi p + \left(\frac{\xi_x}{J}\right)p_\xi\right] + \eta_x\left[\left(\frac{\eta_x}{J}\right)_\eta p + \left(\frac{\eta_x}{J}\right)p_\eta\right] +$$

$$+ \eta_y \left[\left(\frac{\xi_y}{J} \right)_\xi p + \left(\frac{\xi_y}{J} \right) p_\xi \right] + \eta_y \left[\left(\frac{\eta_y}{J} \right)_\eta p + \left(\frac{\eta_y}{J} \right) p_\eta \right]$$

$$= \eta_x \left(\frac{\xi_x p}{J} \right)_\xi + \eta_x \left(\frac{\eta_x p}{J} \right)_\eta + \eta_y \left(\frac{\xi_y p}{J} \right)_\xi + \eta_y \left(\frac{\eta_y p}{J} \right)_\eta \tag{H-10}$$

Therefore, the conservative form of Equation (H-6) is expressed as

$$\eta_x \left(\frac{\rho u U}{J} \right)_\xi + \eta_y \left(\frac{\rho v U}{J} \right)_\xi + \eta_x \left(\frac{\rho u V}{J} \right)_\eta + \eta_y \left(\frac{\rho v V}{J} \right)_\eta$$

$$+ \eta_x \left(\frac{\xi_x p}{J} \right)_\xi + \eta_x \left(\frac{\eta_x p}{J} \right)_\eta + \eta_y \left(\frac{\xi_y p}{J} \right)_\xi + \eta_y \left(\frac{\eta_y p}{J} \right)_\eta = 0 \tag{H-11}$$

In order to obtain a finite difference equation for (H-11), a second-order central difference approximation for the ξ derivatives and a second-order one-sided difference approximation for the η derivatives is used. Note that the unknowns are the values of pressure at the surface, i.e., $j = 1$. The values at the interior points have already been computed. The grid points involved are illustrated in Figure H-1. With the second-order approximation described above, the FDE is obtained as

$$\frac{\eta_{x_{i,1}}}{2\Delta\xi} \left[\left(\frac{\rho u U}{J} \right)_{i+1,1} - \left(\frac{\rho u U}{J} \right)_{i-1,1} \right] + \frac{\eta_{y_{i,1}}}{2\Delta\xi} \left[\left(\frac{\rho u U}{J} \right)_{i+1,1} - \left(\frac{\rho u U}{J} \right)_{i-1,1} \right]$$

$$+ \frac{\eta_{x_{i,1}}}{2\Delta\eta} \left[-3 \left(\frac{\rho u V}{J} \right)_{i,1} + 4 \left(\frac{\rho u V}{J} \right)_{i,2} - \left(\frac{\rho u V}{J} \right)_{i,3} \right]$$

$$+ \frac{\eta_{y_{i,1}}}{2\Delta\eta} \left[-3 \left(\frac{\rho v V}{J} \right)_{i,1} + 4 \left(\frac{\rho v V}{J} \right)_{i,2} - \left(\frac{\rho v V}{J} \right)_{i,3} \right]$$

$$+ \frac{\eta_{x_{i,1}}}{2\Delta\xi} \left[\left(\frac{\xi_x p}{J} \right)_{i+1,1} - \left(\frac{\xi_x p}{J} \right)_{i-1,1} \right] + \frac{\eta_{x_{i,1}}}{2\Delta\eta} \left[-3 \left(\frac{\eta_x p}{J} \right)_{i,1} + 4 \left(\frac{\eta_x p}{J} \right)_{i,2} \right.$$

$$\left. - \left(\frac{\eta_x p}{J} \right)_{i,3} \right] + \frac{\eta_{y_{i,1}}}{2\Delta\xi} \left[\left(\frac{\xi_y p}{J} \right)_{i+1,1} - \left(\frac{\xi_y p}{J} \right)_{i-1,1} \right]$$

$$+ \frac{\eta_{y_{i,1}}}{2\Delta\eta} \left[-3 \left(\frac{\eta_y p}{J} \right)_{i,1} + 4 \left(\frac{\eta_y p}{J} \right)_{i,2} - \left(\frac{\eta_y p}{J} \right)_{i,3} \right] = 0 \tag{H-12}$$

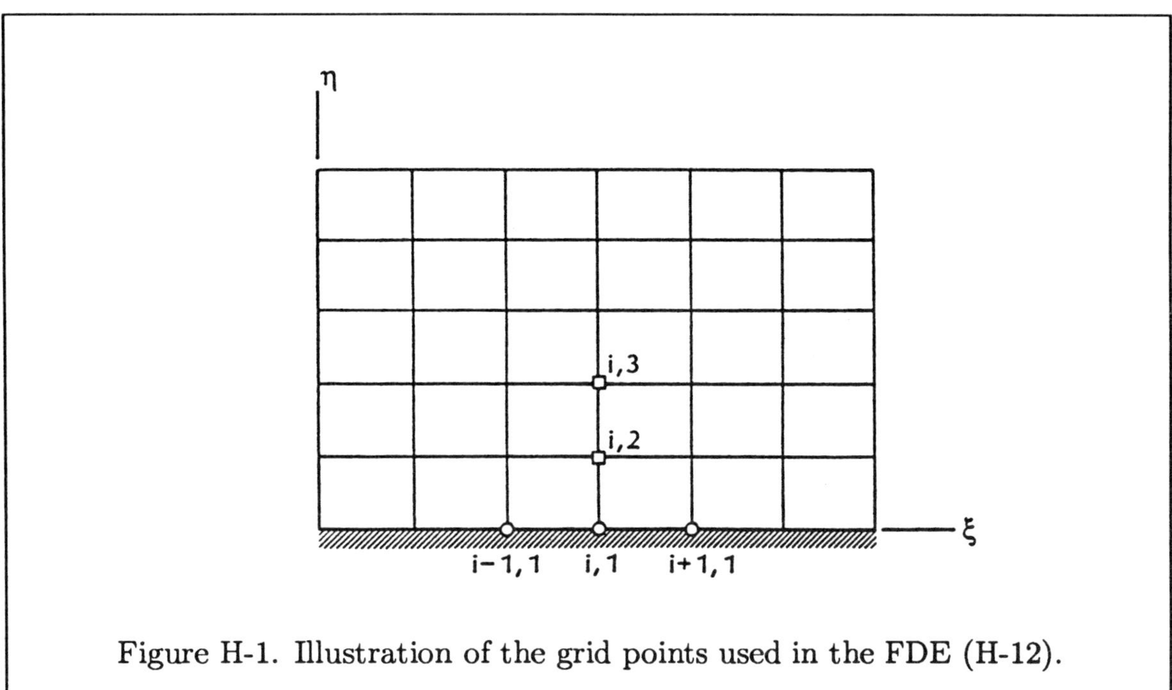

Figure H-1. Illustration of the grid points used in the FDE (H-12).

Note that V is zero at the surface where $j = 1$ and, therefore, those terms are dropped. Now this equation is regrouped so that a tridiagonal system is formed. The rearrangement is as follows:

$$a_i p_{i-1,1} + b_i p_{i,1} + c_i p_{i+1,1} = d_i \tag{H-13}$$

where

$$a_i = -\frac{1}{2\Delta\xi}\left[\eta_{x_{i,1}}\left(\frac{\xi_x}{J}\right)_{i-1,1} + \eta_{y_{i,1}}\left(\frac{\xi_y}{J}\right)_{i-1,1}\right]$$

$$b_i = -\frac{3}{2\Delta\eta}\left[\frac{\eta_x^2 + \eta_y^2}{J}\right]_{i,1}$$

$$c_i = \frac{1}{2\Delta\xi}\left[\eta_{x_{i,1}}\left(\frac{\xi_x}{J}\right)_{i+1,1} + \eta_{y_{i,1}}\left(\frac{\xi_y}{J}\right)_{i+1,1}\right]$$

$$\begin{aligned} d_i = {} & \frac{p_{i,3}}{2\Delta\eta}\left[\eta_{x_{i,1}}\left(\frac{\eta_x}{J}\right)_{i,3} + \eta_{y_{i,1}}\left(\frac{\eta_y}{J}\right)_{i,3}\right] \\ & - \frac{2p_{i,2}}{\Delta\eta}\left[\eta_{x_{i,1}}\left(\frac{\eta_x}{J}\right)_{i,2} + \eta_{y_{i,1}}\left(\frac{\eta_y}{J}\right)_{i,2}\right] \\ & - \frac{2}{\Delta\eta}\left(\frac{\rho V}{J}\right)_{i,2}\left[\eta_{x_{i,1}} u_{i,2} + \eta_{y_{i,1}} v_{i,2}\right] + \end{aligned}$$

$$+\frac{1}{2\Delta\eta}\left(\frac{\rho V}{J}\right)_{i,3}\left[\eta_{x_{i,1}}u_{i,3}+\eta_{y_{i,1}}v_{i,3}\right]$$

$$+\frac{1}{2\Delta\xi}\left(\frac{\rho U}{J}\right)_{i-1,1}\left[\eta_{x_{i,1}}u_{i-1,1}+\eta_{y_{i,1}}v_{i-1,1}\right]$$

$$-\frac{1}{2\Delta\xi}\left(\frac{\rho U}{J}\right)_{i+1,1}\left[\eta_{x_{i,1}}u_{i+1,1}+\eta_{y_{i,1}}v_{i+1,1}\right]$$

When Equation (H-13) is applied to all i at $j = 1$, the following tridiagonal system of equations is obtained.:

$$\begin{bmatrix} b_2 & c_2 & & \\ a_3 & b_3 & c_3 & \\ & & & \\ & a_{IMM2} & b_{IMM2} & c_{IMM2} \\ & & a_{IMM1} & b_{IMM1} \end{bmatrix} \begin{bmatrix} p_{2,1} \\ p_{3,1} \\ | \\ p_{IMM2,1} \\ p_{IMM1,1} \end{bmatrix} = \begin{bmatrix} d_2 - a_2 p_{1,1} \\ d_3 \\ | \\ d_{IMM2} \\ d_{IMM1} - c_{IMM1} p_{IM,1} \end{bmatrix}$$

Since $p_{1,1} = p_{2,1}$ and $\dfrac{p_{IM,1}}{J_{IM,1}} = \dfrac{p_{IMM1,1}}{J_{IMM1,1}}$, therefore

$$\begin{bmatrix} \bar{b}_2 & c_2 & & \\ a_3 & b_3 & c_3 & \\ & & & \\ & a_{IMM2} & b_{IMM2} & c_{IMM2} \\ & & a_{IMM1} & \bar{b}_{IMM1} \end{bmatrix} \begin{bmatrix} p_{2,1} \\ p_{3,1} \\ | \\ p_{IMM2,1} \\ p_{IMM1,1} \end{bmatrix} = \begin{bmatrix} d_2 \\ d_3 \\ | \\ d_{IMM2} \\ d_{IMM1} \end{bmatrix} \tag{H-14}$$

where

$$\bar{b}_2 = a_2 + b_2$$

and

$$\bar{b}_{IMM1} = b_{IMM1} + c_{IMM1}\frac{J_{IM,1}}{J_{IMM1,1}}$$

APPENDIX I: Transformation of Turbulence Models from Physical Space to Computational Space

The details of the coordinate transformation for turbulence models are provided in this appendix. Since the terms for most models in each category, e.g., one-equation models, are similar, mathematical steps will be carried out for a typical model only. Extension to similar turbulence models is straightforward.

I.1 Baldwin-Barth Turbulence Model

To better manage the mathematical work, each term in the equation will be considered separately. Again, using the notation R for $\nu \widetilde{Re}_T$, the left-hand side of (17-71) is dR/dt, which is expanded in Cartesian coordinates to yield

$$\frac{dR}{dt} = \frac{\partial R}{\partial t} + \vec{V} \cdot \nabla R = \frac{\partial R}{\partial t} + u\frac{\partial R}{\partial x} + v\frac{\partial R}{\partial y}$$

Now, using the Cartesian derivatives (17-73) and (17-74), the convective term becomes

$$\begin{aligned} u\frac{\partial R}{\partial x} + v\frac{\partial R}{\partial y} &= u\left(\xi_x \frac{\partial R}{\partial \xi} + \eta_x \frac{\partial R}{\partial \eta}\right) + v\left(\xi_y \frac{\partial R}{\partial \xi} + \eta_y \frac{\partial R}{\partial \eta}\right) \\ &= (\xi_x u + \xi_y v)\frac{\partial R}{\partial \xi} + (\eta_x u + \eta_y v)\frac{\partial R}{\partial \eta} = U\frac{\partial R}{\partial \xi} + V\frac{\partial R}{\partial \eta} \end{aligned}$$

The diffusion term in Equation (17-71) involves the Laplacian $\nabla^2 R$ and $\nabla \nu_t \cdot \nabla R$. Each term is treated separately, and details are as follow.

Using the expressions given by (17-75) and (17-76), one has

$$\begin{aligned}
\nabla^2 R &= \frac{\partial^2 R}{\partial x^2} + \frac{\partial^2 R}{\partial y^2} = \xi_x^2 \frac{\partial^2 R}{\partial \xi^2} + 2\xi_x \eta_x \frac{\partial^2 R}{\partial \xi \partial \eta} + \eta_x^2 \frac{\partial^2 R}{\partial \eta^2} + \\
&\quad + \xi_x \left(\frac{\partial \xi_x}{\partial \xi}\frac{\partial R}{\partial \xi} + \frac{\partial \eta_x}{\partial \xi}\frac{\partial R}{\partial \eta} \right) + \eta_x \left(\frac{\partial \xi_x}{\partial \eta}\frac{\partial R}{\partial \xi} + \frac{\partial \eta_x}{\partial \eta}\frac{\partial R}{\partial \eta} \right) \\
&\quad + \xi_y^2 \frac{\partial^2 R}{\partial \xi^2} + 2\xi_y \eta_y \frac{\partial^2 R}{\partial \xi \partial \eta} + \eta_y^2 \frac{\partial^2 R}{\partial \eta^2} \\
&\quad + \xi_y \left(\frac{\partial \xi_y}{\partial \xi}\frac{\partial R}{\partial \xi} + \frac{\partial \eta_y}{\partial \xi}\frac{\partial R}{\partial \eta} \right) + \eta_y \left(\frac{\partial \xi_y}{\partial \eta}\frac{\partial R}{\partial \xi} + \frac{\partial \eta_y}{\partial \eta}\frac{\partial R}{\partial \eta} \right)
\end{aligned}$$

Rearrange terms,

$$\begin{aligned}
\nabla^2 R &= \left(\xi_x^2 + \xi_y^2\right) \frac{\partial^2 R}{\partial \xi^2} + 2\left(\xi_x \eta_x + \xi_y \eta_y\right) \frac{\partial^2 R}{\partial \xi \partial \eta} + \left(\eta_x^2 + \eta_y^2\right) \frac{\partial^2 R}{\partial \eta^2} + \\
&\quad + \xi_x \left(\frac{\partial \xi_x}{\partial \xi}\frac{\partial R}{\partial \xi} + \frac{\partial \eta_x}{\partial \xi}\frac{\partial R}{\partial \eta} \right) + \eta_x \left(\frac{\partial \xi_x}{\partial \eta}\frac{\partial R}{\partial \xi} + \frac{\partial \eta_x}{\partial \eta}\frac{\partial R}{\partial \eta} \right) \\
&\quad + \xi_y \left(\frac{\partial \xi_y}{\partial \xi}\frac{\partial R}{\partial \xi} + \frac{\partial \eta_y}{\partial \xi}\frac{\partial R}{\partial \eta} \right) + \eta_y \left(\frac{\partial \xi_y}{\partial \eta}\frac{\partial R}{\partial \xi} + \frac{\partial \eta_y}{\partial \eta}\frac{\partial R}{\partial \eta} \right) \\
&= \left(\xi_x^2 + \xi_y^2\right) \frac{\partial^2 R}{\partial \xi^2} + 2(\xi_x \eta_x + \xi_y \eta_y) \frac{\partial^2 R}{\partial \xi \partial \eta} + (\eta_x^2 + \eta_y^2) \frac{\partial^2 R}{\partial \eta^2} \\
&\quad + \left(\xi_x \frac{\partial \xi_x}{\partial \xi} + \eta_x \frac{\partial \xi_x}{\partial \eta} + \xi_y \frac{\partial \xi_y}{\partial \xi} + \eta_y \frac{\partial \xi_y}{\partial \eta} \right) \frac{\partial R}{\partial \xi} \\
&\quad + \left(\xi_x \frac{\partial \eta_x}{\partial \xi} + \eta_x \frac{\partial \eta_x}{\partial \eta} + \xi_y \frac{\partial \eta_y}{\partial \xi} + \eta_y \frac{\partial \eta_y}{\partial \eta} \right) \frac{\partial R}{\partial \eta}
\end{aligned}$$

Using previously defined expressions given by (11-210d), (11-211d), and (11-212e) repeated here for convenience, and additional definitions, the Laplacian becomes

$$\nabla^2 R = a_4 \frac{\partial^2 R}{\partial \xi^2} + 2c_5 \frac{\partial^2 R}{\partial \xi \partial \eta} + b_4 \frac{\partial^2 R}{\partial \eta^2} + g_1 \frac{\partial R}{\partial \xi} + g_2 \frac{\partial R}{\partial \eta}$$

where

$$\begin{aligned}
a_4 &= \xi_x^2 + \xi_y^2 \\
b_4 &= \eta_x^2 + \eta_y^2
\end{aligned}$$

$$c_5 = \xi_x\eta_x + \xi_y\eta_y$$

$$g_1 = \xi_x\frac{\partial\xi_x}{\partial\xi} + \eta_x\frac{\partial\xi_x}{\partial\eta} + \xi_y\frac{\partial\xi_y}{\partial\xi} + \eta_y\frac{\partial\xi_y}{\partial\eta}$$

$$g_2 = \xi_x\frac{\partial\eta_x}{\partial\xi} + \eta_x\frac{\partial\eta_x}{\partial\eta} + \xi_y\frac{\partial\eta_y}{\partial\xi} + \eta_y\frac{\partial\eta_y}{\partial\eta}$$

The second term $\nabla\nu_t \cdot \nabla R$ is first expanded as

$$\nabla\nu_t \cdot \nabla R = \frac{\partial\nu_t}{\partial x}\frac{\partial R}{\partial x} + \frac{\partial\nu_t}{\partial y}\frac{\partial R}{\partial y}$$

Now, using the Cartesian derivatives, one has

$$\left(\xi_x\frac{\partial\nu_t}{\partial\xi} + \eta_x\frac{\partial\nu_t}{\partial\eta}\right)\left(\xi_x\frac{\partial R}{\partial\xi} + \eta_x\frac{\partial R}{\partial\eta}\right) + \left(\xi_y\frac{\partial\nu_t}{\partial\xi} + \eta_y\frac{\partial\nu_t}{\partial\eta}\right)\left(\xi_y\frac{\partial R}{\partial\xi} + \eta_y\frac{\partial R}{\partial\eta}\right) =$$

$$\xi_x^2\frac{\partial\nu_t}{\partial\xi}\frac{\partial R}{\partial\xi} + \xi_x\eta_x\frac{\partial\nu_t}{\partial\xi}\frac{\partial R}{\partial\eta} + \eta_x\xi_x\frac{\partial\nu_t}{\partial\eta}\frac{\partial R}{\partial\xi} + \eta_x^2\frac{\partial\nu_t}{\partial\eta}\frac{\partial R}{\partial\eta}$$

$$+ \xi_y^2\frac{\partial\nu_t}{\partial\xi}\frac{\partial R}{\partial\xi} + \xi_y\eta_y\frac{\partial\nu_t}{\partial\xi}\frac{\partial R}{\partial\eta} + \eta_y\xi_y\frac{\partial\nu_t}{\partial\eta}\frac{\partial R}{\partial\xi} + \eta_y^2\frac{\partial\nu_t}{\partial\eta}\frac{\partial R}{\partial\eta}$$

$$= \left(\xi_x^2 + \xi_y^2\right)\frac{\partial\nu_t}{\partial\xi}\frac{\partial R}{\partial\xi} + (\xi_x\eta_x + \xi_y\eta_y)\frac{\partial\nu_t}{\partial\eta}\frac{\partial R}{\partial\eta}$$

$$+ (\xi_x\eta_x + \xi_y\eta_y)\frac{\partial\nu_t}{\partial\eta}\frac{\partial R}{\partial\xi} + (\eta_x^2 + \eta_y^2)\frac{\partial\nu_t}{\partial\eta}\frac{\partial R}{\partial\eta}$$

$$= a_4\frac{\partial\nu_t}{\partial\xi}\frac{\partial R}{\partial\xi} + c_5\left(\frac{\partial\nu_t}{\partial\xi}\frac{\partial R}{\partial\eta} + \frac{\partial\nu_t}{\partial\eta}\frac{\partial R}{\partial\xi}\right) + b_4\frac{\partial\nu_t}{\partial\eta}\frac{\partial R}{\partial\eta}$$

Now, consider the Baldwin-Barth turbulence model given by (17-72). The transformation of the convective term is exactly the same as the previous case, that is

$$\vec{V}\cdot\nabla R = u\frac{\partial R}{\partial x} + v\frac{\partial R}{\partial y} = U\frac{\partial R}{\partial\xi} + V\frac{\partial R}{\partial\eta}$$

The Laplacian in the diffusion term is first expanded in Cartesian coordinates and expressed as

$$\nabla^2 R = \frac{\partial^2 R}{\partial x^2} + \frac{\partial^2 R}{\partial y^2} = \frac{\partial}{\partial x}\left(\frac{\partial R}{\partial x}\right) + \frac{\partial}{\partial y}\left(\frac{\partial R}{\partial y}\right) = \frac{\partial A}{\partial x} + \frac{\partial B}{\partial y}$$

where

$$A = \frac{\partial R}{\partial x} \qquad \text{and} \qquad B = \frac{\partial R}{\partial y}$$

Now relations (17-73) and (17-74) are used to provide

$$\begin{aligned}
\nabla^2 R &= \xi_x \frac{\partial A}{\partial \xi} + \eta_x \frac{\partial A}{\partial \eta} + \xi_y \frac{\partial B}{\partial \xi} + \eta_y \frac{\partial B}{\partial \eta} \\
&= \xi_x \frac{\partial A}{\partial \xi} + \xi_y \frac{\partial B}{\partial \xi} + \eta_x \frac{\partial A}{\partial \eta} + \eta_y \frac{\partial B}{\partial \eta}
\end{aligned}$$

The second expression in the diffusion term $\nabla \cdot \nu_t \nabla R$ is written as

$$\begin{aligned}
\nabla \cdot \nu_t \nabla R &= \nabla \cdot \left(\nu_t \frac{\partial R}{\partial x} \vec{\imath} + \nu_t \frac{\partial R}{\partial y} \vec{\jmath} \right) \\
&= \frac{\partial}{\partial x} \left(\nu_t \frac{\partial R}{\partial x} \right) + \frac{\partial}{\partial y} \left(\nu_t \frac{\partial R}{\partial y} \right) = \frac{\partial C}{\partial x} + \frac{\partial D}{\partial y}
\end{aligned}$$

where

$$C = \nu_t \frac{\partial R}{\partial x} \qquad \text{and} \qquad D = \nu_t \frac{\partial R}{\partial y}$$

In the generalized coordinates, one has

$$\begin{aligned}
\nabla \cdot \nu_t \nabla R &= \xi_x \frac{\partial C}{\partial \xi} + \eta_x \frac{\partial C}{\partial \eta} + \xi_y \frac{\partial D}{\partial \xi} + \eta_y \frac{\partial D}{\partial \eta} \\
&= \xi_x \frac{\partial}{\partial \xi} \left[\nu_t \left(\xi_x \frac{\partial R}{\partial \xi} + \eta_x \frac{\partial R}{\partial \eta} \right) \right] + \eta_x \frac{\partial}{\partial \eta} \left[\nu_t \left(\xi_x \frac{\partial R}{\partial \xi} + \eta_x \frac{\partial R}{\partial \eta} \right) \right] \\
&\quad + \xi_y \frac{\partial}{\partial \xi} \left[\nu_t \left(\xi_y \frac{\partial R}{\partial \xi} + \eta_y \frac{\partial R}{\partial \eta} \right) \right] + \eta_y \frac{\partial}{\partial \eta} \left[\nu_t \left(\xi_y \frac{\partial R}{\partial \xi} + \eta_y \frac{\partial R}{\partial \eta} \right) \right] \\
&= \xi_x \frac{\partial \alpha}{\partial \xi} + \xi_y \frac{\partial \beta}{\partial \xi} + \eta_x \frac{\partial \alpha}{\partial \eta} + \eta_y \frac{\partial \beta}{\partial \eta}
\end{aligned}$$

where

$$\begin{aligned}
\alpha &= \nu_t \left(\xi_x \frac{\partial R}{\partial \xi} + \eta_x \frac{\partial R}{\partial \eta} \right) \\
\beta &= \nu_t \left(\xi_y \frac{\partial R}{\partial \xi} + \eta_y \frac{\partial R}{\partial \eta} \right)
\end{aligned}$$

Finally, the following definitions are introduced which will be used in the formulation of FDE.

$$M^1 = -U \frac{\partial R}{\partial \xi} - V \frac{\partial R}{\partial \eta}$$

$$M^2 = -\frac{1}{Re_\infty}\frac{1}{\sigma_\epsilon}\left(\xi_x\frac{\partial\alpha}{\partial\xi} + \xi_y\frac{\partial\beta}{\partial\xi} + \eta_x\frac{\partial\alpha}{\partial\eta} + \eta_y\frac{\partial\beta}{\partial\eta}\right)$$

$$M^3 = \frac{1}{Re_\infty}2\left(\nu + \frac{\nu_t}{\sigma_\epsilon}\right)\left(\xi_x\frac{\partial A}{\partial\xi} + \xi_y\frac{\partial B}{\partial\xi} + \eta_x\frac{\partial A}{\partial\eta} + \eta_y\frac{\partial B}{\partial\eta}\right)$$

The last term of the Baldwin-Barth turbulence model includes a production term given by

$$(c_{\epsilon^2} f_2 - c_{\epsilon^1})\sqrt{c_\mu D_1 D_2}\ \nu \widetilde{Re}_T\, S$$

where S in the computational space is given by (17-83).

APPENDIX J:

Rate of Formation of Species

Consider the chemical reaction given by Equation (18-11), i.e.,

$$O_2 + M \rightleftarrows 2O + M \tag{J-1}$$

For a five-species model, the rate of formation for O_2 may be written as

$$\begin{aligned} \frac{d[O_2]}{dt} &= [O]^2 \left\{ K_b^1 ([N] + [O]) + K_b^2 ([N_2] + [O_2] + [NO]) \right\} \\ &\quad - [O_2] \left\{ K_f^1 ([N] + [O]) + K_f^2 ([N_2] + [O_2] + [NO]) \right\} \\ &= R_1 \end{aligned}$$

and for O, $\dfrac{d[O]}{dt} = -2R_1$.

From the chemical reaction

$$N_2 + M \rightleftarrows 2N + M \tag{J-2}$$

the rate of formation is expressed as

$$\begin{aligned} \frac{d[N_2]}{dt} &= [N]^2 \left\{ K_b^3 [N] + K_b^4 [O] + K_b^5 ([N_2] + [O_2]) + K_b^6 [NO] \right\} \\ &\quad - [N_2] \left\{ K_f^3 [N] + K_f^4 [O] + K_f^5 ([N_2] + [O_2]) + K_f^6 [NO] \right\} \\ &= R_2 \end{aligned}$$

and $\dfrac{d[N]}{dt} = -2R_2$.

Similarly for $NO + M \rightleftarrows N + O + M$

$$\frac{d[NO]}{dt} = \left(K_b^7 [N] \, [O] - K_f^7 [NO] \right) ([N] + [O] + [O_2] + [N_2] + [NO]) = R_3 \tag{J-3}$$

and $\frac{d[N]}{dt} = \frac{d[O]}{dt} = -R_3$.

and for

$$NO + O \rightleftarrows N + O_2 \tag{J-4}$$

$$\frac{d[NO]}{dt} = \frac{d[O]}{dt} = K_b^8\,[N]\,[O_2] - K_f^8\,[NO]\,[O] = R_4$$

$$\frac{d[N]}{dt} = \frac{d[O_2]}{dt} = -R_4$$

and for

$$O + N_2 \rightleftarrows N + NO \tag{J-5}$$

$$\frac{d[O]}{dt} = \frac{d[N_2]}{dt} = K_b^9\,[NO]\,[N] - K_f^9\,[N_2]\,[O] = R_5$$

$$\frac{d[N]}{dt} = \frac{d[NO]}{dt} = -R_5$$

Therefore, the five species reactions for the rate of formation of the five species are:

$$\frac{d[O_2]}{dt} = R_1 - R_4 \tag{J-6}$$

$$\frac{d[N_2]}{dt} = R_2 + R_5 \tag{J-7}$$

$$\frac{d[NO]}{dt} = R_3 + R_4 - R_5 \tag{J-8}$$

$$\frac{d[O]}{dt} = -2R_1 - R_3 + R_4 + R_5 \tag{J-9}$$

$$\frac{d[N]}{dt} = -2R_2 - R_3 - R_4 - R_5 \tag{J-10}$$

Note K_b^i and K_f^i (for $i = 1, \ldots 9$) are defined in Table J.1.

Chemical Reaction	Equilibrium Constant					Heavy Particle M	Constants for K_f			K_f^i, K_b^i
	A_1	A_2	A_3	A_4	A_5		C	n	E/K	i
$O_2 + M \rightleftarrows 2O + M$	2.855	0.988	−6.181	−0.023	−0.001	N, O	2.90^{23}	−2.0	59750.	1
						N_2, O_2, NO	9.68^{22}			2
$N_2 + M \rightleftarrows 2N + M$	1.858	−1.325	−9.856	−0.174	0.08	N	1.60^{22}	−1.6	113200.	3
						O	4.98^{22}			4
						N_2, O_2	3.70^{21}			5
						NO	4.98^{21}			6
$NO + M \rightleftarrows N + O + M$	0.792	−0.492	−6.761	−0.091	0.004	N, O, N_2, O_2, NO	7.95^{23}	−2.0	75500.	7
$NO + O \rightleftarrows N + O_2$	−2.063	−1.48	−0.58	−0.114	0.005		8.37^{12}	0	19450.	8
$O + N_2 \rightleftarrows N + NO$	1.066	−0.833	−3.095	−0.084	0.004		6.44^{17}	−1.0	38370.	9

Table J.1

REFERENCES

[12-1] Steger, J. L. and Warming, R. F., "Flux Vector Splitting of the Inviscid Gasdynamic Equations with Application to Finite Difference Methods," NASA TM-78605, July 1979.

[12-2] Van Leer, B., "Flux Vector Splitting for the Euler Equations," Lecture Notes in Physics # 170, 8th International Conference on Numerical Methods in Fluid Dynamics, 1982.

[12-3] "MACSYMA Mathematics and System Reference Manual," Macsyma Inc., 1996.

[12-4] Maple V, Waterloo Maple Software and the University of Waterloo, 1981–1994.

[12-5] Yee, H. C., "Numerical Approximation of Boundary Conditions with Applications to Inviscid Equations of Gas Dynamics," NASA TM-81265, March 1981.

[12-6] Thompkins, W. T., Jr. and Bush, R. H., "Boundary Treatments for Implicit Solutions to Euler and Navier-Stokes Equations," NASA CP-2201, October 1981.

[12-7] Bertin, J. J., "Engineering Fluid Mechanics," Prentice-Hall, 1984.

[12-8] Staff, "Equations, Tables, and Charts for Compressible Flow," Report 1135, Ames Research Center, NACA, 1953.

[12-9] Hoffmann, K. A., Chiang, S. T., and Siddiqui, M. S., "Fundamental Equations of Fluid Mechanics," EES, 1996.

[12-10] Hodge, B. K., and Koenig, K., "Compressible Fluid Dynamics," Prentice Hall, 1995.

[12-11] Saad, M. A., "Compressible Fluid Flow," Prentice Hall, 1993.

[12-12] Anderson, J. D., Jr., "Modern Compressible Flow," McGraw Hill, 1990.

[13-1] Vigneron, Y. C., Rakich, J. V., and Tannehill, J. C., "Calculation of Supersonic Viscous Flow over Delta Wings with Sharp Subsonic Leading Edges," AIAA-78-1137, July 1978.

[15-1] Lo, S. H., "A New Mesh Generation Scheme for Arbitrary Planar Domains," *International Journal for Numerical Methods in Engineering,* Vol. 21, pp. 1403–1426, 1985.

[15-2] Spragle, G. S., McGrory, W. R., and Fang, J., "Comparison of 2D Unstructured Grid Generation Techniques," AIAA Paper 91-0726, 1991.

[15-3] Weatherill, N. P., "A Method for Generating Irregular Computational Grids in Multiply Connected Planar Domains," *International Journal for Numerical Methods in Fluids,* Vol. 8, pp. 181-197, 1988.

[15-4] Sloan, S. W., "A Fast Algorithm for Constructing Delaunay Triangulations in the Plane," *Adv. Eng. Software,* Vol. 9, No. 1, 1987.

[17-1] Marvin, J. G., "Turbulence Modeling for Computational Aerodynamics," AIAA-82-0164, 1982.

[17-2] Kral, L. D., Mani, M., and Ladd, J. A., "On the Application of Turbulence Models for Aerodynamic and Propulsion Flowfields," AIAA-96-0564, January 1996.

[17-3] Rumsey, C. L., and Natsa, V. N., "Comparison of the Predictive Capabilities of Several Turbulence Models," *Journal of Aircraft,* Vol. 32, No. 3, May–June 1995.

[17-4] Kern, S., "Evaluation of Turbulence Models for High-Lift Military Airfoil Flowfields," AIAA-96-0057, January 1996.

[17-5] Baldwin, B. S. and Lomax, H., "Thin-Layer Approximation and Algebraic Model for Separated Turbulent Flows," AIAA-78-257, 1978.

[17-6] Granville, P. S., "Baldwin-Lomax Factors for Turbulent Boundary Layers in Pressure Gradients," *AIAA Journal,* Vol. 25, No. 12, December 1987.

[17-7] Baldwin, B. S., and Barth, T. J., "A One-Equation Turbulence Transport Model for High Reynolds Number Wall-Bounded Flows," NASA-TM-102847, August 1990.

[17-8] Spalart, P. R., and Allmaras, S. R., "A One-Equation Turbulence Model for Aerodynamics Flows," AIAA-92-0439, January 1992.

[17-9] Jones, W. P., and Launder, B. E., "The Calculation of Low-Reynolds-Number Phenomena with a Two-Equation Model of Turbulence," *International Journal of Heat and Mass Transfer,* Vol. 16, 1973, pp. 1119–1130.

[17-10] Wilcox, D. C., "Reassessment of the Scale-Determining Equation for Advanced Turbulence Models," *AIAA Journal,* Vol. 26, November 1988, pp. 1299–1310.

[17-11] Menter, F. R., "Assessment of Higher Order Turbulence Models for Complex Two- and Three-Dimensional Flowfields," NASA-TM-103944, September 1992.

[17-12] Cebeci, T. and Bradshaw, P., "Physical and Computational Aspects of Convective Heat Transfer," Springer-Verlag, 1984.

[17-13] Kays, W. M. and Crawford, M. E., "Convective Heat and Mass Transfer," Second edition, McGraw-Hill, 1980.

[17-14] Libby, P. A., "Introduction to Turbulence," Taylor and Francis Publishers, 1996.

[17-15] Landahl, M. T., and Mollo-Christensen, E., "Turbulence and Random Processes in Fluid Mechanics," Second Edition, Cambridge University Press, 1992.

[17-16] Wilcox, D. C., "Turbulence Modeling for CFD," DCW Industries, 1993.

[18-1] Hirschfelder, J. O., Curtiss, C. F., and Bird, R. B., "Molecular Theory of Gases and Liquids," John Wiley and Sons, Inc., 1954.

[18-2] Stull, D. R., "JANAF Thermodynamical Tables, National Bureau of Standards," NSRDS-NBS 37, 1971.

[18-3] Tannehill, H. C. and Mugge, P. H., "Improved Curve Fits for Thermodynamic Properties of Equilibrium Air Suitable for Numerical Computation Using Time-Dependent or Shock-Capturing Methods," NASA CR-2470, October 1974.

[18-4] Vincenti, W. G. and Kruger, C. H., "Introduction to Physical Gas Dynamics," Robert E. Krieger Publishing Co., Inc., 1965.

[18-5] Anderson, J. D., "Hypersonic and High Temperature Gas Dynamics," McGraw-Hill Book Company, 1989.

INDEX

The numbers 1 or 2 preceding the page numbers refer to Volumes 1 and 2, respectively.

Acoustic equation, *1(429)*
Advancing front method, *2(312)*
Alternating direction implicit (ADI)
  for diffusion equation, *1(78)*
  for incompressible Navier-Stokes equation, *1(305)*
  for Laplace's equation, *1(165)*
Algebraic grid generator, *1(365)*
Amplification factor, *1(125)*
Approximate factorization, *1(85)*, *2(193, 247)*
Anti-diffusive term, *1(234)*
Area-Mach number relation, *2(124)*
Artificial compressibility, *1(315)*
Artificial viscosity, *1(146)*
Backward difference representations, *1(57, 59)*
Backward time/central space, *1(279)*
Baldwin-Barth model, *2(388)*
Baldwin-Lomax model, *2(385)*
Beam and Warming implicit method, *1(213)*
Beta formulation, *1(67)*
Body-fitted coordinate system, *1(360)*
Boltzmann constant, *2(438)*
Boundary conditions, *1(20)*, *2(276, 286)*
  Body surface, *1(323, 344)*, *2(186)*
  Far-field, *1(325, 346)*
  Symmetry, *1(325, 346)*, *2(189)*
  Inflow, *1(326, 347)*, *2(190, 192)*
  Outflow, *1(326, 348)*, *2(190, 192)*
Boussinesq assumption, *2(381)*
Branch cut, *1(395, 401)*
Buffer zone, *2(379)*
Bulk Viscosity, *1(452)*
Calorically perfect gas, *1(461)*
Cebeci/Smith model, *2(312)*
Cell-centered scheme, *2(339)*
Cell Reynolds number, *1(131)*
Central difference approximations, *1(32, 58, 59)*
Characteristics, *1(4, 431)*
Characteristic variables, *2(455)*
Compatibility equation, *1(434)*
Conservative form, *1(23)*
Conservative property, *1(23)*
Consistency, *1(23)*
Consistency analysis, *1(88)*
Continuity equation, *1(274, 450)*
Continuum, *1(445)*
Contravariant velocity, *2(39)*
Convective derivative, *1(447)*
Convergence, *1(23)*
Courant-Friedrichs and Lewy (CFL) number, *2(126)*
Courant number, *1(119)*

Crank-Nicolson method,
for diffusion equation, *1(66)*
for hyperbolic equation, *1(188)*
for incompressible Navier-Stokes equation, *1(321)*

Damping term, *1(216, 228), 2(249, 304)*

Delaunay method, *2(320)*

Delta formulation, *2(100)*

Difference operators, *1(33)*

Diffusion number, *1(70)*

Dilatation, *1(311)*

Dirichlet boundary condition, *1(20)*

Dirichlet tessellation, *2(321)*

Discretization error, *1(13)*

Discrete perturbation stability analysis, *1(114)*

Dispersion error, *1(143, 196)*

Displacement thickness, *2(385)*

Dissipation error, *1(143, 193)*

Dissociation, *2(437)*

Driven cavity flow, *1(348)*

DuFort-Frankel method, *1(64, 277, 285, 333)*

Dynamic instability, *1(122)*

Eckert number, *1(353)*

Eddy viscosity, *2(381)*

Eigenvalues of flux Jacobians, *2(103, 166, 167)*

Eigenvector, *2(104, 166, 168)*

Elliptic equation, *1(6, 152)*

Elliptic grid generators, *1(383)*
Simply-connected domain, *1(385)*
Doubly-connected domain, *1(395)*
Multiply-connected domain, *1(401)*

Energy equation, *1(275, 352, 454)*

Enthalpy, *1(449), 2(142)*

Entropy condition, *1(243)*

Equation of State, *1(462), 2(101)*

Equilibrium flow, *2(435)*

Error analyses, *1(141)*

Essentially non-oscillatory schemes, *1(238)*

Euler equation, *1(276), 2(97)*
Quasi one-dimensional, *2(99)*
Two-dimensional planar or axisymmetric, *2(162)*

Euler BTCS, *1(188)*

Euler FTFS, *1(186)*

Euler FTCS, *1(186)*

Eulerian approach, *1(449)*

Explicit formulation, *1(39, 61), 2(269, 295)*

Finite difference equation, *1(22, 37)*

Finite volume method, *2(336)*

First backward difference approximation, *1(31)*

First forward difference approximation, *1(30)*

First upwind differencing method, *1(186, 218)*

Implicit first upwind, 1(188, 218)

Five species model, *2(439)*

Forward difference representations, *1(57, 58)*

Forward time/central space (FTCS), *1(64, 276, 282)*

Flux corrected transport, *1(233)*

Flux Jacobian matrix, *1(317)*

Flux limiter functions, *1(242, 245), 2(113)*

Fractional step method, *1(87)*

Freestream boundary condition, *2(243)*

Frozen flow, *2(434)*

Gas Constant, *1(462)*

Gauss-Seidel iteration method, *1(157)*
  Point Gauss-Seidel, *1(160, 387)*
  Line Gauss-Seidel, *1(162, 388)*

Global time step, *2(146)*

Grid Clustering, *1(369, 404), 2(144)*

Heat conduction equation, *1(10)*

Heat of formation, *2(436)*

Holograph characteristics, *1(5)*

Heat transfer coefficient, *1(355)*

Hyperbolic equations, *1(8,185)*

Hyperbolic grid generation, *1(411)*

Implicit flux-vector splitting, *2(116, 194, 279)*

Implicit formulation, *1(39, 63), 2(277, 303)*

Incompressible Navier-Stokes equations, *1(302), 2(85)*

Integral formulation, *1(446)*

Inviscid Jacobians, *2(36–38, 89, 90, 93)*

Irregular boundaries, *1(92)*

Jacobi iteration method, *1(157)*

Jacobian, *1(209)*

Jacobian of transformation, *1(364), 2(25)*

k-ϵ model, *2(399)*

k-ω model, *2(401)*

Kronecker delta, *1(451)*

Laasonen method, *1(66)*

Laminar sublayer, *2(379)*

Lagrangian approach, *1(449)*

Lax's equivalence theorem, *1(23)*

Lax method, *1(186, 207)*

Lax-Wendroff method, *1(187, 189, 208)*

Linear damping, *1(228)*

Linear PDE, *1(3)*

Linearization, *1(90, 317), 2(31, 100, 163, 225)*
  Lagging, *1(90)*
  Iterative, *1(91)*
  Newton's iterative, *1(91)*

Local derivative, *1(447)*

Local time step, *2(146)*

LU Decomposition, *2(291)*

MacCormack method, *1(190, 211, 278, 286, 287), 2(270)*

Marker and cell (MAC) formulation, *1(330)*

Metrics of transformation, *1(363), 2(23)*

Midpoint leapfrog method, *1(133, 186)*

Minmod, *1(246), 2(114)*

Mixed boundary condition, *1(20)*

Mixed partial derivatives, *1(51)*

Modified equation, *1(143)*

Momentum thickness, *2(385)*

Monotone schemes, *1(236)*

Multi-step methods, *1(189)*

Navier-Stokes equations, *1(274, 455), 2(28, 267)*
  Two-dimensional planar or axisymmetric, *2(69)*

Newtonian fluid, *1(452)*

Neumann boundary condition, *1(20)*

Neumann boundary condition for pressure, *1(323)*

Nodal point scheme, *2(340)*

Nonequilibrium flow, *2(321)*

Nonlinear PDE, *1(3)*

Numerical flux functions, *1(241)*

Orthogonality at the surface, *1(407)*

Outflow boundary condition, *2(120, 190)*

Parabolic equation, *1(6, 60)*

Parabolic grid generation, *1(418)*

Parbolized Navier-Stokes equations, *2(60, 218)*
  Two-dimensional planar or axisymmetric, *2(81)*

Partial pressure, *2(434)*

Poisson equation for pressure, *1(310, 311)*

Prandtl mixing length, *2(382)*

Prandtl number, *1(463), 2(25, 371)*

Production of turbulence, *2(399)*

Reaction rates
  Backward reaction, *2(437)*
  Forward reaction, *2(437)*

Real gas, *2(433)*

Recombination, *2(437)*

Reynolds Averaged Navier-Stokes Equation, *2(383)*

Reynolds number, *2(25)*

Richardson method, *1(64)*

Richtmyer method, *1(189)*

Riemann invariants, *1(435), 2(191)*

Robin boundary condition, *1(20)*

Rotational energy, *2(435)*

Round off error, *1(113)*

Runge-Kutta method, *1(219)*
  Modified Runge-Kutta, *1(225, 290), 2(112, 180, 275)*

Spalart-Allmaras model, *2(393)*

Species continuity equation, *2(442)*

Specific heats, *1(461)*

Splitting methods, *1(189)*

Shock fitting, *2(250)*

Shock tube, *2(152)*

Staggered grid, *1(328)*

Stability, *1(23)*

Static instability, *1(123)*

Steger and Warming flux vector splitting scheme, *2(107, 116, 170)*

Stokes hypothesis, *1(452)*

Stream function, *1(307)*

Stream function equation, *1(308)*

Streamwise pressure gradient, *2(223)*

Structured grids, *1(92, 358)*

Successive over-relaxation method, *1(164)*
  Point SOR, *1(164)*
  Line SOR, *1(165)*

Sutherland's law, *1(459)*

System of first-order PDEs, *1(11)*

System of second-order PDEs, *1(16)*

Thermal conductivity, *1(460)*

Thermally perfect gas, *1(461)*

Thin-Layer Navier-Stokes equations, *2(57, 268)*

Total derivative, *1(447)*

Total variation diminishing (TVD), *1(237), 2(112)*
  First-order TVD, *1(239)*
  Second-order TVD, *1(244, 292)*
  Harten-Yee Upwind TVD, *1(245), 2(181)*

Roe-Sweby Upwind TVD, *1(247), 2(183)*
Davis-Yee Symmetric TVD, *(250), 2(185)*
Translational energy, *2(435)*
Tridiagonal system, *1(63, 438)*
Turbulent boundary layer, *2(369)*
Turbulent Prandtl number, *2(383)*
Turbulent kinetic energy, *2(388)*
Turbulent shear stress, *2(381)*
Turbulent viscosity, *2(381)*
Two-equation turbulence model, *2(398)*
Universal gas constant, *1(462)*
Unstructured grids, *1(92, 358), 2(307)*
Vibrational energy, *2(436)*
Viscosity, *1(452, 459)*
Viscous Jacobians, *2(46–57)*
Viscous stress, *1(452)*
van Leer flux vector splitting scheme, *2(108, 178)*
von Karman constant, *2(384)*
von Neumann stability analysis, *1(124)*
Vorticity, *1(307)*
Vorticity-stream function formulations, *1(307)*
Vorticity transport equation, *1(308)*
Wall boundary conditions, *2(186, 235)*
Zero equation model, *2(379, 384)*
Zero point energy level, *2(436)*
Zone of dependence, *1(5)*
Zone of influence, *1(5)*
Zone of silence, *1(9)*